Anthony Cronin

is a comic novelist, poet and broadcaster, and the celebrated biography of Flann O'Brien, *No Laughing Matter* (1989). He has been editor of a number of literary publications, received awards for his services to Irish literature, was cultural adviser to the Taoiseach, and instrumental in establishing Aosdána, an Irish state body that funds writers. Among his other books are *The Life of Riley* and *Dead as Doornails*.

By the same author

Poetry

COLLECTED POEMS 1950-1973
REDUCTIONIST POEM
R.M.S. *TITANIC*
NEW AND SELECTED POEMS
THE END OF THE MODERN WORLD

Fiction

THE LIFE OF RILEY
IDENTITY PAPERS

Non-Fiction

DEAD AS DOORNAILS
HERITAGE NOW
A QUESTION OF MODERNITY
AN IRISH EYE
ART FOR THE PEOPLE?
NO LAUGHING MATTER: THE LIFE AND
TIMES OF FLANN O'BRIEN

ANTHONY CRONIN

Samuel Beckett

The Last Modernist

Flamingo
An Imprint of HarperCollins*Publishers*

Flamingo
An Imprint of HarperCollins *Publishers*
77–85 Fulham Palace Road,
Hammersmith, London W6 8JB

Published by Flamingo 1997

First published in Great Britain by
HarperCollins *Publishers* 1996

Copyright © Anthony Cronin 1996

Anthony Cronin asserts the moral right to
be identified as the author of this work

ISBN 978-0-00-733004-1

Set in Linotype Postscript Bembo

Preface

❧

Writing this book has been a journey into my own past as well as Samuel Beckett's. Many who helped me with it were already friends or acquaintances of long standing and to them my thanks are now due. Many others who helped me are no longer here to thank; for though I was not engaged in writing his biography at the time I knew them, Beckett was often enough mentioned or discussed between us.

I first met Tom MacGreevy around 1948, when, as a representative of that peculiar genus, 'young poet,' I was brought to the flat where he lived with his sister and nieces. He had not yet become Director of the National Gallery of Ireland; and, as I understand it now, his position in Dublin was the ambiguous one of the returned exile. He was said to have tried and, the assumption was, failed in literary London; and though I found his occasional references to the figures he had known abroad intoxicating, he could be characterised by malicious Dublin as a 'name-dropper'.

There were still unsold copies of his only collection of poems, published fourteen years before, on the shelves in Combridges of Grafton Street (where no book was ever moved, or, it seemed to me, sold). Yeats's inclusion of two of MacGreevy's poems in the *Oxford Book of Modern Verse* was dismissed by those in Dublin who adverted to it as one of the notorious eccentricities of that anthology; and his poetry was never discussed in Ireland, where the only poets spoken of tended to be latter-day Irish revivalists rather than such modernists as MacGreevy or his friends Beckett, Coffey and Devlin. Since MacGreevy was the first fully and properly published poet I ever knew I would like to say here that he was a gentle and wonderful representative of the vocation to know and that I treasure the memories I have of him.

There was not much talk of Beckett's prose then either, though he

v

was the author of three books: and he was still virtually unknown and untalked of in Dublin when, around 1950, A.J. Leventhal called the attention of those of us who were associated with the magazine *Envoy* to his work and arranged for us to see part of the then unpublished novel *Watt*.

I knew 'Con' Leventhal from various contexts, not least important or enjoyable among them the racecourse, for he was a keen student of form and we often went to Leopardstown or the Phoenix Park with our mutual friend, John Jay. Ethna McCarthy, now married to Con, I first met some time in the fifties. She was then practising as a doctor and at our first meeting in O'Neill's pub on Merrion Row she prescribed red meat for an ulcer I had, or thought I had. This unorthodox cure worked, or seemed to work; and I pass it on now for what it is worth.

It was about the time *Envoy* published an extract from *Watt* that I shared a house with Sam's composer cousin John in Hatch Street and lived for a memorable few months with Ralph Cusack and Nancy Sinclair at Uplands in County Wicklow; about this time too that I first met Peggy Sinclair's other sister, Deirdre Hamilton. I remember talking of Beckett at this time with Ralph and Nancy, with John, with the painter Sean O'Sullivan, who admired him deeply, and a little later on with Arland Ussher.

After I had read and reviewed the Olympia Press edition of *Molloy* for an English journal, *Time and Tide*, I took more interest in the man and his radical view of human life than I had before and from then on I began to discuss him more often with people I knew, including Mary St John Hutchinson and others connected with the review, *X*.

Godot had been produced in London but I had not seen it when, as I say in this book, I wrote the piece about his novels in the *TLS* which led to our first meeting. Some of those who played a part in his life that I came to know at this time are happily still with us; in writing about others, such as the actor Pat Magee, I have been writing about people I knew very well; indeed I myself woke up more than once in the rather splendid digs in West Cromwell Road where Beckett stayed with him in 1958.

Among the living, it is my first pleasure to thank Beckett's nephew Edward Beckett for his kindness and co-operation as well as for permissions to quote from the works, letters and notebooks. Jérôme Lindon, Beckett's other executor, was also very helpful and encouraging. I also thank Messrs Faber and Faber for permission to quote from the plays. In a more personal sense I would like to thank my former publisher, John Calder, who discussed Beckett's life and attitudes at length with me on several occasions and has also been generous with permissions. To Barney Rosset, for so many years Beckett's friend and the American publisher of his works, I am grateful for many things, not least the perennial pleasure of his company; and I would like to thank also his partner, Astrid Myers, for many kindnesses. For permission to use photographs as well as help and information I have to thank Caroline Beckett Murphy. I have had many conversations which included Beckett but ranged much further with Georges Belmont, his friend of four decades; and with Barbara Bray, whom I have known since the 1950s. If I list other people in no sort of order and according to no set schema, I hope they will forgive me.

Louis Le Brocquy and Anne Madden Le Brocquy; Derek Mahon; Ulick O'Connor; Aidan Higgins; Adrian Kenny; John Montague; Harold Pinter; Hugh and Mary Anne Kenner; Deirdre Hamilton; Richard Kearney; Mary Manning Howe; Cyril Cusack; Desmond O'Grady; John Manning; Robert Pinget; Eoin O'Brien; Denis Donoghue; Suzanne Pegley; Jean Demelier; Pierre Chabert; Marion Leigh; Bill Cunningham; Deirdre Bair; Francis Evers; Isabelle Jandard; Patrick MacEntee; Josette Hayden; Kate Meynell; Michael Walsh; Ludovic Janvier; Bernadette O'Leary; Mary Ryan; Adrian Munnelly; Dorothy Walker; Juliette Girodias Kahane; Vivien Igoe; Therese Cronin; Patrick Connolly; Randall Green; Gerry Dukes; Ian MacMillan; Gordon Aston; Noël de Chenu; Jocelyn Herbert; Françoise Bibolet; Tom Bishop; Mark Mortimer; Patricia King; Eliza O'Grady; Aengus Fanning; Stan Gontarski; Terence Brown; Georges Sandelescu; Trevor West; Richard Dauthuille; Ronan Farren; Lois Overbeck; Martha Fehsenfeld; Colm Tóibín; Hosney Emam; Brian O'Doherty; Columba Quinn; Noel Pearson; Billie Whitelaw; Bruce Stewart; John Reilly; Melissa Shaw-Smith; Benedict Kiely; Michael Kane; Brian Fallon; Walter Asmus, Patricia Collins; Ute Norden; Peter Hayes; Melanie Le Brocquy; Ivor Browne; John Minihan; Val Rice; John Wyse Jackson; Sean ÓMordha; Patrick Cronin; Johnny Granville; Avigdor Arikha; Francis Stuart; Susan Schreibman; Mike Rudd; Norah Lever; Tom Cox; Andy O'Mahoney; Jean Martin; Anthony Carroll; Terri Garvey; Inge Eulitz; Margaret Rogers; Jonathan Williams; Rosemary Rowley; Sean Mac Réamoinn; Michael Colgan; Desmond Guinness; Patricia Noone; Jack Waters; David McConnell; Maurice Craig. And finally some friends who helped me but who have died since I began work on this book six years ago: Eilis Dillon, whose husband, Vivian Mercier, wrote one of the best critical books about Samuel Beckett; Stan Gebler Davies; and Patrick Gallagher, who was, as usual, a fount of information about many things.

I have to thank also Bernard Meehan and the staff of the Library and Manuscript Collection, Trinity College Dublin, which, through the generosity of Elizabeth Ryan and Margaret Farrington, holds, among other important Beckett material, his letters to Tom MacGreevy; Tom Staley and the staff of the Harry Ransom Humanities Research Center in the University of Texas at Austin, Texas; Philip R. Grunematt, Curator of Manuscripts and Chief of Special Collections and his staff at Dartmouth College, Hanover, New Hampshire; James Knowlson and the staff of the library and Beckett Archive at Reading University; Catriona Crowe and the staff of the National Archive, Dublin; Patricia Donlon and the staff of the National Library, Dublin; Robert O'Neill, Burns Library, Boston College; Mme Boulez and the staff of the library of the École Normale Supérieure; Andrée Wuczlz and the staff of the Hôtel de Ville, Troyes; Tadhg O'Sullivan, formerly Irish Ambassador to France; Donal Kelly, First Secretary and the Embassy and Consular Staff; and Philippe Grand and the staff of the Archives de Paris.

Beckett has now been the subject of hundreds of books. Most of them are essays in criticism of one kind or another; but some are concerned with aspects of his biography and a few are godsends to a biographer. Anybody working in this field must owe an immense debt of gratitude to Eoin O'Brien and his splendid *The Beckett Country: Samuel Beckett's Ireland*, a mine of information about many matters stretching far beyond its ostensible theme. The late Vivian Mercier's awkwardly titled *Beckett/Beckett* is full of insights, critical and biographical and, as one would expect, excellently written. In spite of its sometimes clotted style Lawrence Harvey's *Samuel Beckett: Poet and Critic* is an essential sourcebook. Deirdre Bair's 1978 biography, *Samuel Beckett*, has come in for a good deal of criticism in professional Beckett circles, much of it unwarranted and some of it unjust. It was a pioneering effort and the achievement in my view outweighs its faults. Since the hardback edition of this book was finished the publication of James Knowlson's thorough and painstaking *Damned To Fame* has added to our store of knowledge about some of the details of Beckett's life. Patrick Cronin's two monographs on Foxrock history published by the Foxrock Historical Society supplied much useful background information. Though published in 1970 Raymond Federman and John Fletcher's *Samuel Beckett: His Works and His Critics* is still indispensable. The special issues of *Cahiers de L'Herne* and *Revue d'Esthétique* devoted to Beckett are very useful.

And finally heartfelt thanks to those who have helped me with research, editing and other matters: Rodney Sharkey of Trinity College, whose meticulous research illuminated so many areas of Beckett biography and who so often discussed the results with me with such keen penetration; Anne Haverty, novelist, French scholar, and herself a biographer of distinction; John Fletcher, Beckett's illustrious bibliographer, the first Beckett expert to read this book, who made many helpful suggestions; Sophie Nelson, whose careful copy-editing has undoubtedly saved me from many errors; and my publishers at HarperCollins, Philip Gwyn Jones, who cheerfully and intelligently spurred me on, and Mandy Kirkby, whose patience and understanding were exemplary.

Anthony Cronin
Dublin, 1996.

Chapter One

Biographies often begin with a date of birth, the date on which the subject's experiences can be said to begin. In the case of Samuel Beckett there are two difficulties about adopting this simple procedure. One is his claim to have been born on Good Friday, 13 April 1906. The other is his repeated insistence that he had memories of life in his mother's womb.

The idea that he had been born on Good Friday, the day of the Saviour's crucifixion, pleased him, more especially since Good Friday happened in 1906 to have been Friday the thirteenth. What better birth-date could there be for someone so conscious of the suffering which underlies human existence; and conscious also that misfortune, in comic or tragic guise, awaited every venture and departure? In a late work, *Company*, which is highly autobiographical, the coincidence of his birth-date with the day of the Saviour's death is emphasized. 'You were born on an Easter Friday after long labour . . .' And: 'You first saw the light and cried at the close of the day when in darkness Christ at the ninth hour cried and died.' Not only was Beckett pleased with the Christ connection involved in having been born on a Good Friday, but he was never averse to introducing analogies and comparisons between Christ's life-story and those of his degraded characters.

It is a little unfortunate therefore that his birth certificate shows him to have been born on the 13 *May* 1906, which was a Sunday. The birth was not registered until a month later, on 13 June. Since it was Irish custom to allow a clear four weeks to elapse between birth and registration, this fact would if anything tend to confirm the May rather than the April birth-date, but of course there is always the possibility of error – even an error of immediate recollection on his father's part, or a slip of the tongue or the pen, neither of which was rare where registration

I

was concerned. Certainly the tradition of his family was that 13 April was his birthday and this is what he had been led to believe in childhood. In fact, though he may not have known it, an announcement of his birth had appeared in the appropriate column of the Irish Times on 16 April 1906. Though no date was included, this would appear to put the matter beyond reasonable doubt.

The other difficulty about beginning on Friday the thirteenth is Beckett's claim to have had memories which preceded this date. He told more than one person that he had such memories; and in an interview given when he was sixty-four he said, 'Even before the foetus can draw breath it is in a state of barrenness and of pain. I have a clear memory of my own foetal existence. It was an existence where no voice, no possible movement could free me from the agony and darkness I was subjected to.'

On an earlier occasion he had told Peggy Guggenheim that he had, in her words, 'retained a terrible memory of life in his mother's womb. He was constantly suffering from this and had awful crises, when he felt he was suffocating.' And with his close friend Geoffrey Thompson, a psychiatrist who tried to help him with subsequent difficulties, he also discussed his pre-natal memories and the feelings of entrapment and suffocation they brought with them.

But not only did he have these generalized memories; he had, he said, a particular memory of being at the dinner table in his mother's womb shortly before birth. There were guests present and the conversation was, perhaps needless to say, of the utmost banality.

In any case Samuel Beckett, son of William – known as Bill in family circles, but generally to friends and associates as Willie – and his wife Maria – known as May – was born at nightfall, in the family home at Foxrock, a prosperous suburb on the outskirts of Dublin. At the time of his birth Bill and May Beckett had been married almost five years. A first child, Frank, had been born within eleven months of their marriage; the second, Samuel, was not conceived for three more years.

The place of May's accouchement was the bedroom in which the conception had most likely taken place, a big room with a bow window facing the Dublin mountains. It was a difficult labour, the pains lasting all day; and as soon as it began in the morning Bill absented himself, going for a tramp in the mountains with a packet of sandwiches and a flask of whiskey. He returned to find his wife still in labour, and being, like most fathers of the time, neither welcome nor anxious to hear or see anything of the delivery, he took himself off to the garage, where he sat in darkness in the driver's seat of his high De Dion Bouton until a maid came to tell him it was all over and that his wife had been happily delivered of a son. In later life Beckett was fond of the verse in the Book of Job, verse 3, chapter 3, 'Let the day perish in which I was born and the night in which it was said, there is a man child conceived.' Mr Tyler in the radio play *All That Fall* apologizes for cursing, in the presence of a lady, God and man and 'the wet Saturday afternoon' of his conception, while Neary, in the novel *Murphy*, curses first the day he was born, 'and then, in a bold flashback, the night he was conceived'.

Whereas Bill and May's first son, Frank, had been healthy and placid, Samuel was a sickly, thin baby who cried constantly.

The Becketts were prosperous people and the house in which the birth took place reflected that prosperity, as did its situation in Foxrock, then a very exclusive suburb of Dublin. Willie Beckett's father, also William, was a building contractor and what nowadays would be called a speculator who had acquired both money and house property in Dublin on a considerable scale. Among the contracts awarded to his firm were those for the new National Library and National Museum, flanking the Royal Dublin Society's headquarters in Leinster House in Kildare Street, which subsequently became the home of the Dáil or Irish parliament. These had been begun in 1885 and finished in 1891. There was fierce competition for the contract and some of William Beckett's enemies were pleased when it was established that the Mount-charles sandstone employed had been wrongly used: face-bedded instead of end-bedded, so that it began to crumble almost immediately, though this was never clearly established as the builder's mistake and may well have been the architect's.

The Becketts vaguely supposed themselves to be descended from Huguenot refugees who had come to Ireland from France after the revocation of the Edict of Nantes in 1685, fleeing their homeland to escape persecution as Protestants and settling in a country where persecution of Catholics was just about to begin on a systematic scale. The Huguenots had supplied Ireland with much of its industrial and business energy, bringing with them a knowledge of the linen and poplin trades as well as of banking; and they fitted in easily enough in a country which was in process of transition after the Williamite Wars, a place where much was in the melting-pot and most business activity represented a new departure. Within a generation or two the Huguenots generally were as happy to forget their French origins as many other members of the land-owning and business classes were to forget their English; and by the early twentieth century no trace or definite knowledge of a former French connection remained. Samuel Beckett's adoption of France as his homeland would have nothing to do with the French origins of his family: and in fact there is doubt that the

Becketts really were Huguenots. The name does not occur in early listings of Huguenot refugees and is not commonly regarded by historians of the Huguenot influx as a likely Huguenot one. The suggestion has therefore also been made that the Irish Becketts were of Norman origin, and those who like aristocratic lineages for their heroes have even claimed that Samuel Beckett's family were descended from the family of Thomas à Becket, the turbulent priest who was Henry II's Archbishop of Canterbury. What makes the Huguenot connection the more likely is that originally the Becketts were silk- and poplin-weavers, a form of manufacture in which Huguenots certainly engaged; and in fact William Beckett senior's father, Samuel's great-grandfather, James Beckett, had been in the silk-weaving business. By the time his son entered the building trade, Irish silk-weaving, which had once been important, was dying out.

In 1869 William the builder had married Fannie or Frances Crothers – Crowther in one family record, but certainly Crothers – daughter of a Dublin merchant, Thomas Crothers, who was also of course a Protestant. A talented and sensitive woman, perhaps more talented and sensitive than William Beckett had bargained for, to some extent she broke the mould of philistine business success and introduced other elements into the make-up of the Becketts. She was an accomplished musician, who played the harmonium and composed settings for poems which she thought sufficiently solemn and religious-sounding, such as Tennyson's 'Crossing the Bar'. In later life she took to the drink and Dublin gossip was pleased to report on her eccentricities, including the fact that she would lock herself in her room, presumably with supplies of alcohol, for days on end. What in the course of time her grandson, Samuel, would remember best about her was the parrot who always perched on her shoulder and flew into a jealous rage when anyone kissed her.

Fannie and William Beckett had four sons, of whom one, Howard, went into the family contracting business, J. and W. Beckett of South King Street; and two, Gerald and James, studied medicine and became doctors. A daughter, also Frances, but known because the boys called her that as Cissie, would have a somewhat less orthodox career, as a painter, member of Dublin's bohemia and wife to an unsuccessful Jewish art dealer. The fourth son, in fact the eldest and christened William after his father, was destined to be the father of Samuel Beckett.

William senior and Frances lived in a house called Earlsfield which was Number 7 Prince William Terrace in the fashionable district of Ballsbridge, and it was there that William junior was born in July 1871. He grew into a burly, athletic youth of a not very studious disposition and left school early to become apprenticed to a member of the trade or profession of quantity surveying. Quantity surveyors were employed to estimate the amount of material likely to be used in any particular building contract, and had begun to appear in the early nineteenth century, when they were known as 'measurers'. Originally anybody could set up who had the right connections and thought he could get the work, but by the late nineteenth century entry was being increasingly if loosely regulated by an apprenticeship and examination system. Willie Beckett was apprenticed to the firm of J. and E. Pannister where he served the requisite five years, after which a partnership was purchased for him in a firm which had offices in Clare Street, near the back gate of Trinity College.

Besides the fact that estimating could prove a lucrative extension of the activities of an already well-established building concern, William Beckett senior, like many successful business people at the time, was anxious that his eldest son should become a member of one or other of the professions. This was a tendency everywhere, but it was especially the case in snobbish Dublin, which was keenly conscious of social gradations. Since William junior showed no interest in a prolonged period of academic study, quantity surveying was a good halfway house. His success, it was felt, would be almost automatically ensured by his connection with the firm of J. and W. Beckett, while at the same time it would be a step up in the world, since quantity surveying took rank as a profession, not a trade. The line between it, civil engineering and architecture was still rather blurred. Quantity surveyors were free to describe themselves as one or the other and frequently did.

Willie Beckett may not have been academically minded, but neither was he a dullard. He had a good, shrewd business brain and he was soon a success in his profession. He was also a socially minded, good-natured, seemingly uncomplicated young man, fond of masculine pursuits, including the game of rugby and the drinking and horse-play that went with it. All the Becketts were athletic and interested in athletics and Willie's brother Jim, one of the two who became doctors, was a famous swimmer and a member of the Irish international swimming team about which later verses would be written:

> Dockrell, Taggart, Beckett, where
> Now are the men I worshipped there?

In manner Willie Beckett was an extrovert, who appeared glad to see people and was prepared to make the best of them. He was, like his brothers, naturally athletic and besides being a rugby player he was a

5

golfer, a tennis player, a great walker and a year-round swimmer. In the course of time he became a member of several men's clubs, including the exclusive Kildare Street. Until he was over thirty he lived at Earlsfield, a mansion substantial enough to house a British embassy in later years, an active young man already inclining to stoutness. It was an uncomplicated life: the office or the building site during the day, nights at the club with a game of billiards after dinner, summer evenings on the golf course and in the club-house bar. But to go with all this masculine activity and joviality there was church-going with his parents at weekends before Sunday dinner at Earlsfield, often followed by a tramp through the Dublin mountains.

Then he fell in love; and unfortunately with a Catholic – the seventeen-year-old daughter of William Martin Murphy, the most successful Catholic businessman in Dublin, owner of newspapers, the tramway company and much else, whom he had met at the Fitzwilliam Lawn Tennis Club, one of the few places where Catholics and Protestants could meet socially. There was talk of marriage and both sides were appalled, the Catholic Murphys even more than the Protestant Becketts. The girl was forbidden to see any more of him and was forced to make a solemn renunciation at her mother's death-bed, the death itself occurring at just the right moment to induce a sense of guilt and transgression. Then, apparently almost before she had time to breathe, let alone weep, she was married off to a distinguished Dublin surgeon, Sir Arthur Chance, a widower with children who happened to be a knight as well as a Catholic and was not only much older than her, but considerably older than Bill Beckett as well: in short, a good match.

Shortly afterwards Willie went down with pneumonia. He appears to have been out of sorts for some time after being parted from his love, with ailments difficult to diagnose and probably evidence of a shock to the psyche greater than was to be expected from his extrovert demeanour; in fact, a breakdown. But the pneumonia was an indisputable reality and he wound up in the Adelaide Hospital, where he was nursed by a tall, thin-faced, serious girl called Maria, or May, Roe, who was exactly the same age as himself.

The Roes had been in Ireland since 1641, before the Cromwellian invasion, and had once been considerable land-owners in County Tipperary, with villages named after them, such as Roesborough and Roe's – now corrupted to Rose – Green. The Tipperary Roes had a reputation for wildness which sometimes overlapped with one for oddity; but by the early nineteenth century a branch of the family was centered in Leixlip, a small town or village not far from Dublin in County Kildare, and was producing principally clergymen. In the mid-part of the century

May's father, Samuel Robinson Roe, son of a clergyman also called Samuel, had possessed a property near Mountmellick in Queen's County, now County Laois. He was also, however, a miller and grain merchant and in 1864 had moved to another property, Cooldrinagh in Leixlip, and another mill in nearby Celbridge. Besides owning the mill he still had land with tenants on it and had therefore not forfeited his right to be called a gentleman; but when he died he left his family in seriously reduced circumstances. Whatever sort of gentleman he may have been, Samuel Robinson Roe was certainly not much of a businessman. He was, as people benevolently say in Ireland, fond of a drop, which means that he was a heavy drinker and he died relatively young. Whatever his failings, however, the religious atmosphere in his household was strong, and was not lessened by his marriage to Anne or Nanny Belas, for the Belases, who were solicitors in Andrew Street in Dublin, were very evangelical. In the days of his prosperity Samuel Roe had rather fancied himself as a speculator and financier, discounting bills and buying up mortgages, but when the mill, which failed to thrive, ate up his capital, including his land, his resourcefulness appears to have been exhausted at the same time. The strong will and character that May Roe exhibited throughout her life seems to have come from her mother, a neat little woman of high – or should one say low? – evangelical principle, which she passed on to her daughter.

The Roes had a number of children; and when Samuel Robinson Roe died his daughter May was only fifteen. Since they had not been well provided for, it was desirable that she should find an occupation and eventually she had, to use a phrase of the day, 'gone nursing'. The burden on relatives of a large family had to be alleviated in any way possible. What made nursing an acceptable option was the advent in Ireland in the late 1880s of the 'lady probationer', the daughter of gentlefolk who entered a hospital for training as a nurse. After this it became quite a fashion for such young ladies to do so. The Adelaide Nursing School was especially popular among Protestants with some social pretensions, but it was also a good place as far as actual training for the profession was concerned, the Adelaide having been among the first hospitals in these islands to institute any sort of training system at all, in 1858, when a colleague of Florence Nightingale's set it up. The financial rewards of nursing were not very great, varying between £12 and £30 a year when the training period was over, but a nurse might be expected to have good marital opportunities, especially if she had the right social background herself. Samuel Robinson Roe's relatives may well have considered nursing to be as good an option as any other. Whether May had any real vocation for it is a different question. There was not much in her subsequent attitudes which suggested a natural nurse, though she was efficient enough at binding up her children's smaller physical wounds.

While it was not a matter of remark that a young man of Willie Beckett's age and type should still be a bachelor at thirty, a girl of that age who was still unmarried was regarded as in danger of being left on the shelf. When she met Willie, or, as she shortly began to call him, Bill Beckett, May was doubtless under some pressure, whether great or small, overt or otherwise, to find a husband. Bill for his part was on the rebound. He had suffered a severe emotional reverse; and, besides, he too would have had urged upon him the necessity of 'settling down'. No doubt all this contributed to whatever process of falling in love or falling for each other there was; but on the surface at least they were a somewhat ill-matched pair. Bill was a comfortable fellow, pleasure loving and emotionally undeveloped; she was moody, questioning and exacting. She was also witty, even sarcastic, and she had a vivacious side, being subject to states of comparative elation as well as depression; so he may have seen her as high-spirited as well as serious.

She had been educated at the Moravian Mission School, Gracehill, outside Ballymena, County Antrim, which does not mean that either the Roes or the Belases were necessarily of the Moravian persuasion. Gracehill had been the first boarding school for girls in Ireland. The Moravians were known as good, thorough-going educators. Low Church Protestantism – which virtually all Irish Protestantism was – would have little difficulty with their non-doctrinal approach. And in any case their doctrines, such as they are, conform to the major Protestant confessions. At school she had frequently been in trouble for indiscretions and breaches of discipline, the sort of high spirits that verge on hysteria; and there had been a threat of expulsion for talking to a boy over the rear wall, an episode to which as a married woman she made occasional reference and of which she seemed rather proud. Probably the extreme high sprits which she occasionally displayed were evidence of a tendency to mood-swings, what a later age would call manic-depressive tendencies, before calling them bi-polar personality traits, but of course that would not have been evident then.

Whatever the degree of attraction between them, the courtship was swift and conclusive. Bill Beckett and May Roe were married according to the rites of the Church of Ireland on 31 August 1901.

Bill's degree of prosperity as a young man of thirty is indicated by the fact that he immediately set about building a substantial new home in the most fashionable suburb of Dublin. Obviously he was good at his job, but he also had family connections; and in truth it was not easy for a Protestant with a good start to fail in business or the professions at the time.

At the turn of the century the grip of the Protestant element on business and professional activity in Dublin was still very strong, though for a long while now they had been conscious of a challenge from upwardly mobile Catholics. This supremacy was largely maintained through the Freemasons and other mutual aid societies. An approach from one mason to another was a recognized way of landing a contract or getting a job or even a bank loan and it was utilized to the full. The Catholics too had their mutual aid societies, such as the Knights of Columbanus, but the masons, with headquarters in a splendid building in Molesworth Street, not far from Bill Beckett's office, still reigned supreme. Both Bill Beckett and his father were masons; and Bill made use of his membership of the order to an extent which his old master Pannister considered to be scandalous, refusing to speak to him because of it.

Like the aristocracy, the Protestant business community of the towns and cities looked down on Catholics as, in general, rather feckless, lazy and dishonest. A sort of right to ownership and control of business as a prerogative of greater thrift and industry, never mind the favour of Providence, was widely assumed. Except perhaps perforce as employers, and to some extent as manufacturers or shopkeepers, they took care to have very little contact with Catholics; and the aim of many Protestant business people as employers was as far as possible to recruit their clerical staff and work force from among their co-religionists. There were then many thousands of lower-middle-class Protestants from among whom to recruit and even a relatively smaller number of working class, of which number John Casey, or Sean O'Casey, was one. Socially too they kept their distance as far as possible. It was a boast among the denizens of Foxrock, the suburb in which the newly married couple were about to live, that one could pass one's day without speaking to any Catholic other than the railway company's employees. As Vivian Mercier has put it:

The males and some of the females of the typical Protestant family took the train every weekday to office, school or university in Dublin. In all these places they were likely to be associating almost exclusively with fellow Protestants. The females who stayed at home spent their leisure time with other Protestant ladies, though their maids and gardeners were usually Catholic. If one preferred to think of oneself as English there was really no reason not to.

9

But to call this class Anglo-Irish and to lump it in with the Protestant land-owning aristocracy – the class to which Yeats affected to belong and to which J. M. Synge and Lady Gregory actually did – is to create considerable confusions. Of course, the two sometimes overlapped: Bernard Shaw's father was, rather in the pattern of Samuel Roe, an unsuccessful Dublin corn merchant who had a bit of land with tenants on it – as John Butler Yeats, a member of the Protestant professional middle class, also had – and therefore had pretensions to gentility. But in general the line between people in trade and those deriving their income from land or a profession was very clearly drawn, as it was at that time in other countries; and though Willie Beckett's professional or quasi-professional qualification and May's descent from land-owners freed them from the opprobrium attached to being in trade, it was the business class to which the Becketts really belonged. Anglo-Irish is a misnomer also because in fact the Protestant Dublin middle classes probably looked to England less often and with less social anxieties than did their landed co-religionists. They did not, for the most part, send their offspring to English schools or take their daughters to London for the season, still less disport themselves at Cowes or Ascot. They did not take it for granted that their cadets would obtain commissions in the British army. Though their loyalty to the Crown and the Union Jack was automatic and unquestioning, their sources of income had not been threatened by the sort of legislation which had brutally loosened the aristocracy's grip on the land – and therefore on Ireland itself – since the 1880s and so their interest in British politics was less fevered and personal. They were content to vote Unionist and hope for the best.

The other ways in which they differed were the ways in which a bourgeoisie differs from an aristocracy everywhere, but with some exacerbations which were peculiar to the Irish situation. Even among aristocracies, the Anglo-Irish was notoriously improvident and devil-may-care in its outlook. Its tradition of rakehelly violence, celebrated in the novels of Charles Lever and Samuel Lover – just as its improvidence is bemoaned in those of Maria Edgeworth – went back to the seventeenth century; and though tamed somewhat by Victorianism, was still being celebrated by W. B. Yeats in the 1930s. Whatever the past history of the Roes, it was far from the outlook of the Becketts and their kind; and far too from the outlook of their offspring, Samuel Beckett, either as a man or as a writer.

As befitted a bourgeoisie the outlook of the Protestant middle class was far more scrupulous, honest and industrious, less eccentric and also less centred on the Vice Regal Court in Dublin Castle with its multifarious snobberies and its petty pretensions. The middle class was snobbish enough, the principal internal manifestation of its attitude being the line drawn between professional people and those in trade, one of the reasons why Bill Beckett called himself a civil engineer in Foxrock when he went to live there. But it was more in touch with reality.

And there was another reason why the term Anglo-Irish for members of the Protestant middle class is misleading. By comparison with Anglicanism the Church of Ireland was a Low Church and had been since Cromwell. But the incumbents of city parishes were expected to be even lower in their practices and their disdain for ritual than their country colleagues and were more jealously examined for traces of Romanism in their ritual or dogma. O'Casey has brilliantly described the violence and obloquy that was visited on the incumbent of an East Wall parish who was adjudged too Anglican in his practices. For the middle class the Bible was the supreme guide and test of religious belief; and in many middle-class homes it was the only reading matter encouraged.

Bill and May Beckett's first home was a substantial late Georgian, rented house in Pembroke Road, fairly near Earlsfield. Three years later they moved to Foxrock, which for most members of the Protestant Dublin middle class represented the pinnacle of social and commercial success. Where he built, inland towards the mountains, new houses were appearing, but nearer the Stillorgan road the merchant princes of Dublin had earlier bought some of the substantial mansions to be found there, each behind its demesne wall with a stretch of driveway, croquet lawns and frequently tennis-courts, some of them houses of considerable architectural distinction. As Foxrock spread west towards the mountains it retained its exclusive character, the new houses being sizeable and standing in fairly extensive grounds. At a slightly later period, when suburbia was growing fast, the great Dublin comedian Jimmy O'Dea would describe the more central districts of Rathmines and Rathgar as 'a purgatory for souls awaiting the heaven of Foxrock'.

The house, which was proudly featured in a supplement to the *Irish Builder* on its completion in 1903, is sizeable but commonplace, with elements of the mock Tudor in its mullioned windows and projecting gables. When it was sold again in the 1970s, its mock Tudor elements were made a virtue: 'This is a charming Tudor style family residence well set back from the road at the corner of Kerrymount Avenue and standing amid totally secluded mature gardens laid out in lawns, tennis court and croquet lawn.' After pointing to the 'extremely high standard of finish' and the 'spacious proportions' of the main rooms, the advertisement concluded: 'Churches and schools are conveniently situated and there are excellent shops situated in Foxrock Village which is five minutes' walk from the property. 'Cooldrinagh' is approached by a sweeping gravelled driveway.'

Except that the implied length of the driveway is something of an exaggeration, it is true enough; and of course Foxrock with its little railway station was even more of a village when Samuel Beckett was born. Inside, the house gave an impression of bourgeois solidity and comfort, with mahogany panelling in the hallway and on the staircase, heavy mahogany furniture and thick velvet curtains. It was a time when most people in most social classes preferred evidences of expense, durability and workmanship to aesthetic considerations – or rather, they equated these things with the beautiful: and the Becketts were no exception. Except for the master bedroom with its big bow window on the first floor the house was rather dark. There were two other bedrooms on that first floor and two more rooms on the top or attic which could be used as servants' or children's bedrooms or playrooms. Downstairs were a drawing room and a dining room. Like the bedrooms, these had bell pulls to which, supposedly at least, a servant would respond. Outside the grounds were, as the advertisement said, spacious enough to accommodate a tennis-court and a croquet lawn, as well as a vegetable garden, flower beds and wilderness areas with larch and pine trees. Besides all these there were outhouses, including a hen-house, a garage and a stable in which, later on, May would keep a donkey. There was also a little wooden summer-house or gazebo with stained-glass windows which is described in the story *Company*.

The house was called Cooldrinagh after the house which the Roes had owned in Leixlip. Some commentators have identified it with the description of Mr Knott's house in the novel *Watt*; and certainly it resembles it in its general situation; but Mr Knott's house is much larger, and its American bar, oratory and dairy are taken from the biggest house in the locality, in fact the original 'big house' which was there before Foxrock became a suburb, the neo-Gothic Glencairn – home, when Samuel Beckett was growing up, to the exotic American, the Boss Croker. Cooldrinagh is, however, the model for Moran's house in *Molloy*, even to the lemon verbena to whose scent he refers, a flower which, by the time Sam was a toddler, already grew in profusion round the hall door, giving forth 'a fragrance in which the least of his childish joys and sorrows were and would for ever be embalmed'.

Chapter Two

May and Willie Beckett differed physically and temperamentally almost
as much as it is possible for two human beings to do; and in most ways
Sam would appear to take after his mother. Whereas Willie was stout
and burly, May was very tall and thin – as Sam would be; and while
Willie Beckett was round-faced, both May and her son would have
strong, bony features with assertive noses and chins. The son would
also inherit his mother's piercing, light blue eyes and fair, faintly reddish
golden hair.

Temperamentally too he would seem rather more like one of his
oddly matched parents than the other. Willie Beckett's nature allowed
him to fit in with his surroundings and make the best of them. Easy-
going, fun-loving and jovial, his was a secular outlook, rejoicing in the
world as he was given it, greeting it with enthusiasm and shrewdly
accepting its values. May, on the other hand, found the world a rather
wicked place which only the strict principles of conduct which sprang
from her own variety of religious puritanism could redeem. Although
she accepted the values of her own class without question and was
blind to its failings, to everything else she brought the all-or-nothing
perfectionism of the essentially religious temperament. Her son would
later claim to have little sympathy with such an outlook, but he would
all the same inherit this extremism rather than his father's adaptability
and moderation. While Willie disliked trouble and fuss, May demanded
that those around her live up to her own high standards and was not
averse to scenes and confrontations if they didn't. Though Sam would
dislike confrontations and argument and, significantly, would always
take pains to avoid them, he too would set himself standards that were
almost impossible to meet.

Discussing the religious aspect of his upbringing later on, Beckett would say that he had been brought up 'like a Quaker', seeming to imply a degree of strictness rather than any particular sectarian outlook. In his other statements about his childhood there is a dry insistence that by orthodox standards it should have been happy enough – he was not beaten by his father or deserted by his mother – combined with suggestions that there was a good deal lacking that the heart of a child might desire. 'You might say I had a happy childhood,' he told one questioner, adding, however, that he had 'little talent for happiness' and that he was 'often lonely'. Referring to his childhood on another occasion, he said that he was 'aware of the unhappiness' of others around him.

During the early years of his marriage Willie Beckett was much more often at home in the evenings and at weekends than he would be later on. He was addicted to the type of literature which could be bought at railway bookstalls for a shilling or two, often in lurid yellow covers, the sort of thrillers and adventure yarns that he called, perhaps misunderstanding the term, 'potboilers'. While May seldom read anything except the Bible and parts of the *Irish Times* – she declared that reading made her nervous – Willie would read for a while every evening. Tired after his day in Dublin he would throw himself into an armchair and sit there 'motionless . . . under the singing lamp', seemingly 'absorbed and null' and bestirring himself only to turn the pages or relight the pipe which continually went out. While he read he was apparently oblivious of what went on around him, but if next day anybody chanced to ask him what the book he was reading was like, he could not tell them.

The little boy's often somewhat tearful and stormy bedtime was usually preceded by the telling of a bedtime story, and as time went by one story in particular was fixed upon as a ritual. While his father sat in his armchair, 'his big belly bursting out of the old cardigan and unbuttoned trousers that rested him from his office canonicals', Sam would climb up on his lap and listen with fascinated attention until he grew drowsy.

The often-repeated story concerned

. . . the adventures of one Joe Breem, or Breen, the son of a lighthouse keeper, a strong muscular lad of fifteen, those were the words, who swam for miles in the night, a knife between his teeth, after a shark, I forget why, out of sheer heroism. He might have simply told me the story, he knew it by heart, so did I, but that wouldn't have calmed me, he had to read it to me, evening after evening, or pretend to read it to me, turning the pages and explaining the pictures that were of me already, evening after evening the same pictures till I dozed off on his shoulder.

14

And after the bedtime story would come night prayers, a recital in which the child was expected to invoke God's blessing on Daddy, Mummy, Frank and Nanny, and to beseech his creator to make him a good boy 'for Jesus Christ's sake, Amen'.

After Sunday dinner, which in accordance with the custom of all creeds and classes was eaten in the mid-afternoon, Willie would retreat to the summer-house in the garden with a cushion and a copy of *Punch* and his little son would go with him. 'The waist of his trousers unbuttoned, he sat on the one ledge, turning the pages. You sat on the other with your feet dangling. When he chuckled you tried to chuckle too. When his chuckle died yours too.'

Little boys are supposed to be physically adventurous and brave, and as soon as he had learned to walk Sam was certainly both; in fact a child psychologist might think that some of his exploits were so dangerous as to suggest a certain level of disturbance. On some of these occasions May beat him severely, either with her large hands or, occasionally, with a stick. From an early age he bore these punishments with the same stoicism and seeming indifference as he bore the bruises and cuts which resulted from the exploits themselves; and far from teaching him caution, they made him if anything more reckless.

One of his diversions was to climb to the top of a fir tree in the garden and, stretching out his arms and legs as if for flight, fall deliberately to the ground, trusting to the lower branches to break his fall.

You are alone in the garden. Your mother is in the kitchen making ready for afternoon tea with Mrs Coote. Making the wafer-thin bread and butter. From behind a bush you watch Mrs Coote arrive. A small thin sour woman. Your mother answers her saying, He is playing in the garden. You climb to near the top of a great fir. You sit a little listening to all the sounds. Then throw yourself off. The great boughs break your fall. The needles. You lie a little with your face to the ground. Then climb the tree again. Your mother answers Mrs Coote again saying, He has been a very naughty boy.

Sam was four years younger than Frank, and his feeling of loneliness is largely attributable to this crucial age gap. But from the time when he was old enough to be tolerated by an elder brother they played together, the effort to keep up with his brother making him apparently tougher and more competitive. In their contests he refused to admit that the difference in physique gave the eldest an insuperable advantage and fought hard to win. Though such a situation almost always induces a latent pessimism of outlook in the younger and necessarily weaker,

15

there was no sign of it in his competitive attitude; and he was certainly more reckless than Frank in many ways. May's anger was especially aroused when her younger son dropped a lighted match into one of the cans of petrol which in those days were an indispensable adjunct of motoring. Finding it standing beside the kitchen door, he decided to test what he had been told about its inflammability. The can was almost empty but the fumes ignited and flared up, singeing his eyebrows and burning his face. His mother attended to his wounds with her usual skill, but her anger was deep.

That May should be the anxious parent meting out punishment while Bill Beckett took a more tolerant and amused attitude to the adventurousness of his young son is not surprising. Bill was not there in the daytime to worry about where the child was and what he was doing. And he was also a sedulously masculine man who was keenly, not to say ferociously, devoted to physical pursuits. Whatever anxiety he may have felt on occasion was probably not unmixed with pride that both his male offspring were showing early signs of physical courage and prowess. A woman in such circumstances often finds herself cast as the one who discourages dangerous games and initiatives while the father tacitly encourages; but May was not only an anxious mother whose fate it was to contend with a small son who was seemingly careless of the risk of physical injury; she also had a fierce and unpredictable temper, so that her punishments sometimes struck onlookers as out of proportion to the crime.

On at least one occasion, however, Bill Beckett's desire to be the father of brave, athletic and masculine sons overstepped the mark. He was accustomed to swim at the Forty Foot, a celebrated swimming hole among the rocks at Sandycove, which in those days was an exclusively masculine preserve where men bathed naked. As soon as Sam could reasonably be expected to learn to swim Willie began to take him there with his brother. The diving board was about ten feet above the water (which in fact is twenty-four feet in depth) and when Sam was still quite small his father placed him on it and ordered him to jump. For the rest of his life Sam was to be visited by a recurring nightmare in which he was expected to dive into a small and distant pool which was often ringed around with dangerous, jagged rocks. 'You stand at the tip of a high board. High above the sea. In it your father's upturned face. He calls you to jump. He calls, Be a brave boy. The red round face... The thick moustache. The greying hair. The swell sways it under and sways it up again. The far call again. Be a brave boy. Many eyes upon you. From the water and from the bathing place' the hero of *Company* is told by the voice which recapitulates incidents in his childhood. Watt, in the novel of that name, is likewise subject to an uneasy sleep, 'lacerated by dreams' in which he has to dive 'from dreadful heights into rocky waters, before a numerous public'. While at the opening of the third act of an unproduced play, *Eleutheria*, Victor is waking from a nightmare and crying out, 'No – no – too high – rocks – my body – papa – be brave – a good little boy – I am brave –'.

The gift of easy and uneventful sleep was never in fact to be his. All his life he suffered from nightmares, many of them, as the poem 'For Future Reference' would record, associated with the fear of high places. Belacqua, in *Dream of Fair to Middling Women* refers to 'dream, with its sweats and terrors'. As Lawrence Harvey, to whom he spoke about such matters, puts it, 'The fear of dreaming is a leitmotif that runs through Beckett's writing as a kind of subterranean analogue to the misery of living.'

His insomnia was probably inherited, for his mother suffered from the same dreadful complaint. She also, as has been suggested, suffered from mood-swings, periods of often acute depression, balanced by occasional and less obviously noticeable elated states. Like other sufferers from this still not fully understood syndrome she appeared to be two different people, and of course those closest to her were more aware of the depressive than others, who tended to see her during her more active and outgoing states. Naturally both her children and her husband found this puzzling and sometimes acutely dismaying. One of the perverse effects of the ailment is that it sets up in those close to the sufferer an acute anxiety to please: partly, of course, a desire to help, but partly also a hope merely to avert the apparent hostility of the depressed, for the change of personality seems like a calculated withdrawing and fickleness. Much of Sam's later anxiety about his mother and desire to please her were probably fostered in childhood by these alternations of personality, the apparent warmth and approachability of one phase being horribly followed by silence and seeming hostility. The length and frequency of her unapproachable periods also created a need for mother substitutes which had an influence on his later sexual development.

One early episode in which May displayed coldness and unapproachability appeared to rankle more than others, finding an astonishingly frequent place in his work. They had been to Connolly's Stores in the neighbouring village of Cornelscourt and were walking back afterwards hand in hand up Cornelscourt Hill Road past Tyler's, a market garden which they also often visited. Sam, who had been looking up at the sky and remarking its blueness, asked his mother whether the sky was not further off than it seemed. According to the version of the story in *Malone Dies*, she replied: 'It is precisely as far away as it appears to be.' In *Company*, written almost thirty years later, the reply is more acidulous: 'For some reason you could never fathom this question must have angered her exceedingly. For she shook off your little hand and made you a cutting retort you have never forgotten.' In the story called 'The End' the small boy's question is slightly different but the mother's retort is indeed cutting: 'A small boy, stretching out his hands and looking up at the blue sky asked his mother how such a thing was possible. Fuck off, she said.' It is unlikely that May retorted with these exact words, but whatever she said, it hurt, as did other evidences of unapproachability. It was probably the circumstance that they were walking home along a familiar route in apparent harmony when the illusion was suddenly shattered that made the rejection memorable, causing Beckett to return to it for the third time when he was nearly seventy.

In *Malone Dies* the question about the height of the sky is related to an air display which took place on nearby Leopardstown racecourse when Beckett was four. Less than twelve months after the first flight ever achieved in Ireland had taken place, the newly formed Irish Aero Club staged an event in which well-known international aviators took part, using, among other aircraft, Farman Biplanes and a monoplane similar to the one in which Blériot had crossed the English Channel the year before. The Becketts were naturally present and the looping of the loop on this occasion is recalled in *Malone Dies*, in which the protagonist recollects his mother holding him by the hand and gazing at the little plane while saying, 'It's a miracle, a miracle.'

Those who saw May only in her more outgoing phases naturally had a somewhat different idea of her personality. With some of them she was charming, kind and, on the subject of gardening, for which she had a passion, both helpful and informative. She was very friendly with Mrs Elvery, who lived less than a mile away at Carrickmines. The Elverys owned a well-known emporium in the centre of Dublin, the Elephant House, specializing in waterproof and sporting goods and were looked down on in Foxrock because they were 'in trade'. Theresa Elvery was intellectual and artistic, which was certainly no ground for intimacy with May, and she was also much older; but nevertheless they became great friends. Her daughter, Beatrice, subsequently Lady Glenavy, believed their intimacy had as much to do with her mother's deafness as anything else. 'This was largely due to the kindness of May Beckett's heart. She took pity on my mother because of her deafness. They were both interested in gardening, and I remember May Beckett coming flying on her bicycle with some new cutting to plant for my mother's garden. I know my mother valued her kindness and companionship very much.'

Another of May's friends was Mrs Susan Manning, whose daughter Mary, subsequently an actress and playwright and a well-known figure in Dublin literary and theatrical circles, was a frequent visitor to the house as a child. Mary remembered May Beckett as 'a very moody person' and suspected even then that she was a recluse by nature who appeared unwillingly in social situations. She also considered her 'very principled and very deeply Protestant' in what Mary thought of as 'a Baptist sort of way'; she would have made a good nun, if her Church had nuns. Mary also thought that she was strikingly mannish – 'virginal, a nun, a manly nun' – and entirely unsuited to marriage and child-bearing.

Protestantism as most people understood it was mainly a system of punishments and rewards, with the Bible as the principal source of authority and at least as much respect accorded to the Old Testament as to the New. May read the Bible to her children every day and continued to do so even after they had begun to attend Sunday school. Her husband attended church and looked upon himself as an orthodox Irish Protestant, but had no attitude to religion except an unreflective agreement with its doctrines and a belief that it was a social and personal necessity. May, on the other hand, was a believer. Even if her Christianity was severely limited by her time and her place, it still had an effect on her treatment of people; and this she tried to transmit to her sons. More than that, she was exceptionally devout and pious by nature and she naturally hoped that something of this piety would be transmitted also.

With Frank she succeeded. He recited his night and other prayers with marked devotion and continued to do so 'as long as he could kneel'. In Sam's case, according to himself at least, the story was otherwise. Whatever the behavioural precepts, the examples of piety had little effect on him because, as he afterwards put it, he had 'no religious feeling' and, except on one occasion, never felt 'a religious emotion'. There is a celebrated photograph which shows the little boy kneeling with his hands clasped in prayer at his mother's knee. May bends over him, 'her huge head hatted in birds and flowers' and appears to be joining his childish hands in prayer.

in a word bolt upright on a cushion on my knees whelmed in a nightshirt I pray according to her instructions

    that's not all she closes her eyes and drones a snatch of the so-called Apostles' Creed I steal a look at her lips

she stops her eyes burn down on me again I cast up mine in haste
and repeat awry

the air thrills with the hum of insects

that's all it goes out like a lamp blown out.

In so far as it suggests native piety on the little boy's part, however,
the photograph is a fake. Beatrice Elvery's sister Dorothy, who was an
art student, wanted to paint a subject picture called 'Bedtime'. Four-
year-old Sam Beckett was 'just the right size and age', so she asked
May if he could pose for her. Realizing the practical difficulty of repeated
sittings, she took a photograph to paint from. This photograph is among
those which the central character in *Film* tears up, dropping the pieces
on the floor beside him.

As has been the case with other people, including Napoleon Bonaparte, the solitary occasion on which Sam felt what he called 'a religious
emotion' was that of his first communion, which he took at the age
of seven in Tullow parish church, the bare, undecorated Gothic little
place of worship, with its steeply pitched roof and small, semi-detached
steeple, where the family attended service. When he spoke of it he
gave no indication of the exact kind of transport he felt, but there may
be a clue, however satirically expressed, in the radio play *All That Fall*,
where Miss Fitt in the same church feels 'alone with her maker' and
after she comes out into 'the sweet fresh air', she says, 'why even then
for the first furlong or so I stumble in a kind of daze.' But whatever
the case on this occasion, those who saw the Beckett family at Sunday
morning services there remembered Sam as having an even glummer
expression than the rest of the congregation. Bea Orpen, whose parents
lived in Carrickmines and worshipped in the same church, used to
think he was frowning at her when she met his glance. It was only
later that she realized his dissatisfaction was nothing personal. It was
directed at the whole of creation.

The extent to which he was impressed by the cadences and spiritual
content of the Authorized Version of the Bible during his mother's
daily readings has perhaps been exaggerated by a generation less familiar
with family readings than his own. Until a comparatively short time
ago the Bible was the first great work of literature or ontology to
which most people were introduced and, as such, it of course proved
memorable, but his frequency of quotation from it is no guide to the
extent to which he was influenced, either theologically or in a literary
sense. When he became a writer he made frequent reference to the
Bible in his works, but reference in Beckett may be ironic or even

20

sarcastic. It may be intended simply to bring about an echo chamber effect, giving resonance to what is being said; or it may be an oblique reminder of how far his personages have travelled from the certainties of the past. Later he would sometimes insist that his use of the Bible and Christian mythology was merely a literary device, rejecting the idea that he had been profoundly affected or influenced by it. This may have been true, but in after-life he always possessed a Bible, at the end more than one edition, and Bible concordances were always among the reference books on his shelves. He certainly knew the book backwards and as a boy he won a prize for knowing it in the diocesan synod examination. The prize was Thomas Hardy's *Tess of the D'Urbervilles*, and if Beckett had been of an age to reflect on such things he might have been struck by how quickly time had promoted Hardy's so recently scandalous novel into a classic which could be given to little boys in the confidence that it would almost certainly be neither read nor understood. But Sam may have read it, for he underlined a sentence which struck him as relevant to his own sleeping difficulties even at this early age and is quoted by Belacqua in *More Pricks Than Kicks*: 'When grief ceases to be speculative, sleep sees her opportunity.'

Later on, Samuel Beckett would be dismissive of Irish Protestantism, saying: 'My brother and mother got no value from their religion when they died. At the moment of crisis it had no more depth than an old school-tie. Irish Catholicism is not attractive, but it is deeper.' He added that as he grew up he found religion 'irksome' and so he 'let it go'. Yet he would also say afterwards that Protestantism, with its 'absurd prohibitions' was 'something you can't escape'.

But that some sort of religious feeling is manifested in his work it would be dangerous to deny; and religious interpretations have not been lacking. It could with some truth be suggested that nobody unaffected by religion could have written much of it; and it could certainly be said that the 'all-or-nothingness' which he may have inherited from his mother is part of the religious temperament.

May struck many people as exceptionally well-mannered and considerate of others, something that may have been partly an expression of Christian charity; and Sam in later years – particularly the famous Sam of the post-war era – would also strike many people as extraordinarily courteous and mannerly. She was also passionately fond of animals and it was probably from her too that he inherited his own fondness for the brute creation – something which is scarcely to be guessed at from his work, or indeed his life later on, since he took care never to own an animal and take responsibility for its well-being or its love. One of the traumas of his childhood is described in *Company*, in which the voice reminds the central character how, as a little boy, he took pity on a hedgehog which he found out in the cold garden and put it in an old hatbox with some worms. He goes to bed glowing with his own good deed and includes the hedgehog in his prayers. 'You glowed easily in those days but seldom for long. Hardly had the glow been kindled by some good deed on your part or some little triumph over your rivals or some word of praise from your parents or mentors when it would begin to cool and fade leaving you in a very short time as cool and dim as before.' In this case the glow was succeeded the following morning by a great uneasiness, 'a suspicion that all was perhaps not as it should be'. Days passed before he could bring himself to return to the box; and he was never to forget what he found there: 'The mush. The stench.'

May had a succession of donkeys over the years, which she drove around the neighbourhood harnessed to a little trap. The most memorable of these was called Kisch, a name bestowed on him by Sam and Frank. Kisch would sometimes stop on the road of his own accord and stand apparently lost in contemplation for lengthy periods. At these times May neither beat nor attempted to persuade him, but sat patiently upright in the little trap until he decided that it was time to go on again. She was constantly on the look-out for cases of maltreatment of donkeys, which, particularly where the tinker people who camped in some of the Foxrock back roads were concerned, were not infrequent; and Kisch was actually bought from a tinker who had been beating him and, perhaps anticipating a deal, refused to desist when May reproached him. There were also dogs – Kerry Blue terriers, representatives of a quarrelsome breed who caused a great deal of trouble when taken for a walk; and finally a very ill-tempered Pomeranian. Animals, like gardening, provided a good deal of the ambience in which she met other people and conversed with them. Nearly always wearing flowered hats, she was a well-known figure at dog and donkey shows or driving round the roads of Foxrock in her little trap. These flowered hats, worn in the garden and frequently round the house as well as out shopping, were one of the best-known things about May in the locality.

Like most families of their time and class, the Becketts had a nanny while the children were small; and one particular nanny, known as Bibby, is mentioned in several of Samuel Beckett's writings, including *Happy Days*, *Texts for Nothing* and *Dream of Fair to Middling Women*, unpublished in his lifetime. This was a young woman named Bridget Bray who seems to have entered the Beckett household when Sam was three or four and remained until he was in his teens. She had a fount of stories, including fairy and ghost stories, deriving from her County

Meath background and, when Sam was small, sometimes played dressing-up games with him and Frank. In the third of the *Texts for Nothing* the hero, who is making up a life in his imagination, promises himself a nanny: '. . . I'll be her sweet pet . . . She'll say to me, Come, doty, it's time for bye-bye. I'll have no responsibility, she'll have all the responsibility, her name will be Bibby, I'll call her Bibby, if only it could be like that.' And in an unusually revealing story, 'A Case in a Thousand', published in the *Bookman* in 1934, the central figure, a Dr Nye, suffers from melancholy, which he regards not as a 'natural and proper' state, but as 'a disorder'. In search of its causes, he meets his former nanny, Mrs Bray. She informs him that 'you were always in a hurry to grow up, so's you could marry me' but, we are told, she 'did not disclose the trauma at the root of this attachment'. The trauma may have been a temporary rejection by his mother, or it may have been something more physical, a revelation of sexuality. Beckett later on took semi-humorous notice of the uniformed nannies he saw with their charges in places like St James's Park in London, and on occasion fantasized about the ideal nanny with whom he would like to 'get off' and perhaps form a deeper relationship. But 'she must have a strawberry nose and suck cloves, or at least peppermints', he told his friend Thomas MacGreevy, evidently referring to a previous conversation about Bibby. Bibby subsequently married a gardener and became 'mother of thousands'. Sam's deep attachment to a particular nanny may have been because she gave him some sort of love he felt his mother did not give. Mary Manning, who knew him well subsequently and at a certain point had a relationship with him which was all but an affair, thought that as a small boy he had been starved of love. 'There wasn't much love going in the Beckett household,' she has said, adding that in her opinion Sam was emotionally 'under-nourished. He suffered from emotional malnutrition.' Perhaps what is at issue here is a general atmosphere of love or lovingness in the household. There is no doubt that May Beckett loved her son fiercely. Later on he would speak of her 'savage loving', but somehow it did not come through to him as a child in the right way.

One of Bibby's rules was that the word 'bottom' should not be pronounced in full, a sort of abbreviated pronunciation, 'btm', being the most that was allowed. This was a common prohibition in genteel households, and of course no ruder term could be used either. That the prohibition had the general effect of calling attention to this part of the anatomy is probable enough, but it does not follow that it accounted for the specifically sexual interest that it would have for Beckett during and after his sexual awakening. In any case, this 'sexual

23

over-valuation', to use a rather amusing Freudian term, would last, to be mentioned several times in his work, for the most part comically, of course. In 'A Case in a Thousand' Dr Nye sees from the hospital window a group of people watching the progress of a canal barge through a lock and under a bridge. He observes that they have crossed over the near side of the bridge, 'with the result, most pleasing to him that where formerly he had seen their faces, now he enjoyed a clear view of their buttocks, male and female'. There was also 'an extension of sexual interest', to employ Freud's terminology again, of a rather strange nature to which seemingly comic reference is made in *All That Fall*. Mrs Rooney in that play goes to hear a lecture 'by one of these mind doctors' in the hope that he might 'shed a little light' on her lifelong 'preoccupation with horses' buttocks'. This preoccupation also figures in the strange prefatory note to the highly autobiographical *Dream of Fair to Middling Women*. It concerns the childhood and early sexual awakening of Belacqua, Beckett's fictional alter ego.

> Behold Belacqua an overfed child pedalling, faster and faster, his mouth ajar and his nostrils dilated, down a frieze of hawthorn after Findlater's van, faster and faster till he cruise alongside of the hoss, the black fat wet rump of the hoss. Whip him up vanman, flickem, flapem, collop-wallop fat Sambo. Stiffly, like a perturbation of feathers, the tail arches for a gush of mard. Ah . . . !

Besides the nanny for the children, the Becketts' domestic staff included various cook-housekeepers. One was called Ludo from her addiction to that game; and another Bridey or Bridgey. The Mannings thought Bridey was insane, perhaps partly because she was convinced that Cooldrinagh was haunted by an old man often to be seen sitting on a chair in the hallway. She eventually convinced May that there was some sort of a presence there; but their relationship was rather of the love-hate variety for Bridey played a large part in May's complaints about life in general, and after the Becketts acquired a telephone her doings and delinquencies were part of the long tales of woe she imparted to Mrs Manning in her daily calls. There was also at one time a gardener called Christy – who may have been the gardener that Bibby married – and two uniformed maids. Foxrock was a very formal place, where afternoon tea was usually served by a maid in uniform; and wherever the denizens of the house might be at the time, the serving of dinner was announced by the striking of a gong in the hall. There is no reason to suppose that the domestic arrangements in Cooldrinagh were any more pretentious than they were in other similar households, but certainly some of May's general strictness of character and demeanour were reflected in them.

But Foxrock was also a village and the familiar characters of a child's world included several village notables. There was the milkboy from the small dairy owned by a Mr Tully, who came to the door singing his tuneless song, 'and went singing away, having measured out the milk, from his can, to the jug, with his usual liberality'. There was the whistling postman, Bill Shannon, who had come from the 'horrible slum of the cottages' and 'by sheer industry, my dear, plus personal charm' had 'raised himself above his station' by his ability to play the violin, an ability which had led to the acquisition of an evening suit of his own and an invitation to play for, and even to 'dance fleetly with', the gentry. Besides delivering the letters and parcels, Shannon kept greyhounds; and as the young boy lay awake 'long after he should have been asleep at the top of the house on a midsummer's night' he might hear Bill Shannon's 'light nervous step on the road as he danced home after his rounds', whistling keenly and loudly 'Roses Are Blooming in Picardy'.

No man had ever whistled like that, and of course women can't. That was the original, the only, the unforgettable banquet of music. There was no music after – only, if one were lucky, the signet of rubies and the pleasant wine. He whistled 'The Roses Are Blooming' and danced home down the road under the moon . . . with perhaps a greyhound or two to set him off, and the dew descending.

Another night sound that the little boy could hear as he lay awake was the barking of dogs up in the hills, which he imagined to come from the stone-cutters' village of Glencullen; and though that seems unlikely now, such sounds travelled much further at a time when the roads were virtually traffic-free and in the daytime the musical clink of the stone-cutters' hammers could be heard from the Glencullen quarry on the roads round Cooldrinagh. Other peaceful sounds included, in Moran's words, 'the clicking of mallet and ball, a rake on pebbles, a distant lawn-mower' as well as the bell of the church. And of course there was bird-song, 'blackbird and thrush, their song sadly dying, vanquished in the heat, and leaving dawn's high boughs for the bushes' gloom'.

The stone-cutters, according to Moran, lived in hovels, 'like generations of stone-cutters before them'; but this and the mention of 'the horrible slum of the cottages' from which the postman had originally come are the only intrusions of ongoing, stable, domestic poverty in relation to Foxrock in Beckett's work. It was impossible even for a well-brought-up child in such a locality, however, not to be keenly aware of the drifting, itinerant poverty which was everywhere in Ireland – the tinker encampments of the Ballyogan Road across the racecourse, and the beggars who were all the more numerous because Cooldrinagh and the other houses round about were prosperous residences at the limit of the city, to whose side or back doors beggars came almost daily.

Speaking later of the half-demented woman in *Not I*, he would say: 'there were so many of those old crones, stumbling down the lanes, in the ditches, beside the hedgerows. Ireland is full of them.' And there is a revealing glimpse of another half-blind old beggar woman, fumbling at a neighbour's garden gate in the story *Company*: 'On the way home from kindergarten on your tiny cycle you see the poor old beggar woman trying to get in. You dismount and open the gate for her. She blesses you. What were her words? God reward you little master. Some such words. God save you little master.'

The kindergarten, Beckett's first educational establishment, run by the Elsner sisters, who lived round the corner on Leopardstown Road, was attended by many generations of Protestant children from well-to-do Foxrock homes. The sisters had a cook, Hannah, who was reputed never to go out, and a little dog, Zulu, who barked at the gate and is remembered in *Molloy*. Their father was a German who had set up in Ireland as a professor of music, and Sam's formal musical education also began here, with piano lessons from Ida Elsner, a mannish lady known as 'Jack', who was something of a Foxrock character and rode a bicycle. The sisters were rigid disciplinarians who imposed silence and stillness as an integral part of the learning process. Ida taught music in mechanical fashion with much scales practice. This may have suited Sam rather than irked him, for a subsequent music teacher, a Miss Skipworth, who gave lessons at her home in Seapoint, found him to be a technically proficient but rather soulless player – though how a young boy could be technically proficient enough to exhibit soul is another matter. Whatever her skills as a teacher and whatever satisfactions they gave her, Beatrice Skipworth needed a glass of sherry at half-hourly intervals to get through sessions with her pupils and this would be dutifully brought in to her by a maid. All his life Sam loved music, however, and the capacity for musical appreciation, descending through the Becketts from Grandmother Crothers, was shared with many other members of the family, though his father seems to have been free of it. Sam would in time learn to play the piano well enough to sightread almost anything with great musicality, style and phrasing and to accompany singers even in difficult pieces which he had not seen before. More importantly, he could play it well enough to solace himself at many periods of his life; and would even occasionally, when in the right company, venture a song, though his singing voice left a lot to be desired.

As he grew older the 'often lonely' child began to explore the locality for himself, becoming familiar not only with the racecourse but with the adjoining spaces known as Croker's Acres. The Boss Croker was an Irish American who had ascended through the murky underworld of New York politics to the control of Tammany Hall. His simple ambition was to retire to the Old World, become a respected figure on the English turf and own a Derby winner. Ireland, the country of his origin, did not figure in his original plans, but a brush or two with the English Jockey Club induced him to remove himself to what he hoped would be a more sympathetic environment. He bought Glencairn, the big house just over the racecourse from Cooldrinagh, and proceeded to lay out gallops on the great flat space in front of it. The white railings of these were still there, just inside the hedge separating Croker's Acres from the Ballyogan Road, until very recently, when a developer bought it and built a middle-class housing estate which he called The Gallops, to which prospective house-purchasers were enticed by a large board advertising THE GALLOPS: WHERE THE GOING IS GOOD – a slogan which might have amused Samuel Beckett. On these gallops Orby was trained, the horse which won the English Derby the year after Sam was born, when The Boss became the only owner of a Derby winner whom the reigning monarch refused to invite to the royal box after the race. On his return to Dublin Orby was paraded through the streets and an old woman burst from the crowd crying, 'Thanks be to God a Catholic horse has won the Derby.'

Catholic or not, though, there was much about The Boss which scandalized his respectable, prudish Protestant neighbours, including the fact that he lived in sin with his housekeeper and that this lady, who died while in office, so to speak, and received a splendid interment in the grounds of Glencairn, was succeeded as The Boss's bed companion by an American Indian princess.

The English sporting press said that Orby had been 'trained on a cabbage patch in Ireland', but in truth Croker's Acres was a wide-spreading area of verdant pasture over which sheep and cattle grazed and was an ideal place for a boy to roam. It figures largely in Beckett's work and there are numerous mentions of the sheep, the lambs in springtime and, typically, the sheep's placentae strewn over the grass in the lambing season: indeed the placentae seem, perhaps naturally enough, to have made a profound impression on him as a little boy

27

and are adverted to several times. When Belacqua, in *More Pricks Than Kicks*, returns to Croker's Acres he regrets the absence of horses, but sees plenty of sheep and lambs.

These latter were springing into the world every minute. The grass was spangled with scarlet after-births, the larks were singing, the hedges were breaking, the sun was shining, the sky was Mary's cloak, the daisies were there, everything was in order. Only the cuckoo was wanting. It was one of those Spring evenings when it is a matter of some difficulty to keep God out of one's meditations.

Further down Leopardstown Road, beyond the Cornelscourt Cross, was a curious structure known as Barrington's Tower. A three-storey, mock Gothic edifice, square and unadorned but for its fake turrets, it had been built as a folly by the said Barrington, but at this period stood forlornly among gorse and bracken in a field. It came to mean a great deal as a familiar bolthole to a boy who was beginning, even then, to value solitude almost as much as the man would later do.

To this place he would retreat with a book, sometimes staying a whole day, until it was dark and searchers had set out from Cooldrinagh to call him home: the punishment for such a long absence was to be sent supperless to bed. Inside the tower was littered with tins and other detritus; outside he could sit hidden from passers-by among the gorse, with a superb view of the sea; but he liked the alternation of dark and light and would move from one to the other. On one clear, sunlit day he was certain he had seen Wales in the shape of a solitary mountain peak from here; Wales, however, is seventy miles away and there is no tradition of seeing it, so when he mentioned the possibility he was laughed at. But he treasured the memory and was convinced that he had seen the same coast on two other occasions.

Safe from intrusion in his hide-out and invisible among the gorse bushes or the giant nettles which grew there, he would sometimes talk out loud to himself, making up conversations between imaginary characters, 'now one voice now another', until he was hoarse and all the voices had begun to sound the same.

As he grew older he would wander further afield, up the Glencullen road perhaps, and sometimes over the mountain slopes, sinking up to his knees in the bog or following the faint sheep tracks, which had been deepened by the rains, hearing nothing but the wind and sometimes lying in a hollow to look at the sky. In *Mercier and Camier*, doubtless with some irony, the heather on the mountain side is suggested as a literal last resting place: 'It is here that one would lie down, in a hollow bedded with dry heather, and fall asleep, for the last time, on an afternoon, in the sun, head down among the minute life of stems and bells, and fast fall asleep, fast farewell to charming things.'

And in addition to these solitary ramblings there were the organized walks up the mountains with his father. All his life Willie Beckett loved walking over these little mountain roads, so much so indeed that when he died after a heart attack Sam was to say that he 'walked himself to death'.

Some of these tramps were of extraordinary length and duration, lasting perhaps from eight in the morning until eight in the evening; and for these longer excursions he would take a flask of whiskey and sandwiches with him. In other than warm weather he would wear a long green raincoat with a velvet collar on these expeditions. Originally a motoring coat of the sort that was necessary before the First World War, it finally became his walking garb; indeed, as it deteriorated over the years, his 'walking rags'. Even when Sam was quite small they would set out hand in hand up the Glencullen Road with the sea behind them gradually coming into view, the Three Rock mountain ahead to the left and the equally rocky eminence known as Prince William's Seat directly in front. His father taught him the names of these mountains and of the islands and promontories of the sea, as well as of the lights and lightships that became visible after dark. As Sam grew older the range of these walks naturally extended and they would go on from Glencullen by the military road built in 1798 towards Glencree. This is the 'still carriageable' road over high moorland described in a consciously bravura passage in *Mercier and Camier*:

It cuts across vast turfbogs, a thousand feet above sea-level, two thousand if you prefer. It leads to nothing any more. A few ruined forts, a few ruined dwellings. The sea is not far, just visible beyond the valleys dipping eastward, pale plinth as pale as the pale wall of sky. Tarns lie hidden in the folds of the moor, invisible from the road, reached by faint paths, under high over-hanging crags . . . Under its heather mask the quag allures, with an allurement not all mortals can resist. Then it swallows them up or the mist comes down. The city is not far either, from certain points its lights can be seen by night, its light rather, and by day its haze. Even the piers of the harbour can be distinguished, on very clear days, of the two harbours, tiny arms in the glassy sea outflung, known flat, seen raised. And the islands and promontories, one has only to stop and turn at the right place, and of course by night the beacon lights, both flashing and revolving . . . End of descriptive passage.

29

Over and over again, in *Texts for Nothing*, in *How It Is*, in *Krapp's Last Tape*, in 'The End', he returns to the bare mountains, the city and the sea below, with the arms of Dun Laoghaire harbour stretched towards the faint horizon and, when night falls, the lights flashing and revolving. In all his work there is a city by the sea, a small coastal plain and mountains behind it. His imagining is different from the literal, meticulous imagining of James Joyce; nevertheless what he says in 'The Calmative' is almost as true for him as it was for his fellow Dubliner. He might imagine and re-imagine, he would include elements of the French landscape – that around Roussillon, for example – or even Paris in his work, 'But there was never any city but the one', the Irish locution making clear which city is meant. In much of his work the abiding image is of an old man and a boy walking the mountain roads. Sometimes the boy is small and they hold hands. In 'The End' the boy wants closer contact but is disappointed: 'It was evening and I was with my father on a height, he held my hand. I would have liked him to draw me close with a gesture of protective love, but his mind was on other things. He also taught me the names of the mountains.' In 'First Love' and elsewhere the father teaches the boy the names of the stars and constellations, including the Wains: 'He had shown me others, but alone, without him beside me, I could never find any but the Wains.'

Many intelligent, sensitive children feel protective of one or other of their parents and apprehensive of what fate may have in store for them, seeing the pathos of their state quite clearly even while admiring them or looking up to them. All Samuel Beckett's writings about his father have this curiously protective quality; and more strangely still he sees this gregarious clubman as essentially lonely and almost beyond communication. He had also in relation to his father the child's wish to instruct. 'One day I told him about Milton's cosmology, away up in the mountains we were, resting against a huge rock looking out to sea, that impressed him greatly'. And, of course, guilt. 'That was always the way, walk all over the mountains with you talking and talking and then suddenly mum and home in misery and not a word to a soul for weeks, sulky little bastard, better off dead.' In 'One Evening' a man in a long green coat with buttons of all shapes and sizes (as Willie Beckett's eventually had) is found dead in a ditch beside what is obviously Croker's Acres. The woman in a long black dress who finds him while 'straying in search of wild flowers' may even be May. Willie Beckett did not die in these circumstances, but the story conveys the essential isolation and pathos that Sam conferred on him.

It would be easy to romanticize or make Wordsworthian Beckett's early relationship with the bare, high places of the Dublin mountains that he and his father tramped over together. We have been trained to see natural scenery, particularly mountains, in the Wordsworthian way as uplifting, inspiring and evidence of some grand but beneficent purpose in creation. In fact Beckett often found the Dublin mountains terrifying. Describing the sensation once to his friend Thomas Mac-Greevy, he told how on a walk across Prince William's Seat, the low mountain between the Glencree and Glencullen rivers, he was 'reduced almost to incontinence by the calm secret hostility' of the place and finally had to run down into Enniskerry. He felt the cold, inhuman quality of such places very keenly, and was by no means reassured or strengthened in any spiritual way by them.

Whatever his son's emotions during them, these long mountain walks do suggest a side of Willie Beckett's character not always apparent to business and other acquaintances; but the need to be out of the house all day probably also reflected the deteriorating state of his marriage. An insomniac who now had her own room, May often roamed restlessly about the house during the small hours, opening and closing doors and sometimes merely sitting silently in the dark. Her temper was uncertain and Bill was naturally finding his wife's angers and silences difficult to deal with. His answer was to take avoiding action. To his inevitable absences at the office he added evenings at his club, as well as golf, racing and these long walks in the mountains. In the early days of their marriage the Becketts had given small dinner parties, mostly for Bill's friends or business associates (it was of one such that Sam thought he had a pre-natal memory), but these were eventually abandoned and when Willie wanted to entertain a contact he did so at the Kildare Street or the Stephen's Green Club; and of course he pleaded business engagements and necessities as excuse for his absences when he did not come home in the evenings. As a result he began to see little of the children except at weekends, though on the occasions on which he claimed to be unavoidably delayed he would sometimes speak to them on the telephone before bedtime. Whatever Bill's desire for continued sexual relations, it was probably not now encouraged by his wife. She had got into the habit of speaking to him, and referring to him in his presence, in tones which varied from outright hostility to mild sarcasm. Though Bill bore this kind of thing with apparent stoicism in company, and though he had the ability to shrug off sorrow and be sociable when he could, he was often quite miserable; and to redress the balance of the life he knew he had called an old imagined life back into existence, brooding over his lost happiness with the girl of his dreams, now married to the elderly surgeon.

By coincidence May's friend Mrs Manning had also been in love with one of the Murphy offspring, a son who had likewise broken off the relationship after being threatened with disinheritance. Bill considered this to be a bond between them and, when out of May's earshot, he would often speak mournfully to Mrs Manning about their cruelly lost loves. Sometimes he would manoeuvre things so that he could take her for a drive on Sundays without May and the children. On these occasions he would drive past William Martin Murphy's house in Rathfarnham, and unfailingly say, 'That's where that man lives who murdered my love.' And turning to Mary, who as a child was sometimes brought on these drives and sat in the back seat, he would tell her that the 'old devil' who lived there had also murdered her mother's first love. And he would add bitterly that the 'old devil' had married his daughter off to a man he 'used to have intercourse with round the dining-room table' – meaning, presumably, to one of his cronies.

In the early summer of 1914 the balance of household relationships, such as it was or had become, was slightly altered by the arrival of the three children of May's brother Edward Price Roe – named after a clergyman uncle – from Africa. This brother had gone out to Nyasaland several years before as a tea planter. Now his wife had died; and though he soon married again, it was thought advisable to send the children, two girls and a boy, back to Ireland for the time being – a not uncommon decision where the children of colonists were concerned. The eldest of the girls, Mollie, was at this time thirteen, or slightly older than Frank; her sister, Sheila, was eleven; and the boy, Jack, was nine, or slightly older than Sam.

By this time May and Bill had separate bedrooms on the first or middle floor; and Mollie and Sheila were given a room, in fact the official guest room, on this floor also. The boys already occupied a large room on the top floor, where there was also an official spare room which they used as a playroom. Since their bedroom was large and they kept a lot of their things in the spare room anyway, Jack Roe was moved in with them.

Although many of those who went out to the colonies made arrangements to have their children boarded and educated at home, it was possibly Edward Roe's original intention to bring up his children in the place in which he had made his home, where the altitude modified the effects and dangers of the tropical climate. But in whatever way the death of his wife might otherwise have changed his plans, the outbreak of war that summer decided that the stay in Ireland should be an indefinite one. There was fighting in Nyasaland between the British and forces from German East Africa; indeed for a while the fate of the colony seemed in the balance. In any case the sea routes were now dangerous. So the children went to boarding schools in Ireland and stayed in Foxrock during the school holidays.

Whatever subconscious reverberations may have been caused by the intruders in the nest, everybody seemed to fit in happily with the new arrangements, and for most ordinary purposes the Roe children now became part of the household. Jack played tennis with Sam and Frank as well as golf and cricket; but since his age was nearer to Sam's than it was to Frank's, he was a slightly more fitting opponent for the younger brother. Mollie and Sheila, both pretty, outgoing girls, were a feminine addition to a household where femininity had previously been represented only by May, becoming surrogates for the sisters the Beckett boys did not have. Sam and the slightly older Sheila were to become quite close. Though not especially fond of Mollie, in later life he was always glad to see Sheila and when he was in England would often call to see her or even come to stay at her house in Surrey.

In spite of this influx of femininity, however, the Beckett boys remained very shy with girls. Mary Manning considered Sam to be 'awfully shy' and thought that both he and Frank were repressed and very much in awe of their mother, who ruled them with an extremely firm hand. May, like other Foxrock mothers, gave tennis parties. These would always begin with a ritual visit to the donkey, who would be found lurking in the garden somewhere and given a piece of cake. Then each boy would be paired off with a girl partner and the tennis would begin and continue until the break for tea. The food and drink provided were always delicious; but Mary found everything so formal and repressed as to make the parties an ordeal. There was hardly any conversation between girls and boys, let alone any frolicking. Once when she played with Sam, he said at the end, 'Game and set, thank God,' and that was all he did say. The donkey she remembered as being 'the only light thing' at 'those terrible, stiff parties'.

But if Mary did not enjoy the Becketts' parties, Sam equally did not enjoy anybody else's. Children's parties were a not infrequent event in Foxrock, and when one was in prospect he would be filled with foreboding. When the day came he would often hide himself away or even lock himself in his room; and when finally forced to go he would be noticeably silent, sullen and unwilling to co-operate in whatever entertainment was organized. The dislike of 'gatherings' of all descriptions which would be such a feature of his maturity was evident early.

The Great War, which broke out in August 1914, did not intrude very much on domestic life in Foxrock and had little effect on the Beckett household except perhaps by deciding that the Roe children should remain a part of it. A cavalry regiment, the South Irish Horse, commanded by Lord Decies, established its base camp on Leopardstown racecourse and remained until the middle of 1915. The mostly aristocratic officers were popular with people of the Becketts' class, as perhaps the private soldiers were with their female domestics; in any case when they were moved to barracks there was widespread regret. In response to the recruiting drive launched by the government with the active support of the Irish parliamentary party, about 200 recruits had been found in the Petty Sessions Area of Foxrock and Cabinteely by the end of 1915; and by then, too, casualties were being reported. Willie Beckett's brother Howard answered the call of king and country some time after the conflict had broken out, but the Becketts were business and professional people. Unlike the Anglo-Irish land-owning families, they had no military tradition or military relatives. Recruitment among their class and among non-aristocratic Protestants generally was only very slightly higher than it was among Catholics; their patriotism was not of the flag-waving kind. The *Irish Times* published a roll of honour, which lengthened as the Irish regiments became involved in the Dardanelles and on the Somme; but in Foxrock life went on much as usual.

In the year following the outbreak of war, however, Sam's freedom was, like that of most people in mid-childhood, curtailed by school. The chosen academy was one to which Frank had already been sent, Earlsfort House School, situated in what was then 3 and 4 Earlsfort Place and is now 63 Adelaide Road. Sam went there when he was eight and remained until he was thirteen.

Earlsfort House School was well known in Foxrock and among well-to-do parents in other areas of south Dublin, though it was not exclusively Protestant. It was run by a Frenchman, Alfred le Peton, and one of its selling points was that it was bi-lingual. French played a big part in the education of the upper middle classes in Ireland at the time: Catholics liked to imagine themselves as part of the Catholic culture of the Continent as opposed to the Protestant culture of England, and Protestants looked upon speaking French as an aristocratic accomplishment.

Because the French language played such a large part in Samuel Beckett's own life it is tempting to imagine his affection for it beginning in M. le Peton's academy at an early age, but the facts would seem to be against any such assumption. This may have had something to do with M. le Peton's teaching methods, which were authoritarian and involved a fairly liberal dispensation of physical punishment, applied to the hands with a cane, but in any case there was no attempt to make French or anything else attractive, and it was taught, of course, without reference to any grace or glamour it might have in use, whether in life or literature. In keeping with his persona as a professor of French, however, M. le Peton was a dapper, somewhat dandified individual and possibly something of a sadist, but since he tended to pick on certain boys, he was not unpopular with the rest. The others were rather cynical about him, recognizing him as a rogue who traded on the gullibility of parents, but they enjoyed his witticisms. Sam incurred his wrath on three or four occasions – once for fighting and once for carving his initials on a desk – but was never one of his special victims, while Frank was naturally earnest and law-abiding.

To reach school the boys rode their bicycles to Foxrock Station and took the train to Harcourt Street, from which they walked straight up Adelaide Road to the Leeson Street end, where Earlsfort Place was. The line by which they travelled was the DSER, the Dublin South Eastern Railway – known, though not in earshot of the children, as the 'Damn Slow and Easy' Railway because of its frequent stops and delays. The line, its terminus at Harcourt Street, and the wayside station of Foxrock all figure largely in Beckett's writings, principally in *Watt* and in the exquisite and profound radio play *All That Fall*. The pretty neo-Doric colonnade of Harcourt Street's façade is remembered in *Texts for Nothing* and the eventual closure of station and line take the narrator by surprise in *That Time*. The efficient if somewhat fussy station master at Foxrock, Mr Farrell, was a well-known Foxrock personage and is commemorated as Mr Barrell in *All That Fall*.

During the Easter holidays of Sam's second year at Earlsfort House School, when he was ten, the trains from Dublin suddenly ceased to run and telegraphic communication also ceased. The cause was the Easter Rebellion of 1916 and some word of what was happening soon filtered through, though the first impressions were of an 'outrageous riot' rather than a rebellion. The rebels had occupied many public buildings and held a large part of the centre of the city, but the army was bringing reinforcements from Britain and had begun to shell the rebel-held strongholds.

As to who the rebels were or what motivated them beyond hooliganism and a general dislike of English rule, however, Foxrock was ignorant. The great upsurge of romantic self-regard which had occurred in Irish literature and had influenced some of the leaders of the rebellion, as well as some of the Protestant middle class's more aristocratic and literary co-religionists, had of course passed it by. Though it was known that Home Rule and Home Rule agitators were an immediate threat, the more virulent and militant aspects of Irish separatism were only dimly apprehended and were generally confused with the labour unrest that had led to a general strike in Dublin three years earlier. The *Irish*

Times, Foxrock's principal source of information on political matters, had not forecast any such outbreak as this; and since it ceased publication for the six days the outbreak lasted, it could provide no help to knowledge either. Even when it resumed and published a handbook giving an account of the events of Easter week it was still somewhat confused.

Meanwhile life went on normally enough. Whatever individuals felt or came to feel as the week progressed, most people in Dublin were antagonistic in their attitude towards the rebels, and those of the Becketts' religion and class were even more so. After the initial shock had worn off, confidence in the British army's ability to handle things returned. One night Bill took the two boys up the Glencullen road to a spot where the flames which now lit the night sky could be clearly seen. A knot of onlookers were gathered at this place and there was a certain amount of jocularity, in which Bill and Frank shared, though Sam later recalled the night with horror.

When it was all over the British decided to execute a number of the leaders; and to make matters worse they did it in a secretive, prolonged way so that the shootings went on for two weeks, the names of those killed being announced only after the event. James Stephens said it was like watching blood seeping from under a locked door. A better way of ensuring sympathy for the rebels could hardly have been found and soon there was talk of the gallant fight they had put up and of how their cause was the old one, though not in the *Irish Times* and of course not in Foxrock. The guerrilla conflict which eventually ensued was a war of outrage and reprisal, often murder and counter-murder. As it progressed members of the Protestant minority in places like Foxrock grew increasingly nervous of showing their true colours. The usual response was a nervous silence and a withdrawal from politics rather than any outward manifestation of opposition to the nationalist militants. Outrage would be condemned and events commented on behind closed doors, but often not even in front of one's own servants. When the Irish Free State was established and the British withdrew from a large part of the island, Protestants of unionist sympathies had usually no one to vote for except the more moderate of the Irish nationalist parties, and were with a few exceptions more concerned to show their loyalty to the new regime than to assert their views. The process of de-politicization, begun during a war of terror and counter-terror, thus continued through its aftermath. The Beckett household, like most of their neighbours, had no politics except a silent and unexpressed loyalty to a regime which had vanished for ever, and they thus lived in a sort of political vacuum. No doubt this had some influence on Sam's later attitudes, including his dislike for all forms of political argument or even discussion and his feeling that they were unavailing anyway.

But the household in which he lived was in any case not interested in the world of ideas, political, philosophical or otherwise. Apart from their respect for musical accomplishment and insistence on piano lessons, his parents were philistine in outlook and, except for the Bible and Bill's 'potboilers', did not own or read any books. In fact the only books in the household were Sam's, ranged on a little shelf beside his bed in the room he shared with Frank and his cousin Jack. He would refer in later life to the sort of absorption in literature which one experiences as a child, and which in his, as in most other people's experience, would never come again – remembering reading *Treasure Island* and *Oliver Twist* with this childish absorption. The late Victorian and Edwardian eras were the great ages of 'style'; and even boys' authors like Weyman, Henty and, of course, Stevenson were, by the standards of today, highly conscious stylists. So was the mature Beckett, though he used the stock locutions of the English stylists for his own ironic purposes; and the pleasure he takes in mocking while imitating the syntactical locutions of the more obvious stylists would suggest that there may have been a certain degree of influence. Sam's boyhood was also the great age of boys' weeklies, journals such as the *Union Jack* and the *Boy's Own* paper, which, whatever their ethos, were at least considerably better written than the comics of a later age. As to how much he absorbed from the sterner authors of the eighteenth century whose books have become boys' classics, Defoe and Swift, we can only speculate; but it was not the sort of household in which classic, or any authors, are forced on children at the first sign of bookishness.

Nor indeed was Sam the sort of boy who becomes bookish at the expense of everything else. He seemed, if anything, an outdoor type rather than an indoor one. He enjoyed games and was good at them. He roamed by himself as well as with his cousin and brother; and though he often retreated to his tower with a book and was already noticeable in the family circle for a certain moodiness and taciturnity, he could on the whole have passed for an athletic, extrovert little Protestant middle-class boy with excellent manners when forced to be sociable. Nor, on a deeper level, did he despise his parents for their lack of interest in literature. His bond with his father was strong and cemented by the hours they spent walking together over a landscape which encourages both silence and communion. His love for his mother was both deep and painful, informed by a knowledge of how alike they were as well as how different. In *Watt* many of the sights and sounds of Beckett's childhood are listed in a passage whose tone becomes increasingly ironic as it progresses.

The crocuses and the larch turning green every year a week before the others and the pastures red with uneaten sheep's placentas and the long summer days and the new-mown hay and the wood pigeon in the morning and the cuckoo in the afternoon and the corncrake in the evening and the wasps in the jam and the smell of the gorse and the look of the gorse and the apples falling and the children walking in the dead leaves and the larch turning brown a week before the others and the chestnuts falling and the howling winds and the sea breaking over the pier and the first fires and the hooves on the road and the consumptive postman whistling The Roses Are Blooming in Picardy and the standard oil-lamp and of course the snow and to be sure the sleet and bless your heart the slush and every fourth year the February debacle and the endless April showers and the crocuses and then the whole bloody business starting all over again.

It is, like so much in Beckett, a passage that undercuts itself; but the larch tree in the grounds did indeed turn green a week before other trees and brown a week before them in the autumn. It still does.

Boarding schools on the English or 'public school' model are meant to serve as some sort of termination of childhood and introduction to the adult world, with its hierarchies and its systems of reward for achievement and conformity. That they do not serve this purpose as well as they might is because they are medieval in their value systems, setting a higher value on physical prowess and even chivalry and knightly bearing than the world outside really does.

Portora Royal School, to which Samuel Beckett was sent in 1920 at the age of fourteen, was no exception in this regard. Because Beckett turned out to be a notable athlete he had a better time there than might otherwise have been the case; and his moodiness and taciturnity were more readily overlooked. To pursue the medieval analogy, in the feudal society of Portora he was a baron with a following; and he even inherited some of the status he enjoyed, for Frank, who went before him, was a natural athlete with a following too.

Portora was one of the very limited number of Irish Protestant boarding schools to which he might have been sent. Frank was already captain of the cricket eleven when Sam arrived. It was a small school, with about 120 pupils, splendidly situated above the juncture of upper and lower Lough Erne in what in 1920 was about to become the six-county statelet of Northern Ireland. It drew its pupils from all over Ireland, however, and when partition came most of them would continue to be from the twenty-six counties of the Free State. The cost of sending a boy there was considerably less than it would have been to send him to an English public school.

The main buildings have a certain classical distinction and the school itself is over 300 years old. In the austere entrance hallway is a scroll commemorating academic prize-winners, but few of the names on it mean anything to the average visitor. One that might have meant something to Samuel Beckett on his arrival in 1920, though, had been removed; for Oscar Wilde's name had been thought unfit for mention or remembrance after his conviction for homosexual offences in 1895 and has only been put back in recent years. Apart from a former headmaster, the Reverend Robert Burrowes, appointed in 1798, who wrote the famous, macabre Irish ballad poem, 'The Night Before Larry Was Stretched'; and a former pupil, Henry Lyte, the author of 'Abide With Me', the school was otherwise innocent of literary traditions.

Douglas Seal, later a clergyman and headmaster of Portora, who was a contemporary of Beckett's there, thought it was at that time a 'natural, unconventional' sort of place, but conventionality is somewhat in the eye of the beholder. Portora did not encourage rebels or revolutionaries who rejected its value-system. The virtues to which the boys were supposed to aspire were the 'manly' ones of truth-telling, trustworthiness and fair play – in other words, the virtues inculcated in English public schools, a handy set from the point of view of those in charge of turbulent schoolboys. The ideal past pupil would be a bishop, a general or a colonial administrator; and among those accounted distinguished there were rather more army officers than there were churchmen. The school song was *Floreat Portora*, plagiarised alas from *Floreat Etona* and sung to the same air.

Like most such institutions it was spartan in its outlook. It was a point of honour to take a cold bath every morning and since there were only two baths per dormitory that meant that there was a terrific scramble for them, and for much of the year a long period of shivering while waiting. After rugby everybody took a cold shower in an inadequate washroom cum lavatory. For someone used to the comforts of Cooldrinagh these introductions to barrack life probably took a little getting used to, but the ascetic tradition which was strong in Portora possibly even suited Beckett's temperament. In any case he took it in his stride. The food at Portora was plain but adequate for even the keenest appetite; and when not playing games the boys had the freedom of the countryside for as far as they liked to go.

Portora was not on the whole as barbaric an institution as some others, Catholic and Protestant. The alternative to the medieval emphasis on valour in its garb of physical distinction on the playing field and knightliness expressed through the concept of 'fair play' in one's dealings with one's fellows, is, in the school system as it was in the Middle Ages, barbarism and the rule of the strongest. A certain amount of bullying went on, and a certain amount of persecution of eccentrics; but on the other hand some respect was accorded to bright scholars and even to those who distinguished themselves in intellectual pursuits, the 'clerks' of medieval society. The then headmaster, E. G. Seale, gave a gold medal each year to the winner of an essay competition, and the fact that Beckett won it three times added quite a lot to his eventual stature.

Seale was a headmaster of traditional type, a clergyman, a classicist and a stern disciplinarian, white-haired, stern and distinguished looking. He seems to have been one of those headmasters who make a deep impression on their pupils and this in spite of certain obvious disadvantages – a club foot, a partially paralysed right arm, a nervous twitch in his cheek and a peculiar voice which made words come out as a sort of bark. In religion the school was ostensibly Church of Ireland; in practice strongly evangelical. Every weekday morning the headmaster read prayers and a passage from the Bible at Assembly; and at bedtime there would be prayers and passages from scripture again in the study hall, to send the boys to their slumbers in the proper frame of mind. On Sunday morning the school paraded to St McCartan's Cathedral in the town, wearing straw boaters in summer. The Sunday evening church parade was, however, to a country parish church near by; in inclement weather, evening service was held in the dining hall. In addition to these exposures to the King James version, there were daily scripture classes for everybody up to the fifth form.

At a crisis point in his life later on Sam would look back on the Portora years as the last time of happiness; yet in spite of the popularity and the glittering prizes he attained, he does not give the impression of having been an outgoing or happy schoolboy. He was moody, taciturn and, with most, uncommunicative. Though of course schoolboy photographs can be misleading, those of Beckett at Portora do not suggest happiness or even contentment. His expression is severe and somewhat distrustful, and his head is already tilted forward in the way that was to become so familiar in later photographs, so that he is looking at the world from under lowered brows. His sight even then, in spite of his general athleticism, was not very good; by his third year there he is wearing glasses of the round, steel-rimmed, utilitarian kind he was afterwards to favour.

Yet there is no doubt that he was popular, always an important consideration at a boarding school. Most people at Portora were familiarly known by their surnames or by a not always very flattering nickname. Beckett was almost the only boy in the whole school who throughout his career was accorded by everybody his full name, being referred to as 'Sam Beckett' and nearly always addressed as Sam. Of course, he had a good start, for Frank was already a notable figure, a prefect and captain of the cricket eleven; but he must also have had a good deal of charm to deploy when he wanted to, and in his first weeks he acquired another valuable asset – a close friend who was also popular and in whom he could confide.

Geoffrey Thompson was, like himself, a middle-class Dubliner, who afterwards became a doctor and then a specialist in psychological medicine. Like Beckett he was good looking, if, like Beckett, also slightly jug-eared. In temperament he was meticulous and trustworthy and, besides being an athlete, he was by no means illiterate. This friendship was, like most of Beckett's friendships, to endure.

Sam soon gave evidence of being able to follow Frank with distinction. In his first year he made the cricket eleven. True, his brother was captain, so, as he said himself afterwards, it was easy to be noticed; but even in a small school it was an achievement. His first match for the school was against an adult team, the eleven of the Duke of Cornwall's Light Infantry. Portora actually won, with Beckett contributing 11 in the first innings and 14 not out in the second. He seems to have secured this first place mainly as a batsman, with a notably elegant, left-handed style, though there were adverse comments on his tendency to step out to meet every ball – which may have been an effect of short-sightedness. In his second year the school magazine would say that he had 'some really stylish strokes', but that 'he ought to put more force into them'. He was in fact, however, that cricket rarity, a true all-rounder, who was peculiar in that he batted as a left-hander but bowled as a right. Competitiveness as well as team spirit is suggested by the fact that he was a keen and always alert fielder. Shortly after he made the team he became opening batsman; and since his friend Geoffrey Thompson became the other opener, they were a socially notable pair for the rest of their school lives.

Though he remained opening bat, it was none the less as a somewhat crafty medium-pace bowler with a strong break-back that he eventually made most mark, and when finally he followed in Frank's footsteps to become captain of the first eleven it was as a bowler that he was principally known. Commenting later on Beckett's attitude to cricket, his friend Brian Coffey remarked on the amount of shrewdness, not to say low cunning, a bowler needs in his make-up, as well as mathematics and logic. 'All this is involved in Sam's attitude towards the game, and I think quite possibly in his attitude to life as well.' In 1922 he was first in the bowling averages, having taken 17 wickets in 15.5 overs for an average of 4.7 runs, while he was third in the batting averages, having made a total of 132 runs for an average of 14.69, his highest score being 34.

If there are intellectual and indeed contemplative aspects to the game of cricket which have a strong appeal for intellectual and contemplative people, the same can hardly be said of rugby football, which can of course be played strategically but is very much a matter of immediate and often bruising body contact scarcely suited to the fastidious. Rugby bulked much larger in the consciousness of Portora than did cricket, which had comparatively few followers. A somewhat critical report by the inspectors of the Intermediate Education Board in 1910 had suggested that the then headmaster, A. C. McDonnell, was a special enthusiast for the game; and writing in the *Bell* in 1946 Vivian Mercier, who came to Portora in the 1920s – a few years after Beckett's departure – confirmed that 'The one tradition that the boys themselves take much stock in is the Rugby football tradition . . . There are few enthusiasts for cricket and a few fanatics for rowing . . . but for two out of three terms in the year we played nothing but Rugger and often talked nothing else too.'

Beckett played scrum-half for the rugby team in his final two years, a position where speed of reflex as well as of foot, brains and calculation are more important than among the forwards or even on the wings. Though scrum-halves are usually smaller and supposedly nippier fellows, he was noted for his ability to size up and seize an opening as well as for his clean gathering of the ball; and in 1923, after he had left, the school magazine, *Portora*, remarked on the number of brilliant games he had played. Douglas Graham remembered him afterwards as being 'blind without his spectacles, but as bold as a lion round the scrum'. In 1922–3, his final year, he was captain of the rugby team as well as of the cricket eleven. In that year Portora went through to the final of the Ulster Schools Cup. The opposing side, Campbell's, had a noted scrum-half, Delap, who played for the Ulster schoolboys' team, but the Belfast *Telegraph* remarked on how well Beckett had done against him. Years later Beckett would remember the Portora team going up to Belfast for this match. They were, he said, 'shit scared'.

Beckett also boxed at Portora, being finally, according to the schoolboy weight classifications, a light heavyweight. At this sport, in which aggression counts for so much, he showed speed and footwork, as well as an ability to use both hands to good effect. There is little doubt that he was a determined boxer of some skill and his interest in boxing is shown by the fact that, like cricket, it was engaged in only by a minority and there was no compulsion, either actual or moral, for anybody to take it up.

Homosexuality, certainly not unknown in some Irish boarding schools, does not seem to have been a part of Portora life, but sentimental friendships between older and younger boys, in which there was a greater or lesser element of chivalrous romanticism, were not uncommon and seem not to have been discouraged, either by officialdom or by public opinion. In Beckett's last two years he had such a friendship. It was with a younger boy whose nickname, 'The Skivvy', suggests a certain femininity of appearance or manner as well as a reduction to the status of errand runner or messenger boy – though there was no fagging in Portora, nor was there in other Irish schools – and who seems to have been ragged or bullied to a degree beyond the ordinary until Sam saw him weeping one day, questioned him and, discovering his misery, made friends with him. This may have been no more than charity or practical concern, but whatever his initial motives, there seems to have been a singular degree of attraction between them and, as the friendship developed, they went for walks together and Sam lent him books. Accorded the daily shelter and the glamour of Beckett's own immense prestige, 'The Skivvy' wound up as quite a popular personage in his own right; and it is possible that it was the ending of this happy association which, as much as anything else, contributed to Sam's feeling of desolation when he left Portora for the outer world.

Much of the impression Beckett made on his contemporaries suggests a paragon of schoolboy virtues, both those admired by schoolboys themselves and by their masters, but besides his occasional moodiness and times of withdrawal there was a less ideal side to him, revealing him as – occasionally at least – capable of the sort of cruelties which are common among boys. It happens in many schools that a particular master becomes the butt of jokes and ragging which can often have a cruel edge. It is a horrible situation for a master to find himself in, for it is a slippery slope with almost no way back to respect and authority. Such a master in Portora was an elderly man called Thackaberry, a natural butt who was often made miserable by the contempt in which he was apparently held. Prep was in a large study hall, with the whole school sitting at desks in front of a dais on which sat the master who was supervising that evening. When Thackaberry took prep, Beckett and another boy, Claud Sinclair, subsequently the novelist C. E. R. Sinclair, used to organize what they called 'The Singular Sing Song Singers' and they even issued a programme, so that at a given signal from Sinclair or Beckett the whole study hall could burst into pre-arranged song.

On one particular evening when Thackaberry could stand this flagrant indiscipline no longer, he rushed down from the dais and began beating Beckett about the head with his fists. Beckett, still sitting there, defended himself as well as he could with his hands, then, during a pause, said, 'Hit someone your own size.' It was not a particularly apposite remark, for, after all, he and Sinclair really had the upper hand; but it was calculated to appeal to schoolboys and it stopped Thackaberry, who returned weakly to the dais. What followed shocked everybody, for the boys soon saw to their horror that the elderly master was weeping uncontrollably. Finally, overcoming his sobs, he was heard to say, 'Here I am, reduced to being the piss-pot for the whole school.' This was succeeded by an appalled silence until fortunately the bell rang for supper break and everybody filed out. Though the usual custom was for the same master to take prep for the whole evening, after supper on this evening another member of the staff had taken over and the songsters were silent.

If Beckett had the upper hand where Thackaberry was concerned, the same was not true of his relationship with another master, his science teacher, W. R. Tetley. When Tetley left after thirty-five years' teaching there, the *Portora* magazine declared eulogistically that he had taken with him 'the affection of many and the deep respect of the whole school'. It is more likely that indifference mixed with relief may have been the dominant feeling of the majority if we can judge from the rhyme which circulated, the authorship of which has sometimes been attributed to Beckett:

> Tetley has gone
> To the mountains of Wales,
> Leaving behind
> His balance and scales.

In fact 'loathing' might have been a more apposite word than 'affection' in Beckett's case, though it was perhaps not unmixed with a sort of respect, for their antipathy and the conflict of wills which resulted from it certainly did not lead to a victory for the boy over the master.

Beckett had no aptitude for science. Classes in this subject took place in a dark, unattractive basement and he frequently committed the most elementary blunders in the course of practical experiments. Some of these may have been deliberate, as on the occasion when he poured sulphuric acid from a beaker into the sink, claiming that he thought it was water. Some were due to absent-mindedness and some to his poor sight. Naturally, however, these errors or mishaps led to reproofs from Tetley, which Beckett was inclined to suffer with a very ill grace, perhaps because he was so successful at most of what he undertook and could not bear to be otherwise, perhaps simply because reproof irked him. Tetley also taught mathematics and Beckett spent much of his time in this class drawing caricatures of his detested enemy, who in looks was rather like the notorious murderer Dr Crippen. A number of these drawings were very rude and on one occasion Tetley swooped down and confiscated some of them, after which Beckett had to see the headmaster.

The river Erne played a large part in Portora life, particularly at the part called The Narrows, where it passes from the upper to the lower of the two lakes and where there was a stone parapet from which the boys used to dive. Although Beckett was a keen enough swimmer, who at one period of his school life began every day with a swim across The Narrows, it was an unhappy circumstance for him that Tetley was also in charge of swimming. Though as a junior he distinguished himself by winning both the long and the sprint races in the school sports, as a senior, perhaps because of his antipathy to Tetley, his record was less impressive; and he resented so deeply an occasion on which the master said in the hearing of others that he was 'putting no guts' into his kick that he enshrined it in a poem seven years later. The poem, 'For Future Reference', was published in the Paris magazine *transition* in June 1930:

> But I stilled my cringing
> and smote him . . .
>
> smashed
> mashed . . .
>
> flayed and crushed him
> with a ready are you steady
> cuff discharge.
> But did I? . . .
>
> The hair shall be grey
> above the left temple
> the hair shall be grey there . . .
>
> God blast you yes it is we see
> God bless you professor . . .
> that little bullet-headed bristle-cropped
> red faced rat of a pure mathematician . . .
>
> up he comes surging to the landing steps
> and tells me I'm putting no guts in my kick.
> Like this he says like this.
> Well I just swam out nimbly

46

> blushing and hopeless
> with the little swift strokes that I like and . . .
> Whoops!
> over the stream and the tall green bank
> in a strong shallow arch
> and his face all twisted calm and patient
> and the board ledge doing its best to illustrate
> Bruno's identification of contraries
> into the water or on to the stones?

The poem shows that his hatred for Tetley lasted for a long time after Portora; and it seems also to incorporate elements of the diving nightmare which went back to his experience with his father at the Forty Foot. The final humiliation for Beckett was being forced to give up and climb out of the water before the finish in the long race, either because the fact that Tetley was supervising it affected him psychologically or because he had refused to train sufficiently for it.

Although Beckett made his mark at Portora principally as an athlete, he engaged in other activities too. The test of intellectuality where schoolboys are concerned is often membership of the debating society, and in spite of what later observers would frequently characterize as his shyness, Beckett was prominent in this, becoming a member of the committee in his second year at the school. One debate of which there is record was on the desirability of 'Women's Emancipation'. According to the school magazine he and another advocate of this succeeded in proving themselves capable 'ladies' men'; but although they made 'violent and eloquent speeches' in favour of the motion it was defeated by two votes. That Beckett was prominent in the debating society does not, however, mean that he held or expressed strong opinions about the public and political questions which were now racking his country. In accordance with what was becoming the southern Protestant practice of keeping one's head down where the fiercely divisive questions of nationalism versus unionism and Ireland versus Britain were concerned, the Portora debating society avoided them.

According to Geoffrey Thompson, Beckett had already developed a passion for poetry while at Portora; and he specially remembered his enthusiasm for Keats. It was the custom of the school for boys to pair off together for Sunday walks and Thompson and Beckett often went on these together. On one beautiful summer day as they lay under a tree in a field they decided to learn the whole of Keats's 'Ode to a Nightingale' off by heart instead of just memorizing the short section which their English master had assigned. In his late sixties Beckett was still able to recite the poem and one wonders how much the famous lines on the ills and miseries of life may have done to inform his own outlook:

Fade far away, dissolve, and quite forget
  What thou among the leaves hast never known,
The weariness, the fever, and the fret
  Here, where men sit and hear each other groan;
Where palsy shakes a few, sad, last grey hairs,
  Where youth grows pale, and spectre-thin, and dies;
    Where but to think is to be full of sorrow
    And leaden-eyed despairs; . . .

Thompson also remembered Sam's interest in the great Canadian humorist Stephen Leacock, whose *Literary Lapses* was then, as ever, popular. Leacock's technique, deployed with enormous skill, is to push things to extremes for comic purposes, so that a simple action or course of action leads to multiplied calamity. Whether Leacock had any influence later on or not, there is little sign of the pessimism of Beckett's own later outlook in the lines he wrote in the autograph album of a schoolfellow and friend, Tom Cox; nor, it might be added, is there much sign of the influence of John Keats either. Indeed so at variance with his own mature outlook and so appalling in certain literary respects are they, that one wonders whether they were not inscribed with satirical intent:

When a bit of sunshine hits you
  After passing of a cloud,
And a bit of laughter gets you
  And your spine is feeling proud,
Don't forget to up and fling it
  At a soul that's feeling blue,
For the moment that you sling it
  It's a boomerang to you.

Cox sat beside the author of these lines in class for three years and thought that they were an accurate reflection of his attitude. It is fairly safe to say that no one would ever think so again. What Beckett slung at people as a writer later on were the miseries of living – perhaps in the end a better service to them, 'friending' them, to use A. E. Housman's words, 'in the dark and cloudy day'. Cox, however, had reason to be grateful for encouragement received because Beckett had been the first person to speak a kind word to him when he came to the school as a new boy. He remembered his neighbour in class as being particularly good at English and French, which was the only continental European language taught at Portora. Irish history was not taught there and Beckett rarely showed any but the most minimal interest in it, a fact which makes critical theory relating his work to the guilt he and other Protestants felt about their part in its melancholy course rather dubious. The general feeling among his class would certainly not have been guilt; and most members of it would have inclined to the J. A. Froude view that the Protestants were the civilizing influence, teaching probity, fair dealing and respect for the law to the natives.

At the end of his senior year Beckett sat for the Junior Exhibition, or entrance scholarship examinations to Trinity College. He did well, but not well enough; and, unlike two fellow Portorans, failed to win an Exhibition. He was only seventeen, almost a year younger than the majority in his class. The normal procedure at Portora would have been to keep such a student back for another year to be 'trained like a racehorse' to win a scholarship the following year; but the Becketts were well-to-do people and it was decided that he should go on to Trinity as a fee-paying student that autumn.

Beckett was to commemorate the Portora years principally by the appropriation of names for characters in his works. The surgeon Bor in the uncollected short story 'A Case in a Thousand' got his name from G. T. Bor, who joined with Frank in what the school magazine called 'a fine cheery duet' at the Hallowe'en concert in 1920. There was an E. P. Mahood there during Beckett's first two years at the school whose name would be memorably used in *The Unnamable*. An E. D. Camier arrived just before he left, and a subsequent pupil was the critic Vivian Mercier. These names would be combined in the name of his second novel, *Mercier and Camier*.

While at the school he gave no sign of cherishing values and ideals other than the prevailing ones. He was to all intents and appearances a normal, athletically minded, clean-cut, middle-class Protestant boy, someone who could be singled out by authority as a chap to be looked up to and emulated. There were few hints of complexities or failures, still less of rebellions to come. Yet Beckett would remember the Portora years as the last truly happy ones for a long time. Since one of the themes of his work is the search for the self, a search which becomes a sort of ethical imperative, one is inclined to ponder on the ironies, or even to ask whether the self displayed at Portora was not just as real as any other. But in spite of his success there Beckett sedulously refused to have anything to do with the school in later life. Though oddly loyal to Trinity College, his connection with which brought its share of misery, and prepared to indulge its requests in various ways, it was otherwise with the school which had flattered and rewarded him and where he thought he had been happy. In his first two years at university Beckett would continue to associate with old Portorans, and his friendship with Geoffrey Thompson was to be lifelong. But according to Mrs Mary Rogers, wife of a subsequent headmaster, 'no invitation, no matter how pressing' would induce him to return to Portora. 'All approaches from the school in after years fell on stony ground.' Mrs Rogers, indeed, was of the opinion that he did not remember his schooldays with any great pleasure, 'Perhaps because he was too successful a schoolboy.'

Chapter Four

Beckett entered Trinity in October 1923. During his time in Portora momentous political changes had taken place in Ireland, leading to the collapse of British rule over a large part of it and, with the establishment of the Irish Free State, the partition of the country into British- and Irish-ruled segments.

But boarding schools are worlds of their own and the conflict just ended had not been of such a nature as to intrude into the school's life either by the cutting off of its supplies or the advent of armed men into its grounds. It was an exclusively Protestant school, British in its political outlook, though many of the pupils came from what was now the Free State. Fermanagh, where it stood on its hill outside Enniskillen, was not a very disturbed county; and when things settled down again, it was still in the United Kingdom.

In Portora, as to some extent in Foxrock, the changes would almost have seemed to be taking place somewhere else. The victory of largely Catholic nationalism meant very little in terms of day-to-day Protestant life, even in the Free State. Property rights remained undisturbed, as did religious practices. The grip of Protestants on Dublin business life was only very slightly weakened; and the Free State government immediately began to issue reassuring signals by appointing the distiller Sir Andrew Jameson, the poet W. B. Yeats and other prominent Protestants to the Senate. To a youth who was in any case apolitical the supposed importance of the events and the changes they had led to would have seemed an exaggeration, and may have reinforced a view of politics as an irrelevancy.

Nor, as he walked through the arched doorway in Trinity's chaste and exquisitely proportioned façade and entered its cobbled square, would he have found a world very different from that which had existed before 'the troubles' began. What changes there had been in the previous decade or so owed more to Suffragism and the Great War than to the establishment of Irish independence. Trinity had admitted women students in 1909, being one of the first universities in the British Isles to do so, and there was now a fair sprinkling of them, most in the arts faculty. The atmosphere of what was on the whole a conservative institution was more relaxed and libertarian than it would have been in 1914. Ninety per cent of students were still Protestants, many from Northern Ireland and some from the United Kingdom; but the absence of Catholics was mostly the result of their own hierarchy's decision and had nothing to do with College rules.

Trinity was, however, still quite a snobbish place; and it may have been in deference to this snobbishness that Samuel Beckett told his tutor, A. A. Luce, who liked to have such information on file, that his father was an architect. It may also have been because quantity surveying was considered to be a branch of that profession as well as of civil engineering. Entry into them all was still laxly governed and one description may have seemed as good as the next, especially if he was fed up with explaining to the uninstructed what a quantity surveyor was. If it was snobbishness, he can be forgiven. Almost everybody is a snob at seventeen.

He enrolled in the arts faculty to study modern languages. There was no pressure on him to enter the family business, or to study a subject which would prepare him for the profession of quantity surveying, as his brother Frank, who took civil engineering, had done. As to what he intended to do, he would not be the first, or the last, to enrol for an arts degree without seeing the future very clearly. Some become academics themselves, teaching the subject they have been taught; some become teachers on a lower level. It was quite customary to take an arts degree on one's way to the Irish bar. In addition to the supposed desirability of acquiring a liberal education as well as a legal qualification, the possession of an arts degree shortened the period of legal studies from four years to three. There is some slight evidence, in a note Professor Luce made, that Beckett had a legal career in mind, but it is more likely that he was drifting on towards the acquisition of an arts degree, as so many others have done, without any clear ideas about the future, or whether it was to be in academia, law, journalism or anything else.

When he arrived at Trinity he was an intellectually, emotionally and sexually undeveloped seventeen-year-old who was, like most other students whose parental home was in Dublin, content to live under his parents' roof. At first sight his attitudes seem, if not normal, at least not in need of much explanation. In so far as he had any contact with anybody outside the family circle, it was with a few old Portorans or with social equals from south Dublin. He showed no interest whatever in anything Trinity had to offer outside sports facilities and was not interested either in the lectures he was supposed to attend, occasionally drifting in and out of them until reminded by his tutor that it was necessary to 'keep terms' – to attend a certain minimum number. Unlike others who had to be so reminded, however, he was not lethargic, or dissipated or idle by nature. His main interests, amounting almost to passions into which all other passions were sublimated, were golf, cricket and motor-bikes; and where they were concerned he was very active indeed.

As already remarked, it seems normal enough, but there is evidence that after leaving Portora he found the outer world very difficult to cope with. Later he would speak of a period of extreme unhappiness that began then, and a period of isolation, of withdrawal into the self, of lending himself to 'a crescendo of disengagement' from others. The extreme devotion to sporting activities may even have been a way of disengaging himself and the amount of golf he played was extraordinary by any standards – sometimes thirty-six holes a day or even more – so that one wonders whether he was turning the golf course into some sort of open-air hermit's cell, a place where nothing need bother him except the game itself; or whether his extraordinary dedication to sport generally was not an attempt to prolong the Portora experience, Portora having been a place where a large part of his identity was created by games and where he had been a dominant figure in his little world.

On many occasions he played golf with a fellow student, Bill Cunningham, who was also in his first year at Trinity and lived in Killiney, a couple of miles from Foxrock. They had both become student members of Carrickmines, an exclusive south Dublin club, and they played there as well as at the Royal Dublin and Portmarnock, a windswept links north of the city, which was the 'official' club for Trinity. Since Beckett could play at Carrickmines, which was only about a mile from his home, without ever going into town or near Trinity College, it is no wonder that his record of attendance at lectures was poor. He was an unorthodox golfer, playing mostly with only four clubs and using a number two iron as a putter; but he eventually got his handicap down to six. When in his second year he played Portmarnock in the annual Trinity competition, he had a round of 77, which is pretty good for the course and won him the prize for the best round.

Cunningham, a decent middle-class Dubliner who was not an intellectual but was by no means devoid of intellectual interests, was his fairly constant companion during that first year or two, and they also went on long walks together through the mountains which backed on Killiney as well as Foxrock. One particularly memorable tramp started from Enniskerry. They went up the Glencree valley, across by Lough Bray to the Sally Gap and came down by Lough Tay and Luggala to where they could join the road coming down from Glendalough and walk back into Bray over Calary bog. It is a total distance of about twenty miles, some of it pretty steep walking. Cunningham remembered it afterwards as taking place on a beautiful autumn day, with the heather still in bloom and the grouse rising from the grass beside Lough Bray, though he could not remember what he and Sam talked about.

Beckett also played cricket and tennis, and as soon as he could prevail on his father to buy him a suitable machine he joined the Dublin University Motor-Cycle Club, riding on a competitive basis in trials. Legend has now elevated his status as a cricketer to considerable heights, the statement – often made by intellectuals who know little about cricket – that he is the only Nobel Prize winner to be mentioned in *Wisden's Cricket Almanac* often being accompanied by the suggestion that he played for Ireland. In fact he played for Trinity first eleven during his second, or what would now be called his 'senior freshman' year, and attained his 'pink' or cap in 1926, his third year in college. For the first four weeks of the cricket season of 1925, when he came into the first eleven, the rain poured down and on four successive Saturdays the match scheduled for College Park was abandoned. This must have been rather dispiriting, but on the second occasion when he carried his bat for the college he scored a notable 61 – as it turned out his best of the season – against the Leinster Cricket Club. The official history of the Dublin University Cricket Club describes him as a 'solid rather than spectacular bat who sometimes opened the innings' and who had a 'very respectable career'. As a batsman his first season's average was a highly respectable 18.4. In his second year this average had dropped to 11.5, but as a bowler during the course of ten matches he took 15 wickets.

To assess his true abilities and those of the team he played for is not easy. According to Ulick O'Connor, that particular Trinity side is 'regarded as one of the best ever there'; but it is interesting to look at the scoring in the annual match Trinity played against Northamptonshire. In 1926 this took place at the county ground, Northampton. Beckett opened the bowling for Trinity, but the Northants opening batsmen treated him with disdain and he took no wickets for 47 runs. As a batsman he fared no better, being bowled for 4 by Powell, a Northants fast bowler, in the first innings, and losing his wicket to a leg spinner,

Adams, with the very first ball of the second. All in all this Northants match was a bit of a débâcle for Beckett, though he did secure two catches off the bowling of a team-mate.

When Beckett entered Trinity he already had a motor-bike, a 1914 Villiers-engined, single-speed Royal Ruby. This rather dangerous machine he had inherited from Frank and eventually passed on to Mary Manning's brother John, becoming the proud possessor of a brand new 350 c.c. side-valve AJS, a very good bike his father bought for him which was the envy of John Manning and other motor-cycle enthusiasts. This was the bike he rode in the university novices' trial in March of his second year in Trinity. The trial was on roads well known to him in the Dublin mountains, part of it over the route that he had walked with Bill Cunningham, going up through Dundrum and Stepaside to Glencullen, the Pine Forest, Glencree and the Sally Gap and then coming down Calary Lane and Red Lane to the Glenview Hotel. After this there were water-splashes and a difficult stretch known as the 'Waterfall of Bricks' before part of the route was retraced though Kilmacanogue, the Goat's Pass, Enniskerry, the Devil's Elbow, Stepaside and Dundrum, making sixty-three miles in all. To the disappointment of the organizers there were only seven starters, and of these two failed to finish, so that Beckett's fourth placing was perhaps not a great achievement.

Motor-bikes were the big passion of the day among mechanically minded and adventurous young men in Ireland, where the great era of Stanley Woods and Charlie Manders was just dawning, but whether Beckett was really mechanically minded is open to question. Judging from the crashes he had as a motor-cyclist he was evidently sometimes as absent-minded and careless of consequences as he was to be behind the wheel of a car. Whatever powers of co-ordination and concentration he showed in other physical activities were not always brought to bear on the conduct of mechanically propelled vehicles; but for the time being anyway, the motor-bike was a major interest.

Gradually, however, he was de-cocooning himself, a process not brought about by ordinary student activities as much as by art. He had begun to play the piano again, particularly Debussy, and towards the end of his first year in Trinity he even began to frequent the theatre, going with Geoffrey Thompson to see a revival of Sean O'Casey's *Shadow of a Gunman*, a play whose startling portrait of the artist as an abysmally poltroonish anti-hero struck a new note. This was the only period in his life when he can be said to have been a regular theatregoer. It was an activity which later on he would be at some pains to avoid.

The early 1920s were possibly the Abbey Theatre's best period. O'Casey was in his relatively brief hey-day; Yeats had discovered a more interesting and hard-edged mode; and there were other, not so well remembered dramatists of interest such as T. C. Murray, whose *Autumn Fire* Beckett saw with Bill Cunningham in September 1924, just as his second year in Trinity was beginning. *Autumn Fire* has for its hero a 'strong' farmer who has a young wife but is rendered impotent by an accident. Beckett was fascinated by the central character and his predicament, but it is doubtful whether this reveals anything about his own difficulties in the matter of sexual congress, as perhaps a later fascination with characters who suffered from impotence would. There were as yet no girls in his life and his theatre companions were all male, the most frequent one being his Portoran best friend, Geoffrey Thompson. It was with Thompson also that he saw a revival of John M. Synge's *The Well of the Saints* as well as the other two great O'Casey plays, *Juno and the Paycock* and *The Plough and the Stars*. He and Thompson were at the Abbey for the first night of *The Plough and the Stars* when there was, according to a regular first-nighter, Joseph Holloway, 'electricity in the air' before the curtain went up, though nothing happened in the way of protest. On the next night, Tuesday, there was some hissing and on Wednesday a 'sort of moaning sound' at the point when the flags of the Irish Volunteers and the Citizen army are carried into the pub. Word had got out that O'Casey had written a play which blasphemed against both Irish nationalism and the purity of Irish womanhood, thus going one better than Synge; and on Thursday night the storm broke.

As it happened, Thompson and Beckett had gone again and as they were in the centre of the front row of the balcony they had a good view of the events: the screaming women in the audience, the lumps of coal thrown on to the stage and then the attempt to invade it, resisted by the actors. Yeats had meanwhile arrived and ordered the police to be sent for, telling one of the actors, 'I am sending for the police, and this time it will be their own police,' meaning the police force of Catholic Ireland, a state which he had been to some small degree responsible for bringing into being. Then he paced up and down behind the lowered curtain, composing his speech and smiling to himself, an old war horse delighted at the prospect of battle. Thompson and Beckett, of course, saw nothing of this; but they did see the dramatic moment when, after ordering the curtain to be raised again, Yeats was discovered on stage in solitary glory, with his hand outstretched. What he said was unheard by the audience because of the general hubbub, and he may have improved on it when he gave it to the reporters a little later: 'You have disgraced yourselves again . . . Dublin has again rocked the cradle of a reputation. From such a scene as this went forth the fame of Synge. Equally the fame of O'Casey is born here tonight.' It was high jinks indeed for the two young students to witness but, in keeping with the Protestant mentality of the time, they were mere spectators of the riot as well as of the play.

The main offering at the Abbey was often preceded by a curtain-raiser, sometimes by Sean O'Casey. His were very much dependent on verbal exchanges in Dublinese with a good deal of slapstick, such as *Nannie's Night Out* and *Cathleen Listens In*, the latter making use of the new invention, the wireless. In *Juno and the Paycock* O'Casey had been the first to put a gramophone on stage, such a machine being one of the acquisitions of the supposedly suddenly enriched family. *Juno and the Paycock* remained Beckett's favourite among O'Casey's major plays and the memory of these innovations may have influenced the use of the tape-recorder in *Krapp's Last Tape*.

The music-hall side of O'Casey's work was something Beckett would have clearly recognized because he had also become a regular attender at the places where 'variety', as it was called in Dublin, was on offer; developing a taste for it which never left him. These were the Queen's Theatre near the back gate of Trinity, the Royal near by in Hawkins Street, and the Olympia, which is almost in view of the front gate. Going to a variety show was a more casual matter than going to the straight theatre. There were matinees as well as evening performances and you could slip in without booking in advance. It was also a more solitary matter, such shows being probably beneath the intellectual level of Geoffrey Thompson and his few other friends. Irish variety, like English music-hall, was a form similar to American vaudeville, but with less dependence on the attractions of the girls, who were clad, if somewhat scantily. There were comedians, acrobats and 'specialty' acts, which could include performing dogs or seals. Comedians often came in pairs, of whom one was the 'straight', or puzzled and uncomprehending one, and the other the funny man, who was the wiseacre and delivered the punch lines. The formula could vary slightly, the straight man being sometimes the educated and would-be dominant type, the other the street-wise; and there was often an element of master and man in the relationship. The dialogue between straight man and funny man was known as 'cross-talk' and they often indulged in 'turns', borrowing each other's hats, boots and even trousers, or doing slapstick with ladders and chairs. The influence of such comedians on Beckett's own work is obvious. Dublin variety also consisted of sketches, in which a normal situation between two or more actors is developed into absurdity and chaos, with perhaps collapsing furniture or even, in one famous sketch, a collapsing motor-car.

Beckett referred to the variety theatre aspect of O'Casey's work in a review he wrote for the *Bookman* ten years later, in 1934. O'Casey, he says, 'is a master of knockabout in this very serious and honourable sense – that he discerns the principle of disintegration in even the most complacent solidities, and activates it to their explosion'. He especially praises a one-acter, *The End of the Beginning*, in which the entire set comes to pieces and the two comic characters, Darry Berrill and Barry Derrill, lie supine on the stage 'in an agony of callisthenics, surrounded by the doomed furniture'.

The influence of such theatre-going would remain with Beckett, but on what might be called a more serious or poetic level he was especially moved by W. B. Yeats's *At the Hawk's Well* and John Millington Synge's *The Well of the Saints*, as well as by *Juno and the Paycock*. It is easy to see why the Synge should have affected him, for in this drama about a pair of blind beggars who, when their sight is miraculously restored, are sadly disillusioned, the vanity of human wishes is pretty thoroughly urged. The Yeats, like other Yeats plays he saw, may also have had an effect on his own dramatic practice, in as much as, like Yeats, he reduced action to a minimum and wrote plays which are primarily verbal as well as pictorial compositions.

When, thirty years later, the actor Cyril Cusack asked Beckett for a contribution to the programme for the centenary production of Shaw's *Androcles and the Lion* at the Gaiety, Dublin, he would remember this early theatre-going – the Synge and the Yeats as well as O'Casey's great drama. 'I wouldn't suggest that G.B.S. is not a great playwright, whatever that is when it's at home,' he wrote. 'What I would do is give the whole unupsettable applecart for a sup of the Hawk's Well, or the Saint's, or a whiff of Juno, to go no further'.

Besides this formal theatre-going and the more casual slipping into variety theatres in the afternoons, he was also discovering the cinema, an art form which also had its masters of comedy and pathos, Charlie Chaplin, Harold Lloyd and Buster Keaton. He took to them now as he later would to those comedians who reflected the tradition of variety after the talkies had arrived – Laurel and Hardy particularly, or the Marx brothers. Like the variety theatres the cinema involved no fuss, prearrangement or necessary companionship.

As well as fiercely devoting himself to golf and other sports, Beckett was now at last educating himself in serious and profound ways, but for the time being his interest in a more formal type of education was not great and it was only towards the end of his second year that he began to take it at all seriously. When he did so, the effect of a little work and attendance at lectures was almost spectacular and he finished the year as a prize-winner, being ranked sixth among the senior exhibitioners. Most of this improvement was due to the influence of one professor and the effect on Beckett of attending his lectures and being introduced, through them, to the plays of the seventeenth-century French dramatist, Jean Racine.

The professor, R. B. Rudmose-Brown, was one of those singular, eccentric and dogmatic figures who give an impression of being at odds with academia, but whom only academia can really accommodate. He was physically massive, with a prominent brow, and these characteristics combined with his evident eccentricity to make him impressive to undergraduates. The type is familiar – the sort of professor who fits undergraduate notions of brilliance but is perhaps held in less esteem by his colleagues. Certainly Rudmose-Brown's colleagues did not rate his scholarship impressive enough to elect him a fellow. Granted that there was a certain contempt for mere teaching professors among the Trinity illuminati at the time, some of the references to Rudmose-Brown, familiarly and sometimes rather contemptuously called 'Ruddy', are hostile or dismissive; and on at least one occasion he was treated with scant courtesy in public by the legendary wit and classical scholar, J. P. Mahaffy, who was Provost of Trinity. Mahaffy had received a letter in French, which fact he announced to the common room in Rudmose-Brown's presence, declaring at the same time that he intended to ask that individual to translate it for him. He was, he emphasized, perfectly well able to read it, but 'why keep a dog and bark yourself?'

Though 'Ruddy' was Professor of Romance Languages he did not lay claim to knowledge of any languages or literatures other than French and Provençal, nor was he expected to. The description of the chair was a misnomer and when, in 1937, its title was altered to Professor of French, the change was merely a recognition of a division of responsibilities which had long been a reality. 'Ruddy' is usually credited with the awakening of Beckett's interest in Dante, but this was not the case. He had no direct influence on Beckett's reading of any language other than French; though since it is true that Beckett had not been a serious, or indeed even a seriously enthusiastic reader of any language at all before his second year in Trinity, one might claim that Rudmose-Brown made an intellectual out of a cricket-loving schoolboy.

Obviously he had the supreme and central gift without which all other entitlements to teach are vain: that is, he could convey his own enthusiasm for the subject. What was also unusual about him as a teacher at the time, however, was his interest in the modern. It had only been a relatively short while since vernacular literatures were admitted to the curricula of universities anywhere. Opposition had been strong and the doubts expressed grave. To teach the modern was to take a further step. Even in the early 1920s in such a university as Trinity the 'approach to language learning was essentially philological and to literature historical, not to say archaeological'. Ruddy taught not only Verlaine, which would have been unusual enough, but Larbaud, Fargue and Francis Jammes. His students wrote theses on Valéry, Péguy and Henri Pourrat. In time he would encourage Beckett to think of Pierre Jean Jouve as a subject.

The most important introduction, though, was not to a modern but to a French author who was a classic in both senses of the word. This was Racine, whose influence on Beckett cannot be overestimated. While under Ruddy's tutelage, Beckett read all sixteen of Racine's plays; and though at the time he may have had no ambitions to be a playwright himself, the impression they made on him was indelible.

In his own plays he would observe, or nearly observe, the classical unities of action, place and time, which Racine derived from Aristotle. Reviewing *Godot* over thirty years after Beckett first read Racine, his friend Georges Belmont would say that the three unities are not so much observed in it as 'pitilessly' present. Even unity of time, which, as Vivian Mercier remarks, is the most difficult and the most neglected, is a feature of the play. True, it takes longer than a single revolution of the sun, but Racine and French classical criticism had extended the allowable period to twenty-four hours.

Like *Happy Days*, *Endgame* and, of course, the shorter plays, it has also unity of place and of what might be described as action. To circumvent the restriction on a variety of settings, French classical drama had hit on the idea of the '*lieu vague*', a location which should not be particularized; a setting described simply as a palace, for example, would allow the action to take place in any apartment or any part of the grounds. This device is not necessary to Beckett, but he nevertheless describes his settings in the manner of French classical drama. So the setting for *Godot* is described simply as '*A Country Road. A Tree.*'

More startling is how Beckett's general aesthetic coincides with Racine's. In his preface to *Bérénice*, the French dramatist had declared his conviction that 'all creativity consists of making something out of nothing'. Critics had long complained that in the typical Racine play nothing happened at great length. They would make the same complaint about Beckett.

Essential to the static quality of a Racine play is the pairing of characters who talk at length to each other. This allows for reflective dialogue, confidences exchanged, philosophies expressed, reminiscences indulged in. The companion who can be talked to and indeed at is essential to the classicist's method. A typical cast list is that for *Phèdre*, which, besides Hippolyte, Aricie and Phèdre herself, includes their pairings: Oenone, nurse of Phèdre; Theramène, tutor to Hippolyte; Ismène, friend of Aricie; and Panope, waiting-woman of Phèdre. Beckett too will have his pairs: Vladimir and Estragon, Pozzo and Lucky, Clov and Hamm, Winnie and Willie, who talk to and at each other endlessly. And while it may seem a far cry from this sort of pairing to the cross-talk comedians of the variety stage, the two influences undoubtedly complemented each other. Beckett would become a French classical dramatist with overtones of English and Irish music-hall.

As his range of interests expanded, so did his social life. As a third-year exhibitioner he was now entitled to rooms and he moved into 40 New Square. New Square is the second Trinity quadrangle, new in the eighteenth century. His rooms were on the ground floor and were shared with a fellow Portoran, Gerald Stewart. They were rather gloomy but commodious, with wooden floors, old and decent, plain wooden furniture. Beckett had a bedroom, shared the sitting room, but of course no bath or shower, and he used the toilet on the staircase.

To begin with, he went home every weekend to Cooldrinagh; but as time went by these visits became less regular and finally they almost ceased. For a while after he moved into New Square his mother visited him in his rooms, arriving at least once a week and bringing a basket of food with her. In spite of the good things she brought these quite natural visits were equally naturally resented by her son; and just as he was displeased by her attentions, so was she by the coolness with which she was greeted and the taciturnity with which her questions were met. Also on the wane at this time was the habit of inviting a small number of his Trinity acquaintances to dinner at Cooldrinagh. These had been mostly old Portorans whose social acceptability in middle-class Protestant circles could be taken for granted and whose background, conduct or general demeanour did not have to be explained or explained away, as would be the case with the intellectuals and others he was now meeting. Such invitations to dinner in Foxrock had been formally issued on printed cards, but in his third year they tended, like other customs established in the first two, to lapse.

He was now beginning to go to student parties in other people's rooms and houses, gatherings at which he did not appear to join very wholeheartedly in the fun, such as it was, but attended just the same — a tall, thin, silent, rather morose figure, who leaned against the wall, head forward, silently observing the goings-on through steel-rimmed spectacles. On a more formal social level he went to a few of the sort of 'at homes' or 'evenings' which were such a feature of Dublin in the earlier years of this century. The practices and rituals at these gatherings would vary, as would the beverages on offer; but the basic principles were that the house in question was open to people on an understood guest list plus whoever they might at their discretion decide to bring, that you came for conversation or decorous home entertainment, that you were reasonably sober, and that you did not outstay your welcome. It would be some time before Beckett became a known and acceptable figure at the more celebrated 'at homes' of Dublin. The ones he attended at this time were principally those given by Rudmose-Brown and his wife and by Mrs Starkie, mother of Walter Starkie, whose lectures on Italian literature he was attending.

Walter Starkie was, like Rudmose-Brown, something of a licensed eccentric among his Trinity colleagues and students, but he was much more of a figure in literary Dublin generally than Ruddy was. Member of a well-known Dublin family and, incidentally, brother of Edith, biographer of Baudelaire and Rimbaud, Starkie had become a Trinity lecturer three years before Beckett enrolled as an undergraduate. Unlike Rudmose-Brown he was made a fellow within a fairly short space of time and became Professor of Spanish and lecturer in Italian literature in 1926, the year in which Beckett began to take Trinity seriously. Starkie was a talkative, rubicund, cheerful person, with something of the innocent enthusiast about him. He claimed to have had an adventurous First World War and to have developed during service with the Red Cross a special affinity with the gypsies of Europe. It was known in Trinity that he spent his holidays among these gypsies, playing his fiddle and studying their language and folklore. The fruit of these sojourns was to be *Raggle Taggle*, an account of his travels with gypsy bands in the Balkans published in 1933 and once a well-known book. It was followed some years later by *Spanish Raggle Taggle*, which described his adventures as an itinerant violinist, begging his bread in Spain. When Starkie's name came up Gerald Brenan told the present writer that he knew nothing about the Spanish people or their attitudes – 'or gypsies either', he added; but however that may be, *Spanish Raggle Taggle* was followed by *Gypsy Folk Lore and Music*; and meantime Starkie had become, in 1927, a director of the Abbey Theatre.

Walter Starkie lived with his mother and his younger sisters, Muriel and Chou Chou. At their 'at homes' he was not loath to oblige with a csárdás or two on the fiddle, nor did his sisters need much encouragement to sit down at the piano. Several prominent Dublin literary figures occasionally came to the Starkies', including W. B. Yeats, but there is no evidence that Beckett met any of them at this period; and in fact he was never to meet the greatest poet of the age, there or anywhere else.

He visited the Starkies largely because, as a bright pupil of Walter's, he was expected to; and he eventually ceased to go when he felt he could risk it. Like, it seems, almost everybody else in Dublin at this time, Starkie had some reputation as a conversationalist and he was one of the most lively lecturers on literature that Trinity could boast. It would be rash to say he had much influence on Beckett, but in one direction he did have some at least. Though Sam now had some interest in the theatre, like others in Dublin he knew little of what was going on on the Continent, for it was not until 1928 that Micheál MacLiammóir and Hilton Edwards founded the Gate and presented a number of continental plays, notably those of the German expressionists. Starkie, however, had some knowledge of the continental theatrical avant-garde and Beckett was to remember later that he introduced him to the work of Pirandello, which he read with some pleasure.

The influence of Rudmose-Brown is an entirely different matter and it probably extended far outside the classroom and indeed outside literature. Later in life Ruddy, as Beckett and others called him, or Ruddy-Nose, as students generally referred to him, wrote an autobiography. It remained, as one might expect, unpublished; and as one might expect also, it has a certain egotism and provinciality of tone which jars on the reader. It makes no mention of Beckett, perhaps because its author simply did not think of the younger man as sufficiently important in his own life. But though Beckett would have expressed himself differently, some of the stances adopted might not have been antipathetic to him. Thus Rudmose-Brown declares himself to be against any form of organized religion. He also rejects ideology, being, he says, 'neither Fascist nor Communist, Imperialist nor Socialist'. He will not have anything to do with nationalism or even patriotism. 'If I have loved passionately certain towns and landscapes of Scotland, England, France or Ireland, I loved them for themselves only . . . As far as I can go back in my memory I find no patriotism, no nationalism.' Above all, he says, he is for 'the self-realisation' of the individual, though he couches the sentiment in terms which Beckett then or later might not have approved: 'To impede the self-realisation of any individual is to impede the self-realisation of the Divine.'

On most Sunday afternoons Beckett made the long trek out to Malahide, a distant northern suburb where Rudmose-Brown lived with his wife, formerly Anne Shirrefs-Gordon, to attend the gatherings to which students and others were invited. 'Furry', as Ruddy called her, was, like himself, accounted 'a character' by students and others. She was said to have 'an artistic temperament' and to make life difficult for him. Both of them had affairs; both were over-fond of a drop, though the really deleterious and destructive effects of this weakness would, as is usual, not show themselves for some time; but theirs was one of those marriages which survive in spite of scenes and difficulties and are indeed partly dependent on them. Beckett would be polite to Furry, as he was polite to most other people, for he was a well-brought-up south Dublin Protestant, but his main interest was in Ruddy and he usually contrived to spend most of the afternoon in his company, sitting sometimes on the arm of his host's deep armchair while its occupant discoursed. Less welcome to him perhaps were Rudmose-Brown's encouragements to his students to pair off at some parties, which occasionally included turning out the lights. It would almost certainly be wrong to say that Rudmose-Brown influenced his attitudes or beliefs in the sense in which one person may influence another by forcefully arguing him out of certain positions or into certain others. But the attitudes on which Ruddy prides himself in his autobiography – a distrust of ideologies and isms and collective emotions, a belief in the individual's truth as the only truth – were the attitudes which Beckett was discovering in himself; and there is no doubt that this large, excitable, red-nosed man reinforced them in one way and another and thus helped him along the road he was to take. He would try to help Beckett, in other more practical ways, towards the academic career he expected him to follow; and when at length Beckett decided that his career could not be an academic one after all, the decision would cause some pain to both parties. 'I have, I am fully aware, failed to make good, as man, as scholar, as thinker, as artist,' Ruddy would write of himself; but he leaves out the category of teacher or mentor.

Both Rudmose-Brown and Walter Starkie have been given the credit for arousing in Beckett an enthusiasm for the poetry of Dante Alighieri, one of the most important influences in his life. But although Beckett attended Starkie's lectures on the great Italian poet, he had already discovered Dante with the help of a privately employed Italian teacher, Bianca Esposito, a discovery which came with the force of a revelation.

Bianca was the daughter of a well-known Dublin figure, the pianist and music-teacher Michele Esposito. An Italian immigrant who also composed occasionally, Esposito had a high reputation in a city in which, except for vocal performance – and even that was on the wane – there was little professional musical talent to be found. There is a tenuous link across the generations in Beckett's acquaintance with the

Espositos, for twenty years earlier James Joyce's talents had come to the notice of the Italian and his two attractive and sympathetic daughters. At that time Joyce had hopes of becoming a professional singer and Esposito had offered to help. The connection might have been a fruitful one had Joyce not arrived drunk one evening at the Camden Hall, where Frank and Willie Fay's National Literary Society, forerunner of the Abbey, was rehearsing. He was refused admission and lay down in a half stupor in the passageway, forcing the ladies of the cast, among whom was Vera Esposito, Bianca's sister, to step over him when they were leaving. He later composed some verses about the episode which went:

> But I angered those brothers, the Fays,
> Whose ways are conventional ways,
>   For I lay in my urine
>   While ladies so pure in
> White petticoats ravished my gaze.

But he had angered more than the Fays, for he had angered Vera also and, being as conventional as anybody else, she had ceased to regard him with the same sympathetic interest as before.

Beckett knew Michele Esposito as well as his daughter and Joyce's name must have cropped up then or later, for in Paris a few years afterwards he was able to tell Joyce that Esposito had compared his voice to that of Jean de Rescke, a famous Pagliacci of the 1900s, a piece of information which pleased the author of *Ulysses*, who was always immensely flattered to have his singing voice praised.

When Beckett took lessons from her, Vera's sister Bianca was a still very attractive single woman in her forties. In his story 'Dante and the Lobster' she is called Adriana Ottolenghi. The hero of the story is learning Italian and we are told that 'there subsisted as much of the Ottolenghi as might be expected to the person of a lady of a certain age who had found being young and beautiful and pure more of a bore than anything else'. Beckett thought highly of her intelligence, indeed we are told, perhaps a shade ambiguously, that Belacqua 'did not believe it possible for a woman to be more intelligent or better informed that the little Ottolenghi'. Once she had introduced the young man to his works, Dante became a shared enthusiasm and Bianca drew his attention to questions inherent in the text, suggesting that he study, for example, 'Dante's rare moments of compassion in hell'; but she was rather cool and ironic about some of his gushings and enthusiasms. Altogether the portrait Beckett later drew is an attractive one. Bianca Esposito played a not at all unimportant part in his literary development. The tall, blue-eyed, athletic youth had begun to grow up. He was no longer an adolescent, and if, in the approving sense of the word, he was not an adult either, he had at least begun that attempt to escape from Foxrock, Portora and family which would absorb his energies and emotions – sometimes painfully – for a very long time to come.

In 1926, at the end of his third year in Trinity Beckett was awarded a Foundation Scholarship in modern languages. A Foundation Scholarship was the most important prize that an undergraduate could win. Scholars had commons, or meals in the dining hall, free. They paid only half the usual rate for their rooms, very little in fees and received £5 per quarter as an emolument. But the real importance of a scholarship to a not needy student like Beckett was that it marked him out academically. Favoured by his professor, proven in a considerable test, his feet were now on the first rung of the academic ladder.

In the course of time, little would accrue to him from this except a considerable amount of guilt, but in the summer of 1926 he was probably quite pleased with himself. In so far as he was anything, he was a budding academic. He had not contributed to the school magazine while at Portora nor had he published anything in *TCD*, then the principal outlet for students with literary ambitions, in his first three years in college. In fact he would not seek to publish anything there for a clear two years after ceasing to be an undergraduate. So far as anybody knew – so far as he knew himself – he had no literary ambitions, though he was beginning to acquire genuine literary enthusiasms as distinct from the stirrings of academic ambitions. As Vivian Mercier has said, he was 'that rare, almost unique creature – a literary artist whose first ambitions were scholarly rather than creative'.

After the examination Rudmose-Brown suggested he should go to France for the summer. He was, after all, a student of French literature, but he had never set foot on that particular sacred soil. His parents proving agreeable, it was decided that he would take a bicycle to the Loire valley, beginning at Tours, where he stayed at a pension on the outskirts of the city. There he met a young American, Charles C.

Clarke, who was himself in search of French culture and who also had a bicycle. They visited numerous châteaux together; but the most memorable of their trips was to the burial place of Ronsard at the Prieuré de Saint-Cosmé near Tours. The supposed place of interment was actually being excavated in at attempt to find the poet's remains. Ronsard's chapel had been turned into a barn, sometimes used as a stable, with hay on the floor and pitchforks and other implements stacked against the walls. This experience would provide material for a poem some years later, a typically obscure and referential contribution to the *New Review* in 1931:

> dirt in a dirt floor
> in a chapel barn
> by a stifled stream.
> Zoroaster
> politely factorised
> and a hay-rake
> guarantee his siesta
> except during the harvest season when the
> latter is removed.

The reference to Zoroaster presumably means that there was some sort of a sun-dial there, but the whole thing is a fair example of Beckett's poetical manner at the time.

Clarke was the first intellectual friend of his own age Beckett had had, but on his return to Trinity that autumn he was to find another. Trinity had a reciprocal arrangement with the École Normale Supérieure, by virtue of which a French graduate would come to Ireland for two years to be *lecteur* in French, and alternately an outstanding Trinity graduate would go to Paris to be an equivalent in English. In 1926 it was the École's turn to appoint a *lecteur* for two years. The graduate chosen was an elegant, handsome and cultured example of all that was best in French civilization and he was to have a very large influence on Beckett's life.

When he arrived in Trinity that autumn Alfred Péron was a slender, dark young man with an engaging manner and a capacity to be witty and charming in both French and English. He had an intimate knowledge of contemporary French literature and art and he might almost have been chosen to expand Beckett's horizons, introducing an element of playfulness and humour into the discussion of these matters as well as the excitements of a devotee. Though Jewish, he even proved acceptable in Cooldrinagh, to which he was soon invited. If he was an example of what France had to offer, May and her husband had nothing to fear from their son's increasing interest in that country. This reassurance aside, however, they felt that they had cause for disquiet.

A gap was now opening between Sam's world and theirs which they found difficult to come to terms with. When he came home for week-ends there seemed to be no communication of any sort. His few attempts at sharing his intellectual interests or communicating his new enthusi-asms having proved an awkward failure, he soon lapsed into silence, something he was getting quite good at. At twenty he was displaying the symptoms of estrangement from family common in adolescence. When this estrangement occurs some adolescents experience their parents' pain as well as their own. Beckett was one of them and later he would speak with compassion of May and Bill's puzzled and uncom-prehending attitude to the stranger in their midst.

But house devil, street angel; and to others, for the time being at least, Beckett seemed more, not less sociable. He was beginning to imbibe a bit at social gatherings. (He was also beginning to go to the pubs in the Trinity area, but he seems to have preferred to do this on his own.) He had now not only begun to hold opinions on a variety of literary and artistic matters, but seemed ready to express them. Where previously he had been a detached and shy figure at parties, undergoing his 'crescendo of disengagement' from other people, he was now almost animated. Some people attributed all this to the influence of the vivaci-ous and talkative Alfred Péron. It was one of those young men's friend-ships, deeply charged with romantic and erotic emotion, in which the love is partly for the world opened up by the other person; and it was one of those periods when Beckett seemed no longer like a stranger to the world at large.

Previously known to the general run of students only as a cricketer, and perhaps remarked on as he passed by virtue of his tallness and air of self-absorption, Beckett was by now a fairly notable figure in college, and became even more so when he took to driving his father's new motor-car, a Swift tourer, through the back gates. For a student to have the use of a car was itself a rarity and Beckett in a car was every bit as egocentric as he was on a motor-bike, while the tourer, with its bulb horn and the still mandatory arm signals, allowed him more latitude for theatrical display and imperious gesture. In a story set a little later his fictional alter ego, Belacqua, is described as scoring 'a wake of objurgation as he sped through the traffic', leaving pedestrians and cyclists staring after him while civic guards at various intersections took his number. Frequently in trouble for traffic offences and failing to possess a driving licence, he was sometimes forbidden the use of the vehicle by his otherwise tolerant father.

Together with Péron, he was a member of the newly formed Modern Languages Society, which mainly consisted of Rudmose-Brown's favourites. Since college societies had rooms he now had a social focus, equivalent to a club, other leading members being W. S. Maguinness and Oliver McCutcheon, who had been at Portora with him and had finished first in the Foundation Scholarship examinations in which Beckett had come fourth. Since women tended to take modern languages in numbers disproportionate to their total student representation, there were also women members. Among these was Ethna McCarthy, also a Foundation Scholar, whose life was to be entwined with Beckett's in various ways until the end.

Ethna was four years older than him and she was also far more mature psychologically. Depending on how you looked at it, she had a reputation in Trinity for being either 'fast' or 'advanced'. In the former capacity she wore plenty of make-up, smoked cigarettes, even in public, flirted with men and was rumoured to have consented to sexual intercourse. In the latter, she was independent minded, outspoken and had friendships with men which presumed equality between the sexes. She was certainly beautiful, small, with dark hair and luminous dark eyes. According to the narrator of the story 'A Wet Night', which was to form part of Beckett's first collection, 'She had only to unleash her eyes, she had merely to unhood them, as well she knew, and she might have mercy on whom she would.' Before beginning her first year in Trinity she had been friendly with a Cambridge student, another subsequently well-known playwright, Denis Johnston, who saw her home after parties on several occasions, part of their walk being along the Foxrock railway line during the curfew imposed by the authorities in the Black and Tan era. According to Johnston,

Seeing a girl home during the Curfew had a peculiar attraction and hazard about it – slipping into people's front gardens when a Black and Tan lorry could be heard – lying down together behind a wall or a hedge, giggling in each other's arms – what a way to get acquainted! Best of all was that railway line – Deserted at night, inaccessible to lorries except at level crossings where we had to climb the gates. Ethna . . . was a dream on gates.

After which he remarks drily that Beckett 'used to take her out later, though not along railway lines'. Johnston was deeply in love with her and in later life she would confess to having been in love with him, but, as is the way with the young, they never managed to convey their feelings to each other. Though this later account suggests at least a good deal of heavy petting, what was called in Ireland 'coorting', his contemporary diary entries suggest that little enough happened along the railway line and that he went home puzzled and disappointed as often as not. But not only did he manage by innuendo to convey more than this in his published account many years later – where Ethna is identified only by her Christian name – but in a sort of gloss or abstract from his diaries he made a only few years later he also tries to convey that the walks home were more eventful than was really the case. Perhaps he was trying to deceive himself, so that he would remember things differently. In any case their relationship was a stop-go affair because he was a student at Cambridge and was only home during vacations.

Apart from her greater experience, the measure of Ethna's effect on a young man such as Beckett would be increased by the fact that she was highly intelligent, kind, charming and, in spite of the outspokenness, a good listener. When the present writer met her some thirty years later he was struck not only by her intelligence but by the ease and matter-of-factness of her demeanour.

Beckett's first relationship with Ethna began companionably. They were part of the same circle and could meet as such without the strains imposed or the questions raised by dating as a couple. Later he would become, to a degree, infatuated by her and she would respond indulgently, moved principally it would seem by kindness and some understanding of his difficult nature.

In spite of his new-found sociability, at the time when he first encountered Ethna Beckett was an introverted young man with little experience of women outside the family circle. Like most young men he masturbated, but in his case onanism was abetted by strong idealistic and solipsistic tendencies. In many ways other than sexual he preferred the realm of the imagination to that of apparent reality; he had difficulty in relating to the outside world and saw no great or urgent reasons why he should. Whether his feelings about Ethna at the beginning of their acquaintance had much or little sexual content, he was certainly in no hurry to end his sexual solitude and persuade her to accept him as her lover.

At the French Society and at the student parties he was now attending, Beckett discussed his new literary enthusiasms with Péron, Ethna and others. He was reading a great deal, principally along the line of French poetry which anyone interested in French modernism would regard as the main strand: Baudelaire, Rimbaud, Apollinaire and, among contemporaries, Max Jacob. When later on he praised the Irish poet Thomas MacGreevy for his interest in French models, he would add Corbière and Laforgue to this list, thus expanding it to take account of the line Eliot had established as leading to English modernism, but he added also the surrealists.

Towards the close of his degree year in Trinity the unresolved question of Beckett's future career began to present itself again, partly because he had secured a moderatorship, or post-graduate travelling scholarship. To fulfil the conditions of this he would have to produce the fruits of research and he accordingly decided to write on a literary movement, Unanisme, and its principal figures, Jules Romains and Pierre Jean Jouve. Rudmose-Brown had indicated that if this project proved acceptable, the way would be open for Beckett to join the faculty. There was also the very definite possibility that he could be the *lecteur* appointed to the École Normale for the two years beginning in 1928. He does not seem to have been truly interested in either Jouve, Romains or Unanisme, a movement which believed in the primacy of communal preoccupations and public events over individual experience; and, as things took their course, the project became an increasingly theoretical one.

The prospect of an academic career did not displease Beckett; and certainly the prospect of an immediate sojourn in Paris pleased him greatly, as it would have pleased any young Irishman, even one without his special predilections for the French language and literature. Nor were his parents as disquieted by the prospect of an academic career as they might have been by other choices. Trinity College was still a bastion of Protestant respectability in Ireland and its prestige in Foxrock was great. Even a future Trinity professorship was not an impossibility, and from May's point of view it may have seemed that a Trinity appointment of any kind would keep her son closer to her than other career choices might have done. Emigration, though usually to various parts of the British Empire, was as common a prospect in their milieu as departure to the industrial cities of Britain or to America was to Irish people of other classes. Frank was now in India practising as a civil engineer; but Sam was May's favourite and she wanted to keep him at home. When it was realized that a sojourn in France at this stage would actually tend to increase this possibility, his mother raised no objection. Nor were there any to another, shorter educational trip abroad.

At the end of the summer term of 1927, again at his mentor's suggestion, Beckett departed for Florence, where he stayed at a *pension* run by a Signora Ottolenghi and family at via Campanella 14, off the piazza Oberdam. He paid 30 lire a day, for which he got his meals as well as accommodation. The Ottolenghis were cultured people and not only was the Signora a nice person, but she had daughters who flirted mildly

72

with him, affecting occasional great merriment at his Italian locutions, at other times seeming to be greatly impressed; and he conferred the family name on Bianca Esposito when he came to write *More Pricks Than Kicks*. But though he enjoyed these home delights and sometimes, in spite of his shyness, practised his Italian on casual strangers, he seems to have been otherwise solitary, spending much time in the Uffizi and the Palatine gallery of the Pitti until he was joined by the companion of his bicycling tour, Charles Clarke. In the autumn he took his degree, coming first in his class in modern languages. He was awarded a gold medal and £50, the news of which was received with pleasure in Cooldrinagh.

He had now, however, arrived at a hiatus. He would not begin as *lecteur* until the end of October 1928; and meantime he had a whole year to fill, during which he would be without means of support or the status of a student. Again, Rudmose-Brown's intervention saved the day, or appeared to. He used his influence to get his prize pupil a job as a teacher of French in Campbell College, Belfast.

Campbell was the main alternative to Portora for boys of the Northern Irish Protestant middle class. It was a practical, down-to-earth institution, somewhat more mundane and even less literary in its approach to things than Portora. The headmaster, 'Duffy' Gibbon, was by all accounts a decent enough fellow in the ordinary course of events, but Beckett was not an ordinary employee. Many more or less neurotic young men find the transition from university to job difficult. Beckett, who had had four years to acquire bad habits after leaving Portora and had latterly been acquiring them very rapidly, found it almost impossible. As a teacher of French in a place like Campbell he was a disaster.

To begin with, he found it difficult to get up in the mornings and he was constantly late for classes or missed them altogether. He found teaching grammar a supreme bore and tended to neglect it in favour of poetry and literature. Worse, he departed from the curriculum and often wound up reading poems he admired or enjoyed to uncomprehending pupils. Even this kind of conduct might have been acceptable had he remained on good terms with his class. However, the acerbic and sarcastic comments he wrote on their exercises created a good deal of personal antagonism to him among his students. Many of the victims of these sarcasms had parents who took an interest in their progress and frequently read what had been written in their exercise books. These now began to enquire about Beckett's general habits and characteristics as a teacher. The result was a flood of complaints to Gibbon and inevitable angry confrontations between headmaster and French teacher. At one such meeting Gibbon chose to remind Beckett that his pupils were 'the cream of Ulster'. 'Yes,' the latter responded. 'Rich and thick.'

Since the appointment was only until the end of the school year both parties were spared the trauma of a dismissal, and Beckett returned to Cooldrinagh from his first foray into the world of jobs without that final disgrace. He now told his parents that before going to Paris he wanted to spend some time on the Continent, mostly with his aunt Cissie and her family in Kassel in Germany.

Aunt Cissie was Willie Beckett's only sister. She had been the youngest of the family and was to some extent indulged. In spite of her parents' misgivings she was allowed to attend the Royal Hibernian Academy's School of Art in Abbey Street, where Walter Osborne and John Hughes taught and where William Orpen was the star pupil before he went on to the Slade. There she was the contemporary of Beatrice Elvery, whose sister Dorothy later took the famous photograph of little Sam Beckett affecting to pray at his mother's knee, Estella Solomons and Lily Williams. Cissie showed a degree of talent and exhibited at the Young Irish Artists' Exhibition in Dublin in 1903; after which, in spite of her parents' more positive forebodings, she was allowed to go to Paris to continue her studies. This was a surprisingly common thing for Irish young ladies to do at the time, and part of the inducement for William and Frances Beckett was that her friends Beatrice Elvery and Estella Solomons were being allowed to go as well; but perhaps they thought their fears were justified when, shortly after her return to Dublin, Cissie married a friend of Orpen's, William or 'Boss' Sinclair, who not only had the disadvantage of having no money and no real job but brought with him a suggestion of worlds that the solid, middle-class Becketts feared and, in so far as they knew anything about them, distrusted and disliked.

'Boss' Sinclair was the son of a Jewish antique dealer who had premises in Nassau Street. He worked in his father's shop, but rather unwillingly, for he was a natural bohemian, whose ambition was to deal in art on his own behalf; and it was in pursuit of this ambition that, after some years, he had uprooted his wife and family and settled in Kassel in Germany. In Dublin 'Boss' was a well-known, rather flamboyant figure, who was as often to be found in the smoky billiard saloons of the city as at its artistic gatherings or dining with his friends Oliver St John Gogarty, Cecil Salkeld and William Orpen in Dublin's celebrated – and only – French restaurant, Jammet's. He and Cissie lived in a cottage in Howth and gave bohemian parties during which people sat on the floor and afterwards quite possibly slept on it, waking up in the morning to continue the talk and drinking. Lady Glenavy rather

74

primly says that they kept open house there 'for all the poets and painters of Dublin; indeed, for all the freaks and oddities as well'. For the Dublin of the day it was all very advanced, but Glenavy describes it as also being 'lively, warm and human; with children tumbling all over the place'.

Boss had combined desultory attendance at the shop with desultory picture dealing. The move to Kassel was made at the instigation of Salkeld, a poet and painter, who was, like Boss himself, a well-known Dublin character, and who found Germany a good place to escape from his mother to. The idea was that someone with Boss's flair and developing eye for new movements in painting would be able to buy the work of German modernists and sell it profitably in Dublin; but it was never to turn out quite like that. The demand for the work of the German modernists in Dublin was small. What living Boss earned was mostly by shipping antiques, bought in Germany at a favourable exchange rate, back home for his brother to sell.

For the moment, however, the principal attraction of the Sinclair family and Kassel for Beckett was not Cissie or Boss, but their seventeen-year-old daughter, Peggy. He had first met her several months before, when the Sinclairs had come to Dublin on a family visit. Peggy was a vivacious, talkative, green-eyed girl who did not read many books but had, like himself, an interest in music. He had seen the Sinclairs off from Dun Laoghaire at the end of that visit. She had waved her green beret to him from the rail of the mail boat several times and he had pined a little ever since, remembering her face, 'the loveliest little pale firm cameo of a birdface he had ever clapped his blazing blue eyes on'.

In Kassel he saw her in her home environment, the eldest of a family of four children living in a two-bedroom flat in Landgrafenstrasse with parents who had plenty of time for talk and many visitors. He discovered that Peggy was a creature of moods. Sometimes, when she was in good form, she could be extraordinarily loquacious and funny, effortlessly dominating the company and making everybody laugh until her mother would say she ought to be on the music-hall stage. On other occasions she could be unexpectedly snappish and act like the spoiled child of the family, which indeed she was.

The warmth and seeming spontaneity of the Sinclair ménage was a long way from the frigidities and repressions of Foxrock; and Cissie Sinclair was not only a great deal more cultured than May Beckett, but a great deal more motherly as well, which motherliness was now extended to the young man on the couch in the living room. He could stay in bed in the mornings, look into Peggy's green eyes while they had long, semi-serious conversations, and go drinking with Boss when he chose. The house was full of books and paintings and good music was played on the gramophone. There was an assumption that these things were important which was not made in Cooldrinagh, as well as comings and goings and a general sense of cultured disorder, which pleased him. Nobody for the moment was making any demands. It was no wonder that he began to regard the crowded flat in Kassel as a second home and the Sinclairs as a second family, in many ways preferable to the real one. Cissie soothed him and encouraged him to feel a part of things. Boss was a bohemian man of the world, *au fait* with music, literature and painting, a drinker and a good fellow who treated him as a contemporary and an equal. And of course there was Peggy. Even at this early stage it became understood among the Sinclairs that Peggy and Sam were special friends and when she went back to her school in Vienna he was allowed to accompany her on the journey there by train. The school was one of those Germanic gymnasia for girls which flourished at the time, 'very vanguardful . . . very callisthenic and cere-bro-hygienic and promotive of great strength and beauty'. Music and dance played a large part in the curriculum. The school was in the Altes Schloss of the Laxenburg Palace, and Beckett stayed not far away in the big, blue Hof or pension, known as the Blauer Hof, where his 'high dark room, smelling of coverlets' opened off a courtyard. Peggy visited him there in her blue woollen gymnasium cloak, walking across a park to do so, and they drank tea with lemon juice and smoked Macedonian cigarettes. When he came back to Kassel again it would be partly at least to be with her. But it was now the end of October and the time had come to take up his lecteurship at the École Normale, so, travelling third class in the continental carriage with its wooden seats, he made the long journey from the Westbahnhof by train to Paris.

Chapter Six

Beckett's first period of residence in the city that was to be his home
for so long dates from the first day of November 1928, when, under
the arrangement with Trinity, he became a *lecteur* at the École Normale.
The city to which he came had recovered well from the shock of the
First World War, and to many in literary circles who experienced them
the Parisian 1920s seemed a happy time. In 1932 Julian Green was to
record what he believed to be 'the end of a happy era' in his journal;
and Claude Mauriac, too young to participate directly in the literary
life of the decade, wrote nevertheless of the nostalgia everyone felt for
'those wonderful years'.

This, of course, was in retrospect. In fact at the time, as in England,
despair was a fashionable mode. As in England too, it was often
associated with the experience of the Great War, but numbers of those
who cultivated despair had had no direct experience of conflict at all.
One factor in the fashionability of pessimism was to be operative
again in the 1950s, when Beckett's work first came to international
notice.

Among the freedoms which are almost inevitably lost in war-time is
the liberty to be despairing about the future of the belligerent country
in which you happen to reside, or even of the human race. Pessimism
becomes a sort of treason and compulsory optimism prevails. For four
crucial years everything that had appeared in French, as in British,
newspapers had to be accompanied by an optimistic prognosis of the
end result. Even large-scale retreats or unforeseen and unwished-for
disasters had to be presented as part of a brilliantly conceived overall
strategy. It had to be suggested that in the post-war era political forms
and structures would be found to ensure everybody a measure of prosperity, justice and happiness. And, by a ghastly paradox, it had to be

insisted over and over again that if so many lives were to be lost, life itself was worth living.

All this optimism, all this enthusiasm, all this positive thinking had to be expressed in language that was at best bombastically eloquent and at worst coarsely bellicose, modes which came not unnaturally to senior men of letters in France because of the fashion for eloquence represented by such as Barrès and Claudel, Péguy and Romain Rolland before the outbreak of war. The result, among the younger writers of the post-war period, was what one critic called 'an irresistible taste for sincerity, even a brutal sincerity, a contempt for over-statement, and the desire of not permitting themselves to be duped either by words or by hollow sentimentality'. And all this, he added, 'to the great scandal of more aged citizens, who had imbibed the milk of rhetoric in their cradles'.

After the war, as the same critic put it, 'it was not a question of beginning again, it was a question of beginning'. Among ordinary citizens, in France as in the other victorious countries, there was a reluctance to return to the ordinary respectabilities and drudgeries from which war, in all its horrors, had been an escape; and this gives the 1920s a false appearance of holiday everywhere. But there was also, among intellectuals, a feeling that continued destruction of some sort was necessary and cleansing. Not only the social hypocrisies, but the very social structures of the belligerent countries had led to the pro-longed horror, and the complacencies of society must accordingly be destroyed. The Dada movement initially billed itself as a 'demolition enterprise' and such indeed it was. The surrealists not only refused to accept the controls and imperatives of society, but even those of the conscious mind. Faced with every imperative or seeming imperative, categorical or otherwise, the younger writers, in common with a great many other people, asked 'why?'. 'Why go on writing?' was a common question, discussed over and over again. 'Why go on living?' was an even more widely asked one. And the answer to both in many cases was negative and pessimistic: 'There is no reason why one should.'

And it was not only the surrealists and their soul-mates who asked such questions. Henry de Montherlant, though influenced by d'Annunzio and Claudel and, above all, by Barrès, who had been his respected mentor as well as his beloved friend and neighbour at Neuilly, asked them also and returned a negative answer. 'One by one', he wrote, 'I have beheld dropping away my reasons for bestirring myself, submerged, each in its turn, by indifference, that rising tide.' And after disposing of religion, fraternity, the longing for fame and even curiosity itself, he declared: 'I long for a time when . . . a disgust with self-scrutiny shall have led me to renounce writing and thus shall have freed me from the last social bond.'

In 1924 the young Marcel Arland began his forty-year connection with the *Nouvelle Revue française* by a celebrated essay in that journal in which he compared the 'new *mal du siècle*' evident among the younger writers to the *mal du siècle* of the romantics, which had likewise had its birth while Europe was at war. In this famous essay, as in much else of the influential writing of the decade, one finds a kind of fore-echo of the preoccupations that would one day be central to Beckett's work. Speaking of the history of Dada and of how the Dadaists had made a principle of daring, he went on, 'But then at once comes the thought: supposing we were to go to the limit in daring? The limit in daring is not destruction, but abstention; a violence greater than saying No is to be found in silence.'

Of course, one does not suggest that Beckett's preoccupations as an artist or as a man were dictated to him by the preoccupations of literary Paris in the 1920s, but it is none the less notable how much of his outlook and how many of his concerns are reflected in the outlook and concerns of the younger writers of the decade and how closely their temperament accords with his own. He was, in a sense, made for twentieth-century France. Finding it, he found his homeland.

In his essay Arland had also discussed self-destruction, remarking that just as the limit of daring was to be found in silence, so 'the real despair lies in acceptance rather than in suicide'. Suicide was much in the air throughout the 1920s. The second number of *La Révolution surréaliste* had posed the question, 'Is Suicide a Solution?' To this the answer, on the whole, was no, though some surrealists did take their own lives in well-publicized fashion from time to time. But for other than French writers too, suicide was in the air. In December 1929 Harry Crosby, the rich young American who had founded the Black Sun Press and was very much a part of the Joyce circle and the *transition* group, shot himself (after abandoning an intention to fly his aeroplane into the sun until it ran out of petrol and crashed) in New York. His act was much discussed in *transition* circles and No. 19/20, to which Beckett contributed a poem, 'For Future Reference', contained commemorative contributions by Hart Crane, who himself jumped off a boat in the Caribbean in 1932, as well as by the surrealist Philippe Soupault and by Kay Boyle, who were both members of the Joyce circle. The suicide of the Russian poet Vladimir Mayakovski a few months after Harry Crosby's was also heatedly debated, as had been that of Sergei Yessenin in 1924. As far as younger literary Paris was concerned, however, the most intriguing suicide was that of Jacques

Rigaut, which took place the year after Beckett arrived in the French capital.

Rigaut was one of those writers who exist more in legend than in their work. He made a practice of destroying most of what he wrote, when he wrote, but at the beginning of the decade, in 1920, had appeared one of the only two articles published in his lifetime. It was an essay in the review *Littérature*, Aragon and Breton's predecessor to *La Révolution surréaliste*, which reprinted it after the author's death in 1929. In it he makes the almost specifically Beckettian observation that 'There are no reasons for going on living, but no more are there any reasons for dying.'

'The only manner left us in which we can evidence our disdain for life is by accepting it,' Rigaut had written.

> Life is not worth the trouble of leaving it. One may, out of charity, ward it off for certain others, but for oneself? Despair, indifference, treasons, felonies, solitude, family, freedom, gravity, money, poverty, love, absence of love, syphilis, health, sleep, insomnia, desire, impotence, platitude, art, respectability, dishonour, mediocrity, intelligence, there are no fish there to fry. We are too well aware of what these things are made of to be taken in by them: good to inspire a few negligible accidental suicides . . .

Rigaut's death was the famous 'pistol shot that ended the decade' and was all the more heatedly discussed because he had, after due consideration of the option, ruled suicide out. Though Beckett, as we shall see, like a great many others whom he knew or knew about, talked much of suicide and the merits and drawbacks of various means of doing it during these years, he would, on the whole, have shared Rigaut's previously expressed attitude. Life was not worth the trouble of leaving it. And he would also perhaps have sympathized with another of that celebrated young man's statements. 'To live, that is to say, to accept life . . . when everything is so wretchedly bound down . . . in this acceptance lies all that there could be in the way of the ridiculous, and our life can only take on a real meaning by setting out from an idea in which it is wholly swallowed up and lost, the moral idea.'

Of course, all this nihilism and despair, this public talk of suicide and self-destruction, was confined to coteries. And though it is important to remember in the case of Samuel Beckett that they were coteries and movements of which he was well aware, some even ones of which he was a member, on the whole what Cocteau characterized as 'sensual and intellectual Paris' appeared to be, in the same author's phrase, 'oblivious of danger, dreaming only of pleasure'. During *la belle époque* Paris had earned a reputation as a place where certain modes of enjoyment had been brought to a pitch of perfection never before attained in any other city; and this enviable reputation seemed, if anything, to be enhanced during the 1920s. And at least as far as surface impressions were concerned, Paris seemed also to be a place in which art and life were no longer enemies, so that each was woven in happy comity into the fabric of the other. Cocteau also said that the Paris of this era 'amused itself with the mad prodigality of a city of genius', thus conferring an artistic temperament and an artistic licence on the city as a whole.

By 1928 the French capital was well established as the home of avant-garde modernism. In the visual arts the seal had been set on this reputation by the Exposition des Arts Décoratifs of 1925, with its cubist and modernist outlook and its adumbrations by Le Corbusier of a modernist future for everybody. As someone said, the hero of the show was the reinforced concrete block. Although the everlasting conservatism of French life still had its powerful voices in literature, with the death of Barrès in 1923 and France in 1924 the pre-war era of old-fashioned, graceful scepticism opposed to traditional but stylish conservatism seemed finally to have come to an end. The election of Paul Valéry to the academy in that year conferred a sort of official status on poetic modernism; but there were many to whom Trotsky's theory of the perpetual revolution seemed to apply to literature, their only fear being that soon there would be nothing left to revolt against. 'We had to get at people any way we could ... we had to be outrageous ... Why not admit it? We were madly in love with being shocking. It's what kept us going. We had to be outrageous, and we were outrageous,' wrote Philippe Soupault, a poet who provided a nexus between the Joyce-*transition* circle and surrealism, whom Beckett was to meet and know quite well and with whom he was to be intimately associated when translating *Finnegans Wake*.

Yet the truth was that to secure the approval of the few hundred people who composed what had come to be known as '*le tout Paris*' elements of modernist innovation had become essential. Novelty provided topics for conversation, but in so far as it was not really likely to disturb the basic equanimity of those who approved, it was becoming a new academicism. In the salons of the Vicomte and Vicomtesse de Noailles, the Princesse de Polignac or the Princesse Marie Murat, the intersection of taste and novelty provided a sort of snobbism of the new which to some extent had begun to govern even the responses of the general public. To these people Diaghilev, the Ballets Russes and the Ballets Suédois, Stravinsky, Picasso and Cocteau provided constant proofs that Paris was the 'city of genius'; and largely under the influence of these ballets there was a ferment of interaction between the various arts – librettos by Cocteau, Cendrars or Claudel accompanying scores by Satie or Honegger in productions with settings by Picabia, Léger and Picasso.

Popular culture too was in its way modernist, the public and the intellectuals alike agreeing on the significance of Cole Porter and Charlie Chaplin and the relevance to modern life of the movies and jazz; while in the approval of the circus and the music-hall, of entertainers such as Maurice Chevalier, Josephine Baker and, of course, Mistinguett, there was the element of modernist condescension to mass culture that had distinguished futurism and, on the other side of the Channel, had been a characteristic of T. S. Eliot's essay on Marie Lloyd. This democratization of taste was also reflected in the fashions on offer, for Chanel had provided a Parisian look – cropped hair, boyish figure, tailored suits and short hemlines – to which both rich women and working girls could aspire.

Looked back on now, the twenties in Paris seem to have been largely a surrealist decade, which is a great tribute to the surprising staying power of the movement, but in fact the surrealists stood somewhat outside this unanimity of enthusiasm. In 1928, the year Beckett arrived in Paris, *Un Chien andalou* was first shown (an English translation of the script would be published four years later in an issue of *This Quarter* for which he would himself translate contributions by Breton and Eluard). The reaction it provoked was not favourable and disapproval even extended to the police. The surrealists' attempts to combine the literary, plastic and musical arts had initially seemed to give them an accord with smart Parisian modernism as propagated by Cocteau and the Ballets Suédois; but many of their outbreaks had genuinely offended members of the intellectual smart set. As Germaine Bree has put it, 'they aimed their 'will to destroy' as much toward the Parisian conspiracy of money, fashion, celebrity-mongering and publicity-seeking as toward the social and literary establishment.' Their well-staged mock trial of Maurice Barrès, their public rejoicings over the death of Anatole France, the sedulously offensive and utterly disrespectful tone of the 'Lettre ouverte à Paul Claudel', and the outrageous Saint-Pol Roux banquet at the Closerie des Lilas, a café much frequented by Beckett a little later, had genuinely set them apart. By 1928 so also did their increasing politicization. Most French writers were not political in the 1920s, but as the decade wore on, the fruitlessly and cruelly waged Riff war in Morocco was to have nearly as profound an effect in terms both of politicization and polarization as Indo-China and Algeria would thirty years later.

From the beginning a cult of violence which was more than just intellectual had been one of the principal weapons in the surrealist armoury. The first issue of their review had published a photograph of Germaine Berton, who had just murdered a prominent right-wing member of the reactionary Action française, defiantly surrounded by all the members of the group. When Beckett arrived in Paris the process of formal politicization of the avant-garde was already well advanced. Aragon, Breton, Eluard and Benjamin Péret had all been members of the Communist Party at least since 1926; all the surrealists had signed the manifesto against the war in Morocco; and although the last issue of *La Révolution surréaliste*, which appeared the following year, also posed the famous question, 'What hope do you place in Love?' it was succeeded almost immediately by a new journal with a significant name change: *La Surréalisme au service de la révolution*. The year after he arrived, 1929, was also the year of the famous Second Manifesto and bitter splits in the movement. Breton was now proclaiming his support for Trotsky; but in 1930 Aragon attended a congress in the Soviet Union and simultaneously fell in love with Elsa Triolet and the Five-Year Plan.

Of course, much of what was going on in Paris passed Beckett by, for three principal reasons. The first was his youth and relative lack of sophistication. The second was that he had a fruitful talent for allowing things to pass him by; and the third that in spite of his knowledge of the French language and his interest in French literature he was drawn almost immediately into the circle surrounding James Joyce and therefore into an expatriate and, on the whole, anglophone, rather than into a French or francophone scheme of things.

The avant-garde reputation and atmosphere of Paris naturally attracted expatriates from the English-speaking world. When we think of them now, we think of Hemingway, Scott Fitzgerald, Edith Wharton and others, as well as of Gertrude Stein, Ezra Pound and, of course, James Joyce. But there is a difference between these two, sometimes overlapping groupings. Hemingway and Fitzgerald were mainstream writers, who looked to the United States or England for publication and fame and lived in France principally because they liked what France had to offer in the way of *la douceur de vivre*; whereas the expatriate literary circle into which Beckett was drawn was the sedulously avant-garde one of the English language magazines published in Paris and read, for the most part, by other expatriates who felt that to be modernist and to live in Paris were one and the same thing. Unlike, say, Scott

Fitzgerald or Edith Wharton, these had many francophone literary acquaintances and, sometimes, models. Beckett's introduction to such circles took place almost immediately.

The École Normale Supérieure is in the rue d'Ulm, in what is still the heart of academic Paris in so far as institutions of learning have remained in the centre of the city, and was then even more so. Nearby was the Sorbonne and the two great Lycées which fed the École with their brightest pupils. An inscription on the classical portico would have told the young Irishman that the École had been founded by a decree of the Directory in 1793, confirmed by one of the Emperor Napoleon in 1807. If he was in a mood to be observant rather than anxious he might have been amused to note that the Emperor's confirmation of its status was dated 17 March – St Patrick's Day – but he was scarcely as yet knowledgeable enough about French history to see that the École owed its origin to the period of middle-class consolidation which marked the later stages of the French Revolution, a period whose values and institutions have largely governed French life ever since. The austere, marble-floored corridors and the general air of studious seriousness with which people went about their business would, however, have confirmed what Rudmose-Brown had already told him, that a place in the École was only to be gained by those who did well in fiercely competitive examinations, and that graduating from it with any sort of distinction ensured an influential place in the French scheme of things thereafter.

Besides being a serious place where academic subjects were eagerly discussed by students even in their spare time, it was also in those days markedly austere and monastic, more like a boarding school than a university. There were no women students and the rooms off the upper corridors to which he was shown after being welcomed were of an exceedingly spartan order – in fact little more than sparsely furnished cubicles. Most of them had no doors, the entrance having only a rather dirty and bedraggled curtain to mask it, and the partitions between the cubicles did not go all the way to the ceiling. Though in accordance with his status as a *lecteur* he at least had a door, his room was far less spacious and considerably less private than his rooms in Trinity. It was on the first floor, to the right of the central stairwell, and it had a window that looked westward to the Luxembourg.

To add to the feeling of slight dismay which its spartan arrangements caused, the room or cubicle which had been allocated to him appeared to be already occupied, and there was a hiatus while the person whose belongings were strewn about was identified and fetched. When he appeared, to apologize and collect his scattered effects, he turned out to be Beckett's predecessor as *lecteur*, a dapper and talkative little Irishman named Thomas MacGreevy.

Tom MacGreevy (originally spelt McGreevy) was thirteen years older than Sam, having been born in 1893 in Tarbert, a village on the Kerry side of the broad waters of the Shannon estuary. And since his parents belonged to the Catholic lower middle class, his social background was very different. But Tom had been a bright youngster and so, at the age of sixteen, he had taken the entrance examination for the British civil service, a not uncommon option for bright Irish youths at the time, whatever their parents' political views, and entered the lower ranks of the administration in 1910, working first in Dublin and then in London.

From whatever fate held in store for him as a minor civil servant, however, he was rescued by the Great War: he volunteered in 1916, moved, like so many others, by romantic notions about the rights of small nations. He was posted to the Royal Field Artillery in which he served with distinction on the Western Front, attaining the rank of lieutenant, being twice wounded and mentioned in dispatches.

The war was an important experience for MacGreevy, though he did not talk much about it. After demobilization it also opened up new options for him, in particular that of going to university. Though a devout Catholic, he opted for Trinity rather than the new National University which had been established in 1908. The particular small nation to which he belonged was at the time involved in a conflict with Britain, the object of which was the attainment of that right to self-determination for which he and others had supposedly fought. Soldiers wearing the uniform which he had just taken off were, as he saw it, committing atrocities against Irishmen whose only crime was the assertion of that right and an affirmation of the ideals but lately claimed by British politicians to have been the object of the war. A certain anti-Englishness was, perhaps understandably, to be part of MacGreevy's outlook from then on, though, since he was a civilized person, it did not influence him in his practical dealings or personal relations with English people.

At Trinity he studied political science and began to write poetry which was influenced by Eliot and Pound as well as by the French models that they recommended. He had attended some lectures of Rudmose-Brown's but had conceived rather less admiration for that individual than Beckett had. His only collection of verse, a slim volume called *Poems*, was to be published by Heinemann in 1934, six years after he met Beckett, but at the time of their first encounter he had been publishing poems regularly in periodicals and was well-known in certain circles in expatriate Paris, both as a poet of some apparent promise and as an art critic.

In spite of Beckett's initial bewilderment, this first meeting between the two ended amicably. MacGreevy explained that Beckett had arrived sooner than was expected and before he had bothered to move. Although he had ceased to be *lecteur* he was staying on in Paris, having got a job on the art magazine *Formes*. After some negotiation he found quarters upstairs and the two Irishmen went out together for a drink.

Tom MacGreevy was to become Sam Beckett's principal confidant and, for a long time, his closest friend. He was talkative, witty and perceptive about literature. His manners were exquisite and their formality was balanced by a charm which was immediately experienced by many of those who knew him (including, incidentally, the present writer, who made his acquaintance in the 1950s), though there were those in Dublin who thought, or affected to think, him pretentious and a name-dropper. He was knowledgeable about both Dublin and Paris, was a gifted story-teller and had considerable experience of the world. Eventually the author of four or five poems which are among the best written by any Irish poet in this century (as well as being comparable with poems written by any English language poet anywhere), he was the nearest to an equal in sensibility and general artistic intelligence that Beckett had yet met. But probably none of these things account for the instant attraction which Beckett felt or the continuing affection it led to; indeed their differences of temperament and outlook may have had something to do with that too. MacGreevy was Catholic, gregarious, socially cheerful and something of a dandy, seldom seen without a bow tie and sometimes a buttonhole. Beckett was Protestant, private, serious and given to gloom, even in company.

In his unpublished memoirs MacGreevy described how, quite early in their relationship, when they were both at the École, he became aware of certain differences in character, outlook and modes of behaviour. Beckett had no difficulties about beginning the day with an academic task, going straight from his morning coffee to his desk and his books. The sociable, gregarious MacGreevy felt, on the other hand, a compulsion to take the morning as it came, 'to go out and make sure that the world was as I had left it the evening before.'

MacGreevy's Catholicism was even then a serious matter, intellectually and otherwise. In Dublin later on he would attract a certain amount of derision as a devotee, a daily mass-goer who associated with priests and was dependent on them for literary employment when the war cut him off from other contacts; but he was a lifelong believer and in a gentle way he would attempt to convert Beckett by trying to show him that his attitude to existence did not differ from those of many who had seen it in the light of eternity.

At this particular period of his life, however, MacGreevy's beliefs may not have influenced his personal life as much as they appeared to do later on, when he returned to Dublin. In what seems to be a detached and broken-off fragment of the autobiographical memoir mentioned, MacGreevy speaks of his attraction to men, though he typically adds that when he first experienced these attractions in adolescence and mentioned them to his confessor, the latter's advice was to kick himself every time he had such thoughts, at which point the fragment breaks off. In Dublin later on, where, as the present writer has said in another context, 'thousands of celibates, lay and clerical, thronged the land, living proof that the state was not an unnatural one and that its difficulties were exaggerated elsewhere', MacGreevy passed as a celibate, a religious devotee who was not particularly interested in sex of any kind or had sublimated his sexuality, whatever it was; but in Paris it was otherwise. He was believed in the École to be a homosexual, as his best friends in École circles were. Among them were the overtly homosexual Jean Thomas, later a higher civil servant and official of UNESCO; and the brilliant philosopher Jean Beaufret, whose later career as teacher in one of the great Lycées was to be darkened by a scandal involving his relationship with a pupil. These now became Beckett's friends in École circles also. An acquaintance of the time thinks that although the sexually inexperienced, highly attractive young man from Ireland may not have responded sexually to whatever attentions were paid to him, he was at least flattered by them. It might also be accurate to say that his relationship with MacGreevy had, though it was not sexual, an element of the homo-erotic in it, as indeed some of Beckett's later relationships were to have.

Outside École circles there was another nexus of acquaintance opened up by MacGreevy which was to be of profound importance to Beckett. In their first conversation Tom happened to mention that the job on *Formes* had been landed partly through the influence of James Joyce. Would Beckett like to meet Joyce? he enquired. It was not the sort of question to which a twenty-two-year-old Irishman who was without acquaintance in Paris was likely to say no, however shy and however little acquainted with the work of the master he might be. Although Joyce had not then visited his native country for almost twenty years, he was nevertheless a planet which shone brightly in its literary sky; and of course he was one of the best-known avant-garde writers in the world. In Ireland especially his writings were a source of scandal and his career a cause of much dismay; but his anti-nationalism and his presumed anti-Catholicism were certainly not likely to give Beckett pause. Even in Foxrock these would not have been disapproved of, though the reputation *Ulysses* had acquired on other grounds might have caused head-shaking had anything much been known about it.

Unlike many who lightly offer introductions to the famous, McGreevy was as good as his word. With Robert McAlmon and Arthur Power he was one of Joyce's *camarilla*, regarded with an affection and trust which were none the less real for having sometimes an element of tolerant amusement in them. He saw Joyce often and telephoned him even more frequently. Joyce was by no means averse to meeting young Irishmen with the right credentials. He was having some people to dinner and he readily assented to MacGreevy's request that he might bring Beckett with him on the night in question.

A snag arose when it was discovered that besides the four members of the Joyce family there would be eight other people at table and that the addition of Beckett would bring the number to thirteen. Since Joyce was excessively superstitious this was a real difficulty, but fortunately somebody else dropped out. MacGreevy had already given the master a description of his new friend: that he was from Dublin; that he was a Protestant who had been to Trinity and lived in Foxrock; that he was a cricketer. It was all, of course, grist to Joyce's mill.

When the night came Beckett wore a French suit he had lately acquired. He felt awkward in it because he had decided it was too tight a fit. Joyce wore his flowered waistcoat and after dinner he sat in an armchair smoking and saying little while others gathered round him. Doubtless he took some note of the tall thin young Dubliner in the ill-fitting French suit who, like himself, peered short-sightedly through small round-framed glasses and, like himself, said little enough throughout. The acquaintance was to develop gradually and eventually to ripen into something like intimacy; but even this initial meeting had a profound effect on Beckett. Years later he would remember walking back to the École that night, exhausted but happy, finding the gates closed and climbing over the high railings to get in. From this point on he began to adopt some of Joyce's tastes and mannerisms, ordering Joyce's favourite white wine in cafés and wearing tight-fitting patent-leather shoes.

His duties as *lecteur* at the École would in any case have been light, but they were rendered especially so by the fact that there was only one *angliciste* among the final year students, the *promotion*, as it was called. This was a highly intelligent, engaging and literate young man from the Jura region in eastern France, called Georges Pelorson. Pelorson (who after the war was to adopt the *nom de plume* Georges Belmont) already had a good deal of English, which Beckett was supposed to improve while they discussed literature together. Their first session was scheduled to take place at eleven one morning shortly after Beckett's arrival in Paris. Pelorson came to his room on the dot and after several knocks decided to open the door. On the bed was stretched a tall, half-naked fair-haired young man, who seemed even more fair because of the sunlight falling across him as he slept. Pelorson, impressed by the spectacle, decided not to wake him; and after some hesitation he left a note to say he had come at the time arranged. Later a meeting in a nearby café was agreed on.

The first session in the café proved agonizing to both parties, since it became apparent that Beckett had no idea how to go about whatever it was they were supposed to do together and the normally quite loquacious Pelorson found himself almost as tongue-tied as his tutor. Then Beckett had a bright idea. They would read an English play together, sharing the parts. Finally he chose *The Tempest*. Pelorson noticed ruefully that Beckett took all the men's parts for himself, leaving the women's to his companion. They both found the exercise very boring but it broke the ice and although they finished the play during their afternoon sessions in the café, the reading of it was gradually replaced by conversation until eventually they became quite jolly and intimate together.

Pelorson, who was three years younger than Beckett, was another intellectual equal and another stroke of luck. Like Beckett he seemed at this time destined for an academic career, but he had literary ambitions, and would eventually decide to make literature and journalism his life, writing poems and novels as well as criticism. At this point he admired the surrealists, who were just at the high meridian of their first wave of activity. Through acquaintance with MacGreevy, Pelorson and, of course, the much-admired James Joyce, Beckett's ambience was gradually shifting from the academic to the literary.

The principal topic of conversation in the Joyce circle in 1928 was the appearance of the master's *Work in Progress* in the magazine *transition*, which had started up the previous year. *transition* was edited by Eugene Jolas, a large French-American with a background which stretched from a childhood spent in bilingual Alsace-Lorraine to a young manhood on American newspapers. Latterly he had worked for the Paris edition of the *Chicago Tribune*, contributing, amongst other things, a causerie entitled 'Rambles Through Literary Paris' to the Sunday edition, in which he discussed the contents of the French reviews, the activities of the surrealists and the work of expatriate writers, principally James Joyce. In 1925 he had met a tall, ebullient, talkative Kentucky girl, Maria McDonald, who was studying music in Paris. The couple's first date was a meeting of the International PEN Club, which had been chaired in very English fashion by John Galsworthy. As Jolas's biographer has remarked, the meeting 'was not a very good way to promote international understanding between authors'; but since James Joyce was present and Jolas was able to introduce the young lady from Kentucky, 'not, however, a bad way to start a courtship'.

Whether in the glow of meeting Joyce or not, the couple decided almost immediately to get married and, shortly afterwards, to start a Paris-based magazine. Joyce and his elaborate and obscure work were, naturally, their trump card. It was serialized in *transition* from that journal's beginning in 1927 until its end in 1938. Given the nature of the work there were more than the usual difficulties attached to proof-reading and these were further aggravated by the fact that Joyce's eyes were now giving him serious trouble. He had so far had nine operations. In September, just as Beckett arrived in Paris, he had had a serious set-back and lost all ability to read. He was now visiting Dr Borsch several times a week for pilocarpine treatment, and when he read at all could do so only with the aid of a magnifying glass. One of the first services which Beckett performed was to assist in the proof-reading of the extracts for *transition*; and, since Joyce was never averse to recruiting helpers, he gradually came to depend on his new-found Irish disciple for the performance of other tasks as well. Like others, Beckett often read to him and marked passages in works that he read himself which he thought might interest Joyce; but he was insistent in later years that he had never taken dictation of letters.

As was apparent on closer acquaintance, he and Joyce had a lot in common. In addition to being Irish and short-sighted they were both polyglots, principally interested in the romance languages. They were both agnostics, though both set a high associative value on the language in which the traditional religions of their forebears had been expressed, and in conversation and writing were not averse to ironic reference to certain metaphysical concepts. They were both apolitical, and took a very cool view of the nationalism which had been the dominant factor in Irish politics during both their lifetimes. They both took an interest in simple numerology; and though Joyce was much more frankly super-stitious, Beckett also kept a wary if Descartian eye on numerical patterns and recurrences. As time would show, they were both to be considerable writers, though Joyce would be dead and gone before Beckett really found himself. And finally, it should be said that they both had a sense of humour, often an important bond between people and one that is rather rarer than is sometimes supposed.

None of this means, however, that Joyce believed he had found a younger soul-mate, or even a disciple of great literary promise. He was self-centred in his greatness and had no need of either. Although admittedly a convivial man, who liked to drink with those he trusted and was quite capable of enjoying himself in their company; and although he needed a circle of people around him to provide admiration and to do things which advanced his reputation and made his life easier, his affections, or rather his need to be involved emotionally, found sufficient object in the members of his family; and even where they were concerned he was by no means a model of perception or real sympathy. He had found in Beckett an obliging, highly intelligent young Irishman, who shared many of his own interests and something of his own outlook on life, who had a sense of humour and who was prepared to provide admiration and sympathy even at some cost to his own identity. He was also, now, prepared to drink with the master on level terms; in fact was becoming, like Joyce, quite a serious drinker, though at this period he adopted, perhaps in imitation of Joyce, a rule about not drinking before a certain hour of the day. In the master's case it was six, in the disciple's five. Unlike Joyce, however, he did not stick to his rule, though he reverted to it with some strictness at times in his life when he felt that for one reason or another he needed to. After an early abstinence at Trinity, he had begun to go to pubs in his third year, when the two pints of stout that represents Belacqua's lunchtime intake in the story 'Dante and the Lobster' was probably not untypical.

Beckett's first impression of the work that was to become *Finnegans Wake* must have been bewildering, though a passage which some commentators have assumed to be a description of his reaction was written before his advent: 'You is feeling like you was lost in the bush, boy? You says: It is a puling sample jungle of woods. You most shouts out. Bethicket me for a stump of a beech if I have the poultriest notions what the farest he all means.' Admiration for the work was, however, incumbent on Joyce's Paris acquaintances; and these were heady days: whatever his initial reaction, it was easy for a young man of twenty-two to get carried away. Not long after the acquaintance had been formed Joyce suggested that Beckett might like to contribute an essay to a symposium on *Work in Progress* which would be published in *transition* and afterwards in book form. Valery Larbaud, he said, was editing this under his own overall direction. There were to be twelve contributors, to parallel the twelve customers in Earwicker's pub; or, more significantly, the twelve apostles of Jesus Christ. Would Beckett like to be one of them? It was an invitation to join the happy and dedicated band; and of course it was accepted.

In November, when Nora Joyce was suspected of having cancer and went into the American hospital for observation, her husband insisted on having a room there also: not apparently for her sake as much as for his own, since he could not bear to be separated from her. This necessitated a good deal of ferrying of letters and other documents between the hospital at Neuilly and the Joyces' apartment on the other side of Paris in square Robiac, much of which Beckett undertook. When finally Nora was sent home, he of course continued to visit the apartment. Nora was not averse to visitors; indeed she had complained to Wyndham Lewis that although Joyce was supposed to be surrounded by admirers, they often sat together in the evening like a couple of old hens with nobody to talk to; and for the time being Beckett was specially welcome because of Nora's fondness for Tom MacGreevy, who had introduced him. Joyce often had tasks for him during the day and he welcomed his admirers in the evening if only because he needed people to drink with. Beckett was gradually drawn not only into the Joyce circle, which was also, to some extent, the *transition* circle, but into the Joyce household, and was coming to know their son Giorgio, soon to marry a rich American divorcée, Helen Fleischman, and their daughter Lucia, who was a year younger than himself.

He had met Lucia the first evening he went to the Joyce apartment with MacGreevy. She was slim, dark-haired and, though almost as shy with men as he was with women, occasionally vivacious. Beckett, who always noticed women's eyes, was struck by the brilliant blueness of Lucia's. He did not seem to notice her slight squint and, whether because of the glow the master cast over her or not, he thought her beautiful.

By the end of the year he could look back on a crowded three months. He had met most of the Joyce circle by now, including Sylvia Beach, a minister's daughter from Princeton, NJ, who declared that the three great loves of her life were James Joyce, her friend and room-mate Adrienne Monnier and her bookshop, Shakespeare and Company, in the rue Dupuytren, which had been the publisher of *Ulysses*. Looking back later, Monnier would remember that she and Sylvia had often seen Beckett at this time, sometimes in company with Joyce, who 'had a great deal of esteem and affection for him'. They were struck by 'his resemblance to Joyce as a young man, of whom Sylvia had photos. He appeared to us like a new Stephen Dedalus – the Stephen of the Proteus episode, who walks alone on the seashore.' She also noted that 'Beckett spoke little' and, as she put it, 'discouraged every advance'. Sylvia Beach wore her hair bobbed and usually a tailored suit and cravat. She had a sharp, alert face and glance and she had of course done Joyce a great service by publishing *Ulysses* under the Shakespeare and Co. imprint, though Monnier and others were agreed that her judgement of authors was better than her actual judgement of literature, and that it was this which had given her a place in literary history.

The Joyces were tolerant of homosexuality, male and female. Robert McAlmon, also an intimate of the circle, was a boastful, witty, hard-drinking homosexual who had made a marriage of convenience with the novelist Bryher, the daughter of Sir John Ellerman, a shipowner and one of the wealthiest men in England. Another American couple, the Crosbys, Harry and Caresse, were also wealthy and were the founders of the Black Sun Press, to which Joyce gave extracts from *Work in Progress, Tales Told of Shem* and *Shaun*. Together with Stuart Gilbert, a retired colonial official who was making a study of the mythic basis of *Ulysses*, and Arthur Power, an Irish writer, land-owner and painter who was, like MacGreevy, a First World War veteran, all these were frequently to be met at the Joyces'. Power remarked, though, that however tolerant Nora and James might be in some ways, they were very strict in others. He noted that one could expect a very cold reception from both Joyce and Mrs Joyce if one brought to their flat a casual girlfriend. 'Your belle amie yes – provided it was always she – but a casual piece, no.'

There was no question of this in Beckett's case. During these first months in Paris he displayed no interest in the women he met socially. He was still in fact fretting after Peggy Sinclair; and when Christmas came was reluctant to return to Cooldrinagh, first pleading that his duties and commitments kept him in Paris, then declaring his intention of going to Germany. He arrived in Kassel to the usual warm welcome from the Sinclair family and the expected puzzling, enticing one from Peggy. Part of his difficulties with her sprang from the fact that, like many another young man – and Beckett was not only young but still immature for his age – he did not know what he wanted. Principally it was simply to be with her, talking, playing piano duets, responding to her flirtatiousness, feasting on the mystery of her green eyes and her delicately beautiful face. Whatever Cissie and Boss might in their somewhat happy-go-lucky way come to expect, there was really no question of marriage. The attachment was already being discussed in Cooldrinagh, where parental opposition to such a project was fierce: perhaps more so on May's part than on Bill's – although he could scarcely be expected to approve very wholeheartedly either, for Peggy was a first cousin and she was Jewish. Of course, immature, romantic

93

young men in the world as it was then often did not think of marriage or, for that matter, seduction either. They were content to live from day to day in the ambience of the beloved, flirting a little perhaps, but basically happy simply to know that their feelings were in some sort returned. Of that, by now, in spite of Peggy's flightiness, he had little doubt; and so for the moment he was as happy as his nature would ever permit him to be.

Chapter Seven

The subject of Beckett's proposed essay for the critical symposium on James Joyce's *Work in Progress* had been characteristically suggested by Joyce himself. He knew, of course, of the young man's interest in Dante. It was one of the bonds they had in common; and he told him of the part the theories and writings of Giambattista Vico and Giordano Bruno had played in the basic structure of the work. In fact he even suggested the title, 'Dante . . . Bruno . Vico . . Joyce', the dots between the names being pedantically meant to indicate the number of centuries that separated one author from the other.

That Beckett was so compliant does not suggest any special sycophancy on his part. Joyce ran a tight ship; he allocated subjects to all the contributors and he decided on their method of approach, the other eleven apostles including Frank Budgen, Stuart Gilbert, Jolas, MacGreevy, Elliot Paul, John Rodker and William Carlos Williams.

Beckett got down to work on his contribution immediately after returning to Paris, his first task being to find out about Bruno and Vico. As is evident from what he wrote, he discovered Vico to be interesting; perhaps less so Bruno, with whom he deals in rather abbreviated fashion, though he is pleased to point out that 'The Nolan', reference to whom had mystified the audience of Joyce's youthful lecture 'The Day of the Rabblement', had been Giordano Bruno Nolano; and that references to the Dublin booksellers Browne and Nolan in *Work in Progress* signified the same individual.

Vico is dealt with more fully. He stresses that Vico's account of history is 'clearly adapted by Mr Joyce as a structural convenience – or inconvenience. His position is in no way a philosophical one.' That little caveat out of the way, however, he is free to give a very clear exposition of Vico's theories: his three ages, Theocratic, Heroic, Human; his 'six termed social progression', from the thunder-clap which produces religion to the anarchy which is succeeded by despotism; and his three fundamental human institutions, Church, Marriage, and Burial, which are adverted to throughout *Finnegans Wake*.

Less successful, if interesting, is his attempt to derive Joyce's method from Vico's theory of language, though it climaxes in the famous statements 'here form is content, content is form. You complain that this stuff is not written in English. It is not written at all.'

Turning to Dante, he advances the rather dubious claim that the Italian language Dante wrote in was as much an artificial construct as the language of *Finnegans Wake*: 'He wrote a vulgar that could have been spoken by an ideal Italian who had assimilated all that was best in all the dialects of his country, but which in fact was certainly not spoken nor ever had been.'

Of more interest to Beckett biography is perhaps his disquisition on the significance of numbers to Dante and to Joyce, and his concluding remarks on the purgatorial aspects of human existence. Here one feels he is asserting the prerogative of genius when it writes criticism: to write about itself. As has been pointed out, numbers fascinated Beckett just as they did Joyce and Dante.

> The death of Beatrice inspired nothing less than a highly complicated poem dealing with the importance of the number 3 in her life . . .
> Why, Mr Joyce seems to say, should there be four legs to a table, and four to a horse, and four seasons and four Gospels and four Provinces in Ireland? Why twelve Tables of the Law, and twelve Apostles and twelve months and twelve Napoleonic marshals and twelve men in Florence called Ottolenghi?

To which those not as obsessed by the significance of numbers as Beckett or Joyce might well reply, 'Why not?'

His last word is 'about the Purgatories'. Dante's, he says, 'is conical and consequently implies culmination. Mr Joyce's is spherical and excludes culmination . . . Sin is an impediment to movement up the cone, and a condition of movement round the sphere.' It is possible, of course, to read too much into this kind of thing. He is comparing Joyce to Dante in certain ways but he wants to make it clear that Joyce does not deal in transcendental absolutes. But the comparison of earthly existence to some sort of purgatory, with its Manichaean overtones, was one that he would make again.

This first venture into theoretical writing was successful in as much as Joyce was impressed and so were the *transition* gang. In later years he felt he had overdone the praise somewhat; but in fact the essay contains hardly any direct praise of Joyce at all. His stature was implied by the comparisons with Dante rather than asserted; and Beckett's was one of the least sycophantic contributions to *Our Exagmination*. It appeared in *transition* 16/17 in June 1929, which was a double number and, in a way, a good one to be in, since it contained the 'Revolution of the Word' proclamation, signed by most of the members of the *transition* inner circle and by one or two outsiders, such as Hart Crane, though he subsequently repudiated his signature, saying he was drunk when he signed, which was quite possibly true.

But *transition* 16/17 contained not only Beckett's first published piece of theoretical writing: it also included a piece of creative or fictive prose entitled 'Assumption'. As a piece of writing this three-and-a-half-page document is probably best forgotten. What is of interest humanly and biographically is that this, his very first published work, establishes a theme which would be his till the end: the simultaneous and contradictory necessity and impossibility of giving utterance.

Like so many subsequent Beckett characters, the central one in this is, to begin with at least, alone in a room: 'In the silence of his room he was afraid, afraid of that wild rebellious surge that aspired violently towards realisation in sound. He felt its implacable caged resentment, its longing to be released in one splendid drunken scream and fused with the cosmic discord.' He dreads the escape of this utterance; but at the same time he longs for it. Then he is visited by 'the woman'. The prolonged contemplation of her beauty, her 'green-flecked eyes' which are 'pools of obscurity', her broad brow, her lower lip, induces a sort of mystic experience.

It would seem that what is involved here is the desire for transcendence of ordinary experience. Beckett's desire for some sort of transcendence through some sort of union with the beloved – one has to be vague – is a topic which will be returned to. When the hero of the story achieves something of this sort – again vagueness is all – we are told that 'at last, for the first time, he was unconditioned by the Satanic dimensional Trinity, he was released, achieved the blue flower, Vega, God'; and that 'each night he died and was God, each night revived and was torn and battered with increasing grievousness, so that he hungered to be irretrievably engulfed in the light of eternity, one with the birdless colourless skies, in infinite fulfilment.'

Finally one night, as the woman is 'contemplating the face that she had overlaid with death', she is 'swept aside by a great storm of sound, shaking the very house with its prolonged triumphant vehemence, climbing in a dizzy, bubbling scale, until, dispersed, it fused into the breath of the forest and the throbbing cry of the sea.

'They found her caressing his wild dead hair.'

It will be seen that the writing leaves a lot to be desired; but the theme is interesting, if inchoate; and so is the description of the green-eyed woman with her broad low brow. This is to be repeated in Beckett's first novel, *Dream of Fair to Middling Women*, and it is, of course, a description of Peggy Sinclair. In the novel the idea of some sort of mystical union resulting from the contemplation of a girl's face is again expressed, though the contemplation of her bodily beauty does not seem to yield the same results.

With the appearance of these two pieces in *transition* Beckett became a writer; and becoming so he crossed at once the gap between the unpublished and the published state and acquired some consequent prestige in expatriate circles in Paris. It was at this point that Adrienne Monnier began to think of him as a 'new Stephen Dedalus'; but Richard Aldington, another expatriate writer and a friend of MacGreevy's, remembered his anything but Dedalus-like talk of suicide. This was a topic that others would remember him discussing both then and later, but it should be emphasized that there are no suicide attempts on record and that, as has been said, suicide was very much in the air in Paris just then. At this time Beckett had made the acquaintance of Walter Lowenfels, who was not a contributor to *transition* but with Henry Miller was a member of a group which for a while called itself the 'Death Movement' and advocated suicide, though not apparently for themselves. With Lowenfels also he discussed the merits and possibilities of taking one's own life.

In February 1929 Nora Joyce again entered the clinic at Neuilly. In spite of her declared opinion that he should have some sense and go home, her husband once again moved into an adjoining room; and again Beckett made daily visits to square Robiac and to the clinic, bringing letters and messages from one to the other. He was now almost part of the family circle; and when Nora was discharged after the hysterectomy which the doctors had considered necessary, he continued to make daily visits to the apartment, to attend evening gatherings, and sometimes to go to restaurants as part of the family circle.

In age he was exactly placed between the two Joyce children, being a year older than Lucia and a year younger than Giorgio, who was (literally) Nora's blue-eyed boy, favoured, as observers noted, in various ways by his mother over his sister. Giorgio was supposed to be pursuing a career as a singer as well as conducting his affair with Helen Kastor Fleischman. She was a supremely elegant, attractive and wealthy

American, daughter of a Jewish New York manufacturer and wife of a certain Leon Fleischman, an American who moved in literary circles in Paris as the representative of an American publishing firm, Boni and Liveright. The Joyces disapproved of Helen Fleischman, who was considerably older than their son, even though she wanted to marry him after her impending divorce; but Nora's illness and the anxiety it caused brought about a rapprochement. When the Joyces went out together now, there would be James and Nora, Giorgio and Helen, Lucia and Beckett. There had already been comment on Beckett's indispensability to the master. Some people described him as Joyce's secretary. Richard Aldington was to say unkindly that he was 'James Joyce's white-boy'. He was friendly with Giorgio who, though not a giant intellect nor even very literate, was a contemporary of an affable nature who liked to drink and with whom he could speak Italian. Now, as the family party was seen about together, there began to be a rumour that he had arrived at some sort of understanding with Lucia. It was of course not so, but a young man who was a bit wiser, more experienced and more observant in certain ways would certainly have been more alert to the dangers of the situation.

In April 1929 he went with the Joyces to hear Giorgio make his public debut as a singer at the Studio Scientifique de la Voix of Professor George Cunelli. Giorgio had a pleasant voice, but it suffered from uncertainty of range, being at this stage more or less bass, though subsequently, after an operation in the United States, it became fixed as a baritone. He suffered from stage fright and had a nervous habit of unnecessarily clearing his throat on stage; but the principal impediment to his career was lack of ambition. He was beginning to drink heavily; he did not work hard enough; and his marriage to Helen Fleischman, whose wealth removed any element of bread-and-butter necessity from his pursuit of a singing career, and who in any case did not really want him to succeed, would not help matters.

The following month a family party including Beckett was assembled again, this time to see Lucia dance in a competition at the Bal Bullier. Up to this point Lucia had been seriously intent on a career as a modern expressionist dancer. She seemed not unfitted for this, being tall and angular, though somewhat large breasted, and moving with a certain grace, as did her mother; and she had worked much harder at it than Giorgio had at his singing, practising or attending classes for at least six hours a day and taking courses in various academies and from various teachers, including Raymond Duncan, Lois Hutton and Margaret Morris. She had also made a number of public appearances, though most of them were student recitals, staged by whatever academy she was attending. Her father and mother had come to these religiously, including one in February at the Comédie des Champs-Élysées, where she appeared in an opéra bouffe, the music of which was by Émile Fernandez, a young man with whom she had been briefly in love. When Raymond Duncan's academy had moved the previous summer from Paris to the Isadora Duncan school near Salzburg, her parents had decided to take their summer holiday in the Austrian city too. This reflected a tendency to treat their now grown-up daughter as if she were still a child, and its effect was probably to still further retard her outlook. In fact she was woefully immature, both in social terms and in her relationship with the opposite sex. She was also suffering from the consequences of being brought up in various countries, speaking various languages and having to make a constant succession of new beginnings at new schools.

Beckett liked her, partly no doubt because she chatted volubly to him on a variety of subjects in her disconnected way while he had the luxury of remaining largely silent. He also liked her looks and no doubt was not averse to going to see her dance clad as a fish in a tight-fitting silver costume which she had designed for herself at the Bal Bullier. She did not win the prize but a large section of the audience, not friends of the Joyces, demonstrated in her favour, calling out, at least according to her father, 'Nous reclamons l'Irlandaise! Un peu de justice, messieurs!'

An evening at the Joyces' was on the whole a jolly, somewhat Irish affair, with Irish people, including Power and MacGreevy, present and impromptu song and dance. If Robert McAlmon was there he would force the pace of the drinking, but Joyce usually confined himself to white wine and Beckett would follow suit. At a certain point of the evening Joyce would almost certainly sing an Irish ballad or two in his light tenor voice and on occasion he was even known to improvise a dance while somebody played the piano.

When Beckett and Joyce were alone together, however, mutual silences were often one of their principal methods of communication – silences, as Beckett put it, 'directed towards each other'. Joyce usually sat in the attitude familiar from photographs, legs crossed, the toe of the crossed-over foot pointing downwards in its tight, patent leather shoe, or twined round the calf of the other leg. Beckett adopted a similar posture, the faithfulness and humility of the imitation being emphasized by the fact that he had also begun to wear similar footwear, even though such natty shoes did not suit his feet and he suffered accordingly.

The silences were occasionally broken by a remark or a question,

Joyce's contribution often philosophical and, as befitted an admirer of Aristotle and Aquinas, disrespectful of most systems of thought other than Catholic scholasticism. In his contacts with Joyce and even Mac-Greevy no less than in his familiarity with Dante and certain aspects of French civilization, Beckett was now becoming aware of the elaborate construct which is Catholic theology. This knowledge was to leave its mark on his work, both in its scepticism and in its logic.

He and Joyce had begun to go for walks together, along the Seine as far as the Ile des Cygnes. In a play, *Ohio Impromptu*, written over half a century later, Beckett would remember his companion of these walks. 'Day after day he could be seen slowly pacing the islet. Hour after hour. In his long black coat no matter what the weather and old world Latin Quarter hat. At the tip he would always pause to dwell on the receding stream. How in joyous eddies its two arms conflowed and flowed united on. Then turn and his slow steps retrace.'

Joyce's conversation was seldom specifically literary, but on one occasion when they discussed poetry he gave it as his opinion that it ought to be rhymed, which Beckett thought rather quaint. He recited Verlaine and said that he could not imagine anyone writing poetry except to a *petite femme*. He seldom adverted to his own work except in terms of the techniques and structure of *Work in Progress*; but years later Beckett was to remember the emphasis with which he said, 'I can justify every word I have ever written.' He also remembered the master saying he thought that *Ulysses* was perhaps 'over constructed'. But Joyce was also puzzled that even sympathetic critics had missed the humour of the book. 'Why does nobody ever say how funny it is?' he asked. When, in the early sixties, Beckett told the present writer about this complaint of Joyce's, it may just have been beginning to be borne in upon him that his own work would suffer the same fate. In 1983 Germaine Bree asserted that 'certain aspects of Samuel Beckett's work' and, in particular, 'his sense of humour' had 'escaped his French critics entirely'.

Joyce and Beckett had in fact much the same sort of sense of humour, the uncompromising, often cruel, Irish kind, which has aspects of gallows humour in it, rather than the gentler, more whimsical English variety; and neither of them had any objection to humour which had a certain obviousness or farcicality about it. Beckett remembered Joyce on one occasion telling a story about a man who was having a bowl of soup in the dining room of a hotel. The waiter, who is looking out of the window, remarks, 'Looks like rain, sir.' To which the diner replies, 'Tastes like it too.'

It was the custom common among Joyce's friends to organize

celebrations which they thought might please him. One such was the 'Déjeuner Ulysse' which Sylvia Beach's friend and partner, Adrienne Monnier, organized to celebrate both the publication of the French version of *Ulysses* and the twenty-fifth anniversary of Bloomsday. It took place on 27 June and the venue, chosen because it boasted a Hôtel Léopold, was the little village of Les Vaux-de-Chermay, which is beyond Versailles. All the Joyce family were there, including Helen Fleischman; and among the guests were Beckett and MacGreevy as well as Philippe Soupault, Éduard Dujardin, Paul Valéry, Jules Romains, Léon-Paul Fargue and Sylvia Beach. It was decided to transport everybody by charabanc from the rue de l'Odéon and return them there afterwards; but unfortunately Beckett fell, as Joyce put it, 'deeply under the influence of beer, wine, spirits, liqueurs, fresh air, movement and feminine society'. On the way back, primed by MacGreevy, he several times asked Joyce to have the bus stopped so that another drink might be had at some suitable café; and, indulgent to an Irish weakness, Joyce several times obliged. Other members of the party, including Paul Valéry, whose suggestion that there should be speeches at lunch had been vetoed by the guest of honour, grew angry at this; and finally the unpopular member of the party was, to quote Joyce again, 'ingloriously abandoned by the Wagonette in one of those temporary palaces which are inseparably associated with the memory of the Emperor Vespasian', in other words while pissing in a urinal. Joyce readily forgave; but given Beckett's admiration for Valéry, it was an unfortunate episode.

Since her operation Nora Joyce had been depressed and frequently tearful. It was thought that a change might do her good and so she and James went to England with Stuart and Moune Gilbert in July, staying first at Torquay – Joyce 'had a weakness for English watering places' – and then going on to Bristol and London. The École Normale was of course closing for the summer and most of Beckett's other acquaintance was also clearing out of Paris. He now had no reason to remain there himself, official or unofficial; and so, after a brief struggle with his conscience about the claims of Dublin and his parents, he decided to go to Kassel. There his relationship with Peggy Sinclair now entered what might be called its second stage or phase.

Chapter Eight

Like many other young men, Samuel Beckett longed, or thought he longed, for a union with another person which would be so complete that it would transcend the conditions of ordinary existence. The lovers would not only be fused together on some plane or other but, in that union, conjoined also with an ill-defined Absolute.

> At last I find in my confused soul,
> Dark with the dark flame of the cypresses,
> The certitude that I cannot be whole,
> Consummate, finally achieved, unless
>
> I be consumed and fused in the white heat
> Of her sad finite essence, so that none
> Shall sever us who are at last complete
> Eternally, irrevocably one,
>
> One with the birdless, cloudless colourless skies,
> One with the bright purity of the fire
> Of which we are and for which we must die
> A rapturous strange death and be entire,
>
> Like syzygetic stars, supernly bright,
> Conjoined in One and in the Infinite.

Of course, it is possible to read this sort of stuff as if it had some reference to sexual congress. Puerile though it is, the European tradition to which it belongs, that of the troubadours and the neo-Platonists, is a subtle one. In fact, though, physical union in the sense of congress has usually little to do with the matter: what is being expressed is a romantic hope of somehow rising above or bringing to an end our poor finite and physical creature state. It usually has much more to do with Thanatos than with Eros. It is not uncommon for young men who express this sort of 'desire' and this sort of 'hope' to have imaginings of erotic acts and conjunctions, but they are usually about other objects of a different kind of desire. It is not rare for them to have or to have had actual erotic experiences, but they usually involve others than the supposed beloved. The sort of 'love' expressed in writings like this is in fact deeply opposed to the physical, on the side of the ethereal or the spiritual.

But it is a mistake to interpret such expressions of longing as religious either, certainly in any orthodox Christian sense. Doubtless the lines could not have been written if the author had not had a religious background which had coloured his sensibility, but the hope is rather for a joint annihilation, perhaps a pantheistic annihilation in nature, 'the blue birdless colourless skies' and 'the bright purity of the fire', than for any sort of salvation. The Church which eventually allowed and even fostered the tradition of feeling and expression to which the twenty-three-year-old Beckett's poem about Peggy Sinclair belonged was recruiting both Eros and Thanatos to its service, linguistically at least.

That Beckett was 'in love' with Peggy Sinclair in some sense or other may as well be accepted, for it is a phrase so vague of import as to cover a multitude of states. Her small face, with its green-flecked eyes beneath a low brow, haunted him and seemed to have an import beyond its own beauty. It reminded him of the Madonna Lucrezia del Fede in the National Gallery, a comparison to which he often returned. To begin with he had seen only her face. That she had a body escaped his notice; when he did become aware of it, he was repelled. So, in the novel *Dream of Fair to Middling Women*, Belacqua to begin with sees only the Smeraldina-Rima's face:

> His fatigue on that fatal occasion making him attentive to her face only . . . he had so far forgotten himself as to cast all over and moor in the calm curds of her bosom which he had rashly deduced from her features that left nothing but death to be desired as one that in default of Abraham's would do very nicely to be going on with in this frail world that is all temptation and knighthood.

In the very paragraph in which he introduces the Smeraldina-Rima, he says it all, however affectedly: the attractions of the face, the rash assumption that the attached body will merely inspire the same emotions, the wish for union which is really a death-wish, like the wish to repose in Abraham's bosom, the opposition in this 'frail world' between ideal love, 'knighthood' and carnal delights, 'temptation'.

Kassel in the summertime was a very different place for a courtship, such as it was, than in the winter. The river which flows through it, the Fulda, meanders gently round bends and siphons itself off into many lakes or inlets and there are also lakes in the park called the Wilhelmshöhe. Just north of the city the Fulda joins the Weser and flows with it into the sea at Bremen. In the summer everybody bathed and everybody boated and the Sinclairs were no exception. They went up the river to a place called Kargenhof, a bathing and boating place where they had in other years rented the whole top floor of a pension; and they went also to a Baltic coast resort. Beckett missed the Baltic holiday, which saddened him, but he and the family and he and Peggy together were often at Kargenhof, which is a short train journey from the Wilhelmshöhe Bahnhof or the Hauptbahnhof.

That summer of 1929 they read together *Effi Briest*, by Theodor Fontane, a German classic about a young woman who is betrayed by her beauty and her desires into adultery and becomes the victim of malice and gossip. Peggy found it highly affecting, being reduced to tears on more than one occasion by the heroine's tribulations; and so to a degree did Beckett, retaining a soft spot for the book ever afterwards – to such an extent that he urged its reissue in English translation on his publishers many years later. One of Beckett's later dramatic alter egos, the writer Krapp in *Krapp's Last Tape*, remembers *Effi Briest*, though the author of the play misspells the eponymous heroine's name: 'Scalded the eyes out of me reading Effie again, a page a day, with tears again. Effie ... (*Pause.*) Could have been happy with her up there, on the Baltic, and the pines, and the dunes. (*Pause.*) Could I? (*Pause.*) And she? (*Pause.*) Pah!'

There had by now been some physical love-making, which Beckett was less eager to continue than Peggy was. Of course he wanted her to respond to him, but not on that level. And he had become aware of her body, which, to put it mildly, did not please him. Most of the descriptions of women in Beckett's early works are psychologically cruel. That of the Smeraldina-Rima is physically pitiless.

Because her body was all wrong, the peacock's claws. Yes, even at that early stage, definitely all wrong. Poppata, big breech, Botticelli thighs, knock-knees, ankles all fat nodules, wobbly, mammose, slobbery-blubbery, bubbubbubbub, a real button-bursting Weib, ripe.

Then, perched aloft on top of this porpoise prism, the loveliest little pale cameo of a birdface he ever clapped his blazing blue eyes on. By God but he often thought she was the living spit of Madonna Lucrezia del Fede.

In fact Beckett, like Western civilization itself, was deeply split between the ideal and the real, a split which had been widened by masturbation and would be acknowledged as well as symbolized by occasional recourse to whores. *Dream of Fair to Middling Women* is largely about this split; and behind the Joycean locutions of his prose is a fierce honesty about it. Several times in the opening pages it is emphasized that Belacqua is 'in love' only from 'the girdle up'. Indeed the assertion is made that this is the only kind of love there is, and it is distinguished by what the troubadour poets valued most: the pang, the love-ache of separation.

Not unnaturally, Peggy was uncomprehending. When, in the novel, Belacqua foolishly attempts to explain that '... it was not when he ... held her in his arms ... but only when he sat down to himself in an approximate silence and had a vision on the strength of her or let fly a poem at her ... that he had her truly and totally according to his God', the poor girl is understandably bewildered and angered, not having heard of the troubadours, nor the poetic tradition that found the summit of bliss in separation from the beloved – and not being in any case very sympathetic to their attitude. In this setting of sun and water and pleasure, she expected some love-making; and one morning, to use the language of the novel, 'she violated him', 'raped him'. Thereafter 'they kept it going in a kind of way, he doing his best to oblige her and she hers to be obliged, in a gehenna of sweats and fiascos and tears and an absence of all douceness.'

But the split went deeper than Peggy suspected. In the novel Belacqua substitutes for physical intercourse with the Smeraldina 'a fraudulent system of Platonic manualisation, chiro-platonism' – in other words, masturbation. That Beckett should prefer masturbation to 'the real thing' was in keeping with his general narcissism and quietism, his preference for what took place in his own mind rather than in the outer, 'real' world, with its contingencies, its disturbances of inner tranquillity, its futile exercises of will and ambition. Some of the most lyrical passages in *Dream* are about the descent into the self, into the tunnel, as it is sometimes called, or the dark or, more overtly, the 'womb-tomb'. That it was in these crepuscular realms, descent into which was often thwarted by one demand or another, including the demands of women – who, paradoxically, are nevertheless pursued –

that any sort of peace or happiness was to be found, was becoming Beckett's basic philosophy, powerfully reinforced, as we shall see, by some of his reading, particularly of Schopenhauer. As far as real women were concerned, their use was as objects of contemplation, whether in the immediate moment or, perhaps better still, when called up in imagination later on. The best music, according to the author of *Dream*, was 'the music that became inaudible after a few bars ... the object that becomes invisible before your eyes is, so to speak, the brightest and best.' When Peggy made no demands of her own, and was at her most bodiless, then he could recapture the emotion he had felt when they first met.

So she had been, sad and still, without limbs or paps in a great stillness of body, that summer evening in the green isle when first she heaved his soul from its hinges; as quiet as a tree, column of quiet ... So he would always have her be, rapt, like the spirit of a troubadour, casting no shade, herself shade. Instead of which of course it was only a question of seconds before she would surge up at him, blithe and buxom and young and lusty, a lascivious petulant virgin, a generous mare neighing after a great horse, caterwauling after a great stallion.

The emotional climax of the play *Krapp's Last Tape* occurs when Krapp plays the tape recounting an incident in a punt on a lake.

I said again I thought it was hopeless and no good going on, and she agreed, without opening her eyes. (*Pause.*) I asked her to look at me and after a few moments – (*pause*) – after a few moments she did, but the eyes just slits, because of the glare. I bent over her to get them in the shadow and they opened. (*Pause. Low.*) Let me in. (*Pause.*) We drifted in among the flags and stuck. The way they went down, sighing, before the stem! (*Pause.*) I lay down across her with my face on her breasts and my hand on her. We lay there without moving. But under us all moved, and moved us, gently, up and down, and from side to side.

This has usually been interpreted in straightforward, simple romantic terms: the affair could not go on for circumstantial reasons – parental opposition or whatever. It was not as simple as that; but, though the couple now bickered fairly constantly, it was not over yet.

Beckett came back from the German interlude with his fair skin considerably sunburned. Before he left on the long journey to Ireland through Ostend there were tears and protestations. In *Dream of Fair to Middling Women* the hero says farewell on this occasion as diplomatically as possible, easing his departure by a present of a powder compact which is suitably chirruped and twittered over. The final leave-taking goes off fairly well, the 'platform absquatulation', in other words, the farewell kiss, being only 'vaguely disagreeable, like an introduction'.

At home in Cooldrinagh his sunburnt appearance was noted and approved; but there was disapproval of the fact that he had gone to Germany in preference to coming first to Ireland; and, in his mother's questioning, an ill-concealed anxiety that he was entering into some sort of entanglement with his cousin. There was one particularly angry altercation about this, during which, in response to repeated questions about his future intentions, he left the house, slamming the door after him and staying the night with a friend. The complicated farcical truth and his utter innocence of any intentions whatever, honourable or otherwise, would of course have been far beyond her comprehension.

His references to his intimacy with the distinguished Irish writer James Joyce were received without much comment. Neither May nor Bill had, after all, much knowledge of who or what James Joyce was; but the production of Sam's contribution to *Our Exagmination* was supposed to be at least evidence that he was producing scholarly essays which would have the effect of furthering his academic career. All in all Beckett was pleased when the visit to Dublin was over, term-time came and he could plead the necessity of returning to Paris to resume his duties at the École Normale.

Back in Paris the daily visits to Joyce were resumed, Joyce's demands and intrusions on his inner contemplative ease not apparently being as onerous as some others. Beckett was the more assiduous in his attentions because Joyce's eyes and his general state of health were still causing serious anxiety. In the month Beckett returned to Paris Lucia decided that she was not fitted for a career as a dancer after all, saying that she was simply not physically strong enough. It was a decision greeted with something less than regret by her father, who had always had his doubts about whether it was seemly for the women of his family to prance about on stages in revealing costumes and had been made uncomfortable by his daughter's insistence on doing so. The decision was arrived at with tears which continued intermittently for several weeks (tears were common in the Joyce household throughout 1929); and when one evening about this time Eugene and Maria Jolas were sitting with a woman doctor friend near the Joyces in a restaurant, the doctor remarked, 'If I were the mother of James Joyce's daughter and saw her staring off into space in that way, I'd be very concerned about it.'

Joyce had begun to call his fellow Dubliner Beckett, instead of Mr Beckett, a step which the recipient of the honour rightly recognized as a mark of increased trust and intimacy. But at the same time Joyce made it clear that his capacity to feel for other human beings, nowadays at least, was limited. 'I don't love anyone except my family,' he said, a remark which Beckett believed could be construed as meaning, 'I don't like anyone outside my family either.'

Loving of his family as Joyce undoubtedly was, he could scarcely be considered observant of them. Lucia was now deeply troubled but neither Joyce nor Nora were aware of this. Nor could Beckett be accused of being over-observant either. He sometimes took Lucia to meals in restaurants or to the theatre. Once or twice she even came to tea at the École Normale where he lived. Many people in Paris believed they were engaged; others that they were lovers; but though he still sometimes thought she was beautiful and it occurred to him to wonder why he was not romantically attracted to her, it was borne in on him only very slowly that she had fallen in love with him. When the realization came, it put him in a quandary. It was inconceivable to him that he should not continue to see Joyce every day; but every day when he came it was Lucia who greeted him and smilingly showed him into the master's workroom. Once when James and Nora were away she invited him to dinner, a meal which she prepared. To some extent trapped in continued association, he wondered what to do.

Shortly after his return to Paris Joyce asked him to undertake a translation into French of the Anna Livia Plurabelle section of *Finnegans Wake*. Since of course he would need a French collaborator, Beckett had suggested Alfred Péron. The suggestion had proved acceptable and during the early winter of 1929–30 the two young men were engaged on the complex, frustrating and often seemingly impossible task of finding French equivalents for Joyce's puns, compound words and jokes involving proper names. They would meet two or three times a week in Beckett's rooms in the École, but often progress was minimal. Writing of any kind in committee, even a committee of two, is an exhausting business, since one seldom gets a sustained inspirational run; but translating the densities and double meanings of Joyce's prose was especially arduous and tiring.

Between day-time visits to square Robiac and evening sessions over the text of 'ALP' (as the Anna Livia Plurabelle section of *Finnegans Wake* was known in Joyce circles) his senior compatriot had now taken over a considerable part of Beckett's life. When he socialized otherwise it was mostly in the cheaper cafés of the student quarter round the École Normale, the Café du Départ on the rue Gay Lussac, where he went with MacGreevy and Aldington, or the American Bar on the rue Mouffetard. Until at the end of the year MacGreevy went back to Ireland they also went frequently to a restaurant, Au Cochon de Lait, in the rue Corneille, near the Odéon Theatre, where there were two good-looking, cheerful young Italian waiters, Mario and Angelo, with whom MacGreevy was friends and to whom he gave English lessons. He was perhaps attracted to them both sexually, but the relationship probably remained quite innocent.

The boulevard Saint-Germain had not yet come into fashion as a resort of intellectuals and the cafés of the boulevard Saint-Michel were mostly patronized by students; but those of Montparnasse – the Dôme, the Coupole and the Select – were popular as places of association and converse for many painters and writers, including Picasso and Chagall. The Coupole had recently been re-designed and modelled in the modernist, Art Deco style of the day and it had an American bar which would retain its popularity throughout the 1930s. The fame of these places would, as is the way with artists' cafés, increase with the fame of those who frequented them, reaching its apogee when many of them were dead or had departed elsewhere, but in the late 1920s the Dôme and the Select were the resort of many quite well-known painters and writers as well as of the smart French people who made up '*le tout Paris*'. When Beckett went to such places it was often to meet Richard Aldington or Aldington's friends.

Richard Aldington was a year older than MacGreevy, which meant that he was a significant few years younger than the great proponents of modernism, Joyce, Eliot and Pound. He had achieved some fame in his early twenties as editor of the *Egoist*, the feminist magazine Harriet Weaver had founded, which was taken over by Ezra Pound on behalf of imagism and had published Joyce's *A Portrait of the Artist as a Young Man*. He had married a fellow imagist, the American Hilda Doolittle, 'H. D.'; and by 1929 he was a fairly well-established 'modern' poet. In that year he scored a commercial and critical success with a war novel, *Death of a Hero*, which had to be expurgated for publication in England, but appeared uncensored in Paris later. In a way, he is one of the first casualties of modernism, born either too early or too late to escape the disastrous influence of Eliot. He was to have a somewhat chequered career afterwards and to be involved in nasty controversy over biographies of both Lawrences – D.H. and T.E. – his book about the latter, in the opinion of his friends at least, causing such hostility in English literary circles that he suffered thereafter; but at this time he was the best established writer of English other than Joyce that Beckett knew. Though not particularly attracted to Beckett, Aldington was fond of MacGreevy and through him was to have an important circumstantial influence on Beckett's career.

Among Aldington's friends and contacts was Charles Prentice of Chatto and Windus and it was through him that Prentice became acquainted with MacGreevy. Charles Prentice was a gentle, cultured man of civilised tastes, a connoisseur of wines, paintings and cigars who could afford to indulge his tastes. He was a bachelor, living alone in his Kensington flat and, according to Aldington, often lonely. Certainly he had plenty of time for his authors. He made friends with MacGreevy and, in the course of time, with Beckett. It was a fruitful connection from MacGreevy's point of view, eventually resulting in the publication of his *Poems* in 1934; and a study of T. S. Eliot which was one of a series of little books on the work of eminent modern writers, under the imprint Dolphin Books, which Chatto were publishing. When Prentice suggested to MacGreevy that he might like to write one on Aldington also, he assented to the proposition. *Richard Aldington: An Englishman* appeared along with the Eliot as a Dolphin Book in 1931.

Chatto's Dolphin Books were monographs of a sensible length: a mere 16,000–17,000 words. The risk to the publishers was not great, so when MacGreevy and Aldington suggested to Prentice that an unknown young Irish friend of theirs, who was *au fait* with modern French literature, might do one on Proust, he was quite ready to entertain the idea. Since Beckett was a graduate in French of Trinity College and since in later years he neither solicited nor accepted casual commissions in the way that many writers are forced to do, it has been assumed that he had at the time a special interest in Proust, of which the little book was the fruit. In fact when the project was first suggested to him he had not read Proust's masterpiece, *A la Recherche du temps perdu*, and when he began to do so he was somewhat dismayed. He took the first volume to Kassel with him but thought it seemed strangely uneven. There were, he agreed, incomparable things, but also much that was 'offensively fastidious, artificial and almost dishonest'. All in all, he found it hard to know what to think of Proust, who was so much the master of his form that he sometimes became its slave; some of whose metaphors lit up 'a whole page like a bright explosion', while others seemed 'ground out in the dullest desperation', who had 'every kind of subtle equilibrium, alarming, trembling equilibriums and then suddenly a stasis'. He accepted that some of Proust's general loquacity was brilliant; at other times it seemed to be 'a meandering false teeth foodle-doodle discharge from a cholic afflicted belly'. And to think, he wrote to MacGreevy, 'I have to contemplate him at stool for 16 volumes.' But in spite of his doubts and the lack of an immediate affinity he persisted and (like many another writer who has taken on a chance commission) he found the experience of reading his subject and, still more, of writing about him ultimately rewarding, largely solving any problems there might be by writing a sort of personal manifesto.

According to Brian Coffey, a Dublin contemporary who came to know Beckett a little later, he did not share the Aldington-MacGreevy admiration for Eliot, remarking on his borrowings and calling him a 'jewel thief'. Yet *Dream of Fair to Middling Women* fully recognizes Eliot's existence by overt, if unacknowledged, quotation, or borrowing – one of Eliot's own borrowings from Julian of Norwich being borrowed again by Beckett. And in 1934, when Beckett wrote an article discussing Coffey, MacGreevy and another Irish poet, Denis Devlin, he praised them for having 'submitted themselves to the influences of . . . Corbière, Rimbaud, Laforgue, the *surréalistes* and Mr Eliot, perhaps also to those of Mr Pound, with results that constitute already the nexus of a living poetic in Ireland'.

His only meeting with Ezra Pound took place at Les Trianons, a restaurant much patronized by Joyce, to which he and other members of the master's coterie were summoned one evening when Pound was in Paris. Beckett remembered afterwards that Pound had 'great trouble with a fond d'artichaut and was very aggressive and disdainful'. It would seem that Pound was scornful of the atmosphere of sycophancy and mutual admiration that he found among the Joyce coterie present; and turning to Beckett, who was his neighbour, he enquired sarcastically whether he was writing a new *Iliad* or a new *Divina Commedia*.

Another modern, or, at least 'modernist', writer of English he occasionally met in 1929 was the American Djuna Barnes, whose novel *Ryder* had appeared in 1928. T. S. Eliot would recommend the publication of her second and most famous novel *Nightwood* to Faber and Faber in 1936, writing to Geoffrey Faber that it expressed 'a sorrow much deeper than personal vicissitude . . . the sorrow of life, the worm unkillable by any of the agents of this world.' In the preface he eventually wrote he said that its prose would 'appeal mainly to readers of poetry.' Barnes wrote of course the sort of prose that is usually referred to as poetic, that is, richly textured, metaphorical and adjectival. It was also syntactically elaborate, as indeed Beckett's own would be later on, though his syntactical elaborations would often be for ironic effect.

Barnes was a striking figure, red-haired, always dressed in black and known to be lesbian, who went often to the Café de la Mairie in the Place St Sulpice, where Paul Eluard, a friend of Soupault's and with him one of the founders of surrealism, was also to be seen. At this point she was only slightly acquainted with Beckett, but later on, when he became famous, she would sometimes intrigue people by saying that she knew things about him that she would never tell. Certainly they corresponded in later years, though his letters, apparently sold to the University of Maryland, seem to have disappeared.

Like many of Beckett's acquaintances at this time, Djuna Barnes was friendly with James Joyce; indeed most of his Paris acquaintances came to him in one way or another through Joyce, Aldington or MacGreevy. Friendly with the latter two was George Reavey, an Anglo-Irishman who was an almost exact contemporary of his. Reavey's father, an Armagh linen manufacturer, had gone to Russia to set up mills there and, since he had been brought up in that country, Reavey himself was a fluent Russian speaker. He was a Cambridge graduate who, like Beckett, had supposedly come to Paris to continue his studies. Like many expatriates in Paris he had a small private income, but he also had a job as a tutor. Small, fair-haired and likeable, he wrote mediocre, mildly avant-garde verse, but in spite of a certain amount of ingratitude and much impatience on Beckett's part he would spend a good deal of time and effort promoting his friend's work, becoming for some years his agent. Under the imprint which he originated, the Europa Press, he would publish Beckett's first book of poems, *Echo's Bones and Other Precipitates*, from two small rooms above a shop at the corner where the rue Bonaparte meets the rue des Beaux Arts, but this would be six years later. Later also, he would acquire some reputation as a translator from the Russian, producing versions of Turgenev and Gogol as well as contemporaries such as Yevtushenko and Vosnosensky. Reavey liked to play billiards and so, all his life, did Beckett, and on many evenings they would play together in the saloon of the Closerie des Lilas, further up the boulevard du Montparnasse from the Dôme and La Coupole and not yet as fashionable a place as it would become.

The festive season of 1929 was accompanied for Beckett by the usual dilemmas about Dublin and Kassel. This time he opted for Dublin first, arriving from Holyhead after crossing both seas with a cold in the head and an appearance so haggard and exhausted that his mother was alarmed. She came to the natural maternal conclusion that he had been neglecting himself in Paris – as, of course, to a large extent he had – but her solicitude and concern for his physical welfare at least provided a substitute for any other form of communication and he was able to leave for Kassel without too much inquisition before the New Year.

The Kassel visit was equally brief and any real communication equally sparse. His cold had now worsened, so he stayed in bed for the first few days, from which vantage point he became resentfully aware that Peggy was leading a full social life and seeing an athletic young German skier, Heiner Stärke, whose calm, amused and self-possessed attitude to her flightiness was in humiliating contrast to his own. On New Year's Eve the entire family, with a miserably silent Beckett in tow, did the round of the bars and night-clubs. There seemed to be nothing to say and no point in staying on, so the day after New Year he boarded a train for Paris, where there were also problems, particularly with Lucia. Small blame to him if he was beginning to think of women as a problem area and, more seriously, of being loved by them as a species of misfortune.

In terms of literary activity at least, the New Year, 1930, would prove rewarding. Through Reavey he had been introduced to another avant-garde American, Edward W. Titus, editor of the review *This Quarter*. Although few of his contributors particularly liked Titus, and although he could not number James Joyce among them, *This Quarter* was on the whole a better 'little magazine' than *transition*, more representative and less professionally experimental. It was financed (from a safe distance) by Titus's lawfully wedded wife, the cosmetics queen Helena Rubinstein, though he kept his contributors, as his wife apparently kept him, on fairly short commons. For a while Titus had Samuel Putnam, an American critic, editor and publisher, as associate editor. The short story 'Dante and the Lobster' would appear in *This Quarter* in 1932, but to begin with Beckett contributed mostly translations from the Italian and the French, and does not seem to have been asked for much else.

Some of his translations from the Italian were prepared with a view to publication in the *European Caravan*, an anthology of post-war European writing which Putnam was preparing for the New York firm, Brewer, Warren and Putnam. The translations were signed S. B. Beckett, but in the England-Ireland section Samuel B. Beckett had four poems, none of which he subsequently collected. George Reavey was originally supposed to be the editor of the Irish section of this still interesting anthology but, having decided that he was more enthusiastic about the Russian, he passed the job on to Jacob Bronowski, another young expatriate whom they all knew. The ebullient and eloquent Bronowski's career as a scientist and, more publicly, a popularizer of scientific subjects on the television, was still in the future; in the meantime he was another young Cambridge poet, or at least Cambridge intellectual, living in Paris. The English and Irish sections were lumped together under his editorship and, as far as the English were concerned, they showed a Cambridge bias, William Empson occupying considerably more space than W. H. Auden.

Beckett subsequently referred to his poems of this period as 'the work of a very young man with nothing to say and the itch to make'. He deplored their self-consciousness and display of erudition, which he called 'showing off'. The reader who is not engaged professionally in the explication of Beckett's minor works may be inclined to agree with him, but on a biographical level the one called 'Casket of Pralinen for the Daughter of a Dissipated Mandarin' may have some interest since 'The Mandarin' was Beckett's name for Boss Sinclair and the poem is therefore for Peggy. The late Lawrence Harvey, the most thorough (and therefore, perforce, the kindest) explicator of Beckett's early poems, related this poem to three pages in *Dream of Fair to Middling Women*, which, he says, illustrate '(1) The conflict in the young poet's mind between love for his Irish fräulein and loyalty to Christ . . .; (2) The conflict between opposed loyalties to art and to Christ . . .; and (3) the dilemma of the existence of evil.' If these topics are there, one can only say they are well hidden.

When the *European Caravan* appeared in the following year, 1931, Samuel B. Beckett was declared to be 'the most interesting of the younger Irish writers' in the biographical note which preceded his contributions. Although Rudmose-Brown had been asked for an essay largely through Beckett's influence, Beckett is described rather oddly as a 'friend of Rudmose-Brown and of Joyce', who 'has adapted the Joyce method to his poetry with original results'. The note further asserted that 'his impulse is lyric, but has been deepened through this influence and the influence of Proust and the historic method', whatever that means. These occlusions apart, however, the *European Caravan* was a fairly decent anthology to be in; and, as has been said, the Anglo-Irish section included Empson and Auden as well as some of Eliot's *Ariel Poems* and extracts from *To the Lighthouse* and *Ulysses*. The reference to Proust is accounted for by the fact that Bronowski knew he had written a study of the French novelist and that this would be appearing from Chatto and Windus later in the year.

Besides the Proust commission there was one more important literary event in the earlier part of 1930. In the early summer of that year Nancy Cunard, another Paris expatriate who had founded a publishing firm called the Hours Press, advertised a prize of 1,000f., or roughly £10, for a poem of less than 100 lines on the subject of time. The announcement, which appeared in literary reviews in England and France, read: 'Nancy Cunard, Hours Press, in collaboration with Richard Aldington, offers £10 for the best poem up to 100 lines in English or American on the subject of time (for or against). Entries up to June 15th 1930.' On the closing day Aldington told MacGreevy that none of the entries he had read so far were worthy of the prize and still less of publication by the Hours Press. That afternoon MacGreevy passed this interesting piece of information on to Beckett who immediately set to work.

For some time he had been reading the French seventeenth-century philosopher, René Descartes, and keeping a notebook in which he made occasional entries. Latterly his reading had embraced Adrien Baillet's *La Vie de Monsieur Descartes*, which had been published in two volumes in Paris in 1691, as well as the *Oeuvres de Descartes* edited by Charles Adam and Paul Tannery in the late nineteenth century, of which volume XII is a life and times by Adam. His notebook entries were, he said, a substitute for work.

Proceeding now to draw on his notes Beckett wrote a poem of ninety-eight lines, finishing it at three o'clock in the morning after some salad and Chambertin at the nearby Cochon de Lait. He gave it the title *Whoroscope*, put it in a folder, walked down to the rue Guene-gaud and put it in the letter box of the Hours Press, which had a shop and office there.

The following morning Cunard and Aldington read the poem and were, apparently, hugely impressed. 'What remarkable lines, what images and analogies, what vivid colouring throughout! Indeed what technique!' she wrote afterwards. 'This long poem . . . was clearly the work of someone very intellectual and highly educated.' When these effusive words were written, Cunard seemed to be under the impression that Aldington and Beckett had not been acquainted before the poem was entered: perhaps her fellow judge deliberately gave her that impression. In any case, Beckett was sent for and informed of his good fortune. His poem had won the prize. It was, however, too short for separate publication. Would he add notes? He would, and did, almost there and then, to nobody's surprise, for notes to erudite poems in free verse had been fashionable since *The Wasteland*, though Beckett outdid Eliot in the matter of proportion, his being half as long as the printed poem. The printing took place almost immediately and consisted of 300 copies, bound in a dull red cover with lettering in black and a banner saying it had won the prize. The blurb also pointed out with truth that the poem was its author's 'first separately published work'. The price was 1s., but 100 copies, which had been signed by the author, were priced at 5s. Beckett spent a good proportion of the £10 that very evening on a dinner for Aldington, MacGreevy, Bridget Patmore and himself at the Cochon de Lait, a jolly occasion with Mario and Angelo waiting at table before sitting down for a glass of wine or two.

Whoroscope is based on Descartes' life, not his theories, so whatever influence the philosopher had on Beckett is best discussed in the context of a later work, *Murphy*. Subsequent critics have been nearly as impressed as Nancy Cunard was by its erudition, but in fact it is all more or less straight out of Baillet. The title relates to the philosopher's refusal to have his horoscope cast, since such a proceeding would compel him to die on the date indicated. Lawrence Harvey found it possible to devote sixty-six pages to the explication of the three-page poem, but a contemporary reader is more likely to be struck by how ably Beckett has, in Vivian Mercier's words, 'caught up with the international avant garde on his first flight'; and to agree with Mercier that it is 'No wonder he had a great deal to unlearn later.'

In Man Ray's photograph taken the year before she discovered Beckett, Nancy Cunard is shown wearing on each arm clusters of the African bangles she affected. In an earlier photograph she wears a silver trouser suit, which admirably shows off her aristocratic height and ultra-fashionable thinness. She is also wearing a mask and her father's top hat and is extending a gracious hand to the Dadaist poet Tristan Tzara, who kneels at her feet in a dinner jacket and black tie. When this photograph was taken they were on their way to a ball at the Comte de Beaumont's. Tzara had been her first artist friend in Paris, or rather the first with whom she had combined the roles of patron and friend. This had been in the earlier part of the decade and since he was then riding high on the crest of the first Dadaist wave, he had opened many doors for her.

Many others would have opened in any case, for she came not only from a very wealthy background, but from a somewhat cultured one. Her father was the shipping millionaire; her mother had also been a patron of the arts and had a long-standing, if only intermittently passionate affair with the great Irish novelist George Moore – indeed Nancy is still reputed in some quarters to have been George Moore's daughter – as well as a wide Paris acquaintance: she too had been photographed by Man Ray.

In Paris Nancy had founded the Hours Press, which at this point had already published collections by Robert Graves and Laura Riding. Later in 1930 it was to publish *A Draft of XXX Cantos* by Ezra Pound and, before she had finished with it, was to bring out volumes by Harold Acton, Brian Howard and others. So when she put *Whoroscope* in the window in the rue Guenegaud, it was in reasonably good company.

Nancy was interested in causes, particularly the cause of what were then called Negro peoples, whether in Europe, Africa or the United States. She lived with a black man, Henry Crowder, who sang at the Dôme and in other cafés and night-clubs, and she had a reputation for promiscuity, not to say voraciousness, where men were concerned, but though the reading of Beckett's poem was the beginning of a long and fairly intimate association, she did not make sexual advances to him. Nevertheless she liked him, describing him in a letter to a friend as 'a lovely young Irishman' and praising his linguistic ability, his modesty and his 'self-containedness', which was a quality about him that was beginning to impress people. For his part he was fond of her, thinking her good-looking but not beautiful and saying he admired not only her energy but her intellect. He also admired, or said he did – or convinced himself that he did – parts of her poem *Parallax*, also published by the Hours Press. They used to meet for a drink at the Café d'Harcourt during the long summer evenings, sometimes driving up to Montmartre with her boyfriend Henry Crowder in the little blue two-seater with a dicky seat which she had bought for Henry, to have dinner together on the *butte*. She was generous and one night she suddenly handed Sam £100, ignoring Henry Crowder's annoyance. Less enjoyable for Sam than some of these more intimate occasions were the long hours he spent sitting silently at night-club tables while she quarrelled or drunkenly danced with Henry and others and he gloomily gazed at the glass in front of him. On one occasion that Harold Acton remembered, he suddenly raised his head to ask, 'Christ in heaven, what am I doing here?'

Beckett's usual passivity where women were concerned had also prevented him from doing anything to discourage Lucia; but by the early months of 1930 it was obvious even to him that some action or other had to be taken. He had picked, perhaps deliberately, a time when Lucia's parents were in Switzerland to invite her to tea at the École and to tell her that he had no romantic interest in her, that his visits to the flat in square Robiac were to see her father and that was all. It was an occasion that called for diplomacy and the facing up to an unpleasant necessity, and when it came to the point he found it extremely difficult, partly because, to his dismay, she seemed to him especially beautiful at the moment he was telling her and he was in consequence totally disarmed. 'What terrible instinct prompts them to have the genius of beauty at the right – or the wrong – moment?' he asked MacGreevy. Later he was to say to Peggy Guggenheim that he had tried to fall in love with Lucia, but that he was dead inside and had no human feelings. By the time he had steeled himself for the encounter he had already told other people, including the American writer Kay Boyle who was friendly with Nora, that he thought James Joyce's daughter was going mad.

In any case his time at the École was drawing to a close and over his last months in Paris lay the shadow of a decision which had to be made. It had always been understood that he would return to Trinity when his time at the École was completed: indeed it had been understood that he would return to take up a post armed with some sort of a thesis on Unanisme in the work of Pierre Jean Jouve. But Jouve had been abandoned. Beckett was now a writer, accepting paid commissions from reputable publishers for works which would appear in book form, not a student composing theses for the eyes of academics. If he returned at all, it would be with *Proust*, which he hoped would suffice.

But he did not want to return; and as the time approached and the offer of a lecturing post was confirmed, he entertained thoughts of escape. MacGreevy, who had returned to Dublin in April 1930 on the death of his father, but wanted to come back to Paris, had suggested that they should move in together. He could, he thought, get Beckett a job of some sort on *Formes*; and they could manage on what they could jointly earn. But Beckett was also considering Italy, which was cheaper, and where he might manage to live for a year or so. Then he thought he might take the job which had been offered in Trinity and resign it after a year. But he had the feeling that if he went back at all his doom was in large part accomplished. It would be very difficult, if not altogether impossible, to escape from Dublin and his family. 'The acceptance of this thing makes flight and escape more and more complicated,' he wrote to MacGreevy, 'because if I chuck Dublin after a year, I am not merely chucking Dublin but my family and causing them pain.'

Meantime he had his two major literary commitments to think about: the translation of 'Anna Livia Plurabelle' with Alfred Péron and the Proust book for Chatto and Windus. Péron, who had now met his future wife, was spending much time in the country. He seemed to Beckett to have changed, and for the worse, a not uncommon reaction for a friend to have in these circumstances and certainly not uncommon with Beckett, who reacted adversely to a friend's marriage in more than one case. He and Péron now usually met in the evening, when they were both tired, and got through a page or so. After a while their attitude became a little more relaxed and they made better progress accordingly; but when Péron left Paris for an extended period Beckett had to face the gruelling task alone. As the summer progressed he felt it was impossible and thought of telling Philippe Soupault, who was editor-in-chief of the translation project, that it could not be done; but in July he was able to give Soupault two copies of the major part of the section, though he had the feeling that Joyce would be 'disgusted by the chasm of feeling and technique between his hieroglyphics and our bastard French'. Soupault and Paul Léon, who was also involved in the project, finally revised and completed the translation of 'ALP'; and it appeared in the *Nouvelle Revue française* in May of the following year.

Proust was also a problem. Beckett declared that he was 'looking forward to pulling the balls of the critical, poetical, Proustian cock' but though he made notes he found it difficult to actually start writing. He wanted to go to Germany again before he left for Ireland, but began to foresee that the German trip would have to be sacrificed and perhaps to no purpose, since he might have to leave the Continent without finishing either Joyce or Proust. Like anybody else with such a task, he was sometimes optimistic and thought it would go very quickly when he actually got down to it; but by early July he still had not put pen to paper. He thought that 17,000 words was 'a hell of a lot' and he simply could not see himself doing so much. Even in August he was saying 'I can't do the fucking thing' and that he didn't know whether to start at the end or the beginning. Still, he was, he said, determined to write the 17,000 words before he left, even though his observations might have 'as little variety and none of the sincerity of Orlando's wood-carvings'. And he did in fact write them, though when he delivered the manuscript in London in September he thought it likely that Prentice and Chatto and Windus would say it was 'not scholarly, primo secundo enough'.

Besides working on *Proust* and 'ALP' he was writing poetry and he sent a poem to George Russell, 'AE', Yeats's friend and a senior figure of the Irish literary revival, who was then editing the fortnightly *Irish Statesman*. Russell sent it back with a note saying, 'I have a copy box stuffed to the brim with poetry sufficient to supply the needs of the *Statesman* for a year to come without taking in a single manuscript and it is no use accepting new manuscripts to add to the pile waiting their turn for publication.' Beckett was both amused and disdainful of this, with its supposed implication that one poem is as good as another, so that trying to sell one on that quarter was like trying to sell manure or bricks to someone who had enough of such commodities. The amusement and disdain were natural enough in a young poet, just as Russell's attitude was a natural and reasonable enough one in an editor.

In the intervals of all these activities he was reading the German philosopher Arthur Schopenhauer. Schopenhauer was an important discovery for him, perhaps indeed the most important literary discovery of his life. Both Péron and Beaufret, the young French philosopher who was a friend of Beckett's and MacGreevy's, were amused by this. Beaufret had, rather unusually for a Frenchman of that era, become acquainted with the work of Wittgenstein and the Viennese, and he was to become, in Geert Lernout's words, 'Heidegger's most important French advocate and teacher of a whole generation of French Heideggerians'. He was a voluble talker who loved discussing philosophy – and a very persuasive one. Faced with the condescending attitude of his friends about his new enthusiasm, Beckett insisted that he was not reading philosophy and had no interest in whether Schopenhauer was right or wrong as a metaphysician. What attracted him in the first place was Schopenhauer's 'intellectual justification of unhappiness – the greatest that has ever been attempted'. This, he thought, was 'worth the examination by one who is interested in Leopardi and Proust'. Bertrand Russell has remarked on the fact that Schopenhauer's appeal '. . . has always been less to professional philosophers than to artistic and literary people . . .'; but, as so often, it is probably on the whole better to speak of an affinity rather than an influence. Schopenhauer probably did no more than confirm the young Beckett in attitudes he was already arriving at; but the resemblances in outlook between the two are certainly striking.

Suffering is for the German philosopher the principal fact of human existence, and he attacks as absurd the idea which underlies almost all metaphysical systems: 'that of explaining evil as something negative'. Suffering, he says, is the positive thing, the norm. Pleasure is the purgative: usually the mere abolition of a desire or cessation of a pain. We are for the most part hardly aware of happiness or satisfaction, but we are acutely aware of pain and deprivation, dissatisfaction and desire, which are with us nearly all the time. In an image worthy of the later Beckett he declares that a quick test of the assertion that pain far outweighs pleasure in this world is to 'compare the feelings of an animal engaged in eating another with those of the animal being eaten'.

For Schopenhauer the world is ruled by Will, inevitably manifested in human existence as striving and ambition, and the more highly developed these are, the more dissatisfaction we feel. But when and if we attain our desires we merely negate for a moment discomfort, pain or yearning; and the probable result of most attainment is satiety, boredom and a fresh set of desires. Instinct, a form of will, urges human beings towards procreation, but this merely brings into existence a new cycle of suffering and death: hence the shame which has always, quite properly, been associated with the sexual act. The real sin, and the inevitable source of guilt, is existence itself. It is not what we do that makes us guilty, but what we are. 'The major sin is the sin of being born,' Beckett himself was to assert later. The only thing that attracts Schopenhauer about Christianity is the doctrine of Original Sin, which brings the Western religion closer to the Eastern faiths which locate guilt in existence itself. However, Christianity is mistaken in postulating that Original Sin was contingent on somebody's action. The sin is the sin of mere being. And, as Iris Murdoch points out, for this most pessimistic of philosophers, 'There is no salvation by works. Schopenhauer regards any emphasis on works as a Protestant aberration.'

The only escape from the egotistical, dissatisfied state into which we are born is through a complete negation of will and desire: in other words through an extreme form of asceticism or quietism. Intellectual and moral betterment, contemplation of the ideal à la Plato will not do for Schopenhauer (though he finds Plato useful in formulating a theory of aesthetics); and whereas Kantian dualism postulates on the one hand a realm of deterministic necessity in which dwells the phenomenal self, and on the other a realm of freedom which can be attained by the use of reason, Schopenhauer does not encourage us to believe in any such prospect. Interestingly enough, though, short of complete negation of will and rejection of desire, there may be a temporary escape through art. As he puts it,

> ... when an aesthetic perception occurs, the will completely vanishes from consciousness ... If it should be objected that the possibility of pleasure would then also be abolished, one should remember that, as I have often demonstrated, happiness, gratification, is of a merely negative nature, namely the mere cessation of suffering, pain on the other hand is positive. Thus when all desire disappears from consciousness, there still remains the condition of pleasure, i.e., the absence of all pain, and in this case the absence even of the possibility of pain, in that the individual is transformed from a willing subject into a purely knowing subject, yet continues to be conscious of himself and of his actions as a knowing subject.

Somebody looking at a painting is, in Schopenhauer's terminology, a 'pure' – that is, an entirely will-less – subject, and it is only through the work of art that there can be a 'pure' object – that is, an object abstracted from its position in space and time. 'For it is this position, always determined by the law of causality, which places this object in any kind of a relationship to me as an individual; so that only when this position is done away with will the object become an *Idea* and I therewith a pure subject of knowledge.'

Murdoch insists that as a theory of aesthetics this is more suited to some kinds of painting than others ('Van Gogh's shoes, Van Gogh's peasant chair which so impressed Heidegger'); and 'It is less easy to apply to literature, though one might instance some poetry and mention

T. S. Eliot's "objective correlative".' This is somewhat arguable. Even the most 'eventful' painting, even a film, succeeds, in Schopenhauer's phrase, in 'fixing forever the fleeting moment and thus extricating it from time' or the realm of will; but it is true that, generally speaking, the more static the work of art, the more he approves of it. *Tristram Shandy* is, for Schopenhauer, one of the great novels, precisely because 'it has as good as no action whatever'; but, he remarks approvingly, 'how little action there is in *La Nouvelle Héloïse* and *Wilhelm Meister*!' Even *Don Quixote*, he goes on, 'has relatively little, and what there is is very trivial, amounting to no more than a series of old jokes.' Yet 'these four novels are the crown of the genre'. While in the case of 'the marvellous novels of Jean Paul' we are asked to wonder at 'how much inner life is set in motion on the narrowest of external foundations', an observation which might apply to the novels Beckett was to write.

Although Beckett rarely attempted to put his outlook on life in propositional terms, was indeed to become famous partly for his refusal to do so, there are striking similarities between the basic feelings about existence that his works convey and Schopenhauer's, and their views on art tend to coincide also. But there are also some (perhaps no more than amusing) similarities of circumstance and character which doubtless reinforced the similarity of view.

Like Beckett, Schopenhauer was more or less indifferent to the claims of nationality. He regarded politics and public affairs very negatively; and though he lived through the German War of Liberation, it aroused no emotions in him. Like Beckett also, he fell very easily into certain habits and routines and was a late riser but a great walker. He had mixed feelings about his mother but considerable affection for his father. Probably as a result of his relationship with his mother, he was a considerable misogynist and for sexual gratification preferred whores or casual acquaintances with whom he was not emotionally involved. Whatever he may have felt or been on a conscious, willed level, Beckett also had strong misogynistic tendencies; and he too tended to divorce the idea of sexual pleasure from obligation or emotional commitment.

Chapter Nine

In the Irish lexicon there ought to be a word for the despair of returning, particularly in youth (returning later on can be a different experience). And if it were a compound word it should include an indication of the particular despair of a forced return for economic reasons, with perhaps a longer word again for the even more acute malaise of a return because of economic and familial reasons. As the years go by and Ireland becomes a more liberal and pleasant place, both in itself and by comparison with the rest of the world, people perhaps feel this less; but between the 1920s and the 1950s a return to the old, obsessive, dull, puritanical, provincial Ireland often created a special sickness in the returning heart.

Paris had been a liberating experience for Samuel Beckett. He had achieved publication and had begun to have a literary identity there. He had become the daily associate of James Joyce and a noted member of the Joyce circle. He had found a sophisticated soul-mate in MacGreevy, perhaps the most intimate friend he would ever have. And in addition to all these things there had been the personal liberation of living in a city where he could be anonymous, where his family identity and background were a matter of indifference to people and where he could enter places of public resort, including the occasional brothel, without fear of being noted or spied on. After all this the return to Dublin would be claustrophobic and frustrating, a long period of confusion, self-doubt and unhappiness.

Seen off at the Gare du Nord by the faithful MacGreevy, he was already in a highly nervous and exhausted condition. He had developed a spectacular facial rash and was recurrently unable to urinate. In London – where he was stopped in the street by a sympathetic old lady who diagnosed his facial complaint as barber's rash – he stayed at Garland's Hotel in Pall Mall, which had been recommended by MacGreevy. The following day he delivered the Proust to Charles Prentice at Chatto and Windus, who was very charming. Aldington had suggested that he go and see J. B. Pinker, the well-known American literary agent who was just then in London, but he did not feel up to such an encounter. Instead he took a taxi out to where he had heard Ethna McCarthy was staying but, failing to find her, he came back on a bus and went out to Hammersmith to see the poet John Lyle Donaghy, then married and living in London with his bride. They arranged to meet later at the Soho restaurant, L'Escargot, where they had dinner and drank a good deal of Chambertin.

Though Beckett did not like the poet's 'silly little wife' he did like Donaghy, a gentle soul, five years older than himself, and did not object to being shown some new compositions. Donaghy was, like MacGreevy, a proponent of *vers libre*. He was already the author of three collections of verse, beginning with *At Dawn Above Aherlow*, which had been published four years earlier by Yeats's Cuala Press, and would produce five more, including a *Selected Poems*, before his untimely death in 1942. He was one of the few Irish contemporaries for whom Beckett had a considerable measure of respect; but though he is one of the few who are still worth reading, he is no longer in print and of course is therefore not read. As Beckett's do, his writings suggest a true self-hood lost amid the contingencies and distractions of the world.

On arrival home Beckett slept for eighteen hours, punctuated by baths and meals. His mother was shocked by his rash and refused to let him leave the house, but she was generally more equable than he had expected. His father suggested that he should see a skin specialist and get his head shaved, but for the moment anyway he was glad to be there, sitting in a comfortable armchair, waiting for the dinner gong and reading the *Strand Magazine*.

Within a few days, however, his contentment, such as it was, began to vanish, and he wrote to MacGreevy that 'this life is terrible and I don't understand how it can be endured.' He was irritated by Rudmose-Brown, who was riding the same hobby-horses and expressing the same attitudes that Beckett had heard so often before, principally anti-romantic, anti-clerical and anti-military. Of course, he had no objection to these anti-isms except that they were drearily familiar.

In and out of Trinity, the thing that irked him most about Dublin were the sort of would-be clevernesses that most people sought to introduce into their talk and which he classed as 'quips', their sedulous pursuit of these supposed marks of wit and sophistication being 'quip-ification'. By comparison with Dublin, life in Paris had been 'an approximation of something reasonable'; but here he was appalled by the combination 'of cheap quip and semi-obscene entirely pointless activity' in the Rudmose-Brown circle and the Trinity Common Room; found no solace in the company of such literary acquaintances as Con Leventhal and Arland Ussher; and was beginning to be irked by the fussy solicitude shown to him at home.

A.J. or, as he was universally known in Dublin, 'Con' Leventhal was Jewish, though he had been educated at Wesley, like Portora a rather exclusive Protestant school. He was another of Rudmose-Brown's bright boys and he would eventually succeed Beckett as lecturer in French, but he was ten years older than his friend and his college career had been interrupted just after the war by a trip to Palestine, where he had worked for the first Zionist Commission and been one of the founders of the publication *Palestine Weekly*. His return to Dublin had been marked by a minor *cause célèbre*, at least in literary circles. Shortly after *Ulysses* appeared in Paris he wrote a review for the *Dublin Magazine*, a literary quarterly which had just been founded by the poet Seumas O'Sullivan. The printers had refused to set this, so to print the review Leventhal founded a journal, the *Klaxon*, which announced itself as hoping 'with an excess of Picabia and banter, a whiff of Dadaist Europe to kick Ireland into artistic wakefulness', from which it can be seen that it was claiming to be avant-garde and in touch with what was going on in Paris. The *Klaxon* lasted for one issue, which was probably all it was meant to last for. In the 1920s Leventhal had a chequered career, which included the proprietorship of a bookshop at the corner of Duke Street and Dawson Street, before a return to Trinity which was made possible by his friend Sam's departure.

Con Leventhal was small, much shorter than the tall Beckett, and, with his large nose and brown eyes, very Jewish in appearance. Partly perhaps because of those remarkably liquid and expressive eyes, he was attractive to women. He was also highly literate, intelligent and sensitive, but his conversation was liberally sprinkled with puns and would-be bon mots, which no doubt Beckett had in mind when he complained of 'quipification'.

Still, Beckett liked him, just as almost everybody else did, in spite of finding some of his attitudes 'sterile' and typical of Dublin. 'I am fond of Leventhal for no reason good, bad or indifferent, which is surely the only possible way of being fond of anybody,' he was to tell MacGreevy a few months later.

The understanding with Rudmose-Brown was that he would become an instructor, or graduate assistant, in the Department of Modern Languages while proceeding to his doctorate; in fact Ruddy was – from his point of view – better than his word, and Beckett was taken on as a lecturer with a three-year appointment and a salary of £200 a year rising by incremental stages of £50. True, he had returned without the thesis on Pierre Jean Jouve which had been promised, but he had a monograph on Proust appearing from Chatto and Windus and Ruddy, who also had ambitions in those directions, was keenly aware of his young favourite's contacts with London publishers and Paris editors.

He began with some mild invigilating for the autumn exams, but then came lecturing. It was a prospect he dreaded, but to begin with he took it very seriously and prepared a course on French literature from the Romantics to Modernism which casts some light on his own future development. Some of his terminology is admittedly confusing. In his schema there were romantics, pre-naturalists, naturalists and post-naturalists. Post-naturalists were really moderns, in poetry beginning with Rimbaud, in prose with Paul Bourget, perhaps an odd choice. Pre-naturalists, chief among them Flaubert, were evidently moderns born out of their time – in other words, born before Zolaesque naturalism swept all before it and, in a sense, delayed the advent of modernism.

His lectures contained a good deal of reference to the philosopher Henri Bergson and his distinction between 'spatial time', which could be measured by clocks, and 'duration' – time as it is really experienced by human beings. Bergson's ideas had an immense influence on all the French writers of the early twentieth century, including Marcel Proust. They seem to have remained with Beckett and it is possible to see later works like *Happy Days*, *Play*, and *How It Is* as being set not in any sort of eternal after-life, as some critics have assumed, but simply as reflecting Bergsonian ideas about time. In the Bergson/Beckett view, the intensities of an experience transcended time.

Most of those who enrolled for modern languages were women, and classes were about thirty strong. Faced with the prospect of almost thirty interested women, Beckett partly froze. Since the classroom was quite small, the front row was squeezed up against the rostrum. He delivered his lectures with his back to the window beside the rostrum and his eyes apparently fixed on a point on the ceiling above the door, speaking slowly and deliberately but with little expression. A little acting, the adoption of a persona would have worked wonders. Nevertheless, his students thought him impressive and classified him as brilliant, while the women found this tall, blue-eyed, seemingly impassive and taciturn lecturer both intriguing and extremely attractive. All in all, he was a better university lecturer than he was a school-teacher and, to begin with at least, he was far more punctual, being nearly always in the lecture theatre before the students arrived and seldom missing a lecture.

As a member of the faculty, Beckett was allotted rooms and he wound up, as before, in New Square, this time in Number 39 (as an undergraduate he had occupied Number 40). Here he had an open fireplace, in which he regularly lit a turf fire, and, on either side of it, two comfortable armchairs which his mother had given him. Georges Pelorson, who had been his pupil and friend in Paris, was now the visiting *lecteur* from the École and they frequently sat together in the evenings, one on either side of the fireplace, slowly emptying the bottle of Jameson set on the floor between them, reading, talking, or silently gazing at the flames. Number 39 New Square also had, like most Trinity rooms, a cooking alcove with a gas stove and here Beckett would prepare the evening meal – usually scrambled eggs – for which Pelorson sometimes joined him. When Pelorson was the host in his rooms they had an omelette instead.

Just as James Joyce had once returned from Paris in a 'Latin Quarter hat', so Samuel Beckett had come back affecting a French beret, which was a cause of some amusement in Dublin. He was also wearing the cheap French suit he had bought in Paris and a tight-waisted French overcoat, and though he sometimes substituted a soft felt hat for the beret there was a good deal of disapproval in Cooldrinagh. His mother and father suggested that when going out he should at least wear a bowler hat, then still considered essential to respectability and the normal garb of the males of his class. After a while he changed his French suit for slacks and a pullover: these now became his constant wear, plus the beret and a raincoat, but he was the more impervious to his mother's suggestions that he should be more respectably clad because he now had no social life except his evening talks with Pelorson, a very occasional visit to Cooldrinagh and the odd visit from Frank.

Like himself the latter had returned somewhat unwillingly to Ireland, the reason in his case being his parents' expectation that he would join his father in the business. Like Sam, Frank found the constraints of family irksome and Dublin rather boring after life abroad – in his case in India. He fretted, like Sam, between rejection and acceptance of what seemed to be his ordained lot and talked of growing old in the shadow of a compromise.

As the year ran out, the only bright spot on Sam's horizon was the possibility of a trip to Kassel during the Christmas vacation. He had been corresponding with Peggy Sinclair, and although he was 'blind to the charms of the mighty steaks and jug dugs' of his Smeraldina, he was still obsessed with her face. In the sample of her side of the correspondence, which he eventually included in *More Pricks Than Kicks* (and which had also been included in *Dream of Fair to Middling Women*), she refers to his 'sweet letters'. Whatever protestations of love he made in them, however, his letters are unlikely to have been as ardent as hers.

After his return to Dublin he had written to MacGreevy in Paris asking him to destroy the two boxes of Peggy's letters which were behind the curtain and the book-case or else in the little attaché case he had left behind; but he had also requested his friend to send on to Dublin the photograph of her which was in one of the drawers. He still liked the photograph because, he said, it could not 'be co-ordinated with any fleshy bloody bedcony in any geranwanen of clumsy copulation', which, so far as it can be made out, means that the young lover liked the photograph because it did not remind him of his beloved's body, particularly of his beloved's body at the time he was enjoying what is supposed to be that summit of bliss which all young lovers desire to attain.

The letters he had kept in Paris seem to have been duly destroyed, but another letter which he received in Dublin he unblushingly used. When it was eventually published in *More Pricks Than Kicks* it was called 'The Smeraldina's Billet Doux'. In the two autobiographical fictions in which he chose to use the letter, the hero is called Belacqua, which the Smeraldina shortens to Bel:

Bel Bel my own bloved, allways and forever mine!!
   Your letter is soked with tears death is the onley thing. I had been crying bitterly, tears! tears! tears! and nothing els, then your letter cam with more tears, after I had read it ofer and ofer again I found I had ink spots on my face. The tears are rolling down my face. . . . Oh! Bel I love you terrible, I want you terrible, I want your body your soft white body nagelnacht . . . Whitch is the greater: the pain of being away from eachother, or the pain of being with eachother, crying at eachother beauty? I soppose the last is the greater, otherwise we would have given up all hope of ever being anything els but miserable . . .

Then she tells him of a 'grand' film she has seen, *Sturm über Asien*, which she seems to have enjoyed very much – partly because there wasn't 'any of the usual hugging and kissing'; of a new girl she has met, 'very beautiful, pitch black hairs and very pale, she onely talks Egyptian'; and of a letter she has had from a man who asked her to go to a dance with him on Saturday evening. Apropos of this invitation she remarks: 'I sopose I will go, I know my beloved dosent mind and it makes the time go round quicker, the man is a bit of a fool but dances quite well and is the right hight for me. A flirt is very amusing but shouldent go further than that.'

129

After recounting a dream she has had, in which Belacqua turned into a baby on a forest path, she continues:

There is no object in me trying to tell you how much I love you because I will never succeed, I know that for sirten. Is he the man I have allways been looking for? Yes! but then why cant he give that what I have been longing for for the last 6 months? I ofen wonder what is on you that makes me love you so greatly. I love you über alles in dieser Welt, mehr als alles auf Himmel, Erde und Hölle. One thing I thank God for that our love is so vast. I ofen wonder who I am to thank that you are born and that we met, I sopose I had better not start trying to find out whose fault it is that you are born.

Because of the references to 'his poor sore face' and the approach of the Christmas visit she is longing for, this letter can be dated to this winter. Though the likelihood is that Peggy's imperfect English was transcribed more or less directly, we may suspect small alterations, if only because the 'whose fault it is that you are born' is, in terms of Beckett's own developing general outlook, almost too good to be true.

Doubtless he was conscious of the Molly Bloom aspect of Peggy's outpourings when he decided to publish them; but he may have overlooked the fact that Molly's monologue comes almost as a sort of *coup de théâtre* after Bloom's rememberings and idealizations. She has already been established as a (to some extent) idealized love-object before we are, as it were, shown the other side of the coin. This process has not taken place with Peggy, who emerges, both in *Dream of Fair to Middling Women* and *More Pricks Than Kicks*, as a merely comic figure. It is easy to overlook the fact that she quotes Goethe, that she is struggling with a Beethoven sonata and that she is something of a poet. And there is real feeling in her letter, both love and passion.

Bel! Bel! Bel! your letter has just come! . . . Oh! God how could you ever say such a thing, for lord sake dont!!! . . . Bel! Bel! how could you ever doupt me? . . . Lord Lord Lord for god sake tell me strate away what agsactly I have done. Is everything indiffrent to you? Evedintly you cant be bothered with a goat like me. If I don't stop writing you wont be able to read this letter because it will be all ofer tears.

The idea of everything being 'indiffrent' to the recipient is again so near the mark that the likelihood is that he improved on the original here also. Calling herself a goat sounds at first like a wry comment on their actual situation, but either in English or German she would be unlikely to use such a metaphor in the sense in which it is often employed for men. That the recipient was conscious of this meaning though, we can scarcely doubt – and conscious of it also when he decided to use the letter for literary purposes later on. He was, as writers so often are, being self-revelatory in a secretive sort of way.

He was to arrive in Kassel on the twenty-third, after spending the first few days of the Christmas term break in Paris, but as the time approached his physical condition, which had not been good since his return to Ireland, became even worse. The eczematous condition of his face had cleared up, but it had been succeeded by an outbreak of boils or cysts, an affliction which was to become almost a commonplace of his life for quite some time. Some of them were in awkward places, such as the groin; but there was one on his neck which would continue to grow throughout his travels, rubbing as it did so against the collar of his jacket, until it reached proportions which disturbed those who saw it. He was suffering from insomnia at this time to an even greater extent than usual. He had constant colds and he now developed diarrhoea, a condition which would last throughout his visit.

There is no doubt that many of the painful and lowering ailments from which Beckett recurrently suffered were psychosomatic in origin. Whatever proportion of blame the diagnostician might decide to allot to physical causes as well, certain situations caused him acute stomach upsets, fevers, colds, heart palpitations, dizziness, boils, cysts, facial rashes and other disturbances. They also aggravated his insomnia and his tendency to nightmare. Though he was athletic and apparently strong, his body let him down regularly and drastically when his mind or emotions were acutely disturbed. Physical maladies and disorders of one kind or another were associated with Ireland and/or a return home to Cooldrinagh, where of course his mother was. They seem also to have been induced by his relationship to Peggy. Though he longed to get away now, and thought he very much wanted to go to Germany to be near the Sinclairs and Peggy, it is pretty obvious that there was also a conflict going on in his subconscious about this.

So bad did his physical condition now become that a couple of weeks before Christmas he wrote suggesting that he might have to postpone his visit. This elicited a reply from Cissie saying that Peggy had become hysterical as a result of his letter and got it into her head that he was either dying or had ceased to care for her. 'For God's sake pull yourself together, throw a bottle and a toothbrush into a bag and come on,' the Smeraldina's mother concludes in the part of her letter also quoted in *Dream of Fair to Middling Women*, where 'the tigress tone of the old multipara's letter' is said to have annoyed the hero considerably.

Multipara means 'mother of many'. In fact Cissie was mother of five; but the references to her in this abandoned novel are tinged with condescension and dislike.

The visit itself was largely, to use a favourite word of his, a fiasco. On the way to Kassel he spent a couple of days in Paris, where he saw Nancy Cunard and others. On arrival in Germany, two days before Christmas, he was met at the station by Peggy, wearing, he noticed, cheap Russian boots and black stockings. Again, it was her face that he found hauntingly beautiful, but even in this first meeting he was conscious that her legs, though the black hose was stretched to breaking point, were not very attractive. In *Dream of Fair to Middling Women* the hero ungallantly observes 'The truly enormous bowel of the hips . . . breaking out and away from the waistroot . . . like a burdocked bulb of Ruffino and the two hoops of the buttocks', which, he says, received 'an almost Rhineline from the dark peltsheath', whatever that may mean.

Perhaps not unnaturally, he went to bed on arrival, gladly accepting the brandy that Cissie offered while assuring her that Peggy looked very well indeed, perhaps more beautiful than before. Over the first few days the supposedly almost betrothed pair were infrequently alone together. Beckett stayed in bed in the mornings and in the afternoons he sat over the stove reading Vasari. In the evenings he drank with Boss and discussed the newly acquired paintings in the flat. On the only occasion they spent an evening alone together there was a tiff and Peggy wound up in tears. As happens at such times, both parties picked on irrelevancies and he commented on her inability to speak proper English.

So passed Christmas, which nobody much bothered about, except that Boss's drinking increased. On New Year's Eve Beckett and Peggy set off for the Ratskeller. According to the account in *Dream*, it was crowded to the doors with revellers and they could not get a drink. He suggested that they go on to the Barbarina, where Boss was to join them, but she objected that if they went they would miss the chimes. Eventually he went to the toilet, gladly absenting himself while the clock of the Rathaus struck the hour and the company joined hands and sang. At this point, if we are to believe the account in *Dream of Fair to Middling Women*, which seems sedulously accurate, one of Peggy's admirers, a local businessman with whom Beckett had sometimes played chess, came over to her and suggested that they join his table; but she refused on the grounds that her companion would not care to. While she waited, another, younger admirer, a glider champion, approached. Then her escort came back and she discovered to her chagrin that on the way he had independently accepted the invitation to sit at the businessman's table. Now it was her turn to say that they should go to the Barbarina; and there, after some further confusion, they went.

According to *Dream* there was dancing in the Barbarina, but when she asked him to dance, he told her that he couldn't, which was something she already knew, and there followed another tiff. What, she demanded, had he come all the way from Paris for? He had done nothing so far but lie about in the house all day reading his old book and making facetious conversation with her father – and this was somebody who was supposed to be in love with her. To all this he gave a succinct and astonishingly truthful reply. He had come to Kassel, he said, so that he could look at her face.

It was, in its way, flattering; but, perhaps understandably unmollified by the brevity and directness of the response, she said firstly that he used to say that he only wanted to look into her eyes, and secondly that in any case he did not look at her face that she could see. He then said that he had come to look at it surreptitiously, a word whose meaning had to be explained to her. What it meant, he declared, was that he liked to steal looks at her face. So, she elaborated, he had travelled all that way, third class, to steal a look at her face. She could, he said, put it that way if she liked, but perhaps now they could drop the subject.

While all this was going on, Boss swaggered in. In the book his arrival is described as pleasing to Sam, but less welcome to Peggy and she eventually got up to dance with the glider champion, who had followed them from the Ratskeller. She danced all wrong, he thought, waggling her large behind, a sight which made him think of canings and lashings. When she returned, she had words with Boss, who wanted to go to a place around the corner called Meisters, where Guinness Export, a strong brew and better value than the beer they were drinking, could be had. Sam was also in favour of the move to Meisters and he now went so far as to suggest that Peggy might have another dance with the glider champion and follow them on. On the way out Sam bought a Goldwasser for the affable if not very pretty barmaid, for whom he often bought drink.

So far, according to the account in *Dream of Fair to Middling Women*, it had not been a very rewarding evening, nor a very pleasant one; but it was not over yet. In Meisters the Mandarin and Belacqua discuss, as might be expected, the perennial topic of women and male attitudes to them. Belacqua insists on the gulf between love and desire and denounces those who are able to feel both for the same person. 'I admit

'. . . Beatrice after the brothel or the brothel after Beatrice, but not Beatrice in the brothel, or rather, not Beatrice and me in bed in the brothel.'

To which the Mandarin, who has already called Belacqua a puritanical Low Church Protestant, asks, reasonably enough, what would be wrong 'with you and Beatrice happy in the Mystic Rose at say five o'clock and happy again in No. 69 at one minute past?'

Belacqua is unable to answer, but at this point in the dialogue Beatrice, in the person of the Smeraldina, arrives with her businessman admirer at heel. She tells them that everybody is invited to the studio of somebody called Sauerwein and the businessman offers to drive them all in his car. She is specially anxious that Belacqua should see the portrait Sauerwein has done of her in her bathing dress. He refuses and she goes off, whereupon the following dialogue takes place:

'In the old town' said Belacqua 'correct me if I am wrong, a certain Fräulein Anita Furtwängler sits by her window'

'Wisdom gleams through me' cried the Mandarin. 'I shudder and kindle.'

'The perfection of her limbs' pursued Belacqua 'has been weighing me up to the peace of Jerusalem. I have the address of Abraham's bosom.'

In other words, Belacqua, having contemplated his Beatrice's face and refused to contemplate her body as delineated by Herr Sauerwein, is now suggesting a visit to a brothel where a whore, whose body he does find desirable, is to be found.

It would appear from *Dream of Fair to Middling Women* that whatever the nature of our hero's feelings about the Smeraldina – his Beatrice – and whatever the disturbance to his soul caused by the beauty of her face, visiting the brothel is by no means incompatible with these feelings and has the effect of assuaging the disturbance. How, the reader is asked,

could a reasonable use of the brothel . . . have constituted the least outrage to the sentiment he entertained for his distant bloom, the light, melody, fragrance and embracement of his inner man? But the inner man, its hunger, darkness and silence, was it left entirely outside the brothel, did it not participate in the shady communion of the brothel? It was not and it did. Inwardly after the act, into the sanctuary whose provision depended on her or on the thought of her, whose assuaging belonged to her or the passional thought, the vision of her,

134

there entered peace and radiance, the banquet of music . . . She ceased to be the bride of his soul. She simply faded away . . . The rare miracle of fulfilment that had been ascribed and referred exclusively to her . . . the gift of magic from her . . . were on tap in the nearest red-lamp . . . Beatrice lurked in every brothel.

All of which may be understood to mean that a young man in love may find relief in visiting the brothel. When he goes to the brothel he forgets about the beloved for a while. What he does there does not affect the inner sanctuary of his spirit where the thought of her is cherished; yet he feels good afterwards, as good as if he had possessed her. To which apologia may be added the consideration that in Beckett's case the body of the beloved herself was not desired, or even found attractive. He was haunted by her face, no more and no less, and he sometimes conjured it up when masturbating.

Beckett's experience of whores had begun in Dublin. When he became a student at Trinity brothels still existed on a large and semi-legal scale in the old Nighttown, which figures so largely in *Ulysses*; and prostitution was carried on so openly that a shocked *Encyclopaedia Britannica* contributor compared Dublin adversely to the south of Europe and Algeria. The brothels catered for all trades and classes, including students, who got a cut rate in some establishments; and, in one of the very select, Becky Cooper's, with astonishing inappropriateness – or perhaps great wit – there even hung in the parlour a reproduction of a painting of Dante encountering Beatrice (possibly Rossetti's). Beckett was certainly acquainted with Becky Cooper's, which survived the sudden official closure in March 1926, carried out with the aid of a military cordon and a fleet of lorries, though after that date business was not supposed to be done on the premises. At the end of the story 'Ding Dong' Belacqua sets off for 'Railway Street, beyond the river', a statement which for anybody acquainted with Dublin could only have one import; and in a letter to MacGreevy later in the 1930s Beckett mentions the reproduction as being still there. The closure had the effect of throwing large numbers of whores on to the street, to ply their trade in alleyways, or to force them to look for clients in public houses and the shabbier kind of hotel. One of these was the Grosvenor Hotel in Westland Row near the back gate of Trinity, which Belacqua also frequents in *More Pricks Than Kicks*. These first encounters were probably furtive enough, though prior to 1926 students went in cheerful gangs to Railway and Mabbot streets. In Paris Beckett was almost certainly acquainted with the famous Sphinx in the boulevard Edgar Quinet behind Montparnasse Station, which was 'a marvel surpassing all others' to his friend

Giacometti, and was indeed a well-decorated and rather jolly place where topless girls mingled with the clients in the downstairs bar, which was more or less open to the public and frequented by many who just wanted to drink and talk in such surroundings. And in Kassel too he went with Boss, the father of his supposed beloved, to the red-light district. The poem 'Dortmunder', which was actually written in Kassel at this time, refers to one such visit. It is a learned affair with references to the minor Jewish prophet, Habbakuk, who dared to confront God with the problem of evil and ask him to account for his administration of the world; and Schopenhauer, who looked upon woman as a creature to be kept in subjection, available for the use of man; but, in spite of the title, which merely refers to a brand of beer, it is about a visit to a brothel in Kassel.

On this particular New Year's Eve Belacqua returns from the brothel to the studio of Herr Sauerwein, to whom he remarks cryptically that he has come from the rosa mundi, the rose of all the world, the rose of the mystics, to the rosa munda – a remark which Herr Sauerwein, who had perhaps some experience of such matters, appeared to understand. Then he accepts the Smeraldina's suggestion that they should go to a *Wirtschaft* on the height above the town together and her businessman friend obligingly drives them there.

At the *Wirtschaft* he refuses soup and other refreshments. She has a *Schokolade* and cookies. When she has finished them she moves closer and begins to caress or, as the book puts it, to 'paw' him. But, we are told, he does not 'want to be pawed. He had got all the pawing he could stand elsewhere, also he had counted on Herr Sauerwein and the Belshazzar, one or the other or both together, to assuage the Smeraldina.' Understandably, having come from a brothel, not wanting to be 'pawed and slabbered on, least of all by the reigning fetish', he moves to the window, ignoring the 'snuffling and muttering' that he hears behind him and, oddly, compares to 'the drip of sanies into a bucket', sanies being a thin discharge from a wound or ulcer and a word that, as we shall see in another context, Beckett was fond of. They both know it is the end. 'She knew and he knew and God knows it was high time.'

Beckett stayed in Kassel for a few days after the New Year, days described in the novel as being accompanied by 'more tears and more recriminations and more tears and more huddling and cuddling and catch-as-catch-can hugger-mugger and more tears and sweats and fiascos, a most painful time'. But in the novel Belacqua and the Smeraldina both know it is the end: 'he knew and so did she. It was all over bar the explanations and the jostling when he stepped off in the Wirtschaft on the Height ... leaving her to her own devices to begin the new year in whatever way she saw fit'.

Belacqua returned via Hamburg and so did Beckett, back to Dublin, where he had no real desire to be, and to the job at which he had no desire to excel. The boil on his neck had now swollen to considerable proportions. The night after his return he sat with Pelorson in 39 New Square. His friend, who knew something of his relationship with the Sinclairs and with Peggy, asked him how it had been. 'It's all over,' Beckett replied. And then, by way of explanation, he added, 'She was too exacting.'

Chapter Ten

Beckett returned to Ireland to face the New Year in no better shape, physically or mentally, than he had left it. The Kassel trip had been an end, but one, typically, without a climax. Now he was back, in a situation which he found almost intolerable. He detested lecturing, and though he was kind to students who consulted him outside the classroom and they found him informative, he already saw a life in academia as a life of misery. The cyst on his neck had now grown to huge proportions, so he had it lanced, which is a painful process, but this was succeeded as the principal locus of physical discomfort by his teeth, which were causing him great agony. He went to the dentist and was told that a number would have to come out and others would have to be filled, a prospect which did nothing to cheer him up.

The new term would not begin until the end of January, but the thought of teaching again paralysed him and he could do nothing in the way of preparation. In fact until it began he did little but sit by the fire in his rooms, sometimes reading, sometimes merely brooding and pitying himself.

He had very little communication with anybody except Pelorson and, to some extent, Frank; but Frank too had problems. Having come unwillingly back from India to enter the family business, he, like Sam, could neither accept nor reject his fate and talked of growing old in the shadow of a compromise. Though Sam sometimes cheered up a little under the influence of whiskey as he sat late by the fire with Pelorson, most people who encountered him at this time found him gloomy and withdrawn. He did not welcome callers at his rooms, but Leventhal or Salkeld or others would occasionally come. Mervyn Wall, subsequently a civil servant and novelist and the author of *The Unfortunate Fursey* and *Leaves for the Burning*, was brought there by someone else around this time and remembered that Beckett spoke most of the time of suicide. 'One left filled with thoughts of dissolution and gloom,' he said.

When Beckett did venture forth the state of his clothes and general appearance caused comment – in college, in Cooldrinagh and among his acquaintance generally. His usual garb was a grey shirt, grey Aran sweater and a pair of grey flannel pants. Both the sweater and trousers showed plentiful traces of food, drink and other matter. On top of these he wore a belted trenchcoat which was also in need of cleaning and sometimes his black beret. His shoes too were permanently dirty and in need of repair. It was all a far cry from the bowler-hatted young man his mother would like to have seen.

The two somewhat brighter spots in the general gloom were the forthcoming appearance of *Proust* and a happy new acquaintance, that with the painter Jack B. Yeats; though the imminence of *Proust* was clouded with regret that he had not been more active about an American edition, which Pinker told him he had left too late, and, of course, with the usual authorial apprehension about the contents.

Jack B. Yeats was six years younger than his then vastly more famous poet brother and, when Beckett met him, just over sixty. The introduction had come about through MacGreevy, who had sung Yeats's praises in Paris, but was a comparatively easy one to manage because like many people in Dublin Yeats had an 'at home' day, in his case Thursday afternoons, when writers such as Joseph Hone, the biographer of Jack's brother, W. B., Austin Clarke, Padraic Colum, and other lesser representatives of what remained of the Irish literary revival might be found in his studio in a Georgian house at the corner of Fitzwilliam Square. A very pale, slight man, with a quiet and deliberately somewhat courtly manner, clad nearly always in black with a flowing tie as the sign of his avocation, Yeats would offer his guests madeira and somewhat desultory conversation, usually about art or literature, but with a faint thread of gossip running through it. He had begun his artistic life in London, as a hard-working illustrator and contributor to many illustrated magazines – how hard-working may be judged from the fact that they included *Paddock Life*, *Judy*, the *Boys' Own Paper* and the *Vegetarian*, and that he also did pen and ink drawings illustrating various stances and ways of avoiding or giving blows for a thesis on the noble art of pugilism. Like his brother he had spent a large part of his boyhood in County Sligo – in fact Jack was there for longer periods than W. B. as a boy – and like his brother also he had conceived a passion for the landscape of that county which never left him.

The re-birth of Irish nationalism in the early part of the century and the literary movement which went with it had affected him deeply. After the turn of the century he began to re-create boyhood scenes in paintings such as 'Memory Harbour' and in 1905 he illustrated J. M. Synge's book on the Aran Islands with drawings which unmistakably reflect his heroic, romantic view of peasants and fishermen.

When, shortly after this, he returned to Ireland, his unfailing romanticism continued to be more than evident in the oil paintings which began to supplant the pen and ink drawings and water colours of earlier years. These oil paintings were to culminate in the atmospheric, poetically titled canvases of his rather late maturity, becoming gradually less realistic and illustrative until the comparatively straightforward, illustrator's draughtsmanship is overlaid with thick impasto and the figures are only a component in the poetically suggested landscape. Many of these paintings evoke parties and gatherings of his Sligo youth. Some of them seem to be telling a story, usually a somewhat sad one and usually of years gone by. He had a single artificial rose attached to his easel in Fitzwilliam Square. 'No one creates,' he told Joseph Hone. 'The artist assembles memories.'

It is this general tendency to romanticism, sometimes elegiac or nostalgic in mood, sometimes even heroic, that makes Beckett's admiration for his work a little surprising, as also, it must be said, does Yeats's fondness for vivid colour. It is true that Beckett took a different view. For him Yeats's figures were sad, solitary beings who inhabited a landscape which cared nothing for them and reflected back none of their feelings. He expressed Beckett's own sense of human alienation – as, incidentally, did Cézanne, for Beckett another painter who painted nature as cruelly indifferent to mankind. But, as far as Yeats is concerned anyway, this is certainly a minority view; and all in all we have to conclude that this admiration was personal in origin; and that, conceived as it was at a time in Beckett's life when he sadly needed someone to admire or look up to, it is a triumph of personal affection over critical or aesthetic considerations. Whereas his sense of affinity with the work of James Joyce, the other senior figure for whom he conceived a loyalty in these years, needs little or no explaining, his admiration for Yeats is largely, one feels, an accident of personal circumstance and acquaintance. He was always inclined to over-praise those with whom he had formed personal attachments of any significant sort, an admirable human trait, if, to critics, an occasionally confusing one. Of course, there are many who regard Yeats as a great painter. What is a little surprising is that Samuel Beckett, whose own vision is so austere and whose own literary palette is so bare, should have been such an early enthusiast.

When, in 1945, MacGreevy published a little book about Yeats,

Beckett reviewed it in the *Irish Times*. MacGreevy praised Yeats as the national painter of Ireland 'in the sense that Rembrandt and Velázquez and Watteau were national painters, the painter in whose work was the consummate expression of the spirit of his own nation at one of the supreme points in its evolution.' Beckett diplomatically said that such statements as this seemed to him to be 'art criticism of a very high order indeed' and an 'affirmation of capital importance'; but at the same time he thought that MacGreevy had exaggerated 'the national aspects of Mr Yeats's genius'. And to some, he went on, 'it may seem that Mr Yeats's importance is to be sought elsewhere than in a sympathetic treatment [how sympathetic?] of the local accident, or the local substance. He is with the great of our time, Kandinsky and Klee, Ballmer and Bram van Velde, Rouault and Braque, because he brings light, as only the great dare to bring light, to the issueless predicament of existence ...'

On the first occasion when Beckett attended an 'at home' at the Yeatses', he took Pelorson with him. They found that Yeats, for one reason or another, was not receiving that day, so the three of them went for a long walk through Ringsend and then out along the sea wall towards the old fort known in Dublin as the Pigeon House Fort. Beckett noted for the exiled MacGreevy that the scene was 'very beautiful and nervous and melancholy and windy. With that livid Dublin evening light on the shallows.'

When, shortly afterwards, he did find the painter officially 'at home', he was, as he said, 'staggered' by the paintings he saw and he wrote effusively to MacGreevy to thank him for the introduction to Yeats and, retrospectively, for that to Joyce, with whom he was already classing the painter. When he returned a few days later (again with Pelorson) they found Yeats in and alone and spent 'two entirely delightful hours' looking at paintings that he had not seen on the first visit. Yeats talked about cruelty and whether its existence could be derived from Original Sin, a proposition to which Beckett, describing the occasion to his friend, gave a cryptic 'no doubt'. He added that he thought that cruelty seldom existed in what might be called its pure state, however. 'Can you imagine a pure act of cruelty?' he asked, presumably suggesting that he thought it always had sexual components, or perhaps circumstantial ones, such as revenge or race. He had evidently clicked with Yeats to such an extent that when he began to avoid the formal 'at homes', both because he wanted to avoid the literary men he found there and because he was not too enthusiastic about having to be nice to Cottie, Yeats's English artist wife of many years, he still found it possible to see him when he called.

Yeats was a writing painter, who wrote plays as well as novels, both somewhat avant-garde in form, but, so to speak, traditional in content. Little happens of an eventful character in either. The novels are often monologues. Roads and people walking on them, singly or in pairs, play a large part in the narrative, such as it is. To this extent they resemble Beckett's. Where they differ is in Yeats's interest in the visual picturesque and picturesque Irish 'characters', his willingness to be Celtic and whimsical, his desire to be charming. When Beckett reviewed one of Yeats's books, *The Amaranthers*, under the heading 'An Imaginative Work' in the *Dublin Magazine* in 1936, he spoke of its irony and 'discontinuity', but contented himself for the most part by saying what it was not. It was not allegory, nor symbolism, nor satire. 'The Island', he said, 'is not throttled into Ireland, nor the City into Dublin . . . The end, the beginning, is among the hills, where imagination is not banned . . .' Thus far it might be thought to resemble one of his own works, but the difference is in the point of view. Yeats is a nostalgic romantic; Beckett is something else; but the probability, given his fierce admiration of the man, is that he was influenced none the less; and of course he also saw Yeats's eventless, somewhat circular plays when they were performed in the Abbey's experimental theatre, the Peacock.

On 5 March 1931, about a month after Beckett's first meeting with Jack Yeats, *Proust* was published, though scarcely to fanfares. The little book is almost exactly the 17,000 words which had been promised but thought impossible of achievement; and even after publication Beckett continued to have his doubts. 'I read the book through and really wondered what I was talking about,' he told the immediately responsive and enthusiastic MacGreevy. Confronted with sentences like the following, some readers may have had the same feeling:

> So, unlike Miranda, he [the narrator] suffers with her whom he had not seen suffer, as though, for him as for Françoise, whom Giotto's charitable scullion in childbirth and the violent translation of what is fit to live into what is fit to eat leave indifferent, but who cannot restrain her tears when informed that there has been an earthquake in China, pain could only be focussed at a distance.

Such lapses apart, however, the writing is vigorous and, if sometimes over-rhetorical, often both eloquent and pithy. As for the content, the book, he told MacGreevy, was 'at its best a distorted steam-rolled equivalent of some aspects or confusion of aspects in myself'. It is somewhat more than that, but it is that as well; and this is a characteristic of it that has struck some present-day critics – A. Alvarez, for example, saying that 'Although the essay contains some shrewd and original critical insights, Proust is, above all, an excuse for Beckett's diagnosis of his own problems.'

It begins with a disquisition on time and habit which provides a sort of first course in Beckett's outlook, as it might be expressed in propositional prose at least, rather than embodied in or subsumed into a work of art. 'There is no escape from yesterday because yesterday has deformed us, or been deformed by us. The mood is of no importance. Deformation has taken place . . . We are not merely more weary because of yesterday. We are other, no longer what we were before the calamity of yesterday.'

So much for the past. As for the future, 'Lazily considered in anticipation and in the haze of our smug will to live, our pernicious and incurable optimism, it seems exempt from the bitterness of fatality.' But advance into the future is really a continuous process of disillusionment. Even when we get what we want, 'We are disappointed at the nullity of what we are pleased to call attainment', the principal reason being that what we attain was desired by the person we were, not by the person we have become when we attain it. The subject which desires a particular object has died, perhaps many times, on the way and 'For subject B to be disappointed by the banality of an object chosen by subject A is as illogical as to expect one's hunger to be dissipated by the spectacle of uncle eating his dinner.'

The common state of humanity is suffering and if our sensibility were not dulled by habit we would feel it to an almost unbearable extent. Habit 'paralyses our attention, drugs those handmaidens of perception whose cooperation is not absolutely essential'. When the protective screen of habit is pierced, the results are almost intolerable. Fortunately habit quickly re-establishes itself, for, as Proust says, 'Of all human plants, Habit requires the least fostering and is the first to appear on the seeming desolation of the most barren rock.' Nevertheless there will be brief periods when habit and its boredoms will be dispersed, when we will have to adapt and form new habits. We do this very rapidly but 'The periods of transition that separate consecutive adaptations . . . represent the perilous zones in the life of the individual, dangerous, precarious, painful, mysterious and fertile, when for a moment the boredom of living is replaced by the suffering of being.'

It will be seen that the dye which permeates most of this is Schopenhaurean rather than particularly Proustian; but the idea that human life oscillates between the disillusion plus boredom of attainment and the disturbance and pain created by new (and of their nature illusory) objects of desire is, as Schopenhauer also claimed, said to be 'the wisdom of all the sages' from Brahma on; and, whether it is or not, it is certainly projected on to Proust as well.

At a particular point, with, it must be said, some violence to the argument which is proceeding, we are led to infer that Proust's view of the human condition is essentially the same as that of Arthur Schopenhauer and Samuel Beckett, the Spanish poet Pedro Calderón de la Barca being marshalled also to drive the point home:

> Tragedy is the statement of an expiation, but not the miserable expiation of a codified breach of a local arrangement, organized by the knaves for the fools. The tragic figure represents the expiation of original sin, of the original and eternal sin of him and all his 'socii malorum', the sin of having been born.
>
> *Pues el delito mayor*
> *Del hombre es haber nacido.*

But Proust, of course, is the great novelist of romantic attraction. In according him a pessimistic view of this phenomenon Beckett is on somewhat firmer ground. Some of it ties in neatly enough with the general Schopenhaurean view of the vanity of human wishes, as when Proust says, 'One only loves that which is not possessed, one only loves that in which one pursues the inaccessible.' Some of what is said has no doubt a Beckettian and thus solipsistic emphasis:

> The artistic tendency is not expansive, but a contraction. And art is the apotheosis of solitude. There is no communication because there are no vehicles of communication. . .
> The only fertile research [for the artist] is excavatory, immersive, a contraction of the spirit, a descent. The artist is active, but negatively, shrinking from the nullity of extracircumferential phenomena, drawn in to the core of the eddy. He cannot practise friendship, because friendship is the centrifugal force of self-fear, self-negation.

And perhaps some of it is accordingly questionable, as when the jealousy of the Proustian narrator is said to be 'doubtless a form of his domination complex and his infantilism, two tendencies highly developed in Proust' (and certainly, it might have been added, highly developed in the young man who was writing about him). But on the whole what is said is a brilliant summation of the Proustian view, in so far as the Proustian view can be summarized: 'How can we have the courage to wish to live, how can we make a movement to preserve ourselves from death, in a world where love is provoked by a lie and consists solely in the need of having our suffering appeased by whatever being has made us suffer?'

The answer is: through art; but it is an art of a special kind, open to, and reverent before, the operations of involuntary memory, those moments of evocation of the past when the veil of habit is pierced and the true and almost unbearable nature of reality is perceived.

Those moments are listed, beginning with the famous madeleine steeped in an infusion of tea and including not only the musty smell in the public lavatory in the Champs-Élysées and the uneven cobbles in the courtyard of the Guermantes Hotel but the subsequent hearing of the water in the pipes and the opening of George Sand's *François le Champi* in the library. Eleven are enumerated, but the list is said to be incomplete. Through these moments time is overcome and, therefore, death itself. Each one can be described as a mystical experience. 'But if this mystical experience communicates an extratemporal essence, it follows that the communicant is, for the moment, an extratemporal being. Consequently the Proustian solution consists, in so far as it has been examined, in the negation of Time and Death, the negation of Death because the negation of Time. Death is dead because Time is dead.'

In the faithful recovery of these moments through art, we make that 'movement to preserve ourselves from death', for which the loneliness, the futility, the disillusionment we experience may have robbed us of all enthusiasm; and emerge from the 'darkness of time and·habit and passion and intelligence'. Driving to the Guermantes Hotel Proust feels that

> everything is lost, that his life is a succession of losses, devoid of reality because nothing survives, nothing of his love for Gilberte, for the Duchesse De Guermantes, for his grandmother, and now nothing of his love for Albertine, nothing of Combray and Balbec and Venice except the distorted images of voluntary memory, a life all in length, a sequence of dislocations and adjustments, where neither mystery nor beauty is sacred, where all except the adamantine columns of his enduring boredom, has been consumed in the torrential solvent of the years, a life so protracted in the past and so meaningless in the future, so utterly bereft of any individual and permanent necessity, that his death, now or to-morrow or in a year or in ten, would be a termination but not a conclusion.

But in the library there he has several revelations in the form of those overwhelming evocations which abolish time (they actually begin·on

145

the way in when he stumbles on the uneven cobbles) and as a consequence he is free to

> receive the oracle that had invariably been denied to the most exalted
> tension of his spirit, which his intelligence had failed to extract from
> the seismic [sismic in the original Chatto and Windus] enigma of tree
> and flower and gesture and art, and suffer a religious experience in
> the only intelligible sense of that epithet, at once an assumption and
> an annunciation, so that at last he will understand the promise of
> Bergotte and the achievement of Elstir and the message of Vinteuil
> from his paradise and the dolorous and necessary course of his
> own life and the infinite futility – for the artist – of all that is not
> art.

This is such splendid writing that it would seem a pity to suggest
taking it down a peg. It is of course a version of Art for Art's sake.
Like many such it is couched in quasi religious terms. Not only is the
experience said to be 'religious', but what looks like a qualification, 'in
the only intelligible sense of that epithet', actually provides an opportunity
to add two very powerful words from the vocabulary of religion,
'at once an assumption and an annunciation'. A little later on the
vivid evocations of involuntary memory are described specifically as a
'mystical experience'; and the recipient is called the 'communicant'.
To back up these evocations of Christian – in fact rather heavily Catholic
– sacramentalism and dogma there is also an appeal to Platonic transcendentalism,
as when it is said that a physical object so evoked can be
contemplated in its 'essential' or 'ideal' reality, the 'essence, the Idea,
Imprisoned in matter' being released during this sort of contemplation.
But oddly enough perhaps, there is no attempt to discuss Proust's
own debt to Plato in the all-important passage on the death of Bergotte,
where it is said that 'everything is arranged in this life as though we
entered it carrying the burden of obligations contracted in a former
life' and that 'there is no reason inherent in the conditions of life on
this earth' to make us try to be good, or fastidious or polite even,

> nor to make the talented artist consider himself obliged to begin over
> and over again a score of times a piece of work the admiration aroused
> by which will matter little to his body devoured by worms, like the
> patch of yellow wall painted with so much knowledge and skill by
> an artist who must forever remain unknown and is barely identified
> under the name of Vermeer.

Although this opportunity to attribute a transcendental belief to Proust is passed up, however, there is certainly a general impression of an attitude to art which partakes of a sort of religious fervour, or of an attempt to make a sort of surrogate religion of art. This attempt is not uncommon among hitherto religious young people who discover art at the same time as they are in process of abandoning religion. Beckett would deny in later life that he had ever been a believer or that religious belief had ever made much appeal to him. The religious vocabulary of his *Proust* certainly makes an opposite impression. Some of this coloration in his attitude to art or in his general outlook may, however, have been the product of an intensive two-year association with MacGreevy, whose own pronouncements on literature and painting are certainly not free from devotional fervour and who once told the present writer that 'a good poem is an offering to the Holy Spirit'. Besides MacGreevy, there was James Joyce, whose habits of thought and to some extent modes of feeling remained Catholic to the end.

In similar vein Beckett would later say, in a review of MacGreevy's poems, that, 'All poetry, as discriminated from the various paradigms of prosody, is prayer.' Again, this may be mere malleability and the desire to please. But it is impossible to resist the conclusion none the less that the twenty-three-year-old author of *Proust* was engaging in the not uncommon process of finding in art a surrogate for the faith which has been abandoned or lost. And, strangely enough, the religious vocabulary employed is much more Catholic than Protestant. Words like 'assumption' and 'annunciation' are found much more often on Catholic lips than on Protestant, and in the context of the worship of the Blessed Virgin Mary at that.

When his six author's copies arrived via Cooldrinagh Beckett was 'shackled to the fire' in his room in Trinity with his latest ailment, an 'atrocious' cold. A wind of 'terrifying' intensity had been blowing for three days and had made him very nervous. Apart from wondering what he had been trying to say in some places and noting that much of it seemed to be about himself, he thought the work was like 'pale grey sandpaper, stab stab stab without any enchantment'; that the enthusiasm expressed was merely verbal and that the whole thing had 'a flat syllogistic drift like the fan of the long division sum in a Portrait of the Artist'. He recognized that it would not do much to advance his career in Trinity. 'Not that I care,' he added to MacGreevy. 'I don't want to be a Professor (it is almost a pleasure to contemplate the mess of this job).'

Of the writers he had to teach he was most attracted to Rimbaud, whom he was reading constantly, though he hated having to talk about him to his students. As will be seen, his enthusiasm for Rimbaud lasted long enough for him to do a translation of *Le Bateau ivre* in the following year, 1932. Another and new-found literary enthusiasm was Jules Renard, author of *Poil de carotte*, a refreshingly realistic account of an unloved childhood, and *L'Écornifleur*, whose poet protagonist is one of the first anti-heroes of modern literature, a descendant of Stendhal's Julien Sorel without that individual's occasional romantic daring and an outstanding example of a type which much preoccupied W. H. Auden in the 1930s, the 'completely self-conscious man'. It is easy to understand Beckett's liking for Renard. His friend Georges Pelorson thought he had deep affinities with a writer who was 'very pessimistic, even cruel in a way, who really inflicted wounds on himself'. Renard's journals, which Beckett had now discovered, portray a sensibility very close to his own mature outlook in that their gloom and cynicism is yet pervaded by humour. What most particularly attracted him was the realistic, disillusioned self-regard which they display; and he picked out a passage towards the end, in which Renard, confined to bed, is describing the infirmities and indignities of his last illness. Wanting to piss one night, Renard attempts to leave his bed and get to the chamber pot, but his sphincter control is weak and as he sits on the edge of the bed, about to put his feet on the floor, he urinates. So he sits there, watching the piss flow down the side of his leg on to the floor. When Beckett called Pelorson's attention to the passage that night in Trinity, he read it aloud twice, taking great delight in the minuteness of the description. The pleasure and strange satisfaction that describing his predicament gave to Renard had a great appeal for him. It is a passage which might almost have its place in one of his own novels.

It was the custom of the Modern Languages Society to produce plays in the languages they studied and Beckett became involved in one such production before the mid-term break at Easter. He did so with great misgivings, in response to the blandishments of Pelorson, who not only insisted that he should take part, but also that he should choose the play to be done. The production was to be in the Peacock, across the river from Trinity, and was to run for three nights.

When finally made, his choice of play puzzled everybody. It was a French melodrama, reminiscent of *Madame Bovary* but with comic overtones, *La souriante Madame Beudet*. Like Madame Bovary, Madame Beudet murders her husband, but the outcome is inconclusively comic rather than tragic, which is perhaps what appealed to Beckett. On consideration, however, *La souriante Madame Beudet* was thought too short, so it was decided to add Corneille's *Le Cid* to the bill. Beckett agreed to play Le Cid's father, Don Diègue, who at the point of his appearance in the play is given a long, solemn and beautiful soliloquy.

The comparison of Corneille and Racine is endemic in academic criticism and, while admitting the greatness of Corneille, Rudmose-Brown tended to favour Racine. In contrast to Racine, Corneille is a dramatist of simple energies, problems confronted by intelligence and will-power, whose characters arouse a certain admiration by reason of their resistance to circumstance and their will to struggle rather than by their capacity for suffering. His plays are full of energy and movement, crises and outcomes, whereas Racine's are static, with scarcely any outcome other than in the understanding and expressivity that the characters attain. Of the two, Beckett temperamentally preferred Racine. Pelorson was a high-spirited young man with a sense of fun and he did not think much of Corneille's play either. The temptation to guy it was strong. He began to introduce bits of business, he cut the text ruthlessly, and finally he called it *Le Kid*, partly thinking, of course, of Chaplin's famous film.

Like all dramatists of his time, Corneille had to observe the unities of time and place. To emphasize the limitation of time within which he had to work, Pelorson devised a large clock, the hands of which would be moved forward as the action of the play advanced. To gild the lily, the old man, played by Beckett in a long white beard, carried an alarm clock to time his speech, putting it down on the stage in front of him. He delivered the first part of the speech sonorously enough, but when the clock, as it was meant to, went off, he gabbled the rest of it at astonishing speed, in the manner in which Lucky would deliver his soliloquy in *Godot*. When Pelorson saw that play in Paris he was irresistibly reminded of the way in which Beckett had delivered Don Diègue's speech twenty-five years before.

Beckett refused to rehearse with the rest of the company, assuring Pelorson that it would be all right on the night; and so, for the first two nights, it was; indeed his speech and his long beard were both great successes. On the last night Pelorson had to go in search of him as the time of the performance approached and finally found him in his rooms. He was very drunk and that night he was a disaster. Unfortunately that was the night that Rudmose-Brown chose to come. He was not pleased by the general tone of the production or by Beckett's speech and said so in loud tones as he left the theatre. Still, most people thought *Le Kid* on the whole had been a great success. Characteristically though, Beckett told MacGreevy that the vulgarization had left him exhausted and disgusted.

Beckett was now suffering from the onset of certain symptoms which were unmistakably psychological in origin, broadly describable – at least in the terminology of today – as 'panic attacks', though of course not recognized as such then. Chief among these was a feeling of suffocation, which often came on him in his room as night was falling and for which he found some alleviation in rushing out and just walking the streets for an hour or two. He choose to keep quiet about these, but he did consult a doctor about palpitations of the heart which occurred when he was in bed at night, causing him great terror. In spite of his obvious worries, the doctor gave him short shrift, as most doctors would have done at the time; and instead of enquiring into his psychological stresses and condition told him to consume fewer of the Wills Woodbines he smoked. These were small, strong, pungent cigarettes which the poor had recourse to. They were sold loose for a halfpenny each or in open paper packs of five for twopence, but Beckett usually bought the green cardboard packet of twenty, which cost eightpence. Whether he really had anything wrong with his heart or not, Woodbines could hardly be said to be good for the health. Added to all this, the cold from which he had been suffering had turned into what was described as pleurisy, and this ailment still lingered. When he came home to Cooldrinagh to recuperate, it would seem that something about his presence in the house annoyed his mother and there followed a blazing row. So it was no wonder that when he received a printed invitation from Adrienne Monnier and La Maison des Amis des Livres to a '*Séance consacrée à James Joyce*', which was to be held in the bookshop '*à 9 heures précises du soir*' on Thursday 26 March, at which Philippe Soupault was to read from the translation of 'Anna Livia Plurabelle' and Adrienne Monnier was to lecture on it, he decided to accept. The invitation listed the names of the nine translators responsible and Beckett's came first, followed by that of Alfred Perron (sic).

This translation, on which he and Péron had laboured with such anxious zeal, had after his return to Ireland been taken over by a sort of superior committee consisting of Paul Léon, Eugene Jolas, Ivan Goll and Philippe Soupault. These luminaries had met every Thursday for several months in Paul Léon's flat, where they sat for two or three hours at the dining-room table, with the author of the original text lounging in an armchair, smoking cigarettes and making occasional languid interpolations. Though the Beckett-Péron text had been well worked over (their version of the opening pages is now at Yale and may be compared by thesis-writers with the final one), many phrases from it remained.

The proceedings at the *séance* were unremittingly reverend. There were about 200 people present and they sat with attentive expressions while the translation was read in French, after which Joyce's recording of the original English was played. Robert McAlmon, who had been drinking before he came, had just decided it was like being in Madame Tussaud's when suddenly an old man rushed across the room and slapped his face. It was Éduard Dujardin, the elderly French author to whom Joyce had decided should go the credit for having invented the interior monologue. Deciding (possibly with justification) that McAlmon was making jokes about his wife's hugely swollen ankles, Dujardin had reacted as a Frenchman should; but apart from this there was little diversion.

His presence in Paris was of course spoken of in Joycean and *transition* circles, and the Jolases' friend, Wambly Bald, who wrote a Paris column for the *Chicago Tribune*, recorded his presence in the 7 April issue of that paper. In this piece he was described as an Irish poet and an instructor at the University of Dublin whose 'book on Proust has just been published in London'. Bald also said that this young Irish poet had had tea with Gertrude Stein, but this was incorrect. It was George Reavey who had conveyed to Bald the information about Beckett, and it was he who had been to the famous flat in the rue de Fleurus.

When he returned to Dublin he found his ailments returning also and before the summer term was over he had another bout of the 'pleurisy' which had afflicted him before Easter, so when Frank suggested that they should go to the south of France together during the summer holidays, he readily accepted. They went to the Côte d'Azur, staying at Canadel-sur-Mer and called on Aldington who had rented a villa nearby. All the Joyce family had been present at the Easter séance and when Sam and Frank returned to Paris from the south, the first person they met was Giorgio, drinking Vittel on the terrace of the Deux Magots. He had been married the previous December to Helen Fleischman, whom Beckett wickedly called 'the Fleshwoman' and the conversation was about family matters and his future plans. Eventually they dined together and Sam also dined and lunched with his friend, the philosopher Jean Beaufret. He tried to see Nancy Cunard but she was in America with her boyfriend Henry Crowder.

Sam parted from Frank in London, where he called on the Joyces at their new-found abode at 28b Campden Grove – which Joyce called Campden Grave – in Kensington. Joyce had announced to Miss Weaver that the *séance* in Paris would mark the end of his Paris career, just as that organized by Valery Larbaud for the completion of *Ulysses* in December 1921 had begun it; and two weeks later, while Frank and Sam were making their way slowly back to Paris through Grenoble, Dijon and Troyes, the Joyces had moved to London. By insisting, against Nora's wishes, that they give up the lease on the apartment at square Robiac, he had begun with his family what he grimly called their 'fifth Hegira'. In other words, he had effectively, and with some satisfaction, rendered them homeless once again. Part of the reason was his determination that after twenty-six years of cohabitation he and Nora must now get legally married. Since they were both British subjects it would be easier to effect this strange design in England than anywhere else. The ostensible reasons were two. Marriage would settle all testatory problems with regard to Nora. It would also legitimize his children, for in 1926 Britain had come into line with the rest of Europe and a new Legitimacy Act now allowed for the post facto legitimization of children born out of wedlock. Stubbornly bent on his own object, he in fact exaggerated the testatory difficulties that were likely to arise if the marriage did not take place – and which would have been greater if he had been a French subject. Typically, what he did not consider was the effect on his children of discovering that they were bastards. When Giorgio, in spite of his artistic habits and tastes a continental bourgeois, discovered that his parents had not, as they pretended, been married in Trieste, he was horrified. The effect on Lucia was deeper and far more disturbing.

There is an impression that Beckett's attempted démarche to Lucia about his feelings – or rather lack of them – for her had led to a total breakdown in his relationship with the Joyces and that there had been no communication between them for over two years afterwards because of it. In fact no such situation had arisen. He had continued to communicate with the Joyces and they with him; and of course he had continued to work on 'ALP'. Shortly after the bungled conversation with Lucia she had written to him and he noted to MacGreevy that she was 'calm', but this related much more to the now general fears about her state of mind than it did to anything he had said. At this point, June 1930, he sent Joyce père a copy of *Whoroscope*, the necessity to send it rather than bring it round being that the Joyces were in England. Just before they left for Oxford that summer he heard from Lucia again, and imparting that fact to MacGreevy he declared, 'I never think of her now' – a somewhat self-centred remark which suggests a greater degree of emotional involvement on his part than he ever admitted to except to MacGreevy and, of course, implies as much concern for his own feelings as for hers. In July, while the Joyces were still on the English side of the Channel, he sent the master two copies of what had been done on 'ALP'.

In October, after returning to Dublin to take up his job in Trinity, he wrote to the Joyces collectively and Lucia replied. He was struck by the sane and level-headed tone of the letter, but the fears he now shared with others for her continuing sanity are expressed along with his relief in his comment to MacGreevy that, 'That girl will have her reason if she sticks to her guns.' In January 1931, though, he describes her as 'poor Lucia' for this is a point at which concern is being freely expressed among the Joyces' friends and acquaintances. Later in the same month, however, he is able to note, again with relief, that he has had 'another very calm letter from Lucia'. And the fact is that whatever opinion he and others may have had about her nervous state, he was still confiding in her about his own state of mind and his own now miserable situation. In response to his description of these she in fact offered him some very sensible advice, which was that he 'should accept the world and go to parties'.

But however calm Lucia had appeared to Beckett in her letters, she was in fact a troubled soul. Even though he had not succeeded in getting to the point, he had said enough to convey that he, though the focus of her dreams, had no equivalent feelings for her and he had vanished from the scene shortly afterwards. After his departure there had followed for her a period of intense and overtly sexual interest in other young men, to which the disturbed state of her psyche and newly fashionable theories about 'sex starvation' probably contributed at least as much as sexual frustration. To begin with there was the young sculptor, Alexander Calder. Later that same year, 1930, she met Alfred Hubbell, an art critic who subsequently wrote for the *New Yorker*. She seems more or less to have thrown herself at him; and after going at her invitation to a recital, and then dancing together, they went to bed a few times. She told Hubbell that he was her first lover and her inexperience was evident to him. He was fond of her, but he had a wife whose arrival in Paris terminated the brief affair. After him there were others – according to Sylvia Beach's friend, Jane Lidderdale, 'many' – and Lucia began to be regarded as an easy mark.

There were stormy scenes in the Joyce household about Lucia's supposed promiscuity, and Nora now determined that the cure was marriage – a solution that was urged on her by some of her acquaintance such as the writer Mary Colum. Nora, a monogamist by conviction if not by nature, and certainly less advanced in her views than many people were becoming in what was now the 1930s, was horrified by her daughter's apparent sleeping around, and berated Lucia accordingly. Joyce, as usual, preferred not to know and to overlook what he could not help being told about. Now he contributed his positive share to his daughter's mental difficulties by declaring that he must marry her mother.

The discovery that she was illegitimate had had a disproportionate effect on Lucia, which, of course, seems even more disproportionate to our eyes than it would have done to most of her contemporaries, many of whom would have been similarly dismayed. In a short space of time she had suffered several severe shocks and disappointments. There was, first, the discovery that she would never be a dancer of any real note. There was Beckett's failure to reciprocate her feelings. Then there was Giorgio's marriage to the stylish, flamboyant Helen Fleischman, an older woman who had deprived her of her adored brother, the companion of her childhood migrations and miseries.

But she was brave and she had decided to attempt other forms of self-expression, learning to draw and even making a start on a novel, when she learned that they had all to go to London so that her parents might take part in a grotesque ceremony.

From Paris Beckett had written to MacGreevy, who was in Ireland, expressing the hope that he would see the Joyces in London, where he had no reason to be other than sanguine about his reception. Due as much to Joyce's peregrinations as his own, he had now been separated from the admired and imitated master for too long. The slight, sentimental feelings of attraction he had felt towards Lucia were no match for the much more profound spell that the genius and character of her father had cast over him. It was Joyce himself he wanted to see and to be alone with.

It did not turn out like that. Nora and Lucia were present, to his disgruntlement, throughout; and there was no chance of conversation with Joyce other than *en famille*. Nor was the visit free from tensions where Lucia was concerned. Nora was affable, though considerably taken up with her worries about Giorgio's forthcoming marriage to a divorced woman older than himself; but he felt Lucia to be a problem and eventually he fled. 'Went out to Camden [sic] Grove but as usual impossible to see Joyce for Nora and Lucia. Usual fucking complications and flight,' he noted to MacGreevy. Nevertheless the visit was followed by what he called a 'rather miserable dinner'. He reported that 'les époux de deux Mois', as he called the now married couple, were in good form, and if from his point of view the evening was not exactly a success, at least it confirmed the intimacy which he still enjoyed. After his return to Dublin he received a copy of the extract from *Finnegans Wake* which had been published in booklet form as 'Haveth Childers Everywhere', as well as the *Nouvelle Revue française* containing the French translation of 'ALP', both autographed by his friend Joyce. Characteristically, all that this mark of favour did for him was to fill him with doubts about the quality of the translation of 'ALP', for which he had been partly responsible; and he hurriedly and 'foolishly', as he thought afterwards, dashed off a note to the master to say that it was impossible to read the text without thinking of the futility of the translation process. He also received a Whitsun card from all three Joyces.

Before leaving London he had also dined with Charles Prentice of Chatto and called at his office the next day to meet one of the partners in the firm. The Beckett of this period was not by any means the fastidious, above-the-battle, exclusively dedicated author that he would become later on. He was a young, hopefully professional writer who was desperate to escape from the job he was in; was in search of commissions and often glad when he got them. Of course, like many another young, hopefully professional writer, he was also often in two minds as to whether he really wanted them or not. *Proust* had now been published with a reasonable success and so he now proposed another subject for the same series – Dostoevski (to adopt his own spelling) – to Prentice and partner. The partner remained poker-faced, however, and Beckett later confessed to MacGreevy that he was actually relieved, since in his heart he believed that he neither could nor would write it. He also told MacGreevy that he had made the proposition simply for the sake of something to say. Many young authors have experienced the curious compulsion to keep in touch with their publisher while at the same time having nothing concrete in mind to suggest and being dubious about the desirability of suggesting anything or accepting any commission whatever. What he really wanted from Chatto at this stage was an expression of interest in a book of poems, but this was not forthcoming. A number of poems had now appeared in *transition*, in the *New Review* (another Paris journal edited by Samuel Putnam), and still more were scheduled to come out in Putnam's anthology of new European writing, the *European Caravan*. London is a long way from Paris, however, and Messrs Chatto displayed little interest.

Back in Dublin, he resumed his reluctant lecturing. His basic fondness and respect for Rudmose-Brown had not altered, though he knew his faults and failings better and was too familiar with his attitudes to want to see much of him. He also resented being loaded with minor tasks, such as invigilations, which he felt other people should be doing. His opinion of those others was low and so was his respect for the 'comedy' of education. In October he gave a talk to the Modern Languages Society which revealed his contempt. It was about a French poet Jean du Chas, author of a 'Discours de la Sortie', and the school to which he belonged, which expounded a doctrine known as 'Le Concentrisme'. But Jean du Chas, his poetry and the group which surrounded him were all Beckett's own inventions, and those members of the modern languages faculty who had attended and afterwards taken part in a serious discussion on the basis of the existence of du Chas and his movement considered themselves, with some justice, to have been damaged by their exposed gullibility and ignorance. Some of them had even pretended to know about the poet and his group. While all this was indeed comic, there was also something strange in the fact of Beckett, who found it almost impossible to lecture about real writers, insouciantly getting up to deliver a talk about an invented one.

This venture into the world of literary discussion was an isolated incident, for Beckett was now beginning to avoid humanity in his leisure hours as much as was possible for a drinker who habitually went to the pubs. The pubs he went to were mostly those on the northern and eastern side of Trinity. Pub crawls round the Grafton Street or College Green end were fraught with danger, for this was where literary Dublin foregathered. But like many who claim to be averse to company he would sometimes weaken and seek it out around Grafton Street, where he would see the local literary notabilities, 'cadging drinks and stealing bottles in every pub', if we are to believe his jaundiced remarks about them. 'I dislike the whole lot without exception. I'll drink in Phibsboro or in my rooms. Or I'll stop smoking and drinking.' He was in truth beginning to avoid company on a systematic basis, yet some of those mentioned sarcastically in such a context considered themselves to be friends of his. Beckett's innate courtesy, his air of grave attention as other people spoke, his occasional jokes, created an impression of affability greater than he normally felt; and some, including Ussher – who thought himself on fairly intimate terms with him and was to remember the Beckett of these years as 'the best of company' – and the painter Sean O'Sullivan, who was later to say that 'the life went out of' literary and artistic Dublin when Beckett finally departed the scene, were deceived. The truth, of course, is that like many people he was inconsistent, liking company sometimes when he was in it, but not on principle, and rarely admitting to liking it in prospect, or even retrospect.

When he went out determined to avoid it, he would usually emerge by the Pearse Street Gate and his itinerary would take him along Pearse Street and up Westland Row. Since the clientele of the pubs along this line was entirely lower middle and working class, he could be fairly sure of not meeting anybody he knew, except possibly at the Grosvenor in Westland Row, which attracted a mixed bag of clients and where Murphy meets the poet Ticklepenny. Sometimes he would go to MacLoughlin's of College Street; sometimes as far as Merrion Row, to O'Neill's, where the jarveys from the Shelbourne cab rank drank, but the walk to there from Westland Row was perilous, for it would take him up Merrion Street and along one side of Merrion Square, by Government Buildings, an ambulation Belacqua would think beset by 'poets and peasants and politicians'. If he left by the Pearse Street Gate, there were several pubs in Pearse Street itself which he felt to be safe; and near the Lincoln Place Gate was Kennedy's, which was a favourite.

On rare occasions he would complete the night by going to Becky Cooper's which now, in free Catholic Ireland, had deteriorated into little more than a shebeen, or place where drink could be got after the pubs had closed. Prostitutes still went to Becky's to drink and of course to encounter possible clients, but business was not supposed to be done on the premises.

In 'A Wet Night', a story dealing with this period of his life, Belacqua's ideal pub was described. 'A lowly house dear to shawlies where the porter was up and he could keep himself to himself on a high stool . . . and feign interest in the Moscow Notes of the Twilight Herald.' Porter was a weaker brew than stout, drunk by the lower orders, but the term was sometimes used inclusively for humorous purposes and stout is meant here. Beckett often drank Guinness, but on the whole he perversely preferred a Cork brew, Beamish. This could not be obtained everywhere, since it did not have much market in Dublin, and its availability was another reason for preferring certain pubs over others. A 'shawly' was originally a Dublin working-class woman as seen by student and other middle-class types, but at this time the shawl had been largely discarded as the garb of the majority, though still worn by the often humorously voluble ladies who kept open-air market stalls in Camden Street, Moore Street or Thomas Street and were among the stock types of Dublin comedy.

Beckett would return from the solitary pub crawl at closing time and shortly afterwards, at about eleven, Georges Pelorson would open the door of Number 39 to find him sitting in front of the fire, probably, though not always, reading. Since the Frenchman's arrival he had seen more of him than anybody else attached to Trinity. He had found Pelorson charming, though sometimes moody and a little mysterious. Their conversations had usually taken place late at night, when, as we have seen, they would often sit for two or three hours with a bottle of Jameson on the floor between them, sometimes reading, sometimes conversing or just sitting in companionable silence.

Beckett did not go abroad during the summer holidays of 1931. The pull of Kassel, much lessened by the break with Peggy, had been still more diminished by her engagement to Heiner Stärke, the athletic young German whose presence in her vicinity had irked him; indeed it may have been the attraction Peggy had felt for Stärke which, as much as anything else, had caused the uncompetitive Beckett to pull out. The holidays were shortened for him in any case by invigilation at examinations at both ends of it. He was a severe-seeming invigilator, who is remembered as staring expressionlessly into space over the bent heads of his students.

As autumn gave way to winter, Pelorson had begun to call less often and Beckett naturally wondered why. The reason was in fact an attractive young widow, Marcelle O'Connell, whose husband, a brilliant classicist, had been killed when run over by a tram in Stephen's Green. Marcelle, who was both pretty and intelligent, gave little parties occasionally and among those who had attended them sometimes was Beckett.

In the summer and autumn Pelorson had begun to see her on a regular basis and eventually they had fallen in love. Since his time in Trinity was coming to an end they resolved that they should get married and that she should join him in France as soon as possible after his return. It was a bold decision, all the braver since Georges was only a student with vague literary ambitions and Marcelle was nine years older than he was – he was three years younger than Beckett, therefore twenty-two. When his parents learned of the situation they disapproved strenuously, even invoking the assistance of the French embassy to terminate the affair. To his surprise, some of his male, Irish friends seemed to disapprove too. Beckett received the news of the engagement with studied indifference, but Georges sensed a coldness underneath and it dawned on him that his friend was jealous. When he went to dinner in the Bailey with Beckett and Con Leventhal the disapproval was even more apparent. Leventhal made remarks which verged on the insulting. Instead of wishing Georges well, he suggested that marriage to a Frenchman, and particularly one so much younger than herself, might well turn out disastrously from Marcelle's point of view. Everybody, he hinted, wagging his finger in the air in a manner which would have been less surprising had the speaker been a chauvinist Irish Catholic rather than a supposedly francophile and cosmopolitan Dublin intellectual, knew what Frenchmen were after. Beckett remained silent throughout these exchanges, but rightly or wrongly, Georges felt that he was pleased with Leventhal's performance and that by remaining silent he was egging him on. It took the young Frenchman some time to regain his natural fondness for the companion of his late-night exchanges.

If Sam felt any little sense of loss where Marcelle was concerned, he had also reason to feel some about Ethna McCarthy, to whom he had been attracted after his fashion since his student days. For a while after his return from Paris in the early summer he had seen something of her, going to her house more than once and meeting her for drinks in the Shelbourne. On one occasion, commemorated in the poem 'Alba', beginning 'before morning you shall be here . . .', she came to his room in Trinity. The poem describes the solitary scholar in his room, concerned with 'Dante and the Logos and all strata and mysteries' which are banished by the girl's coming. All we learn from it is that they play music together (probably on the piano which Frank, concerned for Sam's state of mind, had had installed in Number 39) and that she was, apparently, bountiful:

> who though you stoop with fingers of compassion
> to endorse the dust
> shall not add to your bounty . . .

Whatever this means, it certainly does not imply that they went to bed together. In the lengthy discussion of Belacqua's relationship with 'the Alba', as Ethna McCarthy is called in *Dream of Fair to Middling Women* and *More Pricks Than Kicks*, it is specifically stated that 'He has not lain with her. Nor she with him'; indeed the point, like most other points in *Dream of Fair to Middling Women*, is emphasized again and again at rather wearisome length.

An *alba* is the Provençal name given to an *aubade*, or 'dawn poem', in other words a poem lamenting the separation of lovers when daylight comes, but the lament does not necessarily mean that they have been torn from each other's arms. The poem was published in the October issue of the *Dublin Magazine*, a literary journal to which Leventhal contributed and to which Beckett would contribute again. After the publication of the poem Beckett would confer the name 'Alba', or 'the Alba', on Ethna herself, a procedure which is doubly confusing for the Alba in this context means, if it means anything, a fair person, whereas, as is again emphasized in the books, Ethna was dark. The portrait of her which emerges from both books is recognizably the same: a cool, intelligent lady, who was humorously fond of our hero but at the same time impatient of some of his more outrageous characteristics. A passage of dialogue in *Dream* seems likely to have been accurately transcribed:

> 'Smoke less' she said 'drink less, brood less.'
> 'Brood?' he was shocked to hear it called that.
> 'You brood' she said 'like a sick hen.'

To describe their relationship as inconclusive might give the impression that Beckett at least desired to hasten towards a physical conjunction of some sort. In fact he seems to have desired nothing of the kind. Ethna was more of an intellectual equal than Peggy and was already a fairly mature woman, whose level of sophistication was a change from Peggy's girlish moods and bewilderments. He preferred her figure to Peggy's and was not disturbed or revolted by it in the same way. One of the best pieces of writing in *More Pricks Than Kicks* is in fact a lyrical description of her bare back on the occasion when she decides to wear a backless evening dress. But what he wanted from her was basically the same thing he had wanted from Peggy. As he had been entranced by Peggy's green eyes, so he was deeply taken by Ethna's black ones. They too seemed to him to be a doorway into something. He describes what he wanted from both women in almost incomprehensible – and tiresome – semi-mystical terms. They are to be in some way the Beatrice to his Dante. But basically it was that they should excite his soul but not his body – or at least not his body unless he chose the excitement of masturbation with their image in mind.

In describing such a tangled, individual psychology, one has to avoid impatient labelling, but this distancing of women – whose proximity he none the less sought – would certainly justify terms such as scopophiliac, infantile and voyeuristic. In *Dream of Fair to Middling Women* Belacqua is more than once said to be a voyeur. In the story 'The Betrothed' the girl to whom he is engaged, Lucy, suspects as much. It is understood that after they are married she will take a lover. Then she is crippled by an accident. In Lawrence Harvey's words, Belacqua, 'can settle down to enjoy undisturbed the semi-mystical experience for which her presence is a pre-requisite'. But from another story, 'What a Misfortune', we learn that two years later Lucy dies; and soon Belacqua begins 'to feel more and more the lack of those windows on to better worlds that Lucy's big black eyes had been'. Like Belacqua, Beckett as a young man had difficulty in responding sexually to individual women. Hence his occasional recourse to whores; and hence also his masturbations with their images in mind. That Ethna did not wish for an orthodox sexual conjunction did not displease him. In fact her coolness where he was concerned was, if anything, a relief after Peggy's importunities. He wanted women to be love-objects of a sort, even sex objects of a sort so long as, even while he was with them, they kept their distance. At the same time he sought, in Ethna's case anyway, a degree of understanding and even an element of mothering. And he rather liked what other people might have regarded as the hopelessness of the relationship. Unlike poor Peggy she seems to have understood him fairly well; and there seems always to have been a slight tinge of amusement in her attitude. Very likely the passage describing the Alba's attitude towards Belacqua in *Dream* is, if somewhat obscure at times, fairly true of Ethna's attitude towards Beckett.

But it would not do. It could not go on. She was beyond the puerile graciousness of such a relation. She had got over the salt-marsh phase, the pretty pretty noli-me-tangere love-wound phase, while she was yet a child, before she put up her hair or sheared it off or did with it whatever was done when she was ceasing to be a child. All this pallor and umbilicism à deux might be the very thing for a certain class of gémisseur, it might be the very thing for him, permanent and pertinent and all the rest of it for him. But it was fundamentally all my eye for her. It might, like a new game, entertain her for a time, but it would never be more than light entertainment, a piece of mildly amusing, and, for a soul whose drifting was not distress, on the contrary, rather tragic codology. She used to say affectionately that he would get over this and that, she bestowed 'nino's' and 'mamon's' on him when she felt like it, but her real opinion the whole time was that there was little hope for him, that he was too irremissibly naive for her altogether, too permanently selfish, faithful to himself, trying to be like himself as he fancied himself all the time, an irretrievable stickler for his own wretched standard, and wretched was what she thought, and wretched was what she meant. He lay coiled up in the shadow, always the shadow, of the dread of leze-personality, his own or another's. Personality! That old bugbear bastard of hell! She thought that he would not get over it, that he did not want to get over it, that he thought of getting over it as the sin against the Belacqua third person ... When she would make up her mind finally that all that was so, that he was inextricably Limbese, then that was where she stepped off. He could rot away in his darling gloom if that was what he wanted. She would not be there to listen.

The sin against the third person is of course the sin against the Holy Ghost, who is the Third Person of the Blessed Trinity: the denial of truth. But the sin here is against the Belacqua third person, so the truth is the truth to oneself, faithfulness to which can only be assured by detachment from a world which inevitably demands compromise and modifications of personality. Limbo is the place in which there is neither salvation nor damnation, where there is no becoming, but only an eternal state of dim being. 'Limbese' was Beckett's word for the state of self-centred, mildly gloomy meditativeness and detachment in which he liked to live, with the outer world shut away, its noises meaningless, its struggles pointless. What he is describing is a psychology common enough in youth, one in which ordinary shyness and introspection of course play their part, but in which a deeper fear of the world, perhaps innate in the personality, is involved. To accept the tests and challenges of the world is to put the self in jeopardy, for the self of work, social encounter or even sexual relationship is not the real self, it is a compromise self and the real self may be permanently damaged by such encounters. The avoidance or retreat is therefore justified by the protection of self-hood, of the pure core of the personality. Those who suffer from varieties of this syndrome feel as if this inner core of the self – the soul, one might say – cries out against the necessity of engagement with others and with circumstance.

But Ethna, besides being an intellectual, was someone with what were called 'advanced' ideas. And so now she took up with Leventhal, a married man, beginning an affair with him which was to last for the rest of her life. Beckett was excluded and felt excluded, in spite of his theoretical musings about Belacqua and Lucy. But they remained friends, as Beckett usually did with both women and men.

Pelorson had departed in October and now Beckett was more alone than ever. He described to MacGreevy how he spent six or seven hours over the fire each evening, 'mooching about between the bed and the dark'. He was beginning to find even the smallest transactions in life, the purchase of a stamp or a tram ticket or a book, almost unbearably irksome and to long for the peace and security of his room. A crisis was now approaching and the reaction would predictably be a complete retreat into the self. He still went to the pubs occasionally, but less than ever did he want to see anybody. He met the writer Francis Stuart, whose book, *Women and God*, dedicated to MacGreevy, had just appeared; but though he rather liked Stuart, he did not want to talk to anybody or have to 'appear au courant'; and, though they went to Davy Byrne's, he was appalled by the company, which consisted of O'Flaherty, Austin Clarke, Arland Ussher and others. Most weekends he shut himself in his room, discouraging callers, lying in bed for long periods hoping to find strength to face the torments of another week's lecturing. He hated 'this terrible Dublin'; could not write, or even 'imagine the shape of a sentence' nor, though he acquired books occasionally, could he read very much.

Relations between him and his students were now very poor. Lecturing was a daily ordeal. His neglect of elementary hygiene was becoming serious and at a concert in the RDS he was conscious of the unwashed state of his neck. The 'feathering' of his heart was a matter of great concern. Finally, as might have been predicted, came collapse or abdication, whichever it should be called. He took to his bed in Trinity, drawing the blinds and lying there for several days without cancelling lectures or arranging a substitute. When after repeated knockings he finally opened the door, it was charitably assumed that he was ill and Rudmose-Brown came to the conclusion that his parents, his nearest relatives, should be informed. The result was a transfer to Foxrock, where, after a week or so, he decided that he had had enough.

He would go away after Christmas, 'somewhere into Germany', perhaps for several months. For the moment he was not 'burning any boats' so he would not tell Trinity, at least until he was safely away. He would resign from a distance if his resolve held. 'Of course I'll probably want back with my tail coiled round my ruins. And maybe I won't . . .' He was conscious that he was letting Ruddy down, but knew that 'some charming little cunt' would be appointed as a replacement for a term, which would be 'a happy surprise' for his students, after which 'some really responsible pleaser' would get the job. He was letting his parents down too, but it was now or never, he thought, and, for the moment anyway, he was determined that it should be now.

Like most important decisions it was arrived at suddenly, in an end-of-the-road situation, rather than calmly and rationally. And, like many decisions, it would be regretted, ambiguously and in varying degree, over the next few years.

Chapter Eleven

Needless to say, he went to Kassel, arriving in the cold weather of the imminent New Year, 1932. Youth is full of indeterminate situations and, in a way, Peggy's engagement, which was now formal, freed him from any grave embarrassment about a broken understanding. Still, going to stay with the Sinclairs shows a slightly chilling degree of emotional detachment. They all lived at pretty close quarters in the little flat in the Landgrafenstrasse and after all there must have been a certain awkwardness. None the less the Sinclairs were his second family; indeed their home provided an ambience in which he could be much more truly himself than he was in Foxrock. So, licking his wounds, he now lay in bed in the forenoon, browsed through half-read books in the afternoon, smoking and drinking more or less to his heart's content. They, above all people, were capable of understanding why a young man with literary ambitions should chuck up a job. They would also understand why such a young man should want to leave Ireland.

But things were changing in the Sinclair household. The money worries which dogged Boss's heels were more acute than ever. He was simply not making a go of it, either through dealing in art or antiques. And the political situation was beginning to suggest a seriously grim future for Jewish people in Germany, foreigners or not. The violent anti-semitism of the Nazi Party had already been expressed in beatings up of individual Jews and damage to their property. The Sinclairs were well aware of what the future might hold should Hitler, as now seemed likely, come to power, and Boss and Cissie were anxiously discussing the possibility of a forced return to Ireland or emigration to another country. Even the youngest children were becoming aware that being Jewish in Germany could be both dangerous and unpleasant.

Sam was of course aware of the political situation, but it would be difficult to exaggerate the depth of his indifference to politics for all of the early part of his life, even such politics as were now ruling in Germany, with their threat to his relatives and friends. His long indifference to and detachment from political events in his own country had contributed to a general detachment from anything that might happen on a public level anywhere.

The only interruption to his complete state of idleness during this visit was the necessity to regularize his position with regard to Trinity. He had left the job not intending to go back to it. But his qualms continued. Finally he decided to do what he had refused to do before Christmas: to burn his boats, a metaphorical course of action which was at length taken by means of a telegram to Rudmose-Brown which explained the sudden decision by advertence to his state of health. Naturally his mentor felt betrayed, and for some years to come Beckett would regard himself as a betrayer; but of greater immediate concern to the authorities of Trinity than the disappearance of a junior lecturer was the fact that the college master key had disappeared along with him. This was an event of serious concern. One could not have college master keys floating around either Dublin or Germany and it was decided that a considerable number of locks would have to be changed. Of course, Beckett had to communicate the fact that he had resigned to his parents too, a somewhat more difficult matter. One of the odd things about youth is that one seldom feels guilty about the action itself. It is nearly always the disappointment occasioned to others' expectations that is the vengeful fury and causes the sweats and night thoughts.

But if Kassel had seemed a good place to lick his wounds, Paris was undoubtedly where he wanted to resume his life, his friendships and, such as it was, his literary career. In terms of the depth of emotion it aroused, one of the most important parts of this life was his relationship with James Joyce. Shortly before Christmas he had roused himself from his general torpor to answer certain queries which the master had made about Dublin, principally about its public statues, the geography of Chapelizod, and Professor Webb, the Dubliner who became President of the Indian National Congress, a task Beckett had performed with much more enthusiasm than he was then able to bring to his own academic chores. Beckett's arrival in Paris, however, coincided with a family crisis. The Joyces had returned there in September 1931. Now, like everybody else in their circle he had to take account of the fact that Lucia's mental state had entered a new phase. Unlike some of the others, he felt, or was made to feel, a degree of personal involvement in the matter.

On the afternoon of 2 February, Joyce's birthday and usually a day of celebration, Lucia's antagonism to her mother expressed itself in violence, which included the throwing of a chair. Nora was terrified, Joyce helpless, but Giorgio, who had now been sent for, decided that his sister was so ill that she must be taken immediately to a *maison de santé*. Eugene and Maria Jolas had arranged a little dinner for Joyce that night at 40 bis rue de Sévigné and when he at length arrived he told the company of what had occurred, so word soon got around Paris. It has been claimed that Lucia in her frenzy had blamed her mother for interfering in her relationship with Beckett, of which she still had hopes, by banning him from their flat. Since he had never been forbidden to call or to communicate with Lucia this cannot have been the case, but Paris opinion correctly diagnosed her disappointment where Beckett was concerned as being part of Lucia's problem, in fact giving it more than its share of responsibility. A few people blamed with some accuracy Giorgio's marriage, which, Lucia felt, had placed the person in the world she was closest to out of her reach. James Stephens's daughter Iris, who was a contemporary of Lucia's and had come to know her quite well, has said that the main reason for the breakdown was the trauma caused by the public revelation of her own illegitimacy. Since she saw this as cruel and unnecessary, it was no doubt well to the fore in her mind. But the Beckett story, with its romantic overtones, proved a better gossip-piece and soon Kay Boyle and others were having a good time passing it on.

The fact was, though, that Lucia was suffering from schizophrenia. All the disappointments, frustrations and betrayals which she felt she had suffered – and there is no doubt that she had had a series of them in quick succession – would have been precipitants, but they had nothing to do with the basic condition, which had quite another origin.

Schizophrenia is a condition which most commonly manifests itself in the sufferer's mid-twenties to thirties and Lucia was now almost within that category. Though Freudians and others fought a long rear-guard action in favour of its psychological origin, the weight of the evidence and of medical opinion now favours a physiological or bio-chemical explanation. No doubt Lucia's mind had been subject to terrible stresses in the months leading up to her overt manifestation of disturbance; and no doubt also her early history and relationship with her parents had something to do with the matter – *Finnegans Wake* is shot through with a father's guilt about sexual attraction to a daughter – but the probability is that she would sooner or later have suffered a breakdown.

Whatever their concern, James and Nora reacted like most relatives in similar circumstances – they were glad of the temporary solution the hospital provided. But Lucia's stay in hospital was short. Her return posed a problem, for another sojourn in England had already been mooted. The lease was running out on the apartment in the avenue Saint-Philibert and Joyce feared that failure to go back to London might invalidate the legitimacy of his children. There was also the fact that they were paying rent on the empty flat in Campden Grove. As he was to realize, the marriage had been a disastrous move and unnecessary from a testamentary point of view; but by the time he became convinced of this it was too late. On 17 April Joyce, Nora and Lucia went to the Gare du Nord with sufficient baggage for a lengthy stay *en Angleterre*. When Lucia refused to board the train on which the luggage had already been loaded, there followed the most public and distressing scene in their family history, with Lucia screaming that she was being made to leave Paris against her will, and railway officials and interested onlookers gathering round. Joyce did his best to seem detached and calm but naturally enough Nora screamed back. At length the baggage was unloaded from the train and the Joyces went to a hotel. Of course, a section of Paris opinion attributed the decision to take Lucia to England again to a desire to remove her from the ambience of Beckett, who had now reappeared in Paris; and her refusal to leave Paris to a desperate attempt to stay near a beloved who was probably also a lover.

There remained the question of her future, of which there was naturally a good deal of discussion. As has been said, Mary Colum had already propounded the extraordinary suggestion that Joyce should consider the good old French solution of marriage as a therapy for all female erraticisms of behaviour. He should, she said, provide his daughter with a dowry and cast around for a suitable husband. Nora was appalled. If Lucia could not live as a single woman, how could she cope with the strain of being a married one? But Joyce and Mary Colum won the day and appropriate enquiries were made.

The trouble was that Joyce's acquaintance did not seem to include suitable, unattached, heterosexual young men who might consider marrying his daughter. One who did consider doing so was Robert McAlmon, a homosexual and a most unsuitable marriage prospect, who had nevertheless proposed to Lucia after his immensely rich English wife, the novelist Bryher, divorced him. He had already married for money and it had possibly entered his head now that he might marry for fame. MacGreevy was, in Brenda Maddox's words, 'a classic, pious, Irish bachelor'. Power had money, but he was too old and in any case had a regular girlfriend who danced in a revue. Calder was now engaged to Henry James's grand-niece. Beckett was obviously not cut out for marriage and his apparent lack of romantic or sexual interest in her seemed in any case to be part of Lucia's problem.

Joyce, however, had a new-found friend, the Russian-Jewish lawyer Paul Léon, who was also a man of letters and the author of a biography of Benjamin Constant; and Léon had a young brother-in-law, Alex Ponisovsky, who, after some persuasion and a few dates with Lucia, agreed to propose marriage and was, with some hesitation on her part, accepted, apparently on the grounds that she wanted to be married before she was twenty-five. Joyce was dubious of him as a putative husband on the utterly irrelevant grounds that he was a Russian; but Nora remarked that he was at least better than some of the young men Lucia was involved with, instancing Beckett and Alexander Calder. Since there was obviously a sexual content to Lucia's derangement, however, exhibited in her talk and in her scribblings, and since in the aftermath of the engagement her demeanour became more tranquil and composed, it may, in terms of the desperate logic that now prevailed all round, have seemed that the marriage solution was likely to work. She had lately begun to accuse all the young men of her acquaintance of having attempted to seduce her, sometimes embellishing these stories with details. Joyce, by no means tolerant of behaviour that concerned his daughter, was shocked by the possible truth of these allegations but also shocked by the thought that Lucia might have made such things up. Torn between belief and disbelief, he was inclined to hope it was all fiction while at the same time being inclined to blame the young men, who, he conceded, might have 'unnerved' her. To be on the safe side, and since she was now a respectable affianced woman, he banned them all from visiting, even the irreproachably pious MacGreevy, who was quite upset by the ban. Except for Ponisovsky, they were all, Joyce now told Miss Weaver, worthless. A formal engagement party was agreed on, and it took place at the Restaurant Drouot. But it was a ghastly occasion, with everyone sunk in gloom and all the banned males who usually were part of the Joyce *camarilla* absent.

Shortly afterwards Lucia went to stay with the Léons, her putative in-laws, and there her condition worsened. She lay on a couch in the Léons' flat in an obviously catatonic condition. A doctor was summoned and gave her injections. These had the effect of making her talkative at least, but sometimes she did not make much sense. At this point she moved to the Colums'; but her condition did not improve. In the day-time she would disappear and once or twice was found wandering in a confused condition. Mary Colum slept with her and even stitched their nightdresses together so that Lucia could not go off in the middle of the night; but Mary was herself physically unwell, and at length she and Giorgio decided that Lucia must be incarcerated once more. This time the doctors diagnosed hebephrenia, a then fashionable form of schizophrenia which involves emotional outbreaks and thought disorders.

For some of this drama Beckett was in Paris, where he arrived at the beginning of February. He moved into the Hôtel Trianon, on the rue de Vaugirard, where MacGreevy, who was in the thick of things, was already staying; and one of the first things he heard about was the disastrous scene on Joyce's birthday, when the chair had been flung at Nora. And of course he also heard that he was being blamed for a variously reported on and speculated about contribution to Lucia's emotional and mental troubles. Naturally enough, he felt he might be unwelcome at the avenue Saint-Philibert; and that in any case to manifest himself at this juncture might be only to aggravate Lucia's mental state. So to begin with he decided to steer clear. Steering clear came easily to him – the inclination to do so in troublous circumstances was deeply part of his nature – but in this case it was also probably the sensible and tactful thing to do.

So he stayed where he was at the Hôtel Trianon, resumed his literary contacts and eventually began work on the novel *Dream of Fair to Middling Women*, in which, significantly enough, although it is heavily autobiographical, neither the towering figure of James Joyce nor the troubling one of his daughter figure at all. The hero of this novel, like the hero of the sequence of stories with which he would follow it, is called Belacqua Shuah. The Belacqua comes from Dante's *Purgatorio*, Canto IV, in which the poet finds his dead friend, noted in his lifetime for indolence and sloth, seated in the shade of a rock with his head between his knees, caring not to go further upward on the path to salvation. Shuah comes from the Bible, and, though a proper name, has the meaning of 'depressed' in Hebrew, so that in our hero's name, combining as it does notorious indolence and remarkable depression, we are already pretty far advanced with the rudiments of a self-portrait.

But this self-portrait of the artist as a young man differs from James Joyce's more celebrated one in that it was written contemporaneously with the state itself, before the confusions, the débâcles and the miseries of youth were over and before they could be put in their place by the shaping weapons of a developed art. *Dream of Fair to Middling Women* would be rejected by Chatto and Windus and would doubtless have suffered the same fate at the hands of others but for the fact that its course of rejection was stopped by dismantling it and extrapolating much of a subsequent book, *More Pricks Than Kicks*, from the ruins. Reading it now, one can only agree that it thoroughly deserved this fate. It may – indeed it does – give evidence of genius. There are remarkable honesties and even remarkable insights in it, passages in which very difficult matters, usually pertaining to the self, are discussed with some insight, eloquence and feeling. But it also contains long passages of thoroughly confused, obscure and unreadable writing. The tone of voice is embarrassingly wrong, at once ingratiating, cocky and would-be Olympian. When Beckett later described *Dream* as 'immature and unworthy' he spoke no more than the truth.

The narrative, such as it is, begins on Dun Laoghaire pier, where Belacqua has just said goodbye to 'a slob of a girl', the Smeraldina-Rima, who is going back to Germany. He has, we are told, just fallen in love with her 'from the girdle up'. Later he follows her, first to Vienna, where she is a attending a progressive girls' academy, then to Germany, where he stays with her and her parents. After the termination of the affair he returns to Dublin. Much of the rest of the novel is taken up with his relationship with the girl known as the Alba, closely modelled on Ethna McCarthy, but there are also portraits of Jean Beaufret, Georges Pelorson, Rudmose-Brown and others.

Beckett kept the manuscript of *Dream* for a long time after he had abandoned all hope of its publication; then suddenly on impulse gave it to the American scholar Lawrence Harvey. For a number of years before it was deposited in the library of Dartmouth College, New Hampshire, he was the only person, apart from John Fletcher, to have seen it. He came to the conclusion that, whatever its literary merits, *Dream* was 'valuable for the insights it provides into the temperament, intellect, talent and interests of the young Beckett'; and he was right. Besides those already commented on, one of the most notable characteristics displayed is the strong vein of misogyny evident in both books. That it is combined with a need for women on various levels of intercourse – mental, spiritual, and even, in an indirect way, sexual – is a not uncommon paradox.

Harvey, who initially set out to write a study of Beckett's poetry, eventually decided to correlate the early poems, published and unpublished, with the contemporaneous prose. Beckett, initially at least, was enthusiastic about him as an interpreter, and obviously gave him *Dream* so that he could do this, but in the end he committed what is still supposed to be a capital sin in some Beckett circles by shamelessly interpreting the work in terms of its author's inner psychology and deducing aspects of Beckett's sexuality from it.

When his study was finally completed in 1969 the preface frankly stated that 'Beckett's poetry turned out to be less 'self-contained' than is some poetry . . . It points both to itself as art and to its maker as man

. . . Perhaps there is something more deeply human than idle curiosity that leads us to look beyond even privileged works of art to the human realities from which they spring.' These human realities, according to Harvey, included 'sexual reluctance' – in other words, a reluctance to engage in physical sex (except, as Harvey acknowledges, with prostitutes), voyeurism, infantilism, a fixation on buttocks and 'autoeroticism', in other words a preference for masturbation, as well as 'nostalgia for the womb': quite an interesting package for any young man. Though Harvey admits that most of these characteristics can be extrapolated from the fiction, he says they are 'not without correlation in the poems' and asserts that 'It would be unusual, even allowing for the prerogatives of imagination and the necessities of fictional elaboration if these persistent themes had no correlatives in the life of the author'. This is a just enough statement; indeed to the reluctance to engage in physical sex which he attributed to his subject he might have added frequent inability. Beckett had difficulty in associating the act of sex with other forms of emotional involvement.

Beckett's misogyny is fully illustrated in some of the passages about the Smeraldina–Rima which have already been quoted; but the especially cruel tone in which she is dealt with may have something to do with the fact that in February 1932, when he began the novel, he had just come back from Germany, where the Smeraldina's model, Peggy Sinclair, now engaged to someone else, went happily singing about the place, while her neurotic former boyfriend indulged his guilt about his departure from Trinity and lay glowering in a state of paralysis on the sofa in the living room.

In March, shortly after beginning the novel, he was a signatory of the famous or infamous 'Poetry Is Vertical' manifesto in the first issue of *transition* to appear for almost two years. The best that can be said for 'Poetry Is Vertical' is that it is a bit better than the 'Revolution of the Word', though the latter, which had appeared in 1929, had a somewhat more distinguished list of signatories. Besides Beckett, the signatories to the new manifesto included MacGreevy and Pelorson, now also back in Paris and moving in much the same literary circles as the two Irishmen.

Scholarly critics have pored over 'Poetry Is Vertical' to assess the degree of Beckett's input into it. In the words of his bibliographers, they have thought it best to 'assume that he does not sign anything lightly'. But this again is to confuse the older, more established and of course more fastidious Beckett with the young man just about to turn twenty-six who had arrived back in Paris, still somewhat demoralized, just as Jolas was preparing to get out another issue of his magazine. The manifesto is largely Jolas's own work, though the two most distinguished signatories, the German art critic and novelist Carl Einstein, a co-editor with Georges Bataille of the review *Documents*, and the sculptor Hans Arp, may have contributed parts of it. They, with Jolas, formed a little trio, which met weekly at Einstein's apartment and discussed, among other things, the distressing recrudescence of German nationalism and the possibility of combating it with a 'new expressionist ethos'. It is probable, however, that, as in the case of the 'Revolution of the Word', the other signatures were solicited after the event from those young writers who dropped in and out of the hotel room in the rue Fabert, near the Gare des Invalides, which served as the magazine's office.

And yet it is possible that the manifesto does betray some influence of Beckett's, either during its composition or otherwise. Both then and later he would certainly have been able to go along with a phrase such as 'the hegemony of the inner life over the outer life'. He would also perhaps have been able to subscribe to the view that 'Poetry builds a nexus between the "I" and the "you" by leading the emotions of the sunken telluric depths upwards . . .' Though whether he would have agreed then or later with the proposition that the end of this process was 'the illumination of a collective reality and a totalistic universe' is another matter. But even if he – like, possibly, their authors – did not know the meaning of such sentences as 'The reality of depth can be conquered by a voluntary mediumistic conjuration, by a stupor which proceeds from the irrational to a world beyond a world', he would none the less have been inclined to sign. The *transition* gang were his gang and they were also James Joyce's.

And besides, he had a story in this issue, in company with such as Franz Kafka, Henri Michaux and Gertrude Stein, though – probably because he simply had not been in Paris – he was not a contributor to its 'Homage to James Joyce' section, which was sub-titled 'James Joyce at the Half Century' and had offerings by Padraic Colum, Stuart Gilbert, Eugene Jolas, Thomas MacGreevy and Philippe Soupault.

The story is 'Sedendo et Quiescendo', later to form a part of *Dream of Fair to Middling Women*. The title is from Thomas à Kempis's statement of the quietist attitude: '*Sedendo et Quiescendo anima efficitur prudens*' – 'In seatedness and quietness the soul acquires wisdom.' Shortly after the publication of this issue of *transition*, he published a poem of some sixty-four lines in the *New Review*. He was putting his stay and the break with Trinity to some use by resuming publication in the emigré magazines of the Left Bank and had already agreed with Edward Titus to translate a number of surrealist poems and prose by André Breton, René Crevel and Paul Eluard for the autumn issue of *This Quarter*, a surrealist number of which Breton was guest editor. These were finished and delivered early in May; but shortly afterwards his situation and that of many other foreigners in Paris suddenly changed.

On the seventh of the month a White Russian, Gorguloff, assassinated the President of the French Republic and the police began to check on the papers of all the aliens in the city. Naturally enough, Beckett, like many others, had no residence permit or *carte de séjour*. Neither had he much visible means of income, a circumstance which would have been looked on with extreme disfavour if he had been arrested in the sweep through the little hotels of the left bank which was now taking place.

Accordingly he left the Hôtel Trianon and spent some nights sleeping on the floors of his acquaintances, finally staying about a week with the painter Jean Lurcat, a strong-minded and witty man who was a friend and contemporary of MacGreevy's and at that time principally known as a surrealist, though later he was to do some remarkably interesting tapestries. But if Beckett had not enough money to demonstrate to the authorities that he could reside in France without becoming a beggar or a criminal, neither had he enough to leave the country. So he went once more to Titus who had an office in the rue Délambre off the boulevard du Montparnasse, but did most of his business in the Rosebud bar just below. In spite of his notorious tight-fistedness and the money he had already parted with for the surrealist translations, Titus agreed to a deal. He and Beckett had already discussed the possibility of a translation of Rimbaud's *Le Bateau ivre* and in fact Beckett had begun to toy with the project. Now he asked for 1,000f. to complete it. After some brief haggling, Titus gave him 700 and with the money in his pocket Beckett picked up his bag at Lurcat's and caught the night train to London.

There were other reasons for going there besides his quasi refugee status and such a move had already been discussed with MacGreevy. The Left Bank of Paris was an anglophone literary centre of sorts, but it was not London. Beckett had now no money other than the minuscule fees paid by the little magazines and the occasional pound or two that his father sent him, often without his mother's knowledge. The twenty-six-year-old Beckett is not to be confused with the Beckett of later years. He had to prove to his parents that he had some capacity. He was, within limits, concerned about money, sufficient money to enable him to survive without the charity of his family. He was also, within the limits imposed by his lethargy, his neurosis, his basic quietism, his knowledge of the falsity of most judgement, literary and otherwise, concerned about literary reputation. In so far as proving his capacity, earning a living or acquiring sufficient reputation to impress anybody at home were concerned, England was a better bet than France, London than Paris; and in England he could at least kill more than one of these ugly birds with the stone of an ordinary job. The difficulties of the Joyce situation, Joyce being one of the principal attractions of Paris, no doubt made a move across Channel seem less of a separation than it might otherwise have been. Beckett had been around to call and had, as usual, been given tasks to perform, but the situation was not an easy one and in a way he was glad to interpret the ban on male callers, about which he heard from MacGreevy, as applying to him as much as to anybody else.

He had already discussed with MacGreevy the recourse which many other Irish writers have over the years adopted: he could teach in England; and to this end he had already taken care to obtain references from Ruddy and, however distasteful he found the need to ask, from Gibbon, the headmaster of Campbell College.

Arriving now off the boat in the early morning, he deposited his bag with the porter of the Shaftesbury Hotel and went straight to the famous old turkish baths attached to the Imperial Hotel in Russell Square, where he sweated and slept until midday. It was unexpectedly hot in London and he found the heat oppressive. He also found that his country was in the news, for de Valera had just arrived in London to attempt to heal the latest, very serious rift in Anglo-Irish relations. The Italian barber at the turkish baths told him that the Irish leader was a great man who would never let his country down; but the Irish porter at the Shaftesbury thought the same de Valera would be the ruin of his countrymen. In response to Beckett's request to his family for money, his father had sent £10, which had already arrived by letter in care of Chatto and Windus. When he went to collect this he also saw Charles Prentice. He had already sent him the novel which, Prentice politely explained, was now being read by someone else. Rightly or wrongly, though, Beckett felt that Prentice himself had already read it and, though he murmured encouragement, was for one reason or another embarrassed to speak of it. A glance at it sufficiently explains why. Of course, it contains brilliant passages; it is psychologically highly revealing; but it would have been regarded as utterly unpublishable from an ordinary commercial point of view – and, it might be added, would still be regarded as such. In so far as he could understand them, Prentice may also have found some of the psychologically revealing passages altogether too much so. He murmured what he could in the way of commendation and adverted to the second reader, but Beckett understood that the novel was as good as rejected. When he mentioned a collection of poems, again Prentice very politely said he would be glad to see them; and before they parted he asked Beckett to have dinner with him a couple of evenings later.

After leaving him Beckett went to see the lodgings which Jacob Bronowski had recommended to him which were off Gray's Inn Road and therefore not very far away from Bloomsbury. Number 4 Ampton Street was one of those privately owned London houses where furnished rooms were available. It was owned by a Mrs Southon and Bronowski had himself stayed there. It was the great era of the furnished room in London and Mrs Southon, who showed it to him, was anxious to please. When he enquired whether the noise of a typewriter would be a disturbance to other tenants she assured him that he could type day or night. After some consideration he took a room with a gas-cooker, the larger of the two he was offered, for which he agreed to pay 17s. 6d. per week.

Furnished rooms, with which from now on he would have considerable acquaintance, were a London phenomenon, being basically single-room apartments with cooking facilities in houses where the landlord usually lived on the premises. The furnishings were seldom more than bed, table and chairs. There was often a gas fire, but bathroom and toilet facilities would be shared with other residents and the floor was usually covered with linoleum rather than carpeting. The room that Beckett took seemed clean, bright and reasonably comfortable and, after taking it and paying a week's rent in advance, he promptly went to bed and slept for ten hours non-stop. To MacGreevy he expressed himself as pleased by the proximity of the Royal Free Hospital but also by that of a large pub at the end of the road. After his fears of the police in Paris he found the sight of London's bobbies actually soothing. When he left Paris MacGreevy had been ill, depressed and, like himself, virtually penniless. Beckett now proposed that his friend should also come to London, perhaps taking the other room in Ampton Street and he offered to send £5 to bring this about.

Ensconced in Ampton Street, Beckett now began to do the rounds or, in other words, perform the necessary penances that residence in London imposed on a troubled and guilt-ridden young literary man. For an Irishman, one of these penances was a visit to Charles Otto Desmond MacCarthy. The well-born MacCarthy was a former Cambridge 'Apostle' who had graduated to Bloomsbury and membership of the Bloomsbury Group. Though he liked to describe himself as a literary journalist, he was some distance removed from the grubbier aspects of that profession. Still, he was chief reviewer for the *Sunday Times* and at this juncture editor of a middle- to high-brow journal,

Life and Letters. Though he had no real Irish connections beyond his name, he was for some reason looked upon as an avenue of advancement for young Irishmen in London and, though his track record as an advancer was anything but good, his name was passed on from one to another. Now MacGreevy suggested him to Beckett as, a little later, Frank O'Connor was to suggest him to Patrick Kavanagh, and with ultimately as little result. In spite of some initial encouragement and a suggestion that he might use his good offices for Beckett with the firm of Jonathan Cape, MacCarthy did nothing, eventually becoming 'that foul fucker MacCarthy'.

The widely acquainted and much-liked MacGreevy also wrote a letter of introduction to Leonard Woolf, who ran the Hogarth Press in collaboration with his eminent novelist wife, and one day Beckett trundled round to Tavistock Square with a letter that he had carefully composed, MacGreevy's letter of introduction and his poems. Mr Woolf, he was told, was out of town, but yes, he would get the packet. About two weeks later the poems came back with a formal rejection slip. Beckett decided that the MS had never left London, but then on reflection decided that Woolf must have got his letter anyway. 'Or perhaps', he commented acidly, evidently knowing something about Virginia Woolf's difficulties, 'it is his turn for the asylum.'

Meanwhile MacGreevy had come and gone, on his way to Ireland, a place to which Beckett felt he too was being inexorably drawn, partly by parental pressure, for his mother was growing quite insistent in her suggestions that unless he found visible means of support he should return to the fold, and partly because he had not in fact any means of support, visible or otherwise, except an intermittent stream of small sums from his father.

He had been round to another well-known station of the cross for aspirant literary young men, Truman and Knightly, an employment agency which placed schoolmasters. There he had left copies of the testimonials he had received from Ruddy and Gibbon and filled in an enormous form, reflecting as he did so that his qualifications, when he had thought of them and got them all down, were really quite remarkable, sufficient to get him a very exalted post.

The poems came back from Prentice with a note saying he was very very sorry, and the novel from Cape with a letter informing Beckett that their reader's report did not encourage them to make an offer for publication rights. Henry Grayson of Cape had been one of MacCarthy's contacts, but when he had gone round there he had failed to see Sir Henry, being offered an interview with his son Rupert Grayson instead. He told MacGreevy that he thought Rupert Grayson 'a proper pudding', but he did take the typescript of the novel and was courteous about the opportunity to consider it. Everyone in London was, even if their courtesy did not result in much.

In addition to the publishers there were the literary editors, from whom he hoped to get reviewing or commissions for articles. The literary editor of the *Spectator* was Derek Verschoyle, a member of a well-known, eccentric Anglo-Irish land-owning family and subsequently a publisher. This young man had been a student at Trinity while Beckett was lecturing there. Now the boot was to some extent on the other foot, with Verschoyle seated behind the desk and Beckett in front of it. He received his old lecturer affably and gave him a cigarette. Beckett gave him his latest three poems, but Verschoyle had no books for review and the poems came back a few days later.

And so on to Ellis Roberts, who occupied a similar position at the *New Statesman*. Like Verschoyle, Roberts was friendly and this time Beckett was given a cup of tea. He offered what he thought might be attractive suggestions for literary articles. As Beckett reported to MacGreevy, Roberts 'thought he might possibly be interested in a statement on Gide, covering all that artist's vicissitudes from Andre Walter to Oedipus in not more than 1,800 words, or one of similar length on the modernity of Vico'. To his friend he added, 'but of course it can't be done. I don't believe I could get a dozen words together on any subject whatsoever.' He is certainly not the only aspiring literary journalist who has ever felt moved to offer articles which in his heart of hearts he felt he could not write, but the declared inability increases the grim comedy of these interviews. Of course, he might have managed a book review or two, but, like Verschoyle, Roberts had no books which he thought suitable to offer the rather tongue-tied Irishman for review.

Confronted with these disappointments, like many another he took to sleeping through the forenoon, 'perfecting my methodology of sleep', as he put it to MacGreevy. Though he had little impulse to go to art galleries, he made an occasional foray to a tourist attraction. St Paul's struck him as 'hideous', as it might well do given his general circumstances, whatever his feeling for architecture, and there is little evidence that he had any. He dutifully went also to see the Tower of London, but it did not seem very interesting. Afterwards, however, he sat on the wharf and watched the steam barges dipping their funnels to pass under Tower Bridge and the bridge itself opening up to accommodate bigger ships. This sight struck him as 'très émouvant'.

Usually he got up around one, went out around two and found somewhere to sit until the pubs opened again at five, a not uncommon routine, then or later. One afternoon he sat in a twopenny deckchair in Saint James's Park and saw a little boy playing at empty buses with his nurse, whom he called nanny. She had, he thought, exactly the same 'mined-granite' expression as Bibby had had before she got married and her features changed. He told MacGreevy that he had been 'moved almost to eyedew'; but he had to run for a piss to the underground station 'and when I came back to the same chair they were gone. Soon I will be calling to my mother to come and kiss me to sleep. Fall in love to write a lot of poems, have a child to engage a nanny.' Evidently thinking of Bibby, however, he said such a nanny 'must have a strawberry nose and suck cloves, or at least peppermints'. He added almost tearfully, 'She carried his big ball in a green bag and they started an apple.'

Like most people who sit around a good deal in public places, he was reading a lot of trashy newspapers, either left behind by others or purchased for a penny. To his friend he guyed his own knowledge of the unchanging concerns of the British public as well as his occasional moments of optimism:

Perhaps I will get proofs of poems returned from Woolf to-morrow morning, or an offer to instruct Princess Elizabeth in the Florentine positioning. To-day is her mother's birthday. I hope the Duke got back from 'under canvas' all right. I'm well up in social spews. Britannia's truck is 171 feet above her waterline (carries 3000 worth of canvas: only six feet lower than the underground offices).

The pound is at 89.

Generally speaking, however, he was depressed 'the way a slug-ridden cabbage might be expected to be'. And all the time there was his mother's reiterated insistence that he come home. One letter, written, he thought, in Switzers, a fashionable department store in Grafton Street, was especially insistent. And in truth he could see little alternative, since he was barely eking out an existence on the sums his father sent. He certainly was not occupied in writing; and the very idea of trying to do so seemed 'somehow ludicrous'. Nor was the ambition to teach any longer very strong. Truman and Knightly sent him notices of jobs in Cornwall, Devon, Derbyshire and other places, as well as one as an English instructor for the Berlitz school in Basel; but he was 'too tired and eviscerate' to apply, even for the one in Basel – though, as he reflected, 'Better Basel where love is not than D.D.D. with sentimental salmagundis and others on the mat', 'D.D.D.' being 'dear dirty Dublin', a phrase much used by the inhabitants of the city at that time; and the 'sentimental salmagundis' being the attentions of his family and their concerns for his lot. To MacGreevy he permitted himself to wonder whether his father would take him into the office if he went home. Frank, he reflected, had gone into the office after three years in India: 'and now look at him. With a car and a bowler hat'; but he thought even this semi-jocular reflection about going into his father's office was so infra-dig that he retracted it in his next letter, saying he was 'not serious'; and even if he was, 'there is no room for another clerk in the office, and even if there were I simply could not do the work'. It would have to be 'private school or else training college or else handy andy in the garage and back garden at home', if, that was, he could even mend a puncture.

He dreaded the idea of Dublin but at the same time thought there was 'no use insisting further here. This month of creeping and crawling and solicitation has yielded nothing but glib cockney regrets.' Of course, there were alternatives. He might have gone teaching in Devon or Cornwall, or at least applied to do so, but his nature, weighted to sink, was pulling him downwards and therefore homewards. He was drifting and home was where he would fetch up. He thought fleetingly of making a dash for Paris, but, he told MacGreevy, 'I am too unbelievably gutless to do anything and my mother would throw a fit. So write to Cooldrinagh.' The last straw was the theft of a £5 note which his father sent him. He put it in his drawer and when he went to look for it, it was gone. He told the landlady, who affected great distress, appealing to the lodger who appeared 'to have his being in the kitchen' to confirm that such a thing had never happened before, never in all the years. Faced with this loss, he told MacGreevy, the 'villegiatura' was finished. He would take to his bed until more money came from 'the blue eyes of home'. It is a phrase he had used in *Dream of Fair to Middling Women*. There the hero, in Paris for the first time, spends the money which comes from that source on 'concerts, cinemas, cocktails, theatre, apéritifs', but not apparently on opera, 'nor, after a bit, on brothels'. In this case he spent it more humbly on the fare home to dear, dirty Dublin, where there were plenty of cinemas, even theatres, where cocktail parties and cocktail lounges were supposed to be coming in, but where there were, at this date anyway, no brothels.

Chapter Twelve

The closing of the brothels, brought about by pressure on the government from an organization called the Legion of Mary, had probably been approved of in Foxrock. But it was none the less one of the triumphs of the onward march of Catholic triumphalism, which was now the ruling ethos of the Free State and culminated this year in a huge Eucharistic Congress. The Congress had been formally opened in June 1932 by the Papal Legate, Cardinal Lauri, its high point being an open-air mass in Phoenix Park at which the great Irish tenor John McCormack sang 'Panis Angelicus' to a congregation of tens of thousands. Another landmark had been the Censorship of Publications Act, which had been passed three years previously. This provided for the banning of books whose 'general tendency' was 'indecent or obscene'. There would turn out to be a very large number of these and almost every Irish writer of any note would have a book or books banned.

For the moment, however, the prodigal was more concerned with his welcome home than he was with the welcome the entity called Ireland might extend to him. As is supposed to be the case with prodigals, it was kind and understanding. He was supplied with little jobs around the house or in the garden. And, in spite of its 'early to bed and early to rise' aspects and the contrast with the life he had been leading, he enjoyed almost everything about it, even his mother's fussiness. His father, he noted, was 'real'. His mother was 'comico-real'. His own need for the 'anaesthetic of caress' was also 'comico-real'. He walked a lot, sometimes with his father, taking the old walks through the foothills of the Dublin mountains and having a couple of pints on the way. Sometimes he went swimming with his father at the Forty Foot, the bathing place where he had been urged into the water before he first learned to swim. His father also supplied him with a little money now and then, pressing the odd pound or two into his hand in his shy, good-natured fashion. And he got paid for his odd jobs; but beyond that he had no income. Fulfilling his own prophecy that he might return with his tail between his legs, shortly after arrival he made contact with Rudmose-Brown in the hope of getting some work. Ruddy promised to recommend him as someone who could give grinds – a kindness which made him regret the portrait he had drawn of his old mentor as the Polar Bear in *Dream*. No grinds however materialized and he had apparently no prospects of any other kind of work.

Rupert Grayson of Cape had promised him an answer about *Dream* within weeks. An acceptance would of course alter his circumstances, providing a little money and justification for his existence. He waited, he told MacGreevy, 'in nervous comfort, clipping privet and sneering at my thirst'. In response to a suggestion made by Grayson he had sent his poems to Edgell Rickword, the poet and critic who was one of the editors of *Scrutiny* and read for the communist publishers Lawrence and Wishart. Grayson had suggested he should go to see Rickword before he left London, but he had neglected to do this and now felt that he might have missed an opportunity – a feeling common among young authors who neglect to make suggested contacts and usually, as in this case, a groundless one, though the idea of Beckett as a Leavisite new critic has its attractions.

As the weeks passed and, in spite of a polite note, no answer came from Cape, he decided to write Grayson a stinger. The result was that *Dream* did come back, with references to its 'circumscribed appeal'. Beckett at once asked to see their reader's report, but of course he was not shown it. Grayson's advice was that he should put the book aside and turn his attention to something else, actually very sound advice, which naturally did not please him. In fact he immediately sent it off to Edward Titus, who, besides editing *This Quarter*, was also the proprietor of the Black Mannikin Press, publishing mostly small books in limited editions, but also such period pieces as the memoirs of the famous prostitute and model Kiki of Montparnasse, with reproductions of her paintings and an introduction by Ernest Hemingway; and a study of D. H. Lawrence by Anaïs Nin with facsimile reproductions of pages from *Lady Chatterley's Lover*. Beckett feared to strain the connection with Titus, who already had 'Dante and the Lobster' and his translation of Rimbaud's *The Drunken Boat*, by burdening him with a full-length novel, but he sent it anyway.

The London aspirations now collapsing, the Paris connection was at least a mental lifeline and gave him a feeling of professionalism which he valued, but there was naturally an absence of communication. Besides

Titus he had expected to hear from Eugene Jolas and from Stuart Gilbert in their editorial capacities, but neither of them wrote. He did hear from Nancy Cunard, who said she had poems by Breton and Eluard, which he immediately offered to translate. The piece on Gide for the *New Statesman* remained unwritten, though he had voluminous quotations from that author in a notebook, dating from the time when he and MacGreevy had considered the possibility of a book on Gide for Chatto and Windus.

He sometimes made forays into Dublin on a bicycle and resumed his acquaintance there, though without any great enthusiasm. He had bumped into Con Leventhal on the boat coming over and when he saw him now he found him pleasant but, he thought, somewhat secretive. Part of Leventhal's secretiveness may have been about Ethna, with whom he was now partly living, but of course Beckett knew they were seeing each other. When he himself saw Ethna, he told MacGreevy, he found he did not have 'the guts to be disinterested'; but he did not pursue matters or try to put things on to any sort of intimate or personal basis. So the renewal of acquaintance lapsed and in October he wrote to MacGreevy to say so and to tell him that he had adopted a Beckettian solution to the problems posed by her existence:

> I never see nor write to nor hear from nor am seen by Ethna McC. now. 'Tis better thus'. I incline to the opinion that when it is impossible to see people simply it is more satisfactory to wait til they turn up in the memory. I can't see her and I can't imagine her. Occasionally it happens that I remember her and then, presto! I had nothing up my sleeve nor she in her amethyst bodice.

This was indeed dependence on Proustian involuntary memory; but few things are more striking about Beckett than his willingness to abandon himself to the life of memory, both in young manhood and later on. Most of the events of life may have been 'occasions of fiasco' as they occurred; but the subsequent remembrance of them was nevertheless more tolerable than present existence could ever be. The vein of simple nostalgia in some of his writings is surprising, but its importance should not be overlooked. However happy or otherwise his childhood might have been, sometimes his childhood memories overwhelmed him.

Partly perhaps because of the memories stirred during his long walks, he found his idleness and solitude pleasant enough. Of course, he considered the possibility of escape, Paris being the preferred option, but without some positive word about his book he failed to see how he could manage it; and as summer passed into autumn the prospect of hibernating in Foxrock became more attractive and the large cold world outside Ireland less so.

Meanwhile, in spite of his poverty, he was making the best of Dublin, culturally at least – such as that best was. He went to Ibsen's *The Wild Duck* at the Abbey and the MacLiammóir-Edwards *Romeo and Juliet* at the Gate, but disliked what he saw in both cases. He played the piano a good deal, conceiving at this period a particular admiration for Mozart. He regaled MacGreevy with Dublin news and gossip – mostly about mutual acquaintances, the painter Sean O'Sullivan, Arland Ussher, Con Leventhal, but also hearsay about Yeats and AE, who played croquet together at Yeats's house in Rathfarnham, AE always winning. He went often to the National Gallery, spending time with the Rubens and the Poussins. He did not, to begin with, seek out Jack Yeats, though he felt he should; but once, having left a piano recital given by Rubinstein at the Theatre Royal, he met the painter, similarly escaping, at the door of the pub known as the Scotch House and had a drink with him and Boss Sinclair's brother Harry, who happened to turn up.

He was also reading a great deal. He admired Fielding's *Joseph Andrews* and *Tom Jones*, but principally the former. He tried to read Mallarmé and failed, but Diderot's reminiscences intrigued him, and he was particularly taken by the way Diderot sometimes observed that a chapter was too short and openly stated that he was going to spin it out. His own narrators would admit to spinning things out later on, giving away the show in similar fashion and frequently observing with satisfaction that a few more pages have been filled, even if with what they describe as rubbish.

In October 1932 Frank had to go down to Galway on a job and offered to take Sam with him. He was entranced by Galway, which he had never seen before: 'a grand little magic grey town full of sensitive stone and bridges and water'. Afterwards they went on to the Achill peninsula, a part of western Mayo where the Irish language was still spoken, and spent a day walking there, 'right out over the Atlantic'. They came back by Lough Mask and through the mountainy stretch between Lough Mask and Lough Corrib which is known as the Joyce country, the Joyces being one of the 'seven tribes' of Galway. From wherever they were, they could see Croagh Patrick, the holy mountain where the saint fasted, standing up with the chapel on its summit wrapped in cloud. He found the trip 'unforgettable', though 'much too short', and lost his sense of alienation from nature, thinking the bog and mountain country they passed through 'somehow much more innocent and obvious than the stealthy secret variety we have here'.

He told MacGreevy that he would like to go back to Galway and spend some time there. Since these are the first observations he ever made about the landscape of any part of Ireland other than that of Dublin and Wicklow, his attitude is interesting. Though he spent much longer in Ireland as an adult than James Joyce did, his forays outside Dublin were almost as few in number.

Back in Dublin he continued to walk 'immeasurably and unrestrainedly, hills and dales all day', coming back down through 'the Homer dusk' after a couple of pints. Often on these walks he would get into mental states which he found rewarding and even exhilarating. But his feeling about the 'calm, secret hostility' of nature had come back and on one of his walks he was so conscious of it that it terrified him. This feeling of the otherness and even the hostility of nature was strong in Beckett. When, a little later, he praised Cézanne to MacGreevy, he said he was the first to see landscape as 'something unapproachably alien', an 'unintelligible arrangement of atoms'. The recurrence of this feeling now was another sign that the present Dublin sojourn was running its usual course. His familiar ailments were once again beginning to beset him, which was always a sign of gathering crisis. 'Pretending to like fresh air and salt water I got the old intercostal rheumatism back, this time on the left side,' he told MacGreevy; and then, with humour which was doubtless meant to conceal a real hypochondria, 'but it is going away and I am denied even the excitement of a little dry pleurisy in safe surroundings.' A more tangible form of ailment was a cyst on his neck which was beginning to be painful.

As winter approached he felt that his situation was becoming more anomalous. 'The fact of myself here in this place and with these people becomes as painful as a stone,' he wrote to MacGreevy. His father wanted him to start teaching again. Failing that he offered to use his influence to get him a job in Guinness's, a good Protestant employer whose clerkships were highly prized and whose senior clerkships were regarded as among the best jobs in Dublin. His mother's disapproval of his idleness and lethargy was becoming more apparent every day. After subsiding for a while, the cyst on his neck had now swollen to painful proportions. 'My parents do their best to be patient but the situation is becoming more remarkable each day for us all,' he wrote on 21 November.

Nothing more than the fact of my presence is required to exasperate my mother in the end – oh the usual old exasperation – so that she can scarcely keep her hands off me. Because I don't go out for long walks, nor work any more in the garden nor do odd jobs about the house nor earn money nor groom the dogs nor prove myself useful in any of the services that they recognize, but only read in the warmest place I can find and write a little and play the piano a little they cannot help feeling that I am somehow behaving very badly and failing to 'pull my weight' and letting myself down and them also. And because I don't eat or talk or laugh as much as they do they must suppose I am unwell or unhappy. Whereas I am just as well and as happy as I ever will be again. And there is no use at all trying to explain myself. Even if I could, not even to father. Because everything I do and neglect doing, say and neglect saying, is an implied rejection of their values, or an indifference, which is more offensive than a rejection. There has been no row and there shall be none, but the ambience will soon be unbreathable and I will have no choice but to go away again. Not that I want to at all. And where could one go? And in this miserable cold.

From what looked like imminent crisis he was rescued by the cyst on his neck, which is perhaps why the cyst was there – which is not to deny that it was a painful reality. The local doctor had advised that he should have what that worthy described as a 'deep-seated cystic system' cleaned up once and for all; and, when he told his parents that this was necessary, arrangements were made for him to go into Merrion Nursing Home in nearby Mount Merrion. There the cyst was removed. Present-day practice would be to lance it and clear up the infection with anti-biotics before actually removing the offending object, but of course there were no anti-biotics then. Somewhat to his satisfaction, Beckett was told that he would have to stay for another week or ten days. His blood, he was told, was in terrible condition and a course of arsenic injections as well as assorted vitamins had been prescribed. That in spite of regular meals at home and regular exercise he was considerably run down is evident. There was not enough muscle for him to have the injections in his arm, so he had to have them in the buttock instead. But for a week or so afterwards he lay in bed in a sunny room, reading and sleeping and, it would seem, occasionally masturbating and feeling both spoiled and victimized at the same time. A comic account of the procedures before operation is given in the story 'Yellow', the second-last in *More Pricks Than Kicks*. In this the hero is terrified of death, but he puts a brave face on it before the nurses, all of whom are described in terms of their comparative sexual attractiveness. In the end, however, he actually does die on the operating table, presumably of a heart attack, since we are told that 'they had clean forgotten to auscultate him'.

The wound was expected to clear fairly rapidly and so, to all appearances, it originally did; but after nearly two weeks it began to suppurate again. Eventually he left the nursing home just before Christmas, still wearing a dressing, and whether the removal had been properly performed or not, he was to be bothered by cystic suppurations on and off until May 1933, when his neck had to be lanced again. Prior to the operation he had also had a very painful corn, which impeded his walking considerably. This had responded to treatment in the nursing home, however, and in the New Year he proclaimed that he felt much better all round, even though his chronic constipation had returned.

With the onset of Christmas his aunt Cissie had invited him to Kassel; and since the habit of going there at that time of year was now strong, he felt the tug of Germany pretty keenly. But he realized that a visit to the Sinclair household was out of the question. Peggy had been coughing and had continuous chest colds. These had been diagnosed as tuberculosis and her condition had now suddenly begun to deteriorate. She needed a room to herself and much nursing. And, as Cissie had revealed on a visit to Dublin, the Sinclairs were now in desperate straits financially, with 'hardly a coat between them'. In late November 1932 the bailiffs invaded the flat and seized everything that could not be shown to have been acquired with Cissie's money. The paintings which Boss had bought and hoped to re-sell were publicly auctioned, fetching very little, while Sam in Dublin bemoaned the fact that one could have had the best of them for a few Reichmarks. The prosperous Beckett element had always disapproved of Boss Sinclair and Cissie's marriage to him, so appeals to them were useless. Those made to Boss's brother Harry Sinclair were almost equally so, since he too was experiencing hard times and in any case had no faith in Boss's abilities as a picture dealer or in the worth of what he had bought.

When he came back from the nursing home, Beckett found that the attitude towards him at home had much improved and the air had been cleared. Before Christmas there had been pressure on him to at least teach French and his mother had gone so far as to arrange classes for neighbouring young ladies. She may have been hoping that he would fall in love with one and that this would result in an onset of ambition. Now he put his cards on the table. He would continue to do odd jobs around the house for pocket-money, but there would be no more French or other teaching. However, he did occasionally give instruction in the piano to advanced pupils, including an old lady whom he rather liked and with whom he played Mozart. Pressure on him to do anything, though, had now virtually ceased and though he remarked to MacGreevy that the atmosphere was 'subtly unbearable', adding characteristically that this was because he felt 'the implied reproach' to be translated into 'an excess of concern and fondling', inertia and comfort made staying where he was seem more attractive than any move he might contemplate, particularly now that some very cold weather had set in, 'too cold to go out the door let alone over the sea'. He felt the humiliation of having no money apart from what he received for occasional odd jobs or was given on a charitable basis, however subtly and discreetly he was given it. He spent most of the cold weather indoors, looking for the warmest place to read. Sometimes he would do a little mild gardening or take the dogs for a walk, but that was all.

Relations with his father were, as ever, more relaxed than those with his mother and he would sometimes go into town with him and be stood a meal or a Turkish bath in Lincoln Place, close to his father's office. There was now a general election in progress, contested principally by the two wings of the old nationalist movement, one, led by W. T. Cosgrave, attractive to Foxrock because it was considerably less nationalist than the other, led by de Valera. De Valera's party, the Republicans of the civil war, had originally refused to enter Dáil Éireann because they would not take the required oath of loyalty to the crown. In 1927, however, Dev had decided that the oath was merely 'an empty formula' and since then they had contested elections. Though they had abjured physical force, they called themselves the 'Republican Party' and their followers translated the programme of constitutional change they offered into the simple slogans 'Up Dev!' and 'Up the Republic!'

It was a violent election, with meetings broken up by strong-arm tactics and fierce passions aroused on both sides. In the Beckett household, as in Foxrock generally, Cosgrave was much admired for his insistence on the oath of loyalty to the British king as head of the Commonwealth and his defence of the British connection. Sam, having decided that he was a de Valera supporter, found this irksome and said so to MacGreevy, under whose influence he had developed mildly nationalist views. In his contribution to the pamphlet 'Authors Take Sides on the Spanish War' four years later he would make witty and ironic use of the simple cry of Dev's followers, translating it into a one-word message with a prior inverted Spanish exclamation mark, '¡Uptherepublic!' Both the pleasures and the pains of political commitment, the degree of release or satisfaction it may offer some people as well as the complex reservations and afterthoughts others may experience, are suggested by Malone's 'Yes, that's what I like about me . . . that I can say, Up the Republic!, for example, or Sweetheart!, for example, without having to wonder if I should not rather have cut my tongue out, or said something else.' This was in *Malone Dies*, written much later. Whatever degree of commitment he felt at this point did not go very deep; and when election day came on 24 January 1933 he ratted and humorously offered to sell his vote for a pound. The offer was taken up by his father: he duly voted for the Cosgrave party; and so it would seem that on the only recorded occasion on which Samuel Beckett exercised the franchise in any country, he accepted a bribe to vote against his inclinations.

By this time the weather had improved, and in March he was a little cheered up by a visit from MacGreevy. His friend, however, was making another foray to Paris from Kerry, where he too had been sojourning at his family home; and after he had gone the thought of their days together in the French capital made Beckett feel his solitude and peculiarly friendless state rather acutely. 'The thought of taking root, like a polypher, in a place is horrible, living on a kind of mucus of conformity. And in this of all places.' He longed to see some of their Paris friends and found himself particularly thinking of Beaufret, but he felt he had squandered all his friendships. On one of his trips into town he saw Ethna McCarthy and Con Leventhal in the street together and returned Leventhal's salute, which was the only one he got, making no attempt to stop and talk. He found the Easter holiday particularly trying, his father and Frank being away together on a little walking tour in Wales. But on Easter Saturday he got out his bicycle and went as far as Malahide in north county Dublin, going round the estuary by the Portrane lunatic asylum and back by Swords, getting quite cheerful as he rode home in the now lengthening twilight. This little jaunt is the basis of 'Fingal', the second story in *More Pricks Than Kicks*, though there he invents a girl to accompany Belacqua on a similar bicycle ride. She bears some resemblances to Ethna and is ditched at Portrane by our hero, who takes off for the city without her while she flirts with one of the doctors. On the Monday he felt he had to make another effort for his mother's sake and they went to the Botanic Gardens in Glasnevin on the north side of the city together. Both felt the strain but a deliberate affability was maintained which got them through the famous orchid houses and along the paths by exotic shrubs and trees.

Titus had published 'Dante and the Lobster' in the December issue of *This Quarter* while Beckett was in the nursing home and this had impelled him to write more stories in the hope of making a full-length book. MacGreevy had urged him to write, but he claimed that there was nothing to write about. In despair over *Dream of Fair to Middling Women* ever finding a publisher, however, he now began to cannibalize it to produce stories. He lifted out some thirty-seven pages, the penultimate section of the book, to make the story of the party at which the

Alba wears the backless red dress, 'A Wet Night'. And once again he took Peggy's letter, which had been transcribed almost verbatim into *Dream of Fair to Middling Women*, giving it separate status as 'The Smeraldina's Billet Doux'. By May he had two more stories, 'Ding Dong' and 'Fingal', written, as he said, 'sans conviction' and 'because one has to do something or perish with ennui'. This gave him five in all, but he felt he 'could not invite a publisher to wipe his arse with less than a dozen', so he continued to hammer them out and by June he had two more, 'Yellow' and 'Draff'. It was, he said, 'all jigsaw', meaning that there was an element of the mechanical in the manufacture of them, and he was not interested. Still, he would continue to manufacture until he felt he had enough for a book.

Partly because they were lifted out of *Dream*, some of the stories describe a life prior to and rather different from the one he was leading now. The reference to the 'rather handsome face' of McCabe, the Malahide murderer, staring from the *Evening Herald* gives a date of 1927 to 'Dante and the Lobster'. The trip to Malahide in 'Fingal', however, is the one Beckett himself had just taken, though without the girl, and Belacqua is about the same age as the twenty-six-year-old author, older than the young doctor. 'Ding Dong' simply describes a pub crawl round the back of Trinity, the sort that Beckett had often taken while he was living there; but in general the later stories, written at this time to fill out the volume, are more inventive than those lifted from *Dream*, the girls particularly being imagined, as are, of course, Belacqua's marriage and death.

While he thought there was some chance of *Dream* appearing he had been somewhat concerned about the use he had made of real people, among them Peggy and Rudmose-Brown. Now he had to make the same choice again; and again he suppressed his misgivings, such as they were, and this in spite of the fact that Peggy's illness now culminated in her death. She had been in the sanatorium at Wildungen near Kassel, but had actually been feeling better, and had been told by her doctor that her condition had improved and that she could return home and lie out in the sun. She and Heiner Stärke had been full of plans for the future after they were married. She died unexpectedly but peacefully on 3 May, with her inconsolable fiancé beside her, in a sleep which had been induced by a sleeping-draught taken after a prolonged fit of coughing. The tone in which Beckett reported this to MacGreevy does not suggest that he was very deeply moved. The letter passes on to other things without any expression of grief and within a paragraph or two he is reporting that after a Sunday outing and a few drinks everything seems 'quite pleasant and pleasantly null'. He encloses a poem about a bicycle ride through north County Dublin, which includes an obscure reference to

> Her who alone in the accusative
> I have ever dismounted to love.

But the context suggests it is Ethna.

On the day Peggy died he had a further operation on his neck with a local anaesthetic. The wound was stitched and seemed all right, but the next day and for two days afterwards pus poured out through the stitches. The stitches were then taken out and the cut began to heal with no further discharges, but he had no confidence that it would not occur again; and the fear of being ill abroad made him more than ever determined to stay where he was, even if he succeeded in placing the stories or the novel and getting some money. At home he was encouraged to drink Guinness in order to build up his strength. 'They' no longer spoke of him getting a job and he was beginning to be indifferent to anything they might feel about his idleness. He decided that 'it was an ill cyst that blows nobody any good'.

'They' included his father, an unwilling partner in his mother's sporadic exhortations, with whom he was still enjoying walks and with whom on these excursions he could happily discuss many subjects. There was one particularly lovely tramp after the fraught Easter recess that he would remember afterwards. Occasionally cursing the terrain and the obstacles, they barged over heathland and stone walls, discussing everything from astronomy to plant life and the techniques of surveying. He thought his father was growing old gracefully and philosophically, and noted his cleverness in stopping to rest under cover of admiring the view.

He had begun to see Leventhal, though without Ethna, occasionally at weekends, when he would ride up the road to Enniskerry to meet him in the Enniskerry Arms. It was not a 'bona fide', as pubs outside the city limits entitled to serve 'travellers' from Dublin on Sunday afternoons were called, but other nearby places in the mountains were, and thither they would repair. He found he liked Leventhal better and better as the months went by; and he enjoyed the ride to Enniskerry and the drinking and the ride home afterwards through The Scalp. It was also quite pleasant to reach home a little drunk and eat a little and go to sleep.

He still ached occasionally to get away and had just decided to ask his father for the money to spend a few months in Spain, when suddenly his life changed. Towards the end of June 1933 Bill Beckett, an active if overweight man of sixty, who still swam regularly, played golf and climbed mountains, had a sudden heart attack. Everybody was very frightened, including Sam himself, but Abrahamson, a well-known Dublin specialist who was called, assured them that with rest and care he would get better.

For a while it seemed that he was right; and from being, in the first few days, a difficult and fretful patient, Bill Beckett himself became a more cheerful and amenable one, joking with the doctors and swearing that when he got better he would never do another stroke of work in his life. He would drive out to Howth Head and lie in the heather and fart. Sam, who nursed him along with his mother, spent much time at his bedside and noted that the sweet-pea which had been cut in the garden was reflected on his face. On the morning of 26 June, before Sam was up, the doctor called. He found, as he thought, the patient much improved and everybody was happy. When Sam went in to him Bill remarked on the beauty of the morning and these were almost the last words he spoke. He suddenly became worse; and after suffering a good deal he died at about four o'clock in the afternoon. His last words were 'fight, fight, fight', almost certainly an injunction to himself rather than anybody at the bedside. As memories of many little things came back, Sam was deeply grieved. He told MacGreevy, 'I can't write about him. I can only walk the fields and climb the ditches after him'; but he would write about him many times. The sense of some ultimate sadness, even injustice in his father's life would never leave him; and he would mourn for that life as well as for his death; but the memory of few fathers has been better served.

Chapter Thirteen

His father's death changed things for Beckett, though on the surface not immediately or dramatically. In the immediate aftermath it was decided that Frank would carry on the business, but he would need to pass his Institute of Surveyor's exams to be fully competent to do so. Until Willie Beckett's affairs were settled it was not clear whether the family would be able to stay on in a house like Cooldrinagh; and Sam's own position was, as he put it, vaguer than ever.

But he was immediately brought into a new relationship with his mother. They were both very strong-willed people, even if Sam's will was, as is often the case with creative artists, working itself out through negatives and refusals. Now they no longer had a buffer state between them to prevent conflict, which was a role that tolerant, shrewd and affable Willie Beckett had been glad to play. For the moment Sam was busy answering letters on his mother's behalf and generally looking after her, while poor Frank struggled to familiarize himself with the affairs of the office and sort out the estate.

But their mother was already proving very difficult. She had decided that full, traditional, high Victorian mourning must be observed and she created an atmosphere of gloom in the house which Sam found depressing and irksome. Even his piano playing had to be suspended as not sufficiently solemn, though an exception was made for the setting of Tennyson's 'Crossing the Bar' which his grandmother Beckett had made and which – or so he claimed anyway – he played nightly before dinner. The blinds were kept drawn, everybody was expected to maintain a solemn demeanour and dinner was eaten in an atmosphere of gloomy silence.

Faced with this daily regimen of grief Sam made an arrangement with his brother to have the attic room of the offices in Clare Street, to which he might repair on a daily basis. Here, under the roof, he was supposed to do his writing, and language pupils were supposed to climb up there by the final little railed staircase to receive lessons. In fact he only had one pupil, the daughter of a schoolteacher, and he travelled over to Clanbrassil Street to instruct her for an hour every morning. He got five shillings for doing this, which allowed him to have a few drinks in the pubs around town. The office was near the centre of things, near the back gate of Trinity College and near the Grafton Street area pubs where literary men (and the very occasional woman) drank, pubs which he often affected to loathe, but where he often enough found himself just the same.

Among those he met at this time were two young poets whose names have been sometimes associated with his and MacGreevy's by Irish critics, partly because of a piece he would write (under the pseudonym Andrew Belis, the maiden name of his maternal grandmother, slightly misspelled) in the *Bookman* the following year. In that piece the poets of the Irish literary revival, or what remained of them, were attacked as 'antiquarians' and 'Mr Denis Devlin and Mr Brian Coffey' were described as 'without question the most interesting of the youngest generation of Irish poets' – MacGreevy had already been praised – partly because of their lineage. They had, the article said, submitted themselves to the influences of Corbière, Rimbaud, Laforgue, 'the *surréalistes* and Mr Eliot, perhaps also to those of Mr Pound'. In placing them in this way, Beckett was claiming that they were in what was then supposed to be the central line of development of modern poetry – interestingly enough, leaving out Rimbaud and the surrealists, the Eliot line. Thus placed, they were among those who 'evince awareness of the new thing that has happened' and were to be contrasted with those who were not aware, 'or in whom the velleity of becoming so was suppressed as a nuisance at its inception'. In Ireland, these were 'the antiquarians, delivering with the altitudinous complacency of the Victorian Gael the Ossianic goods'. The contrast between one kind of poet and the other was not, however, 'peculiar to Ireland or anywhere else. The issue between the conventional and the actual never lapses, not even when the conventional and the actual are most congruent. But it is especially acute in Ireland, thanks to the technique of our leading twilighters.'

The main characteristic of these 'twilighters' was the fact that they were 'in flight from self-awareness'. At the centre of their poems there was no self and therefore no theme. 'Why not? Because the centre is simply not that kind of girl, and no more about it.' But the circumference was 'an iridescence of themes', not only the standard props of the Irish literary revival – 'Oisin, Cuchulainn, Maeve, Tir-nanog, the Tain

Bo Cuailgne, Yoga, the Crone of Beare – segment after segment of cut and dried sanctity and loveliness' – but other 'presentable' and 'accredited' subjects. They believed in other words that 'the first condition of any poem is an accredited theme, and that in self-perception there is no theme, but at best sufficient *vis in tergo* to land the practitioner into the correct scenery, where the self is either most happily obliterated or else so improved and enlarged that it can be mistaken for part of the décor.' Included in this classification was a wide array of poets, Yeats, Colum, James Stephens, George Russell, 'the rev Monk Gibbon' and 'Mr Austin Clarke'. Exempted were those 'kindly noticed' by W. B. Yeats as 'the fish that lie gasping on the shore'. MacGreevy was 'best described as an independent, occupying a position half-way between the above and the poor fish, in the sense that he neither excludes self-perception from his work, nor postulates the object as inaccessible.'

He had a little trouble fitting MacGreevy into his preferred category, for after all MacGreevy was an intensely patriotic man who drew a good deal of poetic sustenance from some central Irish reality which he (convincingly) glimpsed; but he does well enough and he rather surprisingly describes him as an 'existentialist', declaring roundly that his work constituted 'probably the most important contribution to post-war Irish poetry'. It would be nice to claim that this was the first introduction of the word existentialist into English usage. It was just creeping into use in French and Beckett may have known it through Beaufret, who was the first French philosopher to interest himself in Heidegger. Alexandre Kojève's influential lectures began in 1933.

It is quite a manifesto, this survey of the Irish poetic scene in 1934; and much of it is still relevant, for there are still those to whom self-perception is 'at best sufficient *vis in tergo* to land the practitioner into the correct scenery, where the self will be obliterated, or so improved and enlarged as to become part of the décor'; but of course like all other manifesto makers, he needed to suggest that there was some sort of movement afoot and he adopts Devlin and Coffey for this purpose, declaring that Devlin's work at least 'does not proceed from the *Gossoons Wunderborn* of that Irish Romantic . . . combination, Sir Samuel Ferguson and Standish O'Grady, and that it admits – stupendous innovation – the existence of the author.' Ferguson and O'Grady were of course the two forerunners of the Irish literary revival. To some extent he was right to claim that there was a sort of movement afoot and that it included himself, MacGreevy, Coffey and Devlin. It was one of the many false starts of modernism in Irish poetry.

At this point Coffey and Devlin were both just back from Paris and had been urged by MacGreevy to look Beckett up. In fact he did not greatly care for either of them, perhaps because at this point he cared for hardly anybody. 'I'm afraid I didn't get much kick out of Coffey and Devlin, their pockets full of calm precious poems,' he wrote. 'It was pleasant to hear the Paris news, what films were on and the latest 10% of surrealism,' he went on, saying he had showed them a poem of his own, 'Enueg', which 'Coffey seemed to find delighting, amusing, delighting'; but it was a sticky meeting. A visit to the National Gallery was a '*grosse erreur*', and, though they had a drink afterwards, 'I'm a cranky man and I don't like anyone' was his summing up. In a subsequent letter he back-tracked a little. 'If I said I didn't like Devlin I made a mistake. I did not take kindly to either of them but I preferred Devlin to Coffee,' he declared, continuing to misspell the latter's name; but the truth is that he was persisting in disliking almost everyone with the exception of Leventhal, even though his manner when he met them seemed to give an opposite impression.

Devlin and Coffey were new acquaintances, but on the walk to or from his class in Clanbrassil Street he might meet a great many people he knew and he would often have a seemingly affable drink or two when he met them. Francis Stuart remembered him in the old back room of Davy Byrne's around this time, 'rather detached, rather isolated' from the company, but remaining in it or on its fringes. The isolation of the past year in Foxrock, in part enforced by poverty, in part by lethargy, was coming to an end. But he did go to his little half-attic to write and he kept himself at it, so that by September he had ten stories, a 60,000-word book, to send to Prentice at Chatto and Windus. In spite of the 'tippling' and 'stravaging' around town of which he spoke to MacGreevy, it was a comparatively structured existence, which satisfied to some extent his latent sense of order; and it was certainly a relief to get out of the house.

Nevertheless at this time of sorrow and stress, ailments, or fancied ailments, continued to afflict him. He was troubled again by what he thought were palpitations and a racing sensation in his heart; and, more substantively, a painful swelling had appeared on the palm of his hand. Abscess or cyst, this failed to respond to lancing. However, when his cousin Mollie Roe arrived from England a good deal of the pressure to look after his mother was lifted. He did not like Mollie, who, he said, giggled and stuttered. Some of his comments, particularly on the shortness of her skirts, worn, he said, 'on the principle of diverting attention from her face to her legs', suggest that he was sexually disturbed by her, but her presence was useful, even though he had to be religiously home by seven o'clock every evening to partake of the solemn and silent evening meal. Another slightly disturbing advent was that of

Nuala Costello, who had been on the periphery of the Joyce circle and whom he had met at Helen Fleischman's. She had been engaged then; now she was '*disponible*', and this made him uneasy. Shortly after his father's death Cissie Sinclair had arrived with Peggy's younger sister Nancy. They stayed with Rudmose-Brown to begin with; but soon Boss and the rest of the family joined them. The German period of their lives had come to an end, partly because of Boss's failure to earn a living there but also because of the political situation. They had no money, but managed to rent a cottage in Howth, where they had lived in happier days, up to the early 1920s. They had little enough contact with any of the Becketts, but there is no foundation for the suggestion that Sam was 'not received' there at this point. A coolness would come later, resulting from his publication of the Smeraldina's letter in *More Pricks Than Kicks*.

The Beckett family's own finances were now in the process of being sorted out. Willie Beckett had died worth approximately £35,000 pounds, a considerable sum. He had made a will in March 1923, in which everything was left to May for her lifetime, 'she maintaining and educating my two sons until they respectively shall be in a position to support themselves but no longer than their respectively attaining the age of twenty-five years'. After May's death Sam and Frank were to inherit on an equal basis; but provision was made for the trustees – with May's consent in writing – to raise 'any part or parts not exceeding in the whole one half of their presumptive or vested share of either of my sons and apply the same for his advancement or benefit'. There were four executors to the will – Willie's brother Gerald, May, Frank and Sam; and the fact that Sam resigned as executor has given rise to the suggestion that he was a beneficiary under the clause allowing May to give him part of his share immediately. In fact the arrangement arrived at was that May would make him an allowance of £200 a year, but this would be entirely discretionary and voluntary. The arrangement was not formalized for some time, nor the sum decided on; and for a long time afterwards he could not escape the feeling that he was living on his mother's charity.

Still, £200 a year between the wars was not a bad sum to have, and would have been accounted a very good wage or even a reasonably good salary. When his near contemporary, Brian O'Nolan (Flann O'Brien), joined the civil service two years later as a junior administrative officer, which was the cadet rank for those who would be promoted to the higher echelons and by no means the lowest grade, his salary to begin with was £180 pounds a year, plus £72 pounds a year cost of living bonus, and out of this he had to maintain his widowed mother and eleven siblings. Beckett now had almost as much money, more or less in his own right, as a young civil servant who had his foot on the first rung of the ladder of middle-class success; and though of course its value in real terms would decrease in time, the process would not begin to be noticeable until after the war, when he was not too far from financial success as a writer.

And just at this point he was about to take what he might have convinced May and others was a step forward as a professional writer, for in October 1933, scarcely more than a month after receiving them, Prentice wrote to say that Chatto and Windus were prepared to publish the short stories and were happy to offer an advance on royalties of £25. He was surprised because he had thought the fact that, with the exception of 'Dante and the Lobster', the stories had not even been published in magazine form would militate against them, though he did remark that he supposed that if people could read Saki they could read anything. In September, before receiving this news, Sam had been knocked off his bicycle by a car, suffering injury to his right arm and hip. It is difficult to assess the real gravity of these injuries, for, as usual, he made the most of them, but nothing had been broken; and by October, when his mother decided she wanted to leave Cooldrinagh with its many memories and take a cottage at the seaside for a month, he was able to offer some assistance with the loading of mattresses – including the dogs' mattresses – on to a lorry which had been hired with driver for the move.

The cottage chosen was on the coast just south of Dalkey – a pleasant little township a few miles outside Dublin. There mother and son remained until November. Beautiful though the south Dublin littoral is supposed to be, Beckett did not like living there, feeling the same sense of alienation from the sea as he had often done from the mountains. He wondered how people had the nerve to live by the sea and said it moaned in his dreams at night. He was consequently highly relieved to get back to Foxrock.

He badly wanted to get abroad and, since he was now an author on the verge of publication there, London seemed attractive. 'It's a frail life and I want to shiver mine elsewhere,' he told MacGreevy. Its attractions were increased by the fact that his friend was already there. MacGreevy was free-lancing, lecturing in the National Gallery, translating novels from the French for English publishers – among them Montherlant's *Les jeunes Filles* – contributing to art magazines such as the *Connoisseur* and the *Studio* and reviewing for the *Times Literary Supplement*. He was living in the house of a certain Hester Travers-Smith at 15 Cheyne Gardens, a loosely structured establishment where there was much coming and going and which would provide at least a jumping-off place for Beckett. Hester Travers-Smith was a Dublin type, the sort of elderly, semi-impoverished, cultured Anglo-Irish lady who keeps a digs, and she had already kept such an establishment in Dublin at which the younger MacGreevy and the very young Francis Stuart had stayed. With a view to joining his friend in London in October Beckett had applied for a job at £300 per year in London's National Gallery, giving Prentice and Jack Yeats as referees, but he did not even get called for interview.

The possibility of getting a job as a copy-writer in an advertising firm in London had also been much in his mind for some time. He admitted that he knew very little about such jobs, but thought he might find it entertaining. 'I don't know a damn thing about it but it has been in my mind for a long time and I have often been on the point of putting it up to Father. Now I can put it up to Mother,' he said, adding sarcastically, 'There is always someone to whom one can put it up.' And then regretfully, 'If there were only always someone in whom . . .' meaning, presumably, someone in whom one could confide.

The idea of Samuel Beckett as London advertising copy-writer is an intriguing one, more especially since copy-writing there has over the years been the grave of more literary hopes than even the BBC used to be. That he entertained the wish is again not in keeping with the character of the purist and uncompromising avant-garde writer which is a common image of Beckett. No doubt we are lucky that he did not succeed in finding such a job, but it is salutary to remind ourselves that he wanted one.

In November mother and son moved back to the gloom, the drawn curtains and enforced mourning of Foxrock. The change had done May good, though she prayed a good deal and wept in private; but as Sam feared, the return to more familiar surroundings was accompanied by a revival of her unassuagable grief. They made at the moment a fine pair. May was still resolutely inconsolable, they were both depressed, both suffered from insomnia and his middle-of-the-night 'heart attacks' were getting worse. Almost every night now he woke in fear, his heart pounding at a terrible rate and his panic increasing as he stared into the darkness. Sometimes there would be a feeling of imminent suffocation, such as he had experienced in his lecturing days. These symptoms fed on themselves. If he dozed off and partly woke thinking of the possibility of such an attack, he would be likely to have one. The resultant fear would cause an increase in the heart rate and this would be cumulative, until eventually the organ seemed to be racing madly and its thumping could even be heard. Since the difficulty of getting any full or refreshing sleep had been in any case compounded by an acute return of the nightmares – often themselves of suffocating or drowning – from which he had always had a tendency to suffer, he was sooner or later awake every night with a racing heart and a feeling of terror.

The only person apart from MacGreevy in whom he felt he could confide about these attacks was his oldest friend, Geoffrey Thompson, whom he had known since his Portora days. Thompson had now begun to practise psychological medicine and he sensibly decided that these symptoms were in fact psychological in origin, the product of unresolved conflicts – in other words neuroses – leading to an acute anxiety state accompanied by depression. Since like most young men entering psychological medicine at the time he was an enthusiast for analysis, that was what he recommended; in fact he went further and recommended an analyst. The practitioner in question was in London, but this was no obstacle as far as Sam was concerned, for in any case he desperately wanted to get away; and if in some respects London was not as good an option as Paris, in others it was better, for he had a book coming out there and he might get work on the strength of it. It remained to convince his mother, who would have to pay good money for the analyst's fees, as well as for Sam's expenses in London. Analysis was a novelty in both Britain and Ireland, but especially in Ireland, where it was almost unknown in practice if not unheard of. May therefore took some convincing, but it was now clear to her that there was a serious problem and she eventually consented.

On 20 January 1934 Sam crossed over with Frank, who had business reasons for visiting the English capital. Shortly afterwards he found a room for himself at 48 Paultons Square, which is just off the King's Road in Chelsea. It was not the most perfect of abodes, but one of its attractions was its proximity to Cheyne Gardens where MacGreevy was staying, where there was a grand piano and the buzz of company.

The psychiatrist to whom Thompson sent Beckett was himself a remarkable man, destined to leave a mark on psychiatry of an original and controversial nature. Nine years older than Beckett, but then in his first year in psychiatric practice, Wilfred Rupert Bion had come to his true vocation late and by a circuitous route. Born in India, the son of a distinguished civil engineer, he had, like many sons and daughters of Indian civil servants, been packed off to England on his own at a tender age. After education at Bishop's Stortford School he had become a tank officer in the First World War, winning the DSO at Cambrai, the first great battle in which tanks were fully engaged, and serving with bravery and distinction wherever that weapon was afterwards deployed on the Western Front.

When the war ended Bion read modern history at Cambridge and then returned to his old school as a French teacher. Like Beckett, he was good at games and played rugby for Harlequins as well as re-organizing the rugby team at the school and coaching the swimmers. He was a physical fitness enthusiast and a bit of a disciplinarian: old boys at Bishop's Stortford would remember Bion getting them up at six in the morning to practise diving and swimming.

About ten years before his analysis of Beckett began he had left Bishop's Stortford to study medicine at University College Hospital, London, gradually becoming interested in psychiatry and particularly fascinated by the psychology of groups. About two years before, he had abandoned ordinary medicine to practise psychiatry. At the time his sessions with Beckett began, Bion was himself in analysis with John Rickman, who would be succeeded as his analyst by Melanie Klein. He would not formally qualify as an analyst according to the extraordinarily stringent rules of the British Institute for Psycho-Analysis until 1947; by then he would be Chairman of the Executive Committee of the Tavistock and the author of a celebrated series of papers which had been collected as *Experiences in Groups*. But his encounter with Beckett took place early in his career, before he had much psychoanalytic experience either as a patient or as an analyst.

With his military bearing and his somewhat caustic manner Bion was a formidable fellow. According to one writer, all those who met him could testify to 'the enormity of his vitality, to the power of his presence, to the humor of his dry wit, to the incisiveness of his thinking, and to the (sadly) anachronistic gentleness of his politeness and respect for people'.

He had much in common with Beckett: both were tall and spare; both were strictly brought up Protestants with a minor public school background; both were natural athletes, and both were former teachers of French. They were separated by Bion's war experiences, his greater experience of life at the time they met, and by the authority conferred on him by his position as analyst. In the context of the life and work of Samuel Beckett at least one passage from one of the autobiographical works Bion would later write seems worth quoting. It describes the condition of a fellow tank officer after an engagement:

He has lost both eyes, his right arm and both legs. He didn't know who I was though the nurse told him. He's simple – just has a silly grin. The nurse told me afterwards that every now and then he becomes terrified, cowers down in a corner of the room and sucks his thumb. Once he told his doctor that at these times he could see a patch of lawn open up, his mother rise out of her grave and walk slowly towards him. Otherwise he told no one – just went into a corner, scuttling on his stumps with astonishing speed, stuffed his left thumb into his mouth and waited trembling. When the fit was over he would go back to his silly giggling.

Bion's principal legacy to psychoanalysis is still in the area of group therapy and the use of free association within groups. More controversial is his famous 'Grid System', an attempt to schematize symptoms and place them on a grid, which inevitably reminds one of some of Beckett's comic schematizations. In his mature thinking there was, as one would expect from someone associated with Melanie Klein, a heavy emphasis on the failure or otherwise of the patient's earliest associations with his or her mother. He was also interested in what he called 'attacks on linking', writing of patients who attack the analyst for the act of interpretation. Since this act implies a link, a connection between two thoughts, it could lead to the analyst implying a link or connection between two people. As Bennett Simon, who contributed a study of Bion's analysis of Beckett to the *International Journal of Psychiatry*, puts it, both Bion's patients and Beckett's characters reject the idea of a 'link' with others and, to quote Simon's rather curious phrase, seem 'invested in having no union or sterile union'.

Unfortunately those professional psychiatrists who have written of Beckett's analysis have given rather over-lurid accounts of his mother, based on such biographical facts as were available. Typical are references to May 'screaming' at her children and 'recurrent unpredictable rages'. During these years Sam was shouted at on occasion, but none of May's behaviour justifies Didier Anzieu's account of her as 'a foul-mouthed widow and bigoted Protestant with whom Sam had never been able to communicate'.

Nor are the attempts that have been made to identify Beckett with the patients in any of the individual case reports that Bion wrote very rewarding. Of course, it is customary for the psychiatrist to alter the circumstances of the patient in these reports in order conceal his or her identity, but surely not so much as to render the whole exercise meaningless. In one of the favourites among those who wish to identify Beckett, the case of 'The Imaginary Twin', the patient has had many years of psychotherapy, had a 'shocking family history' and a mother who died after 'a complicated and painful illness' when he was seventeen.

Inappropriate though these descriptions of May Beckett may be, Bion nevertheless did concentrate on her son's unresolved relationship with her as the source of his 'narcissistic regression and depressive episodes'. Beckett had three sessions a week with Bion for nearly two years, with some short breaks. He wondered when they came to an end whether they had done him any good. His relationship with his mother was as strong and as far from resolution as ever, he remarked ruefully, and the night attacks had returned.

But London was not all analysis by any means. Urged on by the indefatigable MacGreevy, he engaged again in a desultory quest for literary employment. It resulted in a few book reviews, two that spring in the *Spectator*, of which, as we have seen, an Irishman – Derek Verschoyle – was literary editor, and one in T. S. Eliot's *Criterion*. The first *Spectator* review was of 'Mozart on the Way to Prague' by Edward Mörike, which was dismissed in short order. It is the sort of short, self-cancelling review which highly original writers tend to write before they learn, if they ever do learn, the sad art of spinning it out.

The other *Spectator* review was of a book about Proust by Professor Albert Feuillerat. It is notable principally for its emphasis on a conflict between 'intervention and quietism' as being one of the mainsprings of Proust's work. Whether this is the case or not, it would certainly be one of the mainsprings of Beckett's later on; or rather, in his case, there is, to use a boxing phrase, no contest. Quietism wins hands down. Otherwise it is suggested that the professor's book was an expression of regret that Proust had not superimposed on his work what Beckett in his own little book had called 'the vulgarities of a plausible concatenation'. What the professor, he suggests, would have preferred to find in Proust would have been 'the sweet reasonableness of plane psychology à la Balzac ... the narrational trajectory that is more like a respectable parabola and less like the chart of an ague', and Time 'proclaiming its day of the month and state of its weather' to 'elapse in an orderly manner'. All of which remarks also have a bearing on his own later practice. Samuel Beckett was a slow developer; and he was at this point, in the parlance of today, a very mixed-up person; but in matters literary at least, he already knew what he wanted, or at least what he did not want.

At MacGreevy's urging and through his agency he also met T. S. Eliot in his small office at Faber and Faber, which was in Bloomsbury. The result was a review which appeared in the *Criterion*, the austere quarterly which Eliot edited, in July 1934. Not much record survives of this brief meeting between two people who were both, in the main, polite and reserved; but if Eliot was, as always, polite, he was also kind and the young Irishman departed with J. B. Leishmann's translations of Rilke under his arm. His review, when it appeared, being neither polite

nor kind, may or may not have pleased Eliot, though that scorner of
vague beliefs might have agreed when he said of Rilke: 'He has the
fidgets, a disorder which may well give rise, as it did . . . on occasion,
to poetry of a high order. But why call the fidgets God, Ego, Orpheus
and the rest. This is a childishness to which German writers seem
especially prone.' The fact that Beckett never reviewed for the *Criterion*
again proves nothing either way. He was the opposite of pushy and a
little pushiness is usually needed to maintain connections, even with
such editors as T. S. Eliot.

In Paris he was also being published. Nancy Cunard's anthology
Negro, inspired by her general interest in black people and her particular
interest in Henry Crowder, appeared in January 1934, containing a
considerable number of translations by Beckett, both prose and verse,
mostly from the French. The following month his convoluted, acrostic
tribute to Joyce, 'Home Olga', appeared in the American miscellany
Contempo, for which Stuart Gilbert had been European editor.

> J might be made to sit up for a jade of hope (and exile
>   don't you know)
> And Jesus and Jesuits juggernauted in the
>   haemorrhoidal isle,
> Modo et forma anal maiden, giggling to death in
>   stomacho.
> E for the erythrite of love and silence and the
>   sweet noo style,
> Swoops and loops of love and silence in the eye of the
>   sun and view of the mew

And so it goes on. Lawrence Harvey says it illustrates Beckett's 'intel-
lectualism', and so indeed it does. It is the pedantic referential side of
him at its worst. The general gist is that Joyce is the liberator of Ireland
from sexual repression and Catholicism, which no doubt he partly was.
Shown Beckett's effort by Stuart Gilbert, Joyce commented that it was
'all right', but noted that when he in turn showed it to Eugene and
Maria Jolas the former thought it 'acid and not funny' and the latter
thought it 'poor'.

But the most important publishing event as far as Beckett was concerned was the appearance on 24 May of *More Pricks Than Kicks* from
Chatto and Windus in an edition of 1,500 copies. The blurb spoke of
its 'rare humour', the sort of humour which, it claimed, was 'the last
weapon against despair', a statement which either augurs an unusual
degree of perception on Prentice's part or suggests the hand of the

author. For a first book, and a collection of short stories at that, it was reviewed widely, with a good deal of perspicacity and by no means unfavourably. The novelist Gerald Gould, who was the resident fiction reviewer for the *Observer*, described its manner as 'dry, harsh . . . not untouched by beauty, though betrayed by an artificial whimsicality and unnecessary obscurity', comments which seem just enough today. Since Joyce was much in the literary eye at the time, he was mentioned by several reviewers. Peter Quennell in the *New Statesman* invoked Joyce to dismiss the book as mere pastiche; but Arthur Calder-Marshall in the *Spectator* thought that Ronald Firbank rather than Joyce was the main inspiration for Beckett's prose and, a professional humorist himself, expressed the view that the author was 'capable of coming into the open as a humorist, instead of retiring as he too often does into the allusive shelter of "the really cultivated man".'

A much more substantial and more favourable review was that by the poet, translator and critic Edwin Muir in the *Listener*. Remarking that *More Pricks Than Kicks* was 'a book very difficult to describe', he went on to say that it consisted of a number of what might be called short stories about a young Dublin man. The incidents did not matter, the point being in the style of presentation, which was 'witty, extravagant and excessive'. Mr Beckett, he said, 'makes a great deal of everything; that is his art'; and though he had the 'particularity of both writers', so tnat the tasting of a slice of bread or the purchase and cooking of a lobster could become 'matters of intellectual interest and importance to him', his master was Sterne rather than Joyce. The 'divagations' of which the book was composed were 'in reality an exploration of a subtle and entertaining mind which is carried out with great wit and is very much worth following'.

Even longer, if not quite as enthusiastic, was a 450-word review in *The Times Literary Supplement*. Here Joyce again bulked large and was described as 'a dangerous model'. Aware that some of 'Draff' had already appeared in *Transition* (sic), the reviewer seemed to think Beckett was struggling free of the influence, however, and discerned 'a definite fresh talent at work'. Concluding that his humour, 'with its curious blend of colloquialism, coarseness and sophistication, is unlikely to appeal to a large audience', a remark which proved accurate since eventually about 1,000 copies were left on Messrs Chatto and Windus's hands. But on the whole the reception of the book should have proved gratifying enough; and if Beckett had been the sort of young author destined to have an orthodox literary career he would have been entitled to look back on the reception of his book as the start of it.

Whatever psychoanalysis is supposed to do for people and whatever it did for Beckett in a deeper, more integral way, he certainly thought the immediate effects were good and it did not seem to occur to him that the improvements in his way of relating to other people might have had something to do with the publication of his second book and its on the whole favourable reception. Psychoanalysis, he thought, made him humbler, more tolerant, more considerate of other people. When he returned to Dublin in August he noticed the change in himself. Mollie Roe was still there with her two children, though they were about to depart, and his mother was irritable, exhausted and demanding. But he found it somehow simpler to cope with her moods than he had done in the past. On the day after his arrival he rather uncharacteristically walked over the fields hand in hand with the two children. He was trying hard to find people pleasant and seemed to think that his success in doing so would gratify his analyst. In spite of his mother's best efforts to produce a response by demonstrating her grief, he thought he had grown indifferent to outbursts of emotion, but her wish to grieve had not abated. The house was as much a mausoleum as ever, even the children being discouraged from showing any gaiety, but if these things had not changed, there was a change in him. He was content to take her as she came and ignore her if necessary.

By now he had had about sixty sessions, going to Bion three times a week. The physical symptoms which had originally driven him to seek some sort of aid through Geoffrey Thompson had improved; but how suggestible he was is shown by the fact that he began to imagine that he had gallstones. Geoffrey himself had had an extremely painful attack, and now Sam seemingly succeeded in inducing a similar pain in the same area. Pamping revealed nothing; neither did X-rays; and so he was eventually obliged to accept that his imagination had got the better of him and that his panic had been 'psycho-neurotic', which made him in a hurry to get back and get on with the analysis.

His mother now talked of letting Cooldrinagh and moving to London. Surprisingly perhaps, but certainly awkwardly for those commentators who persist in believing that at this stage Beckett had identified his mother as the source of all his woes, or that Bion had identified her proximity as dangerous to him, he thought this would be a good thing. It would be especially good, he considered, for Frank, who often had a sad time at home and could in the new circumstances take a flat nearer town. He had found Frank in good form but thought that his circumstances induced a certain desperation.

He made overtures to Cissie and her family, who had been estranged by the publication of More Pricks Than Kicks so soon after Peggy's death. The result was that Boss arrived on a bicycle with an invitation to

Howth which he accepted in spite of his mother's stern disapproval. Nothing he could say would convince her that Boss was not an unpleasant person whom the Becketts were unfortunate enough to have as a relative, and with whom her son should not associate. When he went out to Howth Cissie complained about the use made of Peggy's letter, but kept her complaints to a minimum and amity was soon restored.

Besides the Sinclairs he saw Leventhal and Ethna McCarthy, now an established couple in Dublin, in spite of Leventhal's married state. When Sam met them, he was less than sympathetic to the atmosphere of domestic bliss that they exuded, noting sarcastically that their relationship had now become very cosy, as cosy as pie. He ran into Devlin in Nassau Street and they walked and talked for a while. When Devlin subsequently rang the house he was out but he decided anyway that he did not much want to see him again. He went to the Abbey to see Yeats's two latest plays *Resurrection* and *The King of the Great Clock Tower*, but neither made much impression on him. Thinking over the difference between Dublin and London he made an interesting remark to MacGreevy: 'Dublin costumes one's impatience, London one's patience. Which is worse?' In other words, presumably, as far as he was concerned the human condition was the same in both cities but in Dublin one was impatient that things should be otherwise while in London one was resigned.

However that may be, there is little real impatience in the piece he was now reluctantly grinding out for the *Bookman* on 'Censorship in the Saorstat', Saorstat Éireann being simply the official Irish name for the Free State. This, for which he had been paid four and a half guineas, was to hang over him for more than a year, being finally completed only in late 1935, its slow progress certainly not suggesting any white heat of indignation. After describing the provisions of the act and its definitions of indecency and other matters ('the word "indecent" shall be construed as including suggestive of, or inciting to sexual immorality or unnatural vice or likely in any other similar way to corrupt or deprave'), he noted accurately that the 'essence of the act and its exciting cause were the prohibition of publications advocating the use of contraceptives'. The article makes no concession to the idea that any form of censorship might be needed to combat more extreme forms of pornography. Indeed he scarcely enters this philosophic territory at all. The tone is, as might be expected, one of unrelieved sarcasm, embellished with quotations from speeches made in the Dáil and Senate during the passage of the act. Reference to the banning of *More Pricks Than Kicks* during the previous year is rather coyly made, when, after listing some of the eminent authors of the 618 books which had by then been banned, he remarks, 'My own registered number is 465, number four hundred and sixty-five if I may presume to say so.'

The article took him so long to finish that it never appeared in the *Bookman* since, before it was completed and dispatched, that journal had ceased publication. In late 1935 he asked George Reavey, who was in Paris, to try it on *transition*, but though Jolas had said he wanted some prose from Beckett he was not interested in this kind of piece.

Nothing in the end came of his mother's suggestion that they might both move to London, and when he returned there himself at the end of August 1934 he promptly found himself new lodgings. He had never been happy – or, since happiness was not a state he ever admitted to, perhaps one should say particularly content – in Paultons Square with MacGreevy. The place he now found was in Gertrude Street, which is a little further along the King's Road towards Edith Grove and Redcliffe Gardens, territory in bed-sitter land known to many Irish immigrants. Describing this location afterwards he would often say that he lived at the World's End and, in so far as that somewhat amorphous location has any existence other than being the name of a pub, so indeed he did.

Number 34 Gertrude Street was owned by a Mr and Mrs Frost, who, like many proprietors of boarding houses or houses with rooms to let, were retired servants, he having been a chauffeur and she a ladies' maid. Mrs Frost was called Queeney and she came from Ireland, from Athlone. He thought she was like a mother on whom you could draw as a barman pulled beer – you pulled on the lever and tea, Sanatogen, hot water and various other manifestations of mothering came out. She was not at all dismayed when he presented her with some Lapsang Souchong, saying he preferred it to the Liptons tea she normally made.

He also liked Mr Frost and Fred Frost junior, a dental mechanic who had a great deal of mechanical aptitude and fixed him up a reading lamp. He ate most of his meals in the kitchen, which he found preferable to cooking on a gas ring in his room, and while he found the Frost's company reassuring, they soon grew used to his silences. At night he could hear Mr Frost snoring next door. Upstairs were a young couple, he a waiter in the Cadogan Hotel and she a maid in an aristocratic house in Hans Crescent. They made nocturnal noises too, occasionally waking up for a bit of quick love-making. He did not object to these evidences of human presence, finding them indeed, like the time he spent in the kitchen, a defence against panic, which might or might not cast doubt on the relevance of Bion's anti-links theory to him. His own room was large, with linoleum a bit like a Braque seen from far away and plenty of room to pace up and down while he was writing. To add to the human and circumstantial advantages of 34 Gertrude Street there was a piano in the front drawing room. Some notes on this were silent, but he liked to play it all the same.

Settled in Gertrude Street in October 1934, he resumed his sessions with Bion, though he had to put off a couple of sessions because of a stye in his eye and was also suffering from what he thought was some form of impetigo on his lip. He had begun to have hopes of it all going a bit faster and even of getting it over by Christmas, but this did not accord with Bion's ideas of what was necessary; and as soon as the suggestion that the analysis could shortly be terminated was made, it was firmly rejected. Beckett, however, was now suffering the classical symptom of hostility to the analyst and when he went back to Dublin for the Christmas of 1934 this hostility was very strong. It was aggravated by a remark that Frank made to the effect that their mother was exceeding her income and that this was principally due to the cost of Sam's analysis. He now began to listen to people who told him that psychoanalysts were only in it for the money and told MacGreevy, 'every day I hear of unfortunates bled white by psychologists when there is nothing principally wrong with them'.

Nor had he much reason to believe that the analysis was doing him any good, in the matter of physical symptoms anyway. He was again suffering from what he thought was a heart complaint, having various attacks and spasms of palpitation. His heart was behaving like that of the hero of the novel he would shortly begin, *Murphy*, 'one moment in such labour that it seemed on the point of seizing, the next in such ebullition that it seemed on the point of bursting'. The worst 'attack' occurred when he went to visit the Sinclairs at Howth and caused panic in the Sinclair household. Naturally his mother was both disappointed and impatient with the results of his analysis. Naturally too, she spoke the language of ordinary resolution and moral endeavour, telling him he should pull himself together, stand on his own two feet and be a man, etc., all of which, again naturally, he found to be both irrelevant and distracting.

He spent a lot of time playing the piano and walking with his mother's Kerry Blue terrier bitch. One of these walks, over the fields from Foxrock, across the little mountains known as the 'Three Rock and Two Rock' and back again by the Glencullen Road, he described to MacGreevy: 'It was so still that from the top of *Two Rock* I could hear a solitary accordion playing down near the Glencullen river, miles away. I thought of a Christmas morning not long ago standing at the back of the Scalp with father, hearing singing from the Glencullen chapel.' On the way back he noted the pink and green sunset that one found nowhere else and when it grew quite dark he found a little pub to sit and drink gin in.

The Sinclairs told him that their son Maurice, who had got a music scholarship from the Academy, was going off to Spain. Since he also had a tubercular condition they were worried about his going off by himself as he would not bother to eat properly or take the proper rest. Cissie suggested that Beckett might go with him. 'I wish to God I could,' he remarked. Besides the Sinclairs he saw Ethna and Leventhal; but one evening when he called he found them enjoying such romantic intimacy over a dying fire that he was embarrassed and, though they were hospitable enough to suggest that he might like a cup of tea, he hurriedly departed. Jack Yeats rang and, finding him to be out, had a fairly lengthy talk with his mother on the telephone. Shortly afterwards Sam brought Boss to see the painter at his studio in Fitzwilliam Place. They found him alone, contemplating his works and Boss was naturally enthralled. He met one or two other people, including Francis Stuart. Like other returning literati, he concluded that 'Dublin was Dublin, only more so'.

In spite of his new-found hostility to his analyst he had wanted to get back to London and when he did so in February 1935 he felt an enormous relief. He also began to feel much better physically and resumed the analysis happily enough; but neither the relief at being out of Dublin nor the physical vitality lasted long and soon he was depressed and miserable again, feeling himself to be worthless and his life sordid.

Some of this feeling may have been due to the resumption of his half-hearted and in large part unavailing search for literary employment. MacGreevy was now away in Kerry but during the previous year he had succeeded in bringing Beckett and the influential critic, Desmond MacCarthy, face to face and Beckett now felt it incumbent on him to resume the acquaintance. Though he dreaded the thought of MacCarthy's polite solicitude, to which he felt he would have nothing to reply, the feeling of obligation none the less weighed on him. So did a similar feeling of onus and concomitant reluctance to get in touch with Edwin Muir who, according to MacGreevy, might put him in the way of getting translations to do. Also on his list of people to see was Charles Prentice of Chatto and Windus, but he could not make the effort to do that either.

It was very cold and he mostly stayed in Gertrude Street only venturing forth to see and to squabble with Bion, with whom in the first week in February he had his 133rd session. He was reading Jane Austen and felt she had much to teach him, though he also thought it curious how English literature had never freed itself from typifications and simplifications. He had had a long letter from Lucia Joyce from Zurich beginning, 'my father Mr James Joyce, communicates sad news', the sad news being that of Peggy Sinclair's death. He thought her six-page letter was utterly insane, like surrealist automatic writing, except that there was nothing artificial about it; and after thinking it over he realized there was nothing he could do, so he did not reply. Lucia had lately been under the care of the famous C. G. Jung, the twentieth doctor to be consulted about her illness, previous authorities having included one who had prescribed drinking sea water. Under pressure from Maria Jolas, Joyce had agonized about consulting the great Jung, of whose criticism of *Ulysses* he did not think a great deal; but finally he decided that this avenue too must be explored. 'My daughter is not myself,' he said. 'I wouldn't go to him, but maybe he can help her.'

At first Lucia seemed to respond and talked freely to the great man instead of being – as she usually was with doctors – sulky and uncommunicative. Jung thought of her as Joyce's anima and it is probably fair to summarize his conclusions by saying that he thought that she was Joyce without the genius. One had sought a kind of madness, the other had it, but it was uncontrolled and therefore a calamity. Joyce was still resisting the idea that his daughter could be mad and Jung thought that this was because 'his own anima, i.e., unconscious psyche, was so solidly identified with her that to have her certified would have been as much as an admission that he himself had a latent psychosis. It is therefore understandable that he could not give in.'

Meanwhile Lucia herself was penitently declaring that she was spoiled and apologizing for the trouble and expense she was causing. 'Father, if ever I take a fancy to anybody I swear to you on the head of Jesus that it will not be because I am not fond of you. Do not forget that,' she told him. She was also disowning her artistic ambitions, saying that at a previous nursing home, Pranginn, where she had been under the care of Dr Forel, she had seen a number of artists, especially women, who seemed to her all very hysterical:

Am I to turn out like them? No, it would be better to sell shoes if that could be done with simplicity and truth . . . I should like to have a life as quiet as I have now with a garden and perhaps a dog, but nobody is very contented. Isn't that so? So many people were envious of me and of Mama because you are too good. It is a pity that you don't like Ireland for after all, it is a lovely country if I may judge by the pictures I have seen, and the stories I have heard. Who knows what fate has in store for us? At any rate in spite of the fact that life seems fuller life this evening, here, if ever I should go away, it would be to a country which belongs in a way to you, isn't that true father?

The country was Ireland, and this was the beginning of a campaign, ostensibly to get herself there, possibly really to get to London to see Beckett. Jung's ascendancy over her had not lasted very long and he now told Joyce that 'she was a very exceptional case and certainly not one for psychoanalytic treatment, which he said might provoke a catastrophe from which she would never recover'. Later Lucia herself would say 'to think that such a big fat materialistic Swiss man should try to get a hold of my soul'.

Now, with Jung's approval, Joyce decided that she should return to the outer world, and though it was not advisable that she should live at home she was installed with a nurse companion in the villa Élite, annexed to the hotel James and Nora were staying in. At Lucia's request her aunt, Eileen Schaurek, was sent for, Joyce's youngest sister, who had lived – and married – in Trieste and was now an intelligent though somewhat excitable widow of forty-six. It was but a short step from this arrangement to a proposal that she should go to Ireland with Eileen, stopping for a while in London with Miss Weaver. Lucia frankly admitted to her father that she hoped to see Beckett in London and the project seemed to meet with his approval.

Installed in Miss Weaver's house in Gloucester Place towards the end of February, Lucia wrote to Beckett expressing a wish to see him. She told him in the letter that she was in Gloucester Place with her aunt on her way to Ireland. Again he decided that there was nothing he could do and so again he did not respond. Since she did not appear to be staying long in London he hoped that the problem would go away. Then she began to telephone and he instructed the Frosts to say he was not in. However, since she had his address she could obviously waylay him, so he began to make detours, returning to Gertrude Street by circuitous routes and keeping a wary eye out. Eileen had now gone back to Ireland, leaving her niece in Gloucester Place and, much to Miss Weaver's alarm, Lucia's behaviour now became very odd. She spoke of procuring a pistol to commit suicide; and when Miss Weaver took her to a doctor she walked out of the waiting room saying that she was a grown woman who could look after herself. That night she failed to return home. It was an anxious night for Joyce's friend and patron, but when Lucia appeared in the morning she would give no explanation except that she had spent the night in the streets. This was probably true, but she may also have gone in search of Beckett and Gertrude Street.

It has been suggested that Beckett was trying to keep in with Joyce by avoiding further entanglement with his daughter. In fact the surest way to Joyce's heart would have been to meet Lucia and write assuring him that she was sane, as he kept insisting to poor Harriet Weaver that she was, growing quite angry when any suggestion was made to the contrary. (During all this Paul Léon wrote to the distraught Harriet Weaver that Mr Joyce had an 'inner conviction' that he had saved his daughter from schizophrenia, ending 'Mr Joyce trusts one person alone and this person is Lucia. Anything she says or writes is the thing by which he is guided.') No doubt Beckett's behaviour showed a certain coldness, but he knew Lucia was mad and in this awful situation his avoiding action was no more culpable than the actions of the people who went along with her father's fond illusions and treated her as if she was otherwise. When Lucia did leave for Ireland he was oblivious of the fact, though no doubt relieved that the phone calls had ceased.

Meanwhile Beckett's own analysis continued, with the usual oscillations of faith in the process on the part of the analysand. In March he wrote MacGreevy a long, highly revealing letter setting forth his current attitude. He says that he had been unhappy for years past, dating this unhappiness from the end of his schooldays, the time when he left Portora and went into Trinity. It is not unusual for those who have been noted athletes and popular figures at school to find later life a disappointment; but this feeling of let-down usually occurs in young men who have little reason to believe that they will ever be able to replicate their achievements in the outside world. Beckett's case was different and his attitude makes one wonder what it was that caused the shock when he left school. Was it that Portora was an all-male society in which questions of sexuality scarcely arose and homosexual feelings could be sublimated without question? Or was it that when he left he was faced with the problem of constructing an identity for himself, he who had little sense of identity until he found it in his writing? He says he was unhappy 'consciously and deliberately, so that I isolated myself more and more, undertook less and less and lent myself to a crescendo of disengagement of others and myself'. And he goes on:

But in all that there was nothing that struck me as morbid. The misery and solitude and apathy and sneers were the elements of an index of superiority and guaranteed the feeling of arrogant 'otherness' which seemed as right and natural and as little morbid as the way in which it was not so much expressed as implied and reserved and kept available for a positive utterance in the future. It was not until that way of living or rather negation of living developed such terrifying physical symptoms that it could no longer be pursued, that I became aware of anything morbid in myself.

But for the terror caused by these physical symptoms, he says, he would still be

> boozing and sneering and lounging around and feeling that I was too good for anything else. It was with a specific fear and a specific complaint that I went to Geoffrey and then to Bion to learn that the 'specific fear and complaint' was the least important symptom of a diseased condition that began in a time which I could not remember in my 'pre-history', a bubble on the puddle, and that the fatuous comments which I had treasured as indicative of the superior man were all part of the same pathology.

In other words, the physical ailments from which he thought he suffered were symptoms of a neurotic disorder; and the basic premise on which analysis rested was true: such a disorder would have its origins in infancy, at a time he could not normally remember. To account for the fact that the symptoms continued he persists with his metaphor: the puddle had not yet been drained. If they now afflicted him more fiercely than ever it was perhaps because 'the waste . . . splutters most when the vat is nearly empty', a thought from which he derived some consolation.

The letter is also strange and remarkable, however, because it also contains an attack on the quietest attitude which some may think lies at the very core of Beckett's work from beginning to end. MacGreevy, the devout Catholic, had lent him Thomas à Kempis's classic work *The Imitation of Christ*. Beckett had been reading this and he found parts of it seductive, parts that seemed made for him and were unforgettable. 'But they all conduced to the isolationism that was not to prove very splendid. What is one to make of 'seldom we come home without hurting of conscience' and 'the glad going out and sorrowful coming home' and 'be sorry in your chambers' but a quietening of the sparrow alone upon the house-top and the solitary bird under the eaves?' But à Kempis had belief in the Christian message of salvation. In the case of one without belief such as himself, they would merely encourage 'An abject self-defencing quietism . . . beside the alert quiet of one who always had Jesus for his darling . . . the only kind that I who seemed never to have had the least faculty or disposition for the supernatural, could elicit from the text, and then only by means of a substitution of terms, very different from the one you propose.'

In his own case he replaced 'the plenitude that he called "God" not by "goodness" but by a plenum only to be sought among my own patterns and entrails, a principle of self the possession of which was to provide a rationale and the communion with which a sense of grace'. On the whole *The Imitation* had only served to confirm him and reinforce him in his own way of living. But this was one 'that tried to be a solution and failed'. Instead of approaching the hypostatics of *The Imitation* 'meekly simply and truly', he had 'twisted them into a preponderance of self sufficiency'. *The Imitation* had put forward a model of Christian behaviour. Doubtless it had been composed out of the 'long, long experience of unhappiness' of which MacGreevy had apparently spoken – though whether with reference to himself or Beckett is unclear. But the fundamental end of detachment from the world for à Kempis was still a closer contact with 'God', which for Beckett was impossible.

There is nothing surprising about this oft-repeated disclaimer of the possession of a religious sense. What is rather surprising, though, is the rejection of the self-sufficing quietism which was his basic philosophy and which the earlier parts of *The Imitation* with their lesson in detachment from the world as a preparation for communion with the Godhead might have served to reinforce. Psychoanalysis was evidently, at this point anyway, having an effect which one can only describe as bracing; not far from the effect that his mother desired, which was to make him a viable person who could take his place in the world and get on with it – what could be described as the 'be a man' effect. It is also of some interest that, at this point anyway, he was accepting the basic premise of psychoanalysis, which is of course that the origins of conflicts in the unconscious are to be found in earliest infancy. According to Geoffrey Thompson, however, as soon as the analysis was over he would begin to dispute this and he would consistently do so thereafter.

Chapter Fourteen

Towards the end of April 1935 Beckett went home for a three-week visit. Immediately on arrival some of his physical symptoms returned. The night sweats in particular were worse than they had been for some time past while his mental equilibrium was disturbed by long periods of speechless bad temper associated with the proximity of his mother. Boss Sinclair was very ill with the consumption which would cause his death and was in the Adelaide Hospital, where Sam went to visit him, meeting Cissie and Harry Sinclair while he was there. Other Dublin acquaintances whom he saw without much enthusiasm included Ussher, Leventhal and Leslie Yodaiken, subsequently Leslie Daiken, affectionately known in Dublin as 'the Yod', who slightly annoyed him by seeming to know that he had now put together a collection of poems and that MacGreevy was acting as his agent in the search for a publisher.

This perhaps annoyed him, as things usually do, because it was partly true. He did now have a collection of poems. He had tried the idea on Chatto and Windus, to whom he had first been introduced through MacGreevy; but it was another friend who was about to publish them, for he had fallen back on an offer from George Reavey to bring them out from his newly established Europa Press, an imprint which operated from above the Russian bookshop in the rue Bonaparte.

To his extreme disquiet he also now heard that Lucia was in Bray outside Dublin. 'Please God our ways don't cross,' he prayed when imparting the news to MacGreevy. And in fact they didn't. Lucia had been taken to Ireland by her aunt Eileen and installed with Eileen's daughters in a bungalow in Bray. In a distracted sort of way she searched for Beckett, and when she made the acquaintance of a young Irish artist, Patrick Collins, she talked of him. Collins, who was rather taken with her, found her strange but did not think she was mad; to others too her eccentricities seemed a result of an exotic upbringing and even evidence of genius. Yet her behaviour was strange and getting stranger. In Bray she attempted to commit suicide by taking an overdose of aspirin and then tried to set fire to the bungalow. Finally she disappeared for six days and wound up in a police station in a highly dishevelled state, asking to be put in a nursing home.

Now wildly distressed and concerned, Joyce dispatched Maria Jolas to Dublin and she took Lucia back on the boat with her to England, leaving her again in the care of Miss Weaver. Dr W. G. McDonald, a specialist who had had good results with a bovine serum, was now called in. The injections, which were given to Lucia over a period of five weeks in Grosvenor Place, did seem to have a calming effect and McDonald then suggested convalescence in quieter surroundings. So the ever-patient and faithful Miss Weaver rented a bungalow near Reigate in Surrey and hired a trained nurse to help her with Lucia there. Their stay in Reigate lasted till mid-December, at which point Joyce, correctly suspecting that Miss Weaver was giving edited accounts of Lucia's behaviour to save him as much distress as possible, dispatched Madame Jolas northward again. On Dr McDonald's initiative Lucia had been placed in St Andrew's Hospital, Northampton, for blood tests. She could not be kept there against her wish without certification by her parents – to which Joyce's response was that he would not give authority over his daughter to an Englishman. (McDonald was actually a Scot.) Instead Madame Jolas took Lucia back to Paris to stay in her own large house in Neuilly. But she was growing increasingly violent and unmanageable and after two weeks had to be removed from the Jolases' in a straitjacket. It was the beginning of Lucia's permanent institutionalization. She was taken to a clinic at Le Vésinet but the doctor there said she was dangerous and needed to be taken to a special institution, so she was moved again, this time to Dr Baule's clinic at Ivry.

Beckett was happily oblivious for the time being of much of Lucia's distresses in Ireland, having perhaps enough of his own to contend with, though before his stay was over his physical symptoms had improved somewhat, as had his tolerance of his mother's company. It gave him pleasure to see that she seemed to forget the duty of being wretched more often than she had done formerly. He did not consider that he was ready to accept life in Foxrock yet, however, while at the same time he dreaded being back in London. He had a ticket which was valid for one month only and at the end of that time he insisted on returning, although his mother wanted him to stay for a while longer. In a way he preferred the mindlessness of existence in Foxrock to what he called the 'shabby depravity' of London. But he had the feeling that if he delayed any longer relations with his mother might deteriorate again. In any case Bion had emphasized before he left that the analysis was nowhere near its end and that there was still a great deal to do – a point driven home by the recurrence of the palpitations of the heart in the second week of his visit.

Before he left he hired a car and took his mother to visit Bill's grave, on which they planted a sprig of heather in a cruel north-east wind. He also took Cissie to see Boss at Newcastle Sanatorium, to where he had been moved. They stopped at the Lucan Spa Hotel, where the bar was 'a riot of priests tired after their golf, lowering the hard stuff sur le sink, and talking about the girls they had been giving lifts to, but not so far as to make them happy men, neither the talk nor the lifts'.

When he left for London it was arranged that his mother should come to visit him in England. She came in July, a car was hired and they drove to Stratford-upon-Avon, everything about which – including Shakespeare 'Up to the Universelle' – Sam found unspeakably awful; then on to Wells, whose cathedral struck him as magnificent but 'unmoving' and finally to Lynmouth in Devon, from which they made excursion along the Devonshire coast. In spite of the inescapable and interminable discussions of his future, which, as far as he was concerned, could reach no plausible conclusion, the trip on the whole went better than he expected and an atmosphere of consideration and courtesy prevailed throughout. They came back through Gloucestershire and then went on to Rugby, where he put his mother on the train for Ireland. Then he took a detour to have a look at Samuel Johnson's home town of Lichfield before finally coming on down to London to meet MacGreevy on Thursday 31 July in the Six Bells in Chelsea, their favourite pub.

In August 1935 he had been in Gertrude Street a full year, as Mrs Frost reminded him, and looking back he could not say he was pleased. He again felt that the analysis was getting nowhere and was guilty about the expense of it. True, the poems were coming out and he was somewhat complacently awaiting the proofs, but otherwise it had been a year of very little work. The truth was, he remarked to MacGreevy, that he had 'as little to write as to say, or can write as little as say the immensity. As though the brain were full of milk that the least act of interested thinking brought to the boil.' Perhaps the commonest feeling among young writers is not that they have nothing to say but that the totality of experience is so overwhelming that it defies expression. There were in his case, however, the stirrings of a book.

He had become fascinated by the shabby but respectable old men who were to be seen at weekends doing odd jobs in their gardens or flying kites at immense heights near the round pond in Kensington Gardens. One day he noticed a group of these with grandchildren in attendance sitting around waiting for the wind to become strong enough. Their kites were lying around in the grass with their long tails carefully coiled. He noticed that the kites were brought in separate pieces, the sticks and tails rolled up in the canvas. They were flown at a great height and on this particular day the wind was such that they rose almost out of sight over the trees in a completely cloudless evening sky. He found the sight of birds flying beneath the kites quite extraordinary and was fascinated by the way the old men simply sat there watching the kites fly and occasionally fingering the string as a coachman might the reins of his horses, evidently to keep them from losing height.

He found himself rooted to the spot, wondering what was keeping him there, and thought that his next old man, or young man for that matter, would be a kite-flyer, but that he would be in a poem. In fact it was to be in a novel, *Murphy*, for which this image of the old men flying their kites was the imaginative beginning, though it does not appear in the book until the very end.

Another point of departure was the Bethlem Royal Hospital in Beckenham where Geoffrey Thompson now worked. Thompson lived in and Beckett used to go down to visit him. In *Murphy* the Bethlem Royal becomes the Magdalen Mental Mercyseat. Shortly after his first visit Beckett began to accompany Thompson, who had a number of psychotic long-stay patients under his care, on his rounds. Like Thompson, Beckett felt very sorry for them, but he was also imaginatively fascinated, spending time with some and becoming very interested in their condition. One in particular seemed to fascinate him, and though he plays no part in *Murphy* it is perhaps easy to see why Beckett found him of such interest.

This was a young man of about twenty, who was, in Thompson's words, 'frightfully deeply attached to his mother'. He spent every day sitting in a chair dressed in his overcoat as if he was ready to go out for a walk. He was waiting for his mother to come and visit him and expected each day to go for a walk with her when she did. In addition to the overcoat he wore a pair of hob-nailed boots and kept his legs stretched out in front of him with only the heels resting on the ground. When questioned as to why he adopted that posture he would reply that his mother had told him to do it so that he would not wear out the boots. The young man's relationship with his mother was of course of a psychotic kind. It had become fixed at a certain point and could move no further; but Thompson thought he understood the extraordinary interest of this particular case for Beckett, whose relationship with his own mother was a central fact of his psychology.

About one other important source for *Murphy* one can speak, alas, with less certainty than one can about the kite-flying or Thompson's patients. In *Murphy* the kite-flying old man, Mr Willoughby-Kelly, has a grand-daughter, Celia. Celia is 'on the street' – in other words, a street-walker or prostitute – having adopted that profession when she was left an orphan. She meets Murphy in the course of her street-walking, having just turned out of Edith Grove into Cremorne Road, which are round the corner from Gertrude Street. When she is introduced we are told that Murphy 'craved' for Celia or, rather, 'the part of him that he hated', his body, craved for Celia, while 'the part of him that he loved [presumably his mind] shrivelled up at the thought of her'. This seems at first like a surprising reversal of the Belacqua situation, in which the hero has a sort of mental or spiritual need of a particular woman, though not much bodily need, but in fact it is not, for Belacqua, we know, was quite capable of enjoying a physical relationship with various prostitutes. Unlike the poor Smeraldina's body, which was an object of such detestation, Celia's desirable body is celebrated in *Murphy* to such an extent that her vital statistics, Bust 34', Waist 27', Hips 35', etc., are given a page to themselves and are constantly adverted to. The promise of mere bodily bliss that she offers turns out to be false, however, for Celia too makes demands: in particular, she demands that Murphy should earn some sort of a living and it is on this demand that the plot of the novel, such as it is, basically revolves. Having turned down his rather snidely made suggestion that they should both live on her earnings – a suggestion that he means as a tribute – Celia issues an ultimatum. Either Murphy, who is living on a barely sufficient allowance from an uncle, gets a job or she will sever all relations with him. There is a sub-plot involving Murphy's former relationship with a Miss Counihan, now desired by the Cork philosopher Neary, but Celia is the principal female character in the book.

The suggestion has been made more than once that Celia had a real prototype with whom Beckett had a real relationship; and one or two remarks he made himself have seemed to support this. Certainly street-walkers were not rare in the 1930s around the lower ends of the King's and Fulham Roads, where they intersect with Edith Grove, or in the pubs round about, including the World's End, where members of the profession, not averse to either social pleasantries or more business-like pick-ups, might be encountered having an occasional small port or whisky and Apollinaris water. That Beckett had one or two such encounters would not be surprising – one such, in May 1934, is recorded in a letter to Con Leventhal – and neither would its having led on to another or others; but that there was any sort of prolonged relationship is certainly doubtful. Unfortunately Celia in the novel has not sufficient reality to enable us to judge. Certainly she is an improbable street-walker, and though many young men seem to dream of the prostitute who loves them and is physically desirable yet quasi-maternal, the type has not much more than a cliché reality.

In another story of Beckett's, 'Premier Amour' or 'First Love', written in 1945 but not published until 1970, there is also a prostitute, Lulu, who works along the Grand Canal in Dublin, as many prostitutes did then and still do. She takes him to the house she works from and gives him a room there, where she feeds him and he stays for a while, undisturbed, in spite of his need for silence, by the 'stifled giggles and groans, which filled the dwelling at certain hours of the night, and even of the day'. 'First Love' is written in a much less realistic mode than *Murphy*, but Mary Manning, for one, was convinced that the girl along the canal had a real prototype. The two women, the Edith Grove street-walker and the one who worked the canal, seem to have been confused by some commentators, but the actor Patrick Magee, who was more intimate with Beckett than most, certainly also thought that the girl by the canal had a reality and that she had a child which was not Beckett's, but to whose support he at some stage contributed. If she had such a reality, his encounter with her probably belonged to the period after his return to Dublin in 1935.

Once started on *Murphy*, he went ahead very fast, even though he was spending a good deal of time out of doors in the beautiful autumn weather, and by the third week in September had written about 9,000 words. He thought it was poor stuff but was determined to keep on with it and after a further two weeks had reached the 20,000-word mark.

Perhaps the analysis was after all having some effect, but more illuminating than anything Bion said to Beckett proved to be a lecture by C. G. Jung which he attended at the Institute of Psychological Medicine in company with his analyst and Geoffrey Thompson. They had a hurried but enjoyable sole at the Étoile in Charlotte Street beforehand, Beckett finding Bion off the job rather pleasant, though his youth club and tank corps manner none the less struck fear into him. He thought Jung was a bit like the Irish poet AE; although his mind seemed more ample, authoritative and far-reaching, there was the same 'cuttle fish's discharge' which enabled him to escape the issue in the end. In a previous lecture Jung had used a diagram which he now referred to again and which Beckett would appropriate. In this Jung portrayed the mind as made up of concentric circles becoming ever darker till we reach the final darkness of the unconscious. (Murphy, it will be remembered, prefers the dark part of his mind to the light.)

But for Beckett the most illuminating part of Jung's discourse came at the end. The psychologist had been describing the case of a young girl whose condition had baffled him. As Geoffrey Thompson remembered it, 'Jung had rather a dramatic, impressive way of speaking,' and when he had described this girl he paused for a moment and then he said, 'Of course, the truth of the matter is, as I realised afterwards, this young girl had never really been born.' This is close to Beckett's recollection of the lecture by the famous psychologist, which he attributed to the character Mrs Rooney in the radio play *All That Fall*, written over twenty years later. Mrs Rooney goes to hear 'one of these new mind doctors, I forget what you call them', in the hope that he might be able to shed a little light on her 'lifelong preoccupation with horses' buttocks'. (This peculiar fixation was, it will be remembered, also attributed to Belacqua, the male protagonist of *Dream of Fair to Middling Women*.) Mrs Rooney also remembers

> his telling us the story of a little girl, very strange and unhappy in her ways, and how he treated her unsuccessfully over a period of years and was finally obliged to give up the case. He could find nothing wrong with her, he said. The only thing wrong with her as far as he could see was that she was dying. And she did in fact die shortly after he had washed his hands of her.

When Mr Rooney challenges her to say what was so wonderful about that, she replies

> . . . it was just something he said, and the way he said it, that have haunted me ever since . . . When he had done with the little girl he stood there motionless for some time, quite, quite two minutes, I should say, looking down at his table. Then he suddenly raised his head and explained, as if he had had a revelation, 'The trouble with her was that she had never been really born!'

Jung's assessment of the little girl's problem had a huge effect on Beckett and not only found its way into his work but was adverted to often afterwards in conversation with Geoffrey Thompson and others.

There is no doubt that he thought that the diagnosis was a profoundly suggestive illumination of his own case, his sense of alienation from the world and of not being ready or fitted to cope with it, to join in its activities as others did, or even to understand the reasons for them. Bion too was greatly impressed. In later years he wrote some papers in which he argued that there were cases in which psychological birth occurred before biological birth, while the people concerned were still in the womb. The result was that actual biological birth did not necessarily bring mental separation from the mother. There were people who suffered trauma due to a 'mismanaged birth' and who remained throughout their life with a sensation of still being in the womb because no proper caesura had taken place during biological birth. Alternatively the caesura can take place too rapidly, causing much repressed pain and future anxiety. This is now a very well-known theory in contemporary psychology, in which 're-birthing' is a not uncommon practice. There is no doubt that Bion and Beckett discussed Jung's diagnosis of the little girl's case, and Bion's theory was almost certainly elaborated from Jung and of course also from his analysis of Beckett's case.

But in spite of this illumination from a psychoanalytic source of his own basic state Beckett had now more or less determined to bring the analysis to a termination by the end of the year. He was certain that he would get no further with it than he had done already; that from now on his mother's money would be thrown away. Nor did he any longer expect that the physical troubles which he had hoped to get rid of with analysis would be cured by it. His latest conviction of illness related to his stomach, and he spent a great deal of time worrying about the possibility that there was something seriously wrong with it, either in the shape of an ulcer or something else. He had decided that at the end of the analysis he would return to Dublin and settle there for a while. He ruled out Spain or any other foreign destination, partly because of his mother, whom he felt needed him and to whom, he said, he felt he owed enough to make it incumbent on him to be with her. This sounds very like a rationale for wanting to be in Ireland with her himself; and in fact from now on they would play out something of a tragi-comedy with each other, his mother putting up with Sam because he had no career elsewhere, Sam staying in Ireland because his mother 'needed' him. There had been a temporary sojourn in Killiney, which appeared to have done her good, and in deciding to go back, Sam told himself that their relationship would at least be a tolerable one. He also insisted that he had developed an immunity to Dublin and might be able to work there. So when he came back at Christmas 1935 he had already decided to remain, though he did not tell Bion that.

Afterwards he would look back on the London period without any great satisfaction. Everybody knew you were Irish, he said. The taxi-drivers called you Paddy. Many years later, when asked by a singularly uninstructed journalist in Paris if he was English, he replied, '*au contraire*', a reply which, if not exactly an Irish bull, is something of a solecism, but contrives to mean something nevertheless.

But, perhaps predictably, he did not get any great pleasure out of returning to Ireland either; nor did he find things as easy as he had hoped where his mother was concerned. So angry and demanding was she, indeed, that the day before New Year's Eve he very nearly decided to go away again, only the feeling that he must get on with *Murphy* keeping him where he was; and this in spite of the fact that he had developed what he called a pleurisy since his return and had been in bed much of the time, before coming down to sit before the fire and complain that the pleurisy had now transferred itself to the other side of his chest. It did not seem to occur to him that this kind of carry-on might be the very reason why his mother was so impatient. In a different, if not very much subtler way, he was now being just as demanding as she was. No sooner did he begin to feel that the pleurisy had abated sufficiently to allow him to be out and about than he had a 'heart attack' in bed during the night which he felt was the equal of any he had suffered before beginning the analysis. All in all, it was little wonder that he decided that the analysis had done him little good. Acknowledging receipt of the final sum due to him, Bion politely hoped that he had by now taken up his work 'with pleasure and satisfaction', though he then proceeded in somewhat contradictory fashion to say, 'even though all activities feed from neurosis'.

But Beckett did feel that the composition of *Murphy* had been somehow benefiting from his prolonged period of analysis and that it had gone the more freely as a result of it, even though he had done little over the Christmas period. Before leaving London progress had been very fast, so fast that he now had about 40,000 words and he felt that another 12,000 – or four more chapters – would see him through.

But he had now come to Murphy's death, and there was a structural difficulty. Although the climax, which should be presented as such, it was not the end of the book; and apart from anything else, he had so far not even succeeded in introducing the kite-flying with which the whole thing had begun in his mind. Yet after Murphy's death the rest would have something of the character of an epilogue. Eventually he decided, as he would later put it to MacGreevy, to 'keep the death subdued and go on as coolly and finish as briefly as possible'.

Perhaps because he was working, he was now finding being at home less traumatic. In spite of the difficulties with his mother and the nights which he found 'horrible' physically and emotionally, he was settling in and settling for the cocoon which, if not always pleasant, was safer than the outside world. He was having bookshelves and a few reproductions of pictures put up in his room and beginning to find the prospect of being elsewhere rather frightening. Considering all this, the impartial observer is almost forced to conclude that the analysis had had a regressive effect, though doubtless Bion might have said that this was because it had been left unfinished.

Echo's Bones and Other Precipitates had come out before Christmas. It was No. 3 in the projected series Europa Poets and, being Paris-published, was scarcely likely to set the English-speaking world on fire. Though he had been writing poems since 1928, there were only thirteen, or less than twenty pages of verse, in the little book, which Reavey had first offered to publish as far back as 1933. At that stage, however, Beckett had been hoping that Chatto and Windus would prove interested in a collection and had turned the offer down. When his London publishers said no, Beckett, like many another similarly placed poet, fell back on the lesser option and in June 1934 he had agreed that Reavey should bring out a book. At that stage there were only nine poems but Reavey's long delay enabled him to add four more, the last four in the book including the title poem 'Echo's Bones' itself.

The title came from Ovid's *Metamorphoses*, Book Three, and as the title of the book it was an afterthought. Spurned by Narcissus, Echo, it will be remembered, becomes gaunt and haggard. Her body dries and shrivels until at first only her voice and bones remain and then only the voice, for the bones have been turned to stone. She hides in the woods 'and no one sees her now on the mountainside but all may hear her voice'. The poems are assiduously modernist, in the learned, allusive, 'witty' mode established by Eliot as the norm of modernism and, supposedly at least, derived from Laforgue and Corbière, but Beckett's understanding of free verse, like that of many other would-be modernist poets at the time, was somewhat different from Eliot's. His lines, like most others', end always at the end of the phrase and the effect is rhythmically somewhat flat and monotonous:

> there was a happy land
> the American Bar
> in Rue Mouffetard
> there were red eggs there

The allusiveness and the wit are carried to extremes. A Beckett poem is an almost indecipherable tangle of recondite references and allusions, some from literature or myth or religion, some from the poet's own life and the particular urban geography in which it has been lived. As Hugh Kenner says, the poems 'are apt to leave a reader blank' and they simply do not offer enough in the way of verbal attractions or harmonies to compensate that reader for the almost impenetrable obscurity. In later years Beckett was harsh in his judgement of them and very aware of the lamentably self-conscious way in which they carried their erudition.

But that he had a soft spot for these poems all the same is shown by the fact that he was never reluctant to let them be reprinted, as he was with his early prose works. As the present writer can testify, Beckett was always pleased to be characterized as a poet; and there is no doubt that during these years he thought of himself as a poet rather than as anything else, even though he was at this point engaged on his third work of prose fiction. He had begun as a poet and, in a manner commoner in France than in England, continued to think of himself as such even after he had virtually abandoned poetry for prose. The fact is, though, that like his admired mentor James Joyce, he put his poetry into prose. Just as there is more true poetry in *Ulysses* than there is in *Chamber Music*, so there is more in *Watt* or even in the harsher and more humorous *Molloy* than there is in his avowed poetry.

As Lawrence Harvey rightly says, the poems sin in two directions: against art by relying far too heavily on what Harvey calls 'the verbal machinations of conceptual intellect', but also in the opposite direction by presenting in the personal and autobiographical references raw materials from poetry masquerading as the article itself. 'The reader sees the top and bottom of the poet's mind, but the twilight zone between, from which poetry emerges and to which Beckett is so faithful in his later writings is only rarely glimpsed in these three poems.'

Clearly though he saw them later on, however, he was not to be expected to see the defects of these poems at the time and their reception naturally disappointed him. Even the friends to whom he gave copies seemed to maintain a shifty reticence about them. When he met Con and Ethna he sensed immediately their reluctance to enthuse, indeed to say anything at all, and noticed how they got out of the difficulty by remarking that they had already been familiar with most of the poems in the book. Since everybody seemed to be preserving what he called a 'revolted silence', even his aunt Cissie, he had to conclude that the poems must be even more offensive than he had thought. Naturally his mother and Frank did not refer to them and he could find no trace in the house of the three copies that he had given to his mother. He remembered with amusement, however, that the copy of *More Pricks Than Kicks* that he had given the Foxrock household had not at first appeared on any shelf in the house and that only very gradually had it crept into the open so that it could now be found in the drawing room. He thought it possible that before the middle of the summer *Echo's Bones* would be similarly displayed. He was not of course alone in having these difficulties. Many young authors have felt that their work caused embarrassment to their nearest and dearest; but of course the problem is more acute and the embarrassment the greater when the author comes from a narrow, puritanical background, as Beckett did. More importantly perhaps, there were no professional responses by way of reviews. A copy which had been sent to the *Irish Times* seemed to have gone astray, but when he gave them another copy there was still no review. To make matters worse he was finding it almost impossible to get Reavey in Paris to reply to his angry letters. By the end of March 1936 he was beginning to believe that no review copies had been sent out; and of course he was suffering the usual angers and frustrations through discovering his publisher's delinquencies in the matter of distribution. Neither he nor his friends could find any copies in the shops and one shop, Combridges in Grafton Street, assured a friendly enquirer that all their requests for copies to the publisher had been in vain.

A new development in his attitude and in his mother's to Cooldrinagh being his home was the fact that his literary friends were now not discouraged or debarred from visiting him there. Among those who went to Foxrock was Brian Coffey, who had come over from Paris. Though entertaining in his mother's house was difficult, he thought the visit went all right; but of course Coffey, with his metropolitan air and his rather posh accent – not to mention his physician father, who was President of University College, Dublin – was more suitable for presenting to his mother than others might have been.

Coffey was full of T. S. Eliot and had the *Criterion* in his pocket. Beckett was, in a way, impressed by his discourse, but found the febrility of his mind somewhat alarming and was not over-impressed by the poems he was shown, though he thought that on the whole there was more to him than to Devlin. T. S. Eliot himself came to Dublin and spoke twice at the National University. On the second occasion, when Beckett was present, he described *Finnegans Wake* as 'an unconscious tribute to a catholic education acquired at a time when few people were educated at all', a comment which Beckett thought showed far too much respect for the pedagogic aspect of things. But he does not seem to have reintroduced himself to the distinguished visitor either before or after the lecture. When he next saw Coffey, a few months later, he decided that it was impossible to consort with him; but his opinions of people continued to oscillate according to his own psychological outlook, and when he went one Sunday to have lunch with the Coffeys and the distinguished father discoursed on ecclesiastical architecture while Brian talked of Sainte-Beuve they got on better. In general, however, many of the people he was seeing in Dublin would have been surprised and dismayed if they had known his real opinion of them, Jack Yeats being perhaps the only exception. The opinion could vary, but some of his comments to MacGreevy were very sarcastic indeed. The young Sam Beckett was rather different from the eminence of post-war years who apparently never spoke ill of anybody.

Other artistic visitors to Foxrock who proved perfectly acceptable were Jack and Cottie Yeats, who came out specially with W. B.'s biographer-to-be, Joseph Hone, to see his mother and her donkey. The name Yeats would of course have been familiar in Foxrock, as might also have been that of Hone, for the Hones were a distinguished Anglo-Irish family. On the whole Sam thought the visit went quite well, both his mother and the donkey being completely natural. He had been a frequent visitor at Yeats's studio, though he still tried to avoid Cottie and the 'at home' days, and the review of Yeats's *The Amaranthers* already referred to was written at this time. As the review and the friendship generally indicate, he seems to have been able to rejoice, in Yeats's case, in a view of life which he did not exactly share. The bright and happy landscape of *The Amaranthers* appealed to him and he thought it made Ireland seem like Paradise Island.

One small picture called 'Morning', almost a skyscape with, under the sky, a wide street leading into a town, especially appealed to him. The price was £30 and he wanted to buy it on the instalment system; but when he let fall several hints to this effect there was no indication that they would be acted on. Finally he borrowed £10 and offered it as a down payment. It was accepted and he got the picture. Neither his mother nor Frank particularly liked it but he hung it on the wall in his own room, rejoicing in the fact that it was 'always morning, and a setting out without the coming home', as he wrote to MacGreevy, who had by then been commissioned to write a small book on Yeats. It would take MacGreevy, who was not industrious and had serious writer's blocks, longer than it should have done to write the book; and when in 1945 Beckett wrote the review already discussed he would have certain reservations about MacGreevy's discussions of nationalism with reference to Yeats.

Beckett had of course attacked the nationalist-antiquarians in his piece in the *Bookman* about Irish poetry and when he went to his aunt Cissie's one evening and met one of them, the poet Austin Clarke, he was afraid that Clarke would bear him some ill will as a result of the article, in which it was claimed that he employed the whole 'fully licensed stock-in-trade' of the antiquarians. In fact he found Clarke full of hate for other poets he had referred to, and there was no sign whatever that he had read the piece in the *Bookman*. Beckett thought him pathetic yet somehow sympathetic, and he wondered if he himself was not simply starved of literary contact of any kind.

Another poet whose acquaintance he made about this time had also been mentioned in the *Bookman* article. This was Niall Montgomery, whom he recognized as an oddity, which in many respects he was. Montgomery was a sedulous modernist and, like all of his generation at University College Dublin, a fervent admirer of James Joyce. He belonged to a particular period in Dublin, a time when a reputation for knowing about certain representatives of modernism and being able to talk cleverly about them could be the basis of a reputation for genius or near genius. A much-admired ornament of Flann O'Brien's circle at UCD, he was a new type of clever Dubliner, a type which was emerging from the ruins of the Irish literary revival, was in some ways an advance on the nationalist 'antiquarians' of that era but also, in some ways, equally sterile and suffering to an equal degree from the uncreative paralysis which Joyce had identified thirty-six years before as the city's main characteristic. He was interested in film as an art-form and he lent Beckett books and magazines about it. As a result Sam's own interest in the possibilities of film began to grow and he entertained serious ambitions about getting into a studio where he could learn. This was of course impossible in Ireland, but among the heroes of the books and magazines which Montgomery lent him were the Russian directors Eisenstein and Pudovkin; and it was not long before his ambition began to take the specific form of getting to the Soviet Union and learning from them. He wrote to Eisenstein, care of the Moscow State Institute of cinematography, asking if he could be taken on at the Institute if he made his way to Moscow. Whether the letter reached him or not, Eisenstein unfortunately did not reply, so the ambition to become a film director remained a day-dream never to be realized, one of the more tantalizing might-have-beens of Beckett's life.

One old friend with whom he now resumed acquaintance, but who was himself sadly not thriving, was Rudmose-Brown. Both Ruddy and Furry had been betraying signs of serious alcoholism for some time past. The rows between them, once regarded as amusing light relief, had become more frequent and more devastating. The 'characters', as frequently happens, were turning into apparent wrecks. In January 1936 Mrs Rudmose-Brown was taken in for treatment, and in spite of the relief this might be thought to have afforded him, Ruddy himself soon followed. It became apparent to his friends that something would have to be done and other people, including Ethna McCarthy, who like Beckett was of course one of his brightest pupils, took over his classes. After treatment Ruddy seemed to improve but Furry's fondness for whiskey did not diminish. Finally, in April, Furry died, having taken an overdose of Luminol, and was buried in a family plot at Enniskerry. Beckett attended the funeral with his aunt Cissie and felt that on the whole her death was a solution for both Ruddy and herself. He also thought that Ruddy had taken the fact of her death after an intense and stormy relationship reasonably well.

Beckett had not offered to take any classes during Ruddy's illness, nor might he have been acceptable to the powers that were in Trinity if he had. But that January he passed through what he called the 'abhorred gates' of that institution for the first time since quitting the job, on a commission for Ruddy to the library; and shortly afterwards went in again in search of the works of the philosopher Arnold Geulincx, which were not in the National Library.

Geulincx was a Dutch philosopher and a disciple of René Descartes'. The 'beautiful Belgo-Latin' in which he wrote is mentioned in *Murphy* where his maxim, '*Ubi nihil vales, ibi nihil velis*' is quoted. He is also adverted to in a later prose piece 'The End', in which the narrator remembers his tutor who gave him the *Ethics* of Arnold Geulincx. The tutor, whose name was Ward, was said to have been found dead 'in the water closet, his clothes in awful disorder, struck down by an infarctus', which is a blockage of the blood circulation by dead tissue. After imparting this piece of information the narrator remarks, 'Ah what peace.'

'*Ubi nihil vales, ibi nil velis*' has been translated as 'Where you are worth nothing, there you should want nothing'; but it could equally mean 'Where you can do nothing, there wish nothing' and has been so translated by Geulincx's translator Ronald Begley, who translates Geulincx's own gloss as 'Nothing is to be done for naught.'

Geulincx said that this injunction was the 'highest principle of ethics from which easily follows each and every obligation'; for 'if nothing ought to be done in vain', one ought to accept both death and life, not struggling against death when God called one away, nor against life when it was given one. 'And if nothing ought to be done in vain, one ought not to struggle against God's bidding us still to live, detaining us in the human condition.' As far as that condition was concerned, we can only know what it is, not how we arrived at it. Confronted with that question, he says, 'I am at a loss. I don't know, nor do I have another thing to say except, 'I do not know.''

Since we had to live, Geulincx went on, we had to have an

occupation, and we should choose some province that is not only suitable to us, but to which we are equal. From this it seems to follow that we should have some purpose in life. But in regard to this we should remind ourselves that 'God enjoins a purpose for me but not an outcome'; the achievement of the aim 'is not subject to my obligation and not within my power'. The course of life once resolved upon, 'we should do many things, endure many things' in order to serve it 'without interruption and with equanimity'. If the purpose was the pursuit of wisdom 'hard and difficult things' would await the person concerned, 'the criticism of others will lash me, the envy of others will gnaw at me, even the powerlessness of others will restrain and undermine me'.

Geulincx was a Catholic who taught in the Catholic University of Louvain; then became a Calvinist and sought refuge in Holland, eventually getting a job as Professor of Philosophy at Leiden. As with other seventeenth-century philosophers, including the great Descartes, there are contradictions and non sequiturs in his thought, forced on him by theological exigencies. When Beckett discovered him in the Trinity library he wrote to Arland Ussher recommending him. He calls Ussher's attention to Geulincx's definition of humility, which, roughly speaking, may be translated as meaning an attitude to one's own character, life and circumstances which does not value them very highly but does not treat them contemptuously either. It will be seen that there is a rather grim, existentialist tinge to Geulincx which may have caused Beckett to read him with a certain amusement as well as sympathy; but apart from the maxim '*Ubi nihil vales, ibi nihil velis*', which was deeply in tune with his own quietism and seems to have struck him with great force, it is difficult to say how much he was really influenced by Geulincx.

Both the sympathy and the amusement may have been increased, however, by the passage in which Geulincx makes 'relaxing the soul on occasion, lest it be broken in the struggle' his sixth ethical obligation. 'One ought to make propitiation gratuitously once in a while; one ought to walk; one ought to wander or stroll; one ought to indulge in jokes, stories and dancing with friends, one ought to banquet, one ought to drink . . . (I almost said one ought to be silly, but in the proper place).' If one obeys this particular ethical obligation, Geulincx says, letting 'the soul recede now and then . . . one can return to serious matters all the more vigorously and with a longer stride'. Beckett and his admired James Joyce certainly believed in this obligation; perhaps Joyce the more so, but Beckett certainly as well.

Geulincx followed Descartes in adopting the latter's rigid dualism of mind and body, but he was bothered by the evidently ridiculous aspects of the doctrine just the same. The great Descartes had admitted that there could occasionally be a certain interaction between the two. Animals he regarded as complete automata, but human beings had a soul which resided in the pineal gland. There it came in contact with the 'vital spirits', and through this contact there was an occasional interaction between the soul and body. Geulincx abandoned the pineal gland to account for the quite evident interactions of mind and matter which caused the body to behave as if the mind controlled it, and invented instead an elaborate and rather delightful theory of two clocks, both set by the creator to the same time. When both point to the hour both strike. If you saw only one and heard only the other you would think that the one you saw had caused the other to strike. This was the case with mind and body. When you put forward your hand to take something from the table it was because your body-clock had been set so that you would do this. That your mind appeared to will the operation meant simply that the mind-clock had been set to will it.

This improvement apart, Geulincx followed Descartes in regarding mind and body as separate kingdoms. The world *in re* and the world *in intellectu* were two separate domains, a nonsensical proposition but one which Murphy, the hero of the novel Beckett was now writing, believes. He 'felt himself split in two, a body and a mind'. He admits that they have some sort of intercourse, 'but he felt his mind to be bodytight and did not understand through what channel the intercourse was affected nor how the two experiences came to overlap'. The world was divided into two, macrocosm and microcosm, the macrocosm being the external reality and the microcosm the mental one, the kingdom of one's own mind. Murphy, like the patients in the Magdalen Mental Mercyseat, prefers the microcosm to the macrocosm. They have successfully retreated within the microcosm and he attempts to do the same, achieving such a retreat when, the needs of the body for the moment satisfied, he binds himself to his rocking chair. This preference of Murphy's was undoubtedly shared by Beckett, though, as in the case of any artist, one must beware of confusing the use that he makes of philosophers' theories with a coincidence of belief. There is no doubt that Geulincx's quietism, like Schopenhauer's pessimism, had a deep personal appeal for him. But there is still something ineffably comic in the idea of one who was so much at the mercy of his body, and whose body was so much at the mercy of his mind, toying with Cartesian dualism even as a system of belief worth investigating and understanding.

In general too much has been made of Beckett's interest in philosophy and too little of his impatience with it. He told Tom Driver in the early sixties that the language of Heidegger and Sartre was 'too philosophical' for him. 'I am not a philosopher,' he said. 'One can only speak of what is in front of one and that is simply a mess.' Yet he did take some interest in the pre-Socratic Greek philosophers, Zeno the Eleatic, Parmenides, Democritus of Abdera. Heidegger has been largely responsible for twentieth-century interest in these early Greek figures, who were, he claimed, more surprised by being – which to him was the great surprise – than other later philosophers were to be. Beckett's friend Jean Beaufret would have been especially aware of this attitude of Heidegger's and was himself especially interested in Parmenides, whom he translated. But one must beware of reading too much into Beckett's interest. He liked to quote Democritus of Abdera's 'Nothing is more real than nothing', but Vivian Mercier thought it was simply the shapeliness of the aphorism that delighted him. And he told Alan Schneider, the first American producer of his plays, 'I take no sides. I am interested in the shape of ideas. There is a wonderful sentence in St Augustine: 'Do not despair; one of the thieves was saved. Do not presume; one of the thieves was damned.' That sentence has a wonderful shape. It is the shape that matters.' But while it may have been the shape that mattered in an unattributed sentence which he transcribed into a notebook about this time, the sentiments are so near his heart that one is inclined to think otherwise: 'The stoics aspired to Apathia, the repression of all emotion and the Epicureans to Ataraxia, freedom from all disturbance.'

Chapter Fifteen

In these early months of 1936 he continued to work on *Murphy*, sitting before it most of the day on most days, even though he sometimes thought it was going 'from bad to worse'; and at other times it would not move on for him at all and he got 'held up over the absurdest difficulties of detail'. Having been begun in the summer of 1935, it would take about a year to write, and was finally finished in June 1936. When not engaged on *Murphy*, he walked, sometimes feverishly and up to ten miles a day, accompanied by his mother's Kerry Blue bitch. This is the dog referred to in *Krapp's Last Tape*, when Krapp nostalgically tries to imagine himself as being 'again on Croghan on a Sunday morning, in the haze, with the bitch' and stopping and listening to the bells; before adding 'all that old misery. Once wasn't enough for you.' One such long tramp took him far beyond Croghan and along the ridges between Glendalough and Glencullen. Some days he walked from nine or ten in the morning until six or seven in the evening, scarcely seeing a soul. Telling MacGreevy about this, he said it saved masturbation.

Partly as a result of being stuck so often with *Murphy*, he was toying with the idea of going into Frank's office on a regular and permanent basis and asking for a small salary in return for whatever humble work he was permitted to perform. There was a self-immolatory aspect to this idea which perhaps pleased him; for certainly when what many would have thought was a more enticing prospect opened up before him, he promptly turned it down. Mary Manning, now Mrs de Wolfe Howe, wrote from the United States to say she thought her father-in-law, who was a Harvard professor, would be sufficiently influential to get Beckett a post as lecturer. In reply he hurriedly said he thought MacGreevy would be more fitted for such a position and urged him to write to Mary about it. MacGreevy did so but nothing came of it. His unwillingness to pursue this opportunity for himself has been ascribed to the fact that Manning suggested that he should ask Joyce for a recommendation, something which he was loath to do; but there is no evidence for this and when he suggested applying for the job to MacGreevy he also mentioned the possibility of a reference from Joyce.

Murphy was finished in late June and the typescript was promptly despatched to Chatto and Windus who, having sold only two copies of *More Pricks Than Kicks* in the previous year, now, almost as promptly, turned it down. Charles Prentice, who was away from the office due to illness when this happened, suggested another publisher, Heinemann, but by early August they had turned it down as well, saying that they could not justify its presence in their list on commercial grounds. Though Beckett was not enthusiastic at this point about George Reavey who, he felt, had made insufficient effort to distribute the poems, he now decided to avoid further direct humiliations of the sort by letting him handle the book. He would have been conscious of the fact that his mother and Frank were both made aware of these rejections when they saw the package arriving with the postman; and perhaps he was thinking of something Joyce said when Arthur Power consulted him about employing an agent: 'Anything would be better than having the parcel thrown back at you while you were still in bed in the mornings.'

In late June, while *Murphy* was still with Chatto, Mary Manning arrived in Dublin from the United States, bringing with her two American girls for whom, as a married woman, she was supposed to act as some sort of chaperone. One was the subsequently well-known poet, Isabella Gardner. The other was a blonde, very pretty, rather extrovert girl called Betty Stockton, talkative and intelligent, but without any literary interests and in fact already engaged to be married. When, at Mrs Manning's urging, Beckett came to tea at the Mannings', he found the two girls there as well. As was so often the case, he took little part in the conversation, but since everybody else rattled on his silence was not very noticeable. A day or two later Mary, the two girls and Sam all went for a drive in the Dublin mountains together. Betty enjoyed this, but there was no sign that she had aroused any special feelings in Beckett; and it would have been difficult to discern them if she had, because he was again almost silent throughout.

Shortly afterwards he surprised everybody by calling at the Shelbourne Hotel, where Betty was staying, and asking for her. When she appeared, he asked her to have coffee in the big lounge of the hotel, made some rather incoherent conversation and finally presented her with a poem. It was not the first time she had been given poems by young men, but it had seldom happened on such short acquaintance and she was considerably taken aback, even more so when she read through it.

Why were you not simply what I despaired for
an occasion of wordshed

is it better to abort than be barren
the hours after you are gone are so leaden
they will always start dragging too soon
the grapples clawing blindly the bed of want
bringing up the bones the old loves
sockets once filled with eyes like yours
all always better too soon than never
the black want splashing their faces
saying again nine days never floated the loved
nor nine months
nor nine lives

Saying again
if you do not teach me I shall not learn
saying again there is a last
even of last times
last times of begging
last times of loving
of knowing not knowing pretending
a last even of last times of saying
if you do not love me I shall not be loved
if I do not love you I shall not love
the churn of old words in the heart again
love churn of old thud of the plunger
pestling the unalterable
whey of words
terrified again
of not loving
of loving and not you
of being unloved and not by you
of knowing, not knowing, pretending
pretending
I and all the others that will love you
If they love you.

235

When she read the poem Betty was both mystified and embarrassed. Even a girl unaccustomed to reading any but the simplest poems might have come across the sentiments of the last part before – someone saying that he had never known what love was until he met her, that he loved her and her only and would never love anyone else. But what were the references to abortion and nine months about, and why should they be there?

The receipt of the poem put her in a quandary. Should she allow the acquaintance to continue? She was not attracted to Beckett and other than through politeness to Mary had no desire to see him again. However, when Mary suggested other social occasions at which he might be present she decided to go along without saying anything.

Shortly afterwards they all went to the beach together. During the drive there Beckett was moody and silent and made no reference either to the encounter in the Shelbourne or to the poem he had given her. She had in fact thrown the poem away, partly out of anger and partly out of embarrassment, but she did not tell him that. At the beach Beckett sat much of the time apart. As evening drew on and they prepared to go he stood alone, gazing out at the waves and occasionally skimming a pebble along the wave-tops. Betty thought he looked very desolate and in spite of her anger and embarrassment she pitied him a little. When she approached she noticed that he was shivering and asked him if he was cold. He replied brusquely that he was not, adding, 'I just shiver sometimes.'

Some days later he called at the Shelbourne again, said he had a car outside and asked her to go for another drive into the mountains, 'to listen to the silence'. She refused, saying she was going to the theatre but again she consented to have a cup of coffee with him in the Shelbourne lounge. There he told her about Joyce, a great writer with whom he had been intimate in Paris. She had never heard of Joyce and did not understand why he should tell her about him except to make conversation. But she listened politely, hoping he would go. When shortly afterwards she left Dublin with her family to stay with the Somervilles in County Cork she told Mary not to give Beckett her address. She never saw him again. It had all been a somewhat unpleasant and disturbing experience for her.

What was it for him? The poem begins by saying he had hoped she would just provide the inspiration – or the occasion – for a poem. Mary Manning was inclined to think he was simply infatuated but that of course he had, as usual, no clear object in view, either seduction or anything else; but writing from Germany later on he spoke of the hurt

236

what he considered to be a rejection had caused him: 'That is worse than being disliked from the beginning.' The poem was published in the *Dublin Magazine* in the October to December issue of 1936 and did not appear again until Beckett's *Poems in English* was published in New York twenty-five years later in 1961. It is among the simplest and most straightforward of Beckett's poems and suggests a degree of real feeling for Betty Stockton that is unmistakable. Even in its transparencies, though, it is concerned with the problem of utterance and the relationship of art to life, 'the churn of stale words in the heart again', almost as if the importance of the emotion was that it demanded to be uttered – later to be a perpetual Beckett theme.

With *Murphy* beginning its long course of rejection his thoughts turned again to finding some other occupation than writing, in fact some way of escaping his destiny. Besides his cinematic hopes, he began to cherish a strange ambition to become an aviator, even perhaps a commercial airline pilot. Aviation, like film-making, played a part in the modernism which prevailed among a limited circle in Dublin. A rather raffish baronet, Sir Osmond Esmonde, who frequented the Bailey and the back bar in Jammet's, places where Beckett also sometimes went, had an aeroplane, a De Havilland Moth, in which he took his literary friends, Liam O'Flaherty and Francis Stuart among them, up for jaunts. Flying lessons were easily and cheaply available at Baldonnell Aerodrome, and other writers whom Beckett knew, Stuart included, were also taking them. He hoped he was not too old or too stupid about machines to take it up seriously and to qualify as an airline pilot. 'I do not feel like spending the rest of my life writing books that no one will read,' he told MacGreevy. As in the idea of going to Moscow there was an element of day-dream about this, but then he was for long periods alone, had plenty of time on his hands and suffered from the impractical magnification of wish which afflicts those who are so situated.

Another job opportunity of a surprising sort came about in late summer. The *Dublin Magazine* was a review edited by the poet Seumas O'Sullivan, who had been a minor figure in the Irish literary revival. It was a well-produced quarterly publication whose main fault was a lazy eclecticism which led the editor to publish almost anything submitted by any reasonably well-known Irish writer. Both Beckett and MacGreevy had been contributors and for many years Leventhal wrote dramatic criticism for it. It did not have much of a circulation but O'Sullivan, whose real name was James Starkey and who was married to the painter Estella Solomons, a contemporary at art school of Cissie Beckett's, was able to keep it going financially.

In August, shortly after accepting a poem of Beckett's, he sent the proof asking if Beckett would consider making one line out of two in the interests of his pagination and requesting that Beckett should come and see him. When he did, O'Sullivan surprised him by saying that he was anxious to retire and enquiring whether Beckett would be interested in editing the magazine in his place. There would be a small salary attached and O'Sullivan would arrange for the magazine to be financed for at least three years. It is possible that this offer may have come about through Cissie, with whom the Starkeys were on very friendly terms, but in any case Beckett did not accept the post. He may have been instinctively averse to something that would draw him even further into the Dublin literary swim. One could also say that he was avoiding jobs of a para-literary nature while day-dreaming about more esoteric kinds of employment. But this is probably to be too rational about it. His anxieties about employment were desultory and full of contradictions.

And in any case he was being drawn, not altogether against his will, into the literary and artistic scene. He was regularly meeting the painter Sean O'Sullivan, very much a figure in the pubs around Grafton Street, and drinking Jameson with him in his Stephen's Green studio. He went with the biographer Joseph Hone to visit Arland Ussher in his ancestral home in Cappagh, County Waterford, where he stayed for a few days, making enjoyable trips to Ardmore and to Cashel, the old fortress-cathedral of south Tipperary. At Ardmore he very much admired a small stone carving on the west door of the ruined abbey, a stooping figure under a spear which, he thought, represented the conversion of St Declan, the abbey's supposed sixth-century founder. He thought Cashel was wonderful, raised as it was on its rock under a huge sky, which made him feel very minute.

When George Reavey came to Dublin on a visit and expressed a wish to see Francis Stuart, they drove over the Sally Gap to where Francis lived in desolately beautiful surroundings at Laragh. Francis was married to Iseult MacBride, daughter of Maud Gonne and a French Royalist politician, Millevoye – not, as Beckett thought, Clemenceau – and step-daughter of Yeats's 'drunken vainglorious lout', the patriot John MacBride. The Stuarts had a visitor, a man with one arm called Wall. Wall was a member of a wealthy family and had lost the other arm to a passing tram when driving insouciantly along Merrion Road in a touring-car with his elbow over the driver's door. They had tea and there was a conversation about Stalin, with Iseult saying that he reminded her of Kitchener and Wall averring that he probably was Kitchener. Kitchener had of course disappeared at sea when the warship on which he was journeying to Russia was torpedoed in 1916. The historian A. J. P. Taylor is said by a recent biographer to have once propounded the same theory.

Another house that Beckett visited at around this time was the Upper Drumcondra residence of the left wing, revolutionary novelist Peadar O'Donnell. O'Donnell, who had been prominent in the IRA and its breakaway organization the Republican Congress, enjoyed a reputation as a firebrand. His wife was a businesswoman who owned a nursing home, and in middle-class Drumcondra O'Donnell lived a life which was the acme of bourgeois respectability, with holy pictures on the walls and evenings rounded off with the Rosary.

Beside these sociabilities, there were visits to Rudmose-Brown's, where Owen Sheehy-Skeffington, a well-known liberal intellectual and inspirer of various protest movements who had taken Beckett's place as lecturer in Trinity, was holding forth in French – as Beckett rather maliciously put it, *au brogue*. And of course there were the usual visits to Jack Yeats's, whose painting he was still paying for.

His attitude to the poet Brian Coffey had been very fickle. There were times when he felt he could not bear his allusive, would-be witty conversation with its unremittingly esoteric display of philosophical and literary learning. At others he liked him and must have recognized the poet in him, for Coffey was a poet who achieved his own voice more than once, but most notably in the *Missouri Sequence*, part of which is dedicated to MacGreevy. In the latter part of 1936 Beckett found himself on very friendly terms with Coffey and lunched and dined on several occasions at his family home, seemingly making no objection to the brand of right-wing, Catholic politics, embellished with quotations from Maurras, which Coffey's father, who presided over the dinner-table, was propounding. There were also friendly bathes and walks, and when Reavey came over there was a sort of Paris reunion between the three of them. Someone said of Coffey's poetry that 'you needed a degree in Thomas Aquinas' to get anything out of it. In so far as his outlook influenced Beckett, it was another reinforcement of the respect for theology displayed, in their different ways, by MacGreevy and James Joyce. As has been said, the nihilistic, world-despising part of Beckett's own temperament was not far from being a theological stance; and he was at pains to insist to MacGreevy that the attraction of Geulincx's work for him was 'its saturation in the conviction that the Sub Specie Aeternitatis vision is the only excuse for remaining alive'.

When Reavey came to Dublin he brought news of *Murphy*, which was now well embarked on its futile journeys to and from publishers' offices. Chatto and Heinemann had been followed by Dent, and, in the United States, Houghton Mifflin. When Sam relayed the news of these rejections to Frank he was dismayed that his brother should ask bluntly why he could not write the way people wanted. His answer was that he could only write the one way. Frank, who, after all, was supporting the household, replied rather tartly that everybody was very lucky that he did not feel obliged to implement such a policy in 6 Clare Street.

The exchange hurt Beckett, but he could see that to some extent Frank also felt trapped in the life he was leading. Like Sam, he was perplexed by their mother's moods and frequent depressions. He was worried about the upkeep of Cooldrinagh and of course the presence about the place of a thirty-something-year-old brother who seemed to have no prospects and little practical ambition or capacity did not help. On top of all this Frank was now suffering from a mysterious form of neuritis which had robbed him of the use of his left arm, possibly also psychosomatic in origin, and equally possibly, if some psychologists are right, due to the same absence of mother-love in infancy.

To non-literary eyes the rejections Sam's book was suffering were proof positive of failure in his chosen profession, and with Frank as well as his mother now looking askance at him he began to think about going abroad again. There were also renewed discussions about giving up Cooldrinagh, which would mean either sharing a smaller house with his mother or living by himself. 'I think we all feel that the present home situation should not go on much longer,' he told MacGreevy. 'But', he asked characteristically, 'if we give it up, what becomes of mother?' Of course he was reluctant to face the unsettling prospect of such a change in things himself.

Now, in discussion with his friend, he began to think of Germany as a place to escape to, at least for a while. He had begun to consider art criticism and connoisseurship as a possible profession. It was, after all, MacGreevy's way of earning a living. Why should it not be his as well? And why should the German trip not be presented in this light to his mother? She was worried about his future. Such a trip could be looked on as a preparation for it. And once the idea of an expedition to see paintings had been propounded to his mother as a way of softening the blow of departure, it began to take on a life of its own. Flimsy though the excuse may have been to begin with, it developed, no doubt assisted by his inherent work ethic, into a project. When he did get to Germany he would spend most of his time looking at paintings and he would arrange the course of his itinerary so that he would see them. Of course, he had always been interested in painting. The trip would be primarily to study Old Master paintings. What is surprising, though, is how closely he clung to this intention and how strict he was with himself.

However, no sooner had he formed this resolution and announced it to his brother than Frank asked him to postpone his departure for a few weeks longer. He sensed that Frank too had plans, perhaps in regard to the house in which they all lived. He felt bad about his poor brother toiling away in the office, feeling all the time that it was 'strangling his life'. But then, Sam also thought, who did not? It was usually 'only a question of the hold employed'. All in all, Sam felt it would be easier for Frank if his younger brother were to clear off for a while. Mac-Greevy, now working as an art critic in London, planned to go to Vienna for a week or two to visit art galleries there. Perhaps they might even conjoin for a while? At length he decided that in spite of all obstacles, he would go in late September. 'My plans are now simply to get to Germany and then . . . I hope to be away a long, long time,' he told MacGreevy.

There was, rather surprisingly, an air service from Dublin to Amsterdam and he was much attracted by the idea of flying to the Continent instead of going by a tedious sea route. But he abandoned this on the grounds of expense and booked a passage on the American liner the *Washington* which called at Cobh and would sail again on 25 September. The *Washington* also called into Le Havre, where the passengers were free to disembark for the day. In Le Havre Beckett found himself experiencing a measure of nostalgia for France and would gladly have stayed where he was. Germany's claim on his emotions was after all perhaps a mere matter of associations now broken. On arriving at Hamburg he felt sad, because of them, to be there. Indeed he was hardly on German soil before he began to think how pleasant it would be to return to Paris, perhaps in the spring. He was already depressed and rapidly falling into the sad and solitary state of mind which was to be his for much of the trip.

He had not sought introductions from anybody before he left, but Cissie Sinclair had insisted on giving him some. One was to Frau Sauerlandt, widow of Max Sauerlandt, whose book about the art of the last thirty years, *Die Kunst der Letzten 30 Jahre*, had been banned, and he went to a party she had arranged. She was a great admirer of Karl Schmidt-Rottluff, who had been one of the founders of '*Die Brücke*' with Erich Heckel and Ernst Kirchner. As Beckett rather sarcastically remarked to MacGreevy after a later meeting with her in Berlin, 'She sschwarms madly for Schmidt-Rottluff.' He went on to say that if she had spittoons in her Hamburg flat they would be designed by that painter.

One might have expected Beckett to be impressed by the work of the great German expressionists, whose emphasis on human suffering was not very different from his own, but except in one or two instances the truth was otherwise. For some reason, perhaps because of Frau Sauerlandt's description of his work, he characterized Schmidt-Rottluff as metaphysical and spoke sarcastically of his soulfulness. Though he turned down the opportunity of meeting some of the most famous expressionists, Schmidt-Rottluff included, the only painters he met during his stay were a Hamburg set which had been influenced by '*Die Brücke*'. Of these he was most taken by Karl Ballmer, a Swiss; and by Willem Grimm. He liked the reserved Ballmer personally and was deeply impressed by his work, thinking it was like nothing else he had seen, except, in mood, some later Picassos he admired. He would remain faithful to Ballmer and always thereafter included him in his lists of significant twentieth-century painters. As was often the case with Beckett, critical judgement was influenced by personal acquaintance and liking.

He spent a good deal of time in the cheap hotel he had found in Schlütterstrasse, sleeping for much of the morning, then going to the art galleries and eating a cheap meal on his own in the evenings. The city itself disappointed him as there was little of it that pre-dated the nineteenth century and the city art gallery was not much better. The Nazis had now been in power for more than four years, and though the epoch-making exhibition in Munich which portrayed the great German paintings of the twentieth century as decadent and unworthy of the new order would not take place until 1938, the main lines of their policy and their attitude towards expressionism and other forms of German modernism were already becoming clear. In the Hamburg gallery the modern works had already been taken down and hidden from the public view, but it was possible for foreigners to obtain a permit to see them and this Beckett did. He also saw some private collections, including those of Frau Sauerlandt and a merchant prince called Hudtwalcher. Throughout his trip he would accept such evidences of the new barbarism with what almost amounted to apathy, hearing the complaints of critics and artists with occasional impatience and sometimes even sarcasm. He was bored by Nazi propaganda and the interminable Nazi harangues on the radio; but he was equally bored by anti-Nazi complaints. The passages published from a recently discovered diary relating to this period do nothing to change the impression of an almost total apathy, related perhaps to his generally depressed mood.

Hamburg then, as later, was famous for the pleasures its red-light district offered. Beckett had not been above sampling the similar pleasures on offer in Kassel and elsewhere even at the height of his relationship with Peggy Sinclair. He now felt no impulse to do so, though he did note somewhat regretfully that by government decree brothels were allowed in seaports but nowhere else. His original intention on leaving Hamburg had been to walk across the vast and desolate Lüneburg Heath, but he soon found the weather too cold and the Lüneburg Heath itself too depressing, so he retreated to whence he had come and took trains to Hanover and Brunswick. Much later, in the story titled 'The Expelled', the narrator would remark, 'Well, the Lüneburg Heath didn't please me at all but not at all. I came back disappointed, at the same time relieved.'

From Brunswick, where he spent a week, he made an excursion by train to nearby Hildesheim and found it so entrancing that he repeated the experience several times. It was one of those affinities with a place which the solitary traveller, however lonely and otherwise disconsolate, sometimes experiences. By 11 December he was in Berlin, where he stayed first in a sinister hotel near the railway station which reminded him of the one in Dublin called the North Star with which he was familiar, at that time a rather louche place and a resort of prostitutes and their pimps which was opposite Amiens Street Station in Dublin. Then he found another in Budapeststrasse near the Zoo. Berlin he found more sympathetic than either Hamburg or Brunswick. He spent much time in a nearby park, from which skies which reminded him of Dublin's and were nearly as good were visible, and took trains to the lakes and the forest.

The news from home was good. Frank was off on a skiing holiday and so his mother was alone but she seemed to be in excellent spirits, which was a good augury for a future in which neither of her sons might be under her roof. The truth probably was, however, that May Beckett's states of mind were not greatly contingent on the presence or absence of her sons at all. She too was a depressive, though more inclined to periods of elation than he was; and, as is the case with depressives, her mind was subject mainly to its own laws.

The dominant mood of Beckett's own mind during his sojourn in Germany was now established. Like another Irish traveller in Germany almost 200 years before, the poet Oliver Goldsmith, he would be 'remote, unfriended, melancholy, slow'.

In Berlin he resumed contact with Frau Sauerlandt and accompanied her to a party where he found himself disagreeing with her all evening in a room where Heckels, Kirchners and Schmidt-Rottluffs hung on the walls. To MacGreevy he described Frau Sauerlandt's 'voice crackling as a whole group of German voices crackles, raining down curses on the government'. He said that through her he could have made the acquaintance of Nolde, Schmidt-Rottluff, Ernst Heckel and other publicly banned painters, but turned down the opportunity. 'They are all great proud angry poor put upons in their fastnesses and I can't say yessir and nosir anymore.'

It was an extraordinary blindness, to the art and to the situation of the artists alike. The truth was that they were great, they were put upon and they were now being forced to retreat to such lonely fastnesses as they could find. But Beckett was far more interested in the old masters to be seen in Berlin – Signorellis, del Sartos, a Carpaccio Entombment, a roomful of Van Eycks and six or seven Brouwers, though the latter's famous 'Moon Landscape', which had belonged to Rubens, was not there. When he went to Dresden he would enthuse about Giorgione, Raphael, and an 'indescribably lovely Vermeer'. He was less impressed with the Rembrandts, Halses, Titians and Rubenses, 'the immensely competent bullies and brow beaters and highwaymen and naggers' of art.

'Is it a pettiness to move away from the art that takes me by the scruff of the neck?' he would ask himself. The emphasis throughout the trip was on Old Masters, for these were what he had come to see, but in Hamburg he had admired a Degas portrait and found some of the modern Germans, including Nolde, 'interesting'. In Dresden he would see what he described as one of the best private collections of modern art in Germany, the Ida Bienert, and would speak of its 'marvellous' Cézanne; but for the most part it is Old Masters that he is interested in and wants to see; and though he had used his Hamburg permit on the director of the Kronprinzerpalais in order to see the closed modern rooms, he had been more interested in the Munchs and Van Goghs than in anything German. Sometimes his comments sound more like those of an aspirant professional connoisseur than of a writer seeking sustenance, or even of an art critic. Thus in Dresden he becomes interested in the question of whether some of the Giorgione has been painted over in the nineteenth century, is rather proud of the fact that he notices there is something wrong with the painting of a leg before being told this is a disputed passage, and finally beards the Director and gets to see the X-ray photographs, which he finds of little help. He tells MacGreevy that he had hoped to leave Dresden 'with a more definite attitude towards Giorgione but it is the other way round. He must remain to be seen in Castelfranco as exclusively as Grünewald in Colmar.'

In Berlin in the New Year, his depression continued. He had already decided that his trip was a 'failure' and that Germany was 'horrible'.

His journey was merely a 'journey *from* and not *to*', as he had known it would be before he began it. He suffered another sebaceous cyst on his anus which refused either to burst or go away until finally there were only two endurable positions left to him, 'both fully recumbent'. He stayed in bed and eventually, in his sleep, the cyst burst. But he had a gloomy conviction that the relief was only temporary. He was very tired and only wanted to go home and lie down but the idea that he might never be in Germany again kept him going. In his depressed state he thought of the money he was receiving from home every month as charity rather than something that he was entitled to. And though he saw few people he found his aversion to human beings increasing every time he went out and met any. 'I shall never learn what to do with my tether of life,' he remarked gloomily. And to Mary Manning he said that the physical mess was nothing compared to the mental.

He had made the acquaintance of Axel Kaun, a young man whom he described to MacGreevy as a bookseller's improver, who had been taken on by the publishers Ruhwoldt, publishers of Beckett's Dublin acquaintance Francis Hackett in Germany. Kaun turned up at the hotel one afternoon on his way to catch a train. In the end he missed it by half a minute and Beckett tolerantly waited with him until there was another after midnight, being in his company for twelve hours altogether. It was a long twelve hours.

Before the cyst began to trouble him he had managed to get to Potsdam, where he found the Sanssouci Palace exquisite: 'Truly an architecture without care'. Depressed by the thought of facing trains and boats, he was beginning to entertain the hope that he might fly to London if he could save enough from his monthly allowance. But by the end of the month, in a slightly more cheerful frame of mind, he had arrived in Dresden by way of Leipzig, making the acquaintance en route of a stage designer who spoke enthusiastically to him of Mexico and Ben Craven and was of an 'unbelievable kindness'. He was so taken with this individual that he mentions him twice in his letters. He also liked Dr Willi Gröhmann, the Jewish art critic and historian, to whom as well he had been given an introduction by Cissie Sinclair, though in describing Gröhmann's career to MacGreevy he displays his usual lack of emotion about what was going on in Germany, saying merely that he had been dismissed from his post as Director of the Zwinger Gallery by the Nazis 'like all the others of his kidney'. Though the former gallery director had many German paintings Beckett was more interested in his Picassos, Klees, Kandinskys and Mondrians. And there is a suggestion that Gröhmann was in any case a bit of a name-dropper in Beckett's remark to MacGreevy that he 'knows them all, from Picasso to Salkeld', a joke about the status of his Dublin painter friend, Cecil Salkeld, which would not be lost on his correspondent. It was through Gröhmann that Beckett obtained a pass to the Ida Bienert collection, where he was pleased and somewhat surprised to see a portrait of Nancy Cunard by Kokoschka which had been painted in Paris in 1924.

Through Gröhmann also he fell into company with a group of five Russians, 'blue with blood and revolutionary privations', and though he showed little interest in the Germans he met with the ex-director, he found drinking and talking with these Russian émigrés so agreeable that he delayed his departure from Leipzig for a couple of days. His indifference to the general situation of contemporary artists was marked again when he made the acquaintance of a private gallery owner who still dared to exhibit Franz Marc and Emil Nolde. It is not that one expects Beckett to be in a continual ferment of indignation. There is a limit to the indignation one can go on expressing. But he simply does not seem terribly interested in what is going on; and though some of the public manifestations of the new gospel repel him, he seems oblivious of the deeper threat to civilisation itself. It is strident politics he hates; and strident protest too. Though it was art which had ostensibly brought him to Germany, he seems strangely unperturbed by the regime's determination to bring to an end a great age of adventurous modernism in German art and replace it with a brutal and banal academicism. One could attribute this to his endemic indifference to politics and public affairs were it not for the fact that when he got back to Ireland he would display a positive hostility to those who expressed what he variously called 'all the usual sentimental bunk about the Nazi persecutions' or 'the usual bilge about the persecutions'.

Though the weather there was painfully cold, he found Dresden lovely; but the gallery was full of what he called Italian rubbish, badly lit and badly hung. He went to a marvellous performance of *The Marriage of Figaro* at the opera house. It was, he realized, the first time he had ever gone to the opera and been sorry when the performance ended. His anus was better, the cyst having, as he put it, 'more or less settled down', but he knew it was only a question of how often and how painfully it would fill up again before he finished the journey and could get home and have it lanced and cleaned. He was now also suffering from an itch associated, he thought, with the same complaint; but it was something he had had before and was used to.

In Berlin he had had a letter from Reavey telling him that *Murphy* had been turned down by Cobden Sanderson and was now on its way to Faber and Faber, who would in turn be followed by Secker and

Warburg and the Hogarth Press. More gallingly, Reavey enclosed a report by Richard Church for Dent, who had declared himself greatly impressed when Reavey lunched with him. The report gave it as Church's opinion that 'this man is a most remarkable and highly equipped writer. The humour, the sophistication, the sense of structure and the clear originality make me agree with you that he is a man fully worthwhile fostering.' Nevertheless, Church went on, Dent had already lost money on one or two authors of this kind and could not be expected to do so again. Rejected again, he dreaded the thought of going home and hanging about the house with nothing to do and no visible means of support. He had been day-dreaming about having the proofs to correct and the details of publication to attend to, which, he thought, would have proved to his mother and to Frank that he had an occupation like anybody else. Now he saw nothing but aimless moonings about Cooldrinagh and endless discussions of what he might do with himself. He had sold the trip to Germany to his mother as something which would result in a book or other form of advantage, and he thought she would expect immediate results, but he knew very well that he would not be able to make any immediate capital out of it in this way.

By the last week in February 1937 he was on his travels again, this time to Munich by way of Würzburg and Nuremberg. In the latter city he examined a twelfth-century abbey which had been built by Irish monks and possibly, he thought, illustrated the text 'heaven and earth shall pass away, but my word shall not pass'. He was intrigued by the Celtic motifs, but whether this was because of their intrinsic interest or because of some nebulous project he had in mind is not clear. In Munich he found lodgings in the Akadamiestrasse in the Pension Romana, which he described as no worse than any other lodging he could have found. Blocking out the view from his windows was the Akademie building, on whose roof stood a representation of Minerva which he thought looked like a quantity surveyor taking a level.

Wandering among the classical buildings of the city, he did not find it very exciting. But the weather had improved – the morning he arrived had an almost summer warmth: he liked the blue trams, heard a thrush singing in the Maximiliensplatz, and had a sense of relief at having reached a more permanent resting place again after his journey. He was tired of travelling and felt like making a bolt for home. Stuttgart and Frankfurt still remained on his theoretical itinerary after he had 'finished with' Munich. But in the mood he was now in he thought that he would abandon further explorations, moving to an even cheaper billet than the Pension Romana and saving up for the price of an aeroplane to London, though he had doubts about whether the currency regulations would allow him to pay the whole fare in Reichmarks. Still hoping for good news from Reavey, he went frequently throughout his stay in Munich, which lasted three weeks, to the post office to see if there was any. But until close to the end of his stay, none came. There had been a hope that Houghton Mifflin in the United States might take the book. They had suggested cuts, and so desperate was he for publication that he agreed to make some. The principal cut demanded by Houghton Mifflin was the chapter on Murphy's mind. This cut he balked at, but would make others if required, an attitude which of course the older Beckett would not have approved of and which would have shocked many of his later admirers. The Americans also thought the title was, from an 'American point of view', too suggestive of a particular kind of Irishness or stage-Irishness – in other words, that it lowered the tone.

Before leaving Ireland he had set out his attitude to cuts in a lengthy letter to Reavey, asking him to communicate the author's 'extreme aversion to removing one third of his work proceeding from his extreme inability to understanding how this can be done and leave a remainder?' However, he went on, 'if they would indicate precisely what they have in mind and the passages that caused the pain, I should be willing to suppress such passages that are not essential to the whole and adjust such others as seem to them a confusion of the issue'. They had also objected to a lot of the dialogue, which they thought was unreal. Beckett's reply was to say that these passages of dialogue were 'the comic exaggeration of what elsewhere is expressed in elegy, namely, if you like, the hermeticism of the spirit'.

In December, in Germany, so desperate was he for publication, that his attitude had weakened even further; and he ironically told Reavey that he was now prepared to 'go further and change the title, if it gives offence, to Quigley, to Trompetenschlein, to Eliot, or any other name that the publishers fancy'. An English publisher, Stanley Nott, had declared his readiness to publish if Houghton Mifflin in America would take sheets. This was regarded by Reavey as a last resort. Now Houghton Mifflin positively declined to publish and with their refusal the Nott option fell also.

At the same time he received news from home that the Kerry Blue bitch he was so attached to (the one which he mentioned subsequently in *Krapp's Last Tape*) had had to be tapped and was found to have growths. His mother's letter did not say that the dog would have to be destroyed but he understood quite well that this was the position. Also mentioned in two of the stories in *More Pricks Than Kicks*, she had been in the family for twelve years, since the end of his second year in Trinity when the death of another dog plunged him into such deep gloom that he even contemplated suicide.

May's news about the dog, coming on top of the end of the half-expected possibility of Houghton Mifflin publishing *Murphy* in conjunction with Stanley Nott, plunged him, naturally enough, into even deeper gloom. The journey, he now realized, was over, 'mentally as usual long before physically'. He would be glad to get out; and from now on he would simply be hanging around waiting for his flight home.

He did not in any case like Munich, where there was nowhere to go and nothing to do in the evenings. He had written to Mrs Frost asking if he could have a room in Gertrude Street. Geoffrey Thompson had invited him to stay, but he was now married and Beckett could not face the prospect of staying with his friend and a strange wife. On 2 April he flew by Lufthansa to London, changing planes at Frankfurt and Amsterdam, the journey taking seven hours in all. Because he could pay for the whole journey with Reichmarks at a favourable exchange rate it was almost as cheap as the boat and train, most of which would have had to be paid for in Sterling; and, in his exhausted condition, much more pleasant.

So ended the German trip of 1936-7. For over six months he had traipsed around Germany in bitterly cold weather, mostly in a depressed state, often lonely, sometimes ill. He had seen innumerable pictures, paying particular attention to the Old Masters but displaying little curiosity about contemporary art or what was happening to it. He had not sought to enjoy himself in ways that one might expect a young man on the loose in a foreign country to do, particularly one who had been living in his mother's house in Ireland for most of the previous year. Nor had he sought to learn much of politics or the political system which would eventually uproot so many lives, including his own, and destroy so many others, including those of some of his friends. He had not been greatly impressed by anybody he had met, except perhaps the Swiss Ballmer, or to any real extent moved by the dilemma in which some of them now found themselves. About ten years later, in his first essay on the painter Geer van Velde, he would, when discussing the futility of art criticism, include the names Gröhmann and Sauerlandt in a list of four critics whom he described as 'the best', while saying that even they failed to cast much illumination on the work they discussed. This may have been an attempt at making some sort of amends. More likely it was simply one of those displays of mystificatory and pointless erudition he was still so fond of.

Whatever his nebulous assurances to his mother about the benefits of the German trip in terms of an alternative occupation, the results did not justify them. With the exception of Ballmer, Kandinsky and Klee, few of the painters whose work he had travelled so extensively to see would play much part in his scheme of things, and of these only Ballmer was new to him.

Beckett's homecoming in April 1937 was to be his last homecoming to Ireland. When he returned to that country in future, it would be merely as a visitor; and one, moreover, more or less anxious to get away again as soon as possible. In April 1937 he was, if anything, glad to get back. He was weary of his winter travels in Germany and for the moment at least in two minds about going anywhere else. As soon as he was re-established at Cooldrinagh, he found himself going through the days in a sort of stupor which was, on the whole, comfortable and comforting – so much so that he wrote to MacGreevy saying, 'I feel now that I shall meet the most of my days from now on here and in tolerable content, not feeling much guilt at making the most of what ease there is to be had and not bothering very much about effort. After all there has been an effort.'

The atmosphere in Foxrock was still basically one of gloom; and though his mother, like many depressives, had her good days as well as her bad, the illusion of her dependence on him was just as strong as ever. There was now a possibility that Frank would be getting married and setting up a home for himself elsewhere (in fact he got engaged in August), so May's aloneness was even more of a consideration than it had been. But even taking this into account Sam displayed no eagerness to be elsewhere, and by midsummer he was still quite resigned to the fact that his future seemed to be in the country of his birth. It is true that when urged on by Ruddy and others he did agree to apply for a lectureship in Italian in Cape Town and even went so far as to collect testimonials from Trinity professors such as Walter Starkie. But his heart was not in it and on the whole he hoped he would not get it. For a job in Buffalo University about which Mary Manning told him he made even less effort, reluctantly agreeing to meet someone from the

English department there who was coming to Dublin, passing news of the vacancy on to MacGreevy with some haste, and being finally relieved when 'the brute from Buffalo' failed to turn up.

But though not anxious, for the moment at least, to go on his travels again, neither was he very eager for a job in Ireland. During the summer a job as assistant librarian in the National Library at £150 a year was advertised. His mother urged him to apply for this, apparently on the grounds that his grandfather had been the contractor for the building and therefore he would receive preference. When he pointed out to her that the successful candidate would have to have a knowledge of the Irish language, she did not seem convinced. A job in County Waterford as agent for Lord Rathdowne which offered £300 a year and a free house was a rather different proposition. He was forced to give it some consideration because the suggestion came from Frank, but again hastily took care to offer it to someone else. Though it is not clear whether his lordship really wanted a literary man as his estate agent, Beckett first suggested Arland Ussher, who turned it down out of fear that if he took it his father might leave what remained of the Ussher estate to someone else; and then put forward Joseph Hone, also a member of the land-owning class, who unavailingly communicated his willingness to turn estate agent by telegram from Switzerland.

Within a few weeks of coming back from Germany Beckett had seen both of these literary notabilities. He had said before he returned that his only ambition was to sit for months alone in a room he did not have to pay for, but now he slipped back into the Dublin swim with what almost amounted to eagerness, arranging to review Devlin's poems for the *Dublin Magazine*, lunching with Leventhal, noting that Austin Clarke had returned from England and been seen at the Academy with the person whom Dublin called his woman – in other words, his companion for the rest of his life – and that Francis Stuart, who was staying in the Dolphin Hotel while writing a historical novel, was seeing rather a lot of Lady Glenavy. He dined with Francis and lunched on at least one occasion in Jammet's, the solidly furnished, traditional French restaurant off Grafton Street where literary Dublin went. The poet F. R. Higgins was there, lunching with a lady from *Harper's Bazaar*. Higgins, he observed, was growing more and more indistinguishable from his great model W. B. Yeats, so that they now formed a pair with Higgins as the front legs of a rheumatic pantomime elephant. On many nights he was to be found in the back room of Davy Byrne's and Francis Stuart remembered an amusing incident from around this time. There was a competition to see who could stick a postage stamp highest up on the wall, either by reaching or jumping, and several tall men, Stuart,

O'Flaherty and Ussher, had just been surpassed by the heavyweight boxer, Jack Doyle, then in the glow of his brief pugilistic glory, when Beckett walked in and coolly succeeded in placing a stamp at least a couple of inches higher again. To MacGreevy, who was interested in the Catholic clergy, he passed on the (inaccurate) information that 'Father Paddy', in other words Monsignor Padraig de Brun, translator from the Greek, the French and the Irish as well as of Dante's *Inferno*, and a notable figure on the Irish literary-academic-social scene, was to be the next Bishop of Galway. At a party he met some of the young literary turks from UCD, the more nationalist and indigenously oriented of Dublin's two universities, among them Niall Sheridan and Mervyn Wall, associates of the already legendary Brian O'Nolan (Flann O'Brien), soon to be Myles na gCopaleen, who would shortly publish his first great novel, *At Swim-Two-Birds*. It may have been at this party that he incurred the everlasting dislike of Denis Johnston, apart from himself the most talented Irish playwright of the post-O'Casey era and, as Beckett probably knew, the sometime escort of Ethna McCarthy. Johnston also lived in Foxrock and he knew Beckett had a car – in fact his father's – so he asked him for a lift home. The reply was a single word, 'No.' But he was even beginning to be drawn into some of Dublin's little squabbles, the games of liking and disliking which it then, as always, rejoiced in playing. Joe Hone asked him to dinner and then rang to cancel the invitation on the grounds that the Lennox Robinsons who, he understood, were enemies of Beckett's, would be there. As in many other such cases, it was useless for Beckett to say that he was stupid about such things and never knew who liked him or who didn't: he had to content himself with an invitation for another night. Through Joe he also met the well-known painter Evie Hone, whom he found to be agreeable, but noticeably dirty. It was with Joe Hone that he also went down again to stay with Ussher in his country house at Cappagh and he had a pleasant outing with Frank to Clonmel, which included walking in the Blackstairs Mountains together after Frank had finished his business.

On a rather different plane were his visits to Jack Yeats and his conversations with Brian Coffey. Yeats was perhaps his only true friend in Ireland in the sense that he truly rejoiced in the relationship, felt it to be important to him and was grateful for the illuminations he derived from it. He was still often irritated by Coffey's over-allusive conversation – erudite, rapid and often seemingly disconnected. It was tiring and he found himself, like others, unable to keep up with it. Still, Coffey, he felt, educated him; and since he was beginning to think that he would never learn anything more except from friends or enemies, he was willing to put up with the unease he often felt in his company. Coffey's poet friend Denis Devlin was a different matter. Whereas Coffey was jobless like Beckett himself, in his neat shirts and suits Devlin was now the complete young diplomat. Beckett sometimes found him amusing but he also thought him calculating, and though he gave *Murphy* to Coffey to read with a genuine anxiety about his comments, he was not so happy at finding he then had to give it to Devlin as well. A review by Beckett of Devlin's *Intercessions* did not in the event appear in the *Dublin Magazine*, for which it was written, but in *transition* the following year. Though almost incomprehensible, it seems to be reasonably enthusiastic; but it is principally notable for the statement that 'art has nothing to do with clarity, does not dabble in the clear and does not make clear'; and for a somewhat confused assertion of the artist's 'need' to create. In subsequent formulations this would be stiffened in an important way. The 'need' to create would be expressed first as a 'compulsion'; then as a response to a 'commandment'.

Murphy continued to come back to Reavey, though Beckett now felt oddly confident that it would eventually be accepted by some publisher or other. In addition to rejections by Chatto, Heinemann, Dent, Cobden Sanderson and the two Americans, Simon and Schuster and Houghton Mifflin, the list now included Secker, Robert Hale, Boris Wood and Stanley Nott. Hamish Hamilton and Faber and Faber would shortly be added to the roll of honour. Meantime, like many authors who have one book going the rounds, he could not begin another, declaring that he could not work in such circumstances.

He was not entirely idle, though, whatever he might say, for he had had an idea for a play about Samuel Johnson and had begun to work on it, making voluminous notes and writing at least part of the first act. Surprise has been expressed that Beckett should be more interested in Johnson than in Jonathan Swift, seemingly on the grounds that if he needed an eighteenth-century pessimist to be interested in, he should at least have fixed on a fellow Irishman and a fellow graduate of Trinity College, Dublin. Perhaps, though, one of the reasons for his interest lies precisely in the difference in outlook between the two, a difference important to Beckett and implied very clearly throughout Johnson's *Life of Swift*.

The implication underlying Johnson's *Life* is that the famous '*saeva indignatio*', the 'savage indignation' of the epitaph in St Patrick's Cathedral in Dublin, was never very deeply grounded. Johnson portrays Swift as a rancorous and petulant man, disappointed of preferment, who reluctantly 'turned Irishman for life' when he did not get the bishopric in England that he thought he deserved. The grounds of Swift's pessimism, he implies, are merely circumstantial; if his life, and the chances which influenced its turning points, had been different, Swift would have had a different outlook on things. As it was, he became petulant, irritable, fretful about trifles, the sort of man who counted the faults of Lord Orrery's servant and despaired of human nature because it was not as he would have designed it to be. Marooned in Ireland, 'his asperity condemned him to solitude; and his resentment of solitude sharpened his asperity.'

Johnson, on the other hand, believed that human nature was inescapably flawed and the lives of human beings inevitably melancholy. True pessimism, he seems to say, is cooler and steadier than Swift's. Its view is of the whole of human life, that of the supposedly fortunate as well as of the unlucky, and wherever it turns it sees inevitable disappointment, modified and alleviated only by illusions of various sorts, the recurrence of false hopes and briefly welcome deceits. In all this he is much nearer to Beckett – and indeed to Schopenhauer – than Swift is, or perhaps one should say that Beckett is nearer to him than he is to the Dean of St Patrick's.

Johnson, moreover, according to Beckett, experienced a 'need to suffer', or a 'necessity of suffering'. He was 'spiritually self-conscious' and therefore a tragic figure. His dogmatisms were unimportant, a 'mere façade' for his 'consternation' at being. The eighteenth century had many such figures, constructing systems and embracing dogmatisms, but really alone in their solitude and terror, but few of them were so 'completely at sea in their solitude as he was, or so horrifically aware of it – not even Cowper'.

There was, however, another, deeply personal reason for his interest in Johnson. The play concerned Johnson and Mrs Thrale and the fifteen years during which the sage was a regular guest in the Thrale household at Streatham. Its novel premise was that Johnson, though 'absurdly in love' with Mrs Thrale, was incapable of sexual congress with her; in other words, he was impotent, or felt himself likely to be impotent if there was any question of bedding her. Doubtless Beckett was loading the broad back of Samuel Johnson with his own difficulties in regard to women, for he, like his Johnson and his more autobiographical hero Belacqua, though not impotent, had difficulty copulating with the women for whom he had real feeling, though not, it would seem, with whores, mere casual acquaintances or the figments of his imagination.

There was, Beckett admitted, no text for the impotence; but he decided that it fitted the facts – 'the fake rage to cover his retreat from her, then the real rage when he realizes that no retreat was necessary, and beneath both the despair of the lover with nothing to love with'.

255

In other words, when Mrs Thrale becomes a widow and free to marry him, he has to break with her out of fear that he and his protestations of love will be put to the test. Then, when he finds that she is in love with Piozzi, he realizes that this charade was not necessary and is genuinely angry: with himself for the unnecessary break rather than with her for preferring Piozzi, a gloss on Johnson's anger which contradicts the usual assumptions. Beckett really believed this theory, making great efforts to 'document' it and insisting that 'the evidence for it is overwhelming'. When he put it forward to Reavey as a possible script for Charles Laughton (an extraordinary idea in itself) he admitted that there was a difficulty about Johnson 'being so old at the crisis', i.e., 'she could hardly have expected much from him. We will make him younger and madder even than he was.'

He was quite frank about the nature of his interest in Johnson with Mary Manning, who had remained in Ireland after the departure of Betty Stockton and Isabella Gardner for the USA in the previous year, 1936, and who had had a play produced in both Dublin and London in that year. Beckett had helped her with the play, suggesting the novel innovation of a completely silent character to her, a suggestion which she had accepted. Mary was unhappy with her life in the United States at the time, and Beckett was also miserable; so they comforted one another, 'weeping on each other's shoulders', going to several parties together and seeing each other a good deal. There may have been an episode with Mary which both confirmed his interest in Johnson's possible impotence and made the nature of it clear to her. In any case, he discussed the theme of his proposed play with her without any reservation and she was afterwards associated with Mrs Thrale in his mind, to such an extent that he sometimes referred to her as 'the Swan of Streatham'.

He also discussed the nature of Johnson's difficulty with MacGreevy, saying that if the play he was going to write was more about Samuel Johnson than Mrs Thrale, it did not mean that Johnson was in the right 'or any nonsense like that'; and explaining the nature of his own interest in him rather than her on the grounds of his spiritual self-consciousness, which made him 'worth putting down as part of the whole of which oneself is a part', while she, 'being merely physically self-conscious', was 'less interesting' to him 'personally'. He added for good measure that 'She of course didn't get what she wanted either, Piozzi being a poor performer.'

Beckett filled three notebooks with material about Johnson, but the only part of the play that survives is the beginning. He was naturally still a long way from his mature dramatic practice, but in its spareness and economy, its use of pauses, silences and repetitions, the fragment that remains is still recognizably by the author of *Godot*. The scene is a room in Bolt Court where Mrs Desmoulins, Mrs Williams and Miss Carmichael are sitting. Mrs Desmoulins is knitting, Miss Carmichael reading and Mrs Williams meditating. The cat, Hodge, who is on stage, is, the stage direction says, sleeping (if possible). Johnson himself does not appear.

> Mrs D. He is late.
> *Silence*
> Mrs D. God grant all is well.
> *Silence*
> Mrs D. Puss puss puss puss puss.
> Mrs W. What are you reading, young woman?
> Miss C. A book, Madam.
> Mrs W. Ha!
> *Silence*

Beckett read all of Johnson, including his favourite text, the *Prayers and Meditations*, in the course of his research for the play and he continued faithful to the great Englishman throughout his life. With his usual percipience Vivian Mercier has identified the influence of Johnson's prose style on many sentences and passages in the English version of *Molloy*: 'And from the poop, poring upon the wave, a sadly rejoicing slave, I follow with my eyes the proud and futile wake. Which, as it bears me from no fatherland away, bears me onward to no shipwreck.'

Beckett held by his theory about Johnson until as late as 1959, when Mary Manning sent him a new biography and he was forced to abandon it. It is not clear whether he believed that Johnson was altogether impotent at the time he knew Mrs Thrale, or only impotent with her, or whether his impotence was in spite of or as a result of his undoubted love for her; but in any case Beckett equated love with impotence, writing to Mary and saying apropos of Johnson that 'the impotent can only love where he is impotent, the whole aim of loving being impotence, even a moment's impotence, i.e. a moment's love'; and adding, 'The sad animal, the impotent animal, the loving animal.' Some time before he received the biography from Mary Manning he had begun to suspect that his 'Thrale theory was all haywire'. The book confirmed these suspicions. And 'the time I spent on that red herring', he lamented. But the time he spent on Samuel Johnson was not wasted, any more than was the time spent on the other great pessimists he read – Leopardi and Schopenhauer among them.

Up to this point he had tried for his mother's sake to create the illusion that he had some sort of career in front of him, whether as writer, art critic or academic; and to pretend to be actively preparing for if not actually pursuing it. His technique at home had been to keep up a pretence of work, going about the house with a preoccupied look and retiring to his room for long periods. Now he did not even pretend. As always he got up late, but now he mooched about afterwards in open idleness. The effect of this was to increase his mother's anxiety about his future and to ensure that he got on her nerves even more than before. To have a lanky and, as she well knew, in some respects brilliant thirty-one-year-old lounging about the place in moody idleness and drifting obstinately into an indeterminate future would have strained to breaking point the patience of a less irritable and less neurotic mother than May Beckett. In her case it produced frequent extremes of anxiety and irritability. Of course, her son's situation was objectively worrying and would have been to ninety-nine mothers out of a hundred; but like other depressives, she was determined to concentrate the worry and anxiety she would probably have felt anyway on a single focal point; and chose her son's situation to be it, and this made things worse.

Shortly after his return from Germany they had been brought together by the death of the Kerry Blue bitch to which they were both devoted. On the Saturday after he came back, while he was at one of Jack Yeats's studio gatherings, the vet decided that the dog should be chloroformed and it was done straight away, though Sam had wanted to be with her at the end, convincing himself that it would make things easier for her. On his return he found his mother prostrated in bed, where she remained for two days while he tried to persuade her to take a reasonable view of something that he found it impossible to take reasonably himself. Later, in August, after his mother had again been in bed with what was called a flu, he woke up to find the stairs, living room, dining room and furniture of the house copiously smeared with blood from the other dog, who had come home during the night with a severed artery. With grim determination, May immediately set about cleaning up, not an easy thing to do since the dog had crawled from chair to chair in an attempt to find somewhere to lie. On top of this Frank, now engaged, got a septic arm from a blister on his hand and went into a nursing home for treatment just in time to avoid amputation. This was at a juncture when he was actively house-hunting and May was naturally concerned about the forthcoming marriage.

Also beginning to loom between them now was the shadow of the libel action which Harry Sinclair, Boss's brother, had decided to take against Oliver St John Gogarty and the publishers of his book, *As I*

Was Going Down Sackville Street. This somewhat mannered production described

> An old usurer who had eyes like a pair of periwinkles on which somebody had been experimenting with a pin, and a nose like a shrunken tomato, one eye of which swung independently of the other. The older he grew the more he pursued the immature, and enticed little girls into his office. That was bad enough; but he had grandsons, and these directed the steps of their youth to follow in grandfather's footsteps, with more zeal than discrimination.

There was also a set of verses in the book inset into a piece of dialogue:

> 'And one thing more – where can we buy antiques? –
> 'Nassau Street, Sackville street, Liffey Street where Naylor's is, and all along the Quays. Have you not heard?

> > 'Two Jews grew in Sackville Street,
> > and not in Piccadilly,
> > One was gaitered on the feet,
> > The other one was Willy.

> > 'And if you took your pick of them,
> > Whichever one you choose,
> > You'd like the other more than him,
> > So wistful were these Jews.

> > 'They kept a shop for objects brought
> > by Masters famed of old,
> > Where you, no matter what you bought,
> > Were genuinely sold.

> > 'But Willy spent the sesterces
> > And brought on strange disasters
> > Because he sought new mistresses
> > More keenly than old Masters.'

In his statement of claim, Harry Sinclair actually alleged that his grandfather Norris Sinclair, who also had an antique shop, had in fact enticed little girls into the back room of his premises and given them sweets before he interfered with them sexually.

Boss had died in May after a long spell in the Newcastle tuberculosis sanatorium, the family having known for some time that he had passed the point at which recovery might have been possible. Beckett visited him often, using his father's car to travel down the twenty-odd miles from Dublin. Virtually on his death-bed, Boss had expressed a wish that Gogarty and his publishers should be sued. Beckett knew as well as anybody that being a witness in the action would be an unpleasant experience; that dirt of various kinds might be raked up; and that as the author of *More Pricks Than Kicks*, whose very title might make it seem salacious, he would be especially vulnerable. At the same time he thought the proceedings might amuse him and he had a certain amount of respect for Boss's dying wishes. May's attitude to the whole thing was in part the lay person's cautious fear of legal proceedings, but neither she nor her husband had ever liked the Sinclairs or encouraged Sam's intimacy with them. Anti-semitic feelings have been ascribed to her, but there is not much evidence that she had any beyond the vague traditional distaste of the denizens of Foxrock for people not of their kind – and for Jews as members of the business community who, belonging to a different freemasonry, did not, Foxrock thought, play fair – which was precisely what his former master, Pannister, had thought about Willie Beckett.

As if all this was not bad enough, it was now clear to May that Sam was drinking quite heavily, borrowing money from Frank to do so, going out almost every evening and driving back to Foxrock drunk after a heavy session with the sort of people Foxrock did not know but whom she imagined to be in one way or another disreputable. Never a very sound sleeper, she frequently heard him blundering about downstairs on his way to bed in the early hours. The hangover he suffered from was usually pretty severe and combined with his other, familiar and long-standing ailments to make him even more of a trial round the house than usual. His 'heart attacks' had returned in mid-summer and occurred now on an average of about once a week. 'I am quite convinced,' he had told MacGreevy in July, 'with that barren numb conviction of birth having sprung the trap, that at this rate it is only a matter of a few years before a hideous crisis compared to which the last was a cold in the nose and which I shall be as little fit to deal with as a bad calf with its castrators. There are still things, physical things, which I simply *dare not think of*.' Whatever this last statement might have meant in precise terms, he was complaining now more than ever and, drinking heavily, was advancing to meet the crisis with some determination. 'I feel riddled with pains inside and out and keep wondering when the crisis,' he told his friend in an update a few weeks later. It was no wonder May's patience was exhausted and she now made the flat definite suggestion that he should clear out. There was an element of comedy about this. He was living at home partly because he was convinced – or had convinced himself – that in her depressed condition she needed him. Now she was telling him that she would prefer it if he left the country once and for all. There were scenes almost every day but he noted with clinical detachment that they left him comparatively undisturbed in mind, affecting him far less than they had in the past. If there had been times when self-disgust, remorse and fear of his mother had all combined to make him thoroughly miserable, this was no longer the case. He was wrapped in self-pitying indifference. He had said that Cissie's capacity to care about another had worn out where Boss was concerned. His own capacity to care about his mother was at this point similarly wearing thin.

Just before Frank's wedding he went to a party in the Wicklow mountains, given by Charlie Gilmore and Lilian Donaghy – the poet Lyle Donaghy's wife – who were now living together in a cottage which belonged to the poet Joseph Campbell. Charlie was a well-known republican activist – the word terrorist had not yet been invented – who, with his brother George, had been involved in a sensational shoot-out during an escape from Mountjoy Gaol in the twenties. He had, as is sometimes the case with terrorists in Ireland, a wide literary and bohemian acquaintance, but if May had known that Sam kept such company she would have been even more shocked and probably not mollified by being told that the Gilmores, like many republicans of the era, were Protestants. Mervyn Wall, the painter Harry Kernoff and others were present at the party, though when members of both sexes decided to go swimming naked in the little Glencree river, the cautious Wall demurred. Sam, however, was game, but he contrived to cut his head diving into a rock pool and also lost his hat and his watch.

Next morning he had one of his particularly acute hangovers. That evening there was a reception to view the presents for Frank and his bride, a Foxrock formality which depressed him greatly. The whole custom of present-giving had seemed to him symbolic of the way society absorbed the lives of two people and their relationship. Viewed in terms of teapots and tea-trolleys, marriage seemed to involve the smothering out of existence of the human personal element, reducing the love of two people from a relation to a condition, substituting 'cordial fucking and fucking cordiality' for the darker and more turbulent spiritual and physical need that had brought them together.

Still, at the wedding he did his best to be affable to all concerned and drank his share of champagne. He had thought he would feel more at Frank's departure into another life, but in the event he felt little or nothing and realized that all attempts at real contact between him and his brother for years past had failed. His brother had learned the 'deaf and dumb alphabet of convention'; he had not. After the ceremony he came back to Cooldrinagh with his mother and cousin (weddings at that era took place in the morning and it was the custom for the guests to leave when the bride and groom did), and the problem of the rest of the day and evening was solved by the trio going cheerlessly to the Gaiety Theatre to see a variety show starring the great Dublin comedian Jimmy O'Dea. He was depressed, but no more so than his mother, who naturally felt the loss of the most dependable and companionable of her two sons, nor can the sight of the other one going moodily through the motions of social pleasantry with his head bandaged like Van Gogh's have done much to cheer her up. But she bore herself bravely and in the immediate aftermath of the wedding was as nice to Sam as possible. She had now decided definitely to leave Foxrock and find a house near Frank's new abode in Killiney. The first plan was to let Foxrock on a long lease; but such was the general fear of tuberculosis at the time that this was shelved when it was discovered that one of the prospective tenants was tubercular. Sam rather hoped she would drop her plans for a permanent move altogether and simply go away for a holiday, leaving him to his own devices in spacious Cooldrinagh with the cook and the dog; behind this hope was a desire that the status quo should simply continue indefinitely.

But the improvement in relationships was merely temporary. The day after the wedding he drove up to Glencree to recover his hat and watch and started drinking again with Charlie Gilmore. In mid-September he took to his bed for almost two weeks with diarrhoea and general debility which were classified as gastric flu. This may have been an appeal for his mother's sympathy, but if so it did not work. The day he got up he took the car out and crashed into a lorry. He was uninjured, but the car was a write-off. He claimed that the lorry-driver was to blame, but the police took a different view and he was prosecuted for dangerous driving. Relations between him and his mother had now gone far too far to be mended and even Frank, back from his honeymoon, told him point-blank that he was letting everybody down. There seemed nothing left to him but the usual recourse of young Irishmen faced with insuperable difficulties at home, which was to catch the boat. And in his case, as in many of theirs, the resolve was accompanied by the usual abuse of the country being left. Ireland was 'a useless kip of a country' and as for its police force, there was no animal that he 'loathed more profoundly than a civic guard, a symbol of Ireland with his official, loutish Gaelic complacency', a statement in which there is a definite note of Foxrock unionism. A note of realism about his personal situation was struck, however, by the admission that he had been wrong in thinking that he was 'well enough to deal with' his mother and with himself 'in relation to her', which is also one of his few admissions that he had hoped to gain more from psychoanalysis than merely a cessation of his physical ailments and difficulties.

Of interest also is the fact that his first thought was to go to London, not Paris. He knew the kindly Mrs Frost would have a room for him, but he had 'had enough of mothers and surrogate mothers', so he wrote to MacGreevy to ask if there would be room in his place and what it would cost. Paris was attractive on the grounds that the money he would have to live on would go further; and he did enquire also if MacGreevy ever thought of having a stab at that city again, but at least in the first instance he would go to London, where of course Mac-Greevy was, which was always an important consideration for Sam.

The situation at home was now intolerable. He wanted to move into Clare Street, but when Frank told him that he had to go to Waterford on business he decided to go too, staying a couple of nights at the Imperial Hotel. Then, with Frank's agreement that he could move his books and personal effects into the newly-weds' home in Killiney, he came back with him to Dublin, sleeping on a cot bed in Clare Street and dossing elsewhere for a few nights rather than go home. At this point, however, his mother surprised him by leaving Cooldrinagh for a destination that Frank refused to disclose; and so, somewhat flummoxed by this, but relieved to get between clean sheets and into a bath, he decided to delay his departure for a week or so. His mother had told Frank that he need not move his things immediately out of Cooldrinagh; and that if he wished he could lock the study door and give Frank the key. May's decision to offer this concession and make herself scarce has been interpreted as a move to hold on to him; but this does not seem at all likely. True, both were enacting what amounted to a comedy of motive and feeling. May on the whole wanted him to stay, but his presence in the house was such a constant aggravation to her that she expressed herself otherwise; and while she needed him, she did not need him as much as he liked to think. For his part he was beginning to recognize that, given the fact that she was a depressive, her condition might be much the same whether he was there or not, governed by internal mechanisms that had nothing to do with him. He admitted this to MacGreevy as soon as he was back in Foxrock, writing to him about the 'peace in the heart' that 'will never be there anyway, least of all as the fruit of formal reconciliation'. His mother had wanted him to behave 'in a way agreeable to her in her

October of analphabetic feultility [sic] or to her friends ditto or to the business code of father-idealization (whenever in doubt about what to do ask yourself what Darling Bill would have done)'. But the letter shows a growing realization that even if he could conceivably have satisfied all May's wishes, it might not have made much difference to her outlook. 'There are the grey hairs that will go down in sorrow, that want to go down in sorrow, as they came up in sorrow because they are that kind.' As for himself, he was rather surprised that he felt so little. 'Instead of creeping about with the agenbite as I suppose I ought, I am marvelling at the pleasantness of Cooldrinagh without her and could not wish for anything better than to feel the same when I am away.' ('Agenbite' was a medieval word Joyce had adopted to mean the unease of spirit from which Stephen Dedalus suffered.) If these feelings left something to be desired, 'I am what her savage loving has made me and it is good that one of us should accept that finally, as it has been all this time.' And if this all boiled down 'to saying what a bad son I am then amen. It is a title of as little honour as infamy.'

In view of much that has been written and said about it, it is perhaps necessary to emphasize that Beckett's attitude to Ireland and his eventual preference for living elsewhere were not products of his minority Protestant upbringing, still less of the generally anglophile stance of his family and their friends. Since the 1890s many Irish intellectuals and Irish writers, Catholic and Protestant, unionist and nationalist, had preferred to live abroad, partly for career reasons, partly for psychological. England had journals, reviews, publishing houses and a large number of literate people who read and discussed books. France had, they believed, a widely disseminated knowledge of art and literature and a level of intellectual discussion which was absent at home. It certainly had wine, sunlight and a reputation for being liberal about sexual matters. It offered possibilities of cheap but civilized living such as were not available in Ireland. And both England and France had great cities, with all that cities implied in the way of anonymity, freedom and the pleasures of unobserved observation.

After the foundation of the Free State most liberals and intellectuals did, in varying degree, feel the repressiveness of the Church-State alliance and, in varying degree, resented it. But beyond that there was the fact that Ireland was a small, narrow, familial and puritanically oppressive place in ways less easily defined than by talking about censorship and theocracy. Apart from intellectuals, thousands of people who did not care very much about the censorship of the latest novel by Seán O'Faoláin felt that; and they packed the outgoing boats, the feeling often intersecting with the economic reasons which drove them and others into exile. Thousands of people had mothers, as Samuel Beckett had, or at least families, relatives and localities that they saw as oppressive, demanding, intrusive or merely dull; they wanted to escape.

There was a widespread feeling that to go away was to grow up: more, that it might present the only chance of growing up that the individual in question would ever have. Beckett said also that Ireland was the land of his 'unsuccessful abortion'; and of course, he meant this specifically, for after it had come to him the illumination he had derived from the Jung lecture was never too far from his mind; but many, many people who had not had such a specific illumination about their mental state, who were not writers or even, in the professional sense, intellectuals, felt in those years that getting away from Ireland and away from the particular pressures of locality and family might be a form of re-birth and might give them a chance to discover their individual identity.

Some writers and artists also felt that the paralysis which Joyce had identified as having Dublin in its grip militated against the production of the great works of which they felt themselves capable. There was a standing cliché that in Dublin people went to the pubs rather than to their table or desk or easel; and the writer who talked his books away was a recurrent figure in Dublin's own mythology. There was indeed a considerable element of truth in this. In Dublin people did tend to live out their promise until eventually they had great futures behind them. This was partly an effect of the size of the place, partly its sociability. Word-of-mouth reputations were gained easily and often rested on. More businesslike places had a publishing or exhibiting apparatus which separated the sheep from the goats, the workers from those who merely talked about great works; and they had furthermore journals and critics who were seemingly eager to discuss such work when it was written. But even at this point some of Beckett's Dublin friends, among them Arland Ussher, Sean O'Sullivan and Cecil Salkeld, were beginning to be perceived as figures of unfulfilled promise. Acquaintances such as Mervyn Wall and, more notably, Brian O'Nolan (Flann O'Brien) would shortly, rightly or wrongly, suffer from the same perception. It was the feeling that this might easily have happened to him which Beckett was much later to express in conversation with Martin Esslin, who asked him point-blank why he lived in Paris and if he had anything against Ireland. 'Oh no. I'm a fervent patriot and republican,' Beckett replied with, admittedly, a possible degree of over-statement. 'Well,' Esslin queried, 'why do you live in Paris then?' To which Beckett answered: 'Well, you know, if I were in Dublin I would just be sitting around in a pub.'

In the minds of many intellectuals or many less exalted people who had read him or knew something about him, there was the further fact that James Joyce had, in the early years of the century, made exile fashionable. In a famous outburst Stephen Dedalus had excoriated the nets that in Ireland were flung over a soul to hold it back from flight. 'You call these nets language, nationality, religion: I will fly by those nets.' He had spoken in the same breath of trying to express himself 'in some mode of life or art as freely as I can and as wholly as I can, using for my defence the only arms I allow myself to use – silence, exile and cunning.' The number of writers who identified in one way or another with Joyce was large. In keeping with these words, Joyce had created a myth out of his own exile; and though in his case the myth was not quite true to the facts – for it said nothing about coming back to make a career in cinema management in 1909 – it nevertheless had a clear and definite outline which had great force.

If Beckett knew these words (and there is, oddly enough, no hard evidence that he ever read the early prose of James Joyce, either *Dubliners* or *A Portrait of the Artist*), doubtless he probably would, if put to it, have echoed them; cunning or not, he would use both exile and, after he had learned the value of an enigma, silence, as weapons in his defensive armoury. In various interviews later he would say that he left Ireland because of the censorship laws and the dominance of a repressive church. 'You know the kind of thing – theocracy, censorship of books, that kind of thing,' he told Israel Shenker. But the thought of making any sort of Joycean myth out of his own departure was certainly far from his mind in April 1937. Nor would it be in his mind in November 1938, when what turned out to be a permanent exile was embarked upon. Of course, like many others he preferred to be abroad, away from the restraints, futilities and inhibiting circumstances of home. But when he went, it was more or less on the spur of the moment and as much to get away from an immediate and intolerable personal situation as to escape from a theocratic place of censorship. Censorship was scarcely likely to be much of a real worry to a writer whose latest book was in the process of being rejected by thirteen publishers. In all his thoughts or feelings about going abroad, being abroad and finally staying abroad, personal freedom would play a part, as would opportunity and, finally, creative fulfilment; but thousands of other Irish people, Catholic and Protestant, including many who were not anglophile or francophile and did not write books, went for much the same real reasons as he did; and their going, like his, was inconclusive and undramatic. 'I just slipped away,' he said afterwards, with more truth than when he talked of censorship and theocracy. Though there is a Beckett myth of exile, a counterpart of the Joycean – and it is, to an extent, properly part of his legend as a writer – we must beware of seeing him as having conscious motives or making conscious gestures at a time when gestures were far from his mind.

Before he left he had to spend the best part of a day at Shankill courthouse waiting for the dangerous driving charge to be heard. The District Justice who heard it turned out to be Kenneth Reddin, author of more than one Abbey play, several children's books and a couple of novels. Reddin, who had taken part in the 1916 Rising as a young man and been rewarded with what was then regarded as a rather cushy job, was fond of his reputation as a courtroom wit, an easy thing for a judge to acquire. He knew who Beckett was and, describing him as 'one of the most distinguished of Irish writers', declared that he cut corners in literature with a certain finesse, but perhaps did not do so with equal skill when driving (laughter in court). It would, he thought, be a pity if Mr Beckett's services to literature were ended prematurely and, in fining him £1 with 5s. costs, expressed the hope that he would take more care on the road in future. The object of his solicitude was not amused, since the fact that the verdict went against him ruled out an insurance claim against the owner of the other vehicle.

By the third week in October, with May still absent and incommunicado, he was ready to go and, on the night of Saturday 16 October, crossed to England by boat, arriving at Euston from Holyhead at midday on Sunday.

On arrival in London Beckett went straight to Chelsea, where Mac-Greevy, very much the star boarder, was still ensconced in Hester Travers-Smith's guest house and had arranged for him to have a room. There followed a few days in his friend's company, during which he retailed such gossip as he had not already included in letters and they discussed their present and futures. MacGreevy was working on the *Studio*, of which he was the principal critic, lecturing in the National Gallery and contributing to the *TLS*, among other journals. He was also translating French novels for English publishers. His long-term ambition was to get a decent job as a curator, or possibly even director of a gallery, and he felt he had a foot in London and had a better chance of landing such a post there or in Dublin than in France, which was known to be chauvinistic in the matter of gallery appointments. As far as Sam was concerned, though, Paris was cheaper and, in terms of his own likes and habits, pleasanter. Perhaps also the disadvantages of London were emphasized by the fact that Cheyne Walk was in territory which was all too familiar and too evocative of the recent past. He wanted to be with Tom, but felt he could persuade him to move to Paris also. When, in the event, Beckett went on to Paris, arriving there on Tuesday 26 October after just over a week in London, it was with a promise from MacGreevy that he too would cross the Channel and stay at least for a while in Paris, a city where they had not been together for any length of time in almost eight years.

Almost as soon as he arrived Beckett began to worry about money. In search of a place to stay he naturally drifted back to the area he knew best and finally wound up in a *pension* in the rue de la Grande Chaumière, off the boulevard du Montparnasse, almost at its intersection with the boulevard Raspail. This cost 16f. a night and on his first evening an entrecôte, cheese and half carafe in a nearby restaurant cost him 19.75f. Having cleared a few debts before he left Ireland and succeeded in drawing his allowance up to the beginning of December, he had in fact very little left, and began to worry that he would have to retreat to London again, where he could stay with MacGreevy (who had offered to pay for his room in Cheyne Walk) or Reavey (with whom he did not particularly want to stay) – or even as a last resort with Sheila Roe (now Sheila Page) in Surrey. He was worried as well about the imminent necessity of returning to Dublin for the libel action. Harry Sinclair had written to tell him that this could be any day, but at this point could not be more specific. He hoped he might stay there with the painter Sean O'Sullivan, who had a studio in Stephen's Green, or with Frank, who now wrote saying that May was ill and had had to go into hospital for abdominal X-rays. Taking all these things together he certainly could not rejoice in any definitive feeling of release or even plan to meet people for more than a day or two ahead, and the resultant feeling of paralysis limited his desire to contact friends like the Joyces. Brian Coffey was also in Paris and had arranged to meet the train at the Gare du Nord but he did not turn up. The following day, however, they succeeded in making contact and there followed a very drunken night with Coffey and a couple who were members of the Joyce circle, Alan and Belinda Duncan. The next day he went out to Senlis with Coffey and stayed the night; and in the next few days he succeeded in seeing one or two old friends from the École, such as Georges Pelorson and Jean Thomas. Then word came that the action Harry Sinclair had taken against Gogarty and the publishers of his book was imminent. Before Beckett had left Dublin he had already sworn an affidavit in which he said,

I purchased a copy of a book, *As I Was Going Down Sackville Street* from Greens Library, 16 Sheriff Street, Dublin. My attention had been called to it by the many advertisements I had read and the notoriety of its author. On reading the paragraphs complained of I instantly inferred that the lines quoted referred to Mr Henry Maurice Sinclair and the late Mr William Abraham Sinclair and that the words 'old usurer' and 'grandsons' referred to the late Mr Morris Harris and his said two grandsons. I also considered that the words constituted a very grave charge against the said Henry Maurice Sinclair and his late brother.

At the time that this document was sworn out, the case, like most libel actions, seemed a distant enough prospect and of course the plaintiff had hoped, like all plaintiffs, that the defence would settle out of court. Now through the capricious workings of the legal system the action had suddenly been listed for hearing and he had to decide all over again whether he would be a witness or not. To his credit, he decided to return and take the stand. It was to prove a bitter and humiliating experience but also perhaps a fitting enough way to say goodbye to Dublin.

The last thing he wanted was to return there but nevertheless he gallantly set forth, travelling by rail and sea to London. Unfortunately MacGreevy had just left for Paris, but at least he could use the room in Cheyne Walk, so he spent a couple of days there before going on to Dublin to face whatever lay in store for him in the witness box.

Libel actions involving literary figures have always been a popular source of entertainment in Dublin and this one was no exception. Gogarty was a swashbuckling figure known not only as a poet, but famed far and wide as a wit. Besides being a literary man he was a surgeon with a fashionable practice, a country house and a Rolls-Royce. He was well known as a friend of Yeats, but another friendship, the intimacy he had enjoyed in his youth with James Joyce, also added spice to the occasion, for he had already been identified as Buck Mulligan of *Ulysses*, both by the minority who had read that book and the far larger numbers who hadn't. On top of all this he had an involvement in politics, having been appointed a senator by W. T. Cosgrave and distinguished himself as a bitter opponent of republicanism in that capacity. It was no wonder that public interest in the case was so great that queues had formed and many people were turned away from the first day's hearing. A visiting journalist, Charles Graves, thought that 'only the *Pickwick Papers* re-written by James Joyce' could properly capture the atmosphere of jolly anticipation in the Four Courts. Gogarty and his reputation were undoubtedly the drawing cards. Beckett's part in the proceedings was only a tit-bit, a tasty morsel with which to begin.

Though Gogarty was not popular, public interest in the case may also have had a degree of anti-semitism in it. If so, it received an early gratification, for the first witness was Harry Sinclair, who proceeded to put into evidence documents which he said showed conclusively that his grandfather had indeed been guilty of enticing newspaper girls under twelve into his shop and interfering with them – precisely what Gogarty had insinuated the old usurer mentioned in his book had done. Prosecuting Counsel had claimed in his opening address that Gogarty had 'vilified the living and the dead in a pen dipped in the scourgings of a putrid and amoral mind'. It now seemed to many that the dead man's grandson was doing his own share of the vilification.

He was followed by Beckett, who was what is called a 'publication witness'; in other words, it was his job to testify that he had read the book, identified the plaintiff as the person mentioned in it and considered him damaged and defamed by what was said therein. After he had said all this he was cross-examined by the defendant's Counsel, John Mary Fitzgerald, KC. Fitzgerald was one of the leading barristers of the day, known both as a merciless cross-examiner and as a florid and, when he needed to be, sentimental orator with an expansive style of delivery. He was a scholar and a professor of law, who had been Dean of Studies at University College, Dublin. He spoke excellent French, probably almost as well as Beckett, but for the occasion he had decided to play the common man, thickening his native Cork accent and pretending he did not know how French names such as Proust were pronounced. So thick was the accent he assumed that when a witness found difficulty in understanding him, Opposing Counsel, Ernest Wood, remarked amid laughter that it was the Cork accent that made it difficult.

Fitzgerald now began by challenging Beckett as to his relationship with the accused, saying, 'You have made an affidavit in which you have led the court to believe that you were an impartial independent person, but you forgot to tell the court that it was your uncle-in-law who suggested that you should buy the book?' The reply was, in the circumstances, clever: 'I said in my affidavit that my attention was attracted by the book because of the notoriety of the author and the advertisement it had received.'

Fitzgerald's next tactic was to ask Beckett where he lived, to which the reply was 'In Paris.' This was exactly what Fitzgerald wanted to hear, for Paris was a place whose very name had connotations of iniquity and licence. 'Ah Paris,' he said and then, turning knowingly to the jury, declared with great significance, 'He lives in Paris.' Beckett could just as easily have said he lived in Dublin, for he had been in Paris for less than a month at the date of hearing and nobody was to know that he had any intention of taking up permanent residence there.

The next question seemed equally innocuous. What had Beckett written to justify describing himself as an author? 'I have written verse, fiction and literary criticism,' was the reply. 'Have you written a book about a man called Marcel Prowst?' Fitzgerald asked. It was one of the oldest tricks in the business and two decades later John A. Costello, like Fitzgerald a well-educated, fairly sophisticated man, would employ it against the poet Patrick Kavanagh. Beckett fell for it now. He had written such a book, yes, but the name was not 'Prowst'. It was 'Proost' – giving it an impeccably French pronunciation, even to the lisped 'r' sound. Heavily, humbly, and again turning to the jury, Fitzgerald acknowledged that he stood corrected. The gentlemen of the jury would understand that he, a common man like themselves, knew nothing about these decadent French authors, on whom the witness was an expert.

He then read an extract from a book which, since the matter was never cleared up, the jury might well have assumed was the book about Proust. It was part of the conversation between the Polar Bear and the Jesuit which takes place on top of a bus in the story called 'A Wet Night' in *More Pricks Than Kicks*:

> The Lebensbahn of the Galilean is the tragic comedy of the solipsism that will not capitulate. The humilities and retro mes and quaffs of irreverence are on a par with the prestos, arrogance and egoism. He is the first great self-contained playboy. The cryptic abasement before the woman taken red-handed is as great a piece of megalomaniacal impertinence as his interference in the affairs of his boyfriend Lazarus. He opens the series of slick suicides as opposed to the serious Empedoclean variety.

Was it not a blasphemous caricature of our Redeemer? Fitzgerald asked. Beckett may well have decided not to point out that the extract was from a book called *More Pricks Than Kicks* because the title of the book and the fact that it had been banned would both obviously be grist to Fitzgerald's mill. He contented himself with replying that the book from which the extract was taken was a work of fiction; that the characters in it were fictitious and that he could put words in the mouths of characters which he himself did not necessarily share. The next question went back to Proust, whose name Fitzgerald now took elaborate pains to pronounce correctly.

'Was Proust a man who indulged in the psychology of sex?'
'I have not been aware of that.'

271

'But you have written about him. How long did it take before your book was banned by the censorship of Ireland?'

Again there was the confusion of which Fitzgerald may not have been aware, possibly having been briefed and supplied with extracts by Junior Counsel, one of whom was Gogarty's son, Oliver. In any case, Beckett again decided not to make an issue of it. Again probably not wishing to dwell on the banning of his second book, he drily replied that he thought it was about six months after its publication.

'I suggest that it was banned because it was blasphemous and obscene book,' continued Fitzgerald.

'I have never discovered why it was banned,' was the reply.

The next question could perhaps be described as a leg-spinner. 'Do you call yourself a Christian, Jew or Atheist?'

As Fitzgerald may have hoped, Beckett replied, 'None of the three,' which reply was repeated in 'now we know where we stand' tones by Fitzgerald to the jury.

Cross-examined further, Beckett was prepared to admit that he had indeed written another book which was called *Horoscope* but with the letter W prefixed; and that he had had it printed privately and circulated for the enjoyment of his friends. Asked if his two uncles-in-law were among the choice circle of friends for whom the book had been printed, he was prepared to admit that it was likely.

Fitzgerald must have been fairly well-satisfied with the impression which he had helped the witness to produce on the jury, more especially since the other publication witnesses were a fairly motley crew, most of them either relatives or employees of the plaintiff who had also purchased the book at his suggestion and afterwards proceeded to identify him in it. The last was a Mr Seamus Fenning who owned a bookshop in Dawson Street and with whom Fitzgerald pursued the anti-semitic theme which had been a secondary motif of the case. Cross-examined by Fitzgerald, he said he was a Jew who had become a Catholic. Ernest Wood, KC, who appeared for Harry Sinclair, thought it necessary at this point to intervene by asking the witness if he was a well-known billiard player to the extent that he had won a billiard championship, to which Mr Fenning replied, 'I have won about ten championships in this city.' When the judge observed that he did not see any relevance in this, Wood said, 'It is suggested that he came from nowhere, out of the clouds and that he is an apostate as regards religion.'

When Fitzgerald stood up to put the case for the defence he dealt with these witnesses seriatim. There was no need to deal at length with the case of Mr Beckett, who came from Paris, he said. The newspaper report in the *Irish Times*, the only journal that May ever read and one for which, as a Protestant paper, the people of Foxrock had the highest respect, continued in oratio obliqua:

He might well have stayed in Paris, because they would like to know why, of all the respectable people he knew, Mr Sinclair should select that 'baud and blasphemer from Paris' to make an affidavit in the case to lead to the belief that any ordinary reasonable man reading the book would have identified Mr Sinclair. Could they imagine 'that wretched creature' making representations to the High Court as an ordinary reasonable man?

The object of putting certain phrases in quotations in the course of an indirect report was of course to distance the newspaper itself from the phraseology employed and to make it clear that it was Fitzgerald's. Unfortunately it also served to call attention to the mode of description employed. Fitzgerald has incurred a good deal of opprobrium for his manner of dealing with Beckett, being ranked with Sir Edward Carson in his dealings with Wilde and, in Ireland, John A. Costello in his dealings with Patrick Kavanagh. But to be fair to him, it was Ernest Wood who had first used the phrase 'bauds and blasphemers' with reference to Gogarty and the company he kept. In fact Fitzgerald made this clear in his closing address to the jury. 'Mr Wood', he said, 'has referred to the coterie of "bauds and blasphemers".' This was another boomerang. He did not know where Mr Wood got the phrase but it fitted the gentleman who was put forward as the principal witness, Mr Samuel Beckett, the nephew-in-law of Mr Henry Maurice Sinclair. Certainly if anybody could be said to belong to a coterie of 'bauds and blasphemers', it was Mr Beckett. He might well have stayed in Paris. Could they imagine that wretched creature making representations in the High Court as an ordinary reasonable man?

The fact that this sort of thing was reported in the *Irish Times* was a dreadful shock to May, who of course knew that every word of it would be read by her Foxrock friends and by the Protestant business community generally. She may have hoped that some allowance would be made for the fact that this was Defending Counsel speaking and merely the expression of a partisan viewpoint.

The same allowance, though, could not be made for the judge's statements when he came to deal with Beckett's evidence in his summing up. Emphasizing that the credibility of witnesses was a matter for the jury, he nevertheless felt he was bound to say that he himself would not repose any great faith in the evidence of 'the witness Beckett'. The constant references to the Jewishness of various participants with whom Beckett was associated would not have pleased May either; and again, as she knew, it would have been noted in Foxrock. Ernest Wood, who appeared for Harry Sinclair, had attempted to pre-empt prejudice on the opening day by calling attention to this aspect of the case. 'Is there a hope that a Jew will not receive justice from a Dublin jury? Is it the gambler's throw? Throughout all our chequered history one of the great traditions of the city is that it has never persecuted the Jews and I know, members of the jury, that you will preserve that tradition.'

In the event, as far as this was an issue, he was right. The jury were out for an hour and a half, coming back to announce that they had found for the Plaintiff and were awarding him £900 in damages. This was a big sum for those days and, together with the legal costs which he had to pay, it was, in spite of the fashionable practice, the country house and the Rolls-Royce, a ruinous blow to Gogarty, who claimed afterwards that the case had cost him £2,000 in all. This was of little satisfaction to Beckett, however, who was seriously out of pocket for the return trip from Paris.

In the aftermath he was afraid to face May, but he crossed Capel Street Bridge and walked up Nassau Street to Frank's office in order to apologize to him for any distress or annoyance the family might have been caused by the reports of the case. He found his brother far less forthcoming and dismissive of the consequences than he might have hoped. Frank advised him that since he had not so far seen May, it might be as well to go back to Paris without doing so. It was a big step and a serious test of his new attitude towards his mother for Sam to leave Ireland without making any attempt to contact her; or, for that matter, to feel guilty about something without immediately rushing to mollify her. But somewhat fortified by Frank's advice, he took the boat that very night for London and Paris.

It had been a horrible little foray and of course it had its effect on him. With one part of his nature he welcomed humiliation and abasement, but the distress he might have caused to others was another matter. When he got back to Paris he wrote to Frank. Once again he said how sorry he was and once again, as he had done in Clare Street, he repeated that he was ready to write to his mother to apologize for any grief or embarrassment he might have caused her. To his dismay, Frank did not even answer.

Neither May nor Bill had ever been fond of the Sinclair connection, but May found some solace in blaming the Sinclairs for the embarrassment the public castigation of her son as a liar had caused her and she never spoke to them again. When, six months later, Harry Sinclair had still not received a penny of the sum he had been awarded, Sam wondered bitterly what it had all been for. The only result of the case, he said, 'had been a thorough turning over of the family manure'.

Chapter Seventeen

Beckett arrived back in Paris at the end of November 1937, just as the last golden glow of autumn was giving way to the harsher and greyer light of winter. His first problem was to find a place to stay and after some searching he settled on the Hôtel Liberia, which, like Chez Sarrazin, the pension where he had been staying before his return to Ireland for the libel action, was in the rue de la Grande Chaumière, just off the boulevard du Montparnasse at the junction of streets and boulevards which is now called the place Pablo Picasso. He was thus within a few yards of the Dôme and the Select, still very much the centre of things so far as those artists and writers who frequented Montparnasse were concerned. The Liberia, which he found to be both cheap and tolerable, would be his home during one of the most eventful periods of his life.

The France to which he had returned was an increasingly divided country whose intellectual life had continued to be politicized throughout the 1930s. An exception was the surrealist movement, which, in spite of some well-publicized political stances and happenings – the expulsion of Eluard and Dali and the publication of a tract against the Moscow trials of 1936 – had just taken a turn towards art rather than politics, boosted by a major exhibition in London the previous year and the capture of the review *Minotaure* in the month of Beckett's arrival. But there was no denying that politics, with its angers, divisions and, increasingly, despairs, now held the centre of the stage. The process begun in the twenties with the Riff war in Morocco had been completed by the rise of fascism and the bitter divisions it brought. Both right and left had their intellectual spokespersons, often more prominent and recipients of more publicity – sometimes deservedly – than the parties or groupings they were supposed to represent. The newspaper article, the political pamphlet, the political manifesto, the essay which linked cultural and political trends in a propagandist way, whether hopefully or angrily, took up the time and energies of a host of prominent writers, among them Gide, Malraux and Montherlant. New reviews, *Réaction*, *L'Ordre nouveau*, proclaimed their general allegiances and alignments through their very titles as the Moscow trials, the Stavisky riots, the Popular Front, the Spanish civil war and the growing sense of French inability to find a unified stance with which to confront Hitler caused new bitternesses between right and left.

And of course there were the usual heresy-mongerings and attempts to impose a rigid orthodoxy. The tears which had sprung to André Gide's eyes from 'an excess of joy, tears of tenderness and love' during his visit to the Soviet Union were less important to some than the slight reservations he had expressed about life in Stalin's paradise. Bernanos's condemnations of Franco's atrocities caused more bitterness among his former colleagues of the Camelots du Roi and L'Action française than many pronouncements of the left itself had done.

For the great Exposition Internationale of 1937 Paul Valéry had presided over the creation of a Pavilion of Intellectual Co-Operation; and for the Pavilion of Light Raoul Dufy had painted an enormous celebration of the forms of international intellectual co-operation which had flooded the world with electric power and illumination; but Picasso's grey and black painting of the destruction of Guernica in the Spanish Pavilion seemed to some to express the grimmer realities of the time to come rather better than either.

From most of the intellectual conflicts of 1937 the *transition* circle remained aloof, continuing to revolve around the blind titan at its centre, whose occasional references to politics expressed only cynicism about politicians of all sides and their motives, and whose interest in ideologies or philosophical systems did not extend further than the thirteenth century. The last issue of *transition* would appear in 1938, containing Beckett's poem 'Ooftish' and his review of Devlin's book, which had originally been intended for an Irish journal, *Ireland Today*. In it he complained of the politicization of poetry, 'tied to the cart-tail of faction', asking that it should be judged on its own terms and not on those of politicians and zealots; but in an editorial, 'Homage to the Myth-Maker', Jolas wrote:

James Joyce is completing the last pages of his protean book of the night. *Work in Progress*, eighteen fragments of which have been published in *transition* during the last ten years will appear in book form in 1938 and will doubtless attract the attention of the inter-continental world with the electric shock of the thunder-word that epitomises polysyllabically one of its leit-motifs.

But by the time the book was finished the inter-continental world had a number of other things on its mind and the politicians and zealots were in the ascendant. The thunder-words were the words of Hitler's hung-upon speeches; the electric shocks, the sudden invasion of a new country, a brutal new breach of promises and obligations.

Almost immediately after his return Beckett had rung the author of the protean work in question and spoken to Nora while the master was shaving. When he himself came to the telephone Joyce passed on various items of news before inviting him to dine that very evening. At dinner he tentatively asked him if he could work with Giorgio on the galley-proofs of the great book, which was still without a title. Beckett agreed, as he had already agreed to write a piece about it for the *Nouvelle Revue française*, insisting to Giorgio that it would be his last piece of criticism and that there would be no question of writing any more exegesis of it when the book came out. He found the proof-reading stupefying work, but consoled himself with the thought that he would be paid. 'I am so hard up that I have allowed myself to be employed by him as proof corrector in chief,' he told Mary Manning. He was dismayed by having to stay in a hotel, even one so cheap as the Liberia, for which he was asked to pay 480f. a month, and was making tentative and unavailing efforts to find a place of his own.

Besides the Joyces he saw the Jolases and was told that his poem and the review of Devlin's book would be appearing in the issue of *transition* already referred to, which might well be the last; but the big news was from Reavey in London, who informed him by telephone on 11 December that *Murphy* had been accepted by Routledge and Kegan Paul. Once again he had MacGreevy to thank and also Jack Yeats. The Routledge director who was principally responsible for the acceptance was Tim Ragg. The poet and art critic Herbert Read, had also been approached about the brilliant young Irish writer who had written a novel. The news, coming as it did at the end of a long line of refusals, made him very happy, 'no jubilation but bien content quand même', as he put it when he wrote to MacGreevy. His satisfaction must have been deep for he began to worry when he heard nothing more for a few days from Reavey – usually a sign that the recipient regards the news as almost too good to be true; so he was much relieved when he got the actual contract a couple of weeks later, so relieved indeed that he sent it back immediately, forgetting to sign it.

He found the *NRF* article about the 'protean work of the night' – actually never finished – tough going; and was still determined that in any case it would be his last effort at criticism on Joyce's behalf. 'No more slop-emptying', as he put it to Mary Manning. If that meant a break with the Joyces, so be it; at least this time, he comforted himself, it would not have anything to do with Lucia. He was paid for the proof-reading, however, getting 250f. rather unexpectedly, to which Joyce, after a little hesitation, added an old overcoat and five ties. Beckett was hurt by this generous gesture, but decided that he would say nothing about it. It was far simpler and easier to be hurt and be quiet about it than to hurt in his turn by an indignant refusal. Though Wyndham Lewis's *Blasting and Bombardiering* had just appeared he may not have known that Joyce had sulked for days when T. S. Eliot gave him the parcel containing a charitable gift of a pair of old boots from Ezra Pound seventeen years before.

But living as he did near the famous cafés of Montparnasse, Beckett's acquaintance was widening beyond the Joyce circle. One new drinking companion with whom his relationship finally attained a certain intimacy was the sculptor Alberto Giacometti, with whom he had a great deal in common. According to James Lord's splendid biography of Giacometti, 'it was a friendship very gradual in growth, for neither man was looking for solace or reassurance'. They usually met without prior arrangement in the Dôme or the Select fairly late at night, but they were both night birds and both addicted to walking, so often in the early morning hours they would rise from a café table and patrol the boulevards together, sometimes in a companionable silence. Like Beckett's, Giacometti's process of development had been slow and uncertain, and like Beckett too he was yet to do the sort of work that was latent in him. He had had a partial early success as a member of the surrealist group, the creator of such works as his 'Disagreeable Object', a sort of phallus with spikes, and the 'Surrealist Table', at which a woman with long hair contemplates her severed hand. In the 1930s, however, he had become dissatisfied with surrealism; and when word got round that he was attempting to do a simple head from the life, he was formally expelled from the movement by Breton. From then on he found himself an outcast from the circles in which he had previously moved.

When he met Beckett Giacometti was at a difficult stage. According to Lord's book, he was 'trying to sculpt a head or figure which would satisfy him. None did. The heads never seemed lifelike, while the figures kept on getting smaller and crumbling to dust in his fingers. The results baffled him, but he was not discouraged. What he saw had ceased to be as important as how he saw and therefore the whole process of creativity had changed.'

Like Beckett also, Giacometti had a deeply ambivalent attitude towards women. His experience resembled Beckett's in that he too had difficulties in translating feelings of love or affection into sexual desire. He both feared entanglement and found it an obstacle to physical fulfilment. An early illness had conjoined with these difficulties to make Giacometti intermittently impotent; and according to Lord, '. . . he could never be sure of achieving satisfactory completion of the sexual act, no matter how fiercely he desired it.' Perhaps, Lord hazards, 'that's why he preferred to stare at women from a distance and avoid amorous entanglements.' Certainly it was why throughout his life he had recourse to prostitutes. Of his first encounter he told Pierre Schneider: 'I took a prostitute home with me to draw her. Then I slept with her. I literally exploded with enthusiasm. I shouted, 'It's cold. It's mechanical!''

Beckett would often speak of casual, sometimes commercial encounters which had no emotional content or consequences in similarly enthusiastic terms. At rare moments in the right company he would assert that these were the only satisfactory sexual conjunctions there were. Frequent impotence when the emotions are involved, but not in other circumstances – and particularly not with prostitutes – is a by no means uncommon syndrome, Baudelaire and Cézanne being other well-known examples. As in Baudelaire's and Giacometti's cases also, it is frequently combined with a high degree of arousal through mere looking or, as Freud rather quaintly put it, a tendency to excessive lingering 'over the intermediate sexual aim of a looking that has a sexual tinge to it', which he thought commoner among artists than others and to which he gave the rather ugly name scopophilia.

Giacometti's late-night walks often ended in an encounter with a whore, or in a visit to one of the brothels which flourished in the Latin Quarter. Since Beckett was not averse to occasional dealings with prostitutes either, they were both known to the girls of the quarter. Lord instances an occasion later on when a prostitute who knew them both said to a café proprietor, 'It's your luck to have two of the great men of our time sitting together right now on your terrace, and I thought you ought to know it.' Sometimes people who were about Montparnasse early in the morning might see Giacometti and Beckett still together, bleary eyed, neither having been home to bed.

The friendship between them was gradual in its development; but in their creative difficulties as well as in their ultimate achievements they had a great deal in common. As James Lord finely puts it, 'what they eventually found in each other's company was an affirmation of the supreme value of a hopeless undertaking.' According to Lord, 'It was a very private, almost secretive, and secret friendship. However it did not exist in the vacuum of austere speculation. It came to be recognised by others as a confirmation of something which, even if they did not understand it they could recognise as valuable.'

Over Christmas and New Year, however, Beckett was drawn back into the Joyce and *transition* circles as people are into families, even adopted families, at that season; going to the apartment on Christmas Day; dining with them at Fouquet's the following night; and being present at a family row on New Year's Eve. Helen wanted to go back to America to be near her ailing father, but Giorgio was reluctant and Nora sided with him. When he gave in his mother was very angry. On New Year's Eve Beckett had gone round by arrangement. It had been planned that they would all go out together but Nora flatly refused. For two hours Shem, as Beckett and MacGreevy privately called him – after Shem the Penman in *Work in Progress* – tried to persuade her while Beckett sat uncomfortably by with a silent Giorgio and Helen. Then Shem himself refused to go. Finally Beckett, Giorgio and Helen went out together to celebrate the passing of the old year and/or the advent of the new. When Beckett went round the following evening to take James and Nora to Charlie Chaplin's much acclaimed new masterpiece, *Modern Times*, he feared the worst; but they all went to the movie as planned and James and Beckett at least enjoyed themselves.

At the Joyce party at Fouquet's on the night after Christmas, however, a new factor had entered his life in the shape of Peggy Guggenheim, a rich American art collector and friend of many artists, who was also friendly with the Joyces but whom Beckett now met for the first time. What he would have observed was a forty-year-old, slim, well-dressed woman, who was not unattractive in spite of too much make-up and the rather heavy nose which she herself always thought to be a draw-back. What she saw was a 'tall lanky Irishman of about thirty with enormous green eyes that never looked at you', who was excessively polite and rather awkward in manner. She noted his spectacles and his ill-fitting, too-tight French clothes; thought that his mind seemed to be very far away, solving some intellectual problem or other; and observed that though he spoke very seldom he never said anything stupid.

How much he knew about her in advance is not clear; but their fate was already entwined in certain ways, not only through the Joyces but through her close friendship at that point with Herbert Read, who had advised Routledge to publish *Murphy*. Peggy was herself supposed to be very wealthy – if Beckett knew anything, he probably knew that –

and was a patron of many artists, a remarkable number of whom were or had been her lovers. Both her grandfathers had been immigrants into the United States, her Guggenheim grandfather from Switzerland, and both had wound up as multi-millionaires, Colorado silver being the principal source of the Guggenheim wealth. Her father was a famously handsome, debonair Lothario who had perished with the *Titanic*; her uncle, Solomon Guggenheim, the owner of the collection now housed in Frank Lloyd Wright's wonderful building in New York. Peggy had been married when quite young to one of New York's best-known bohemians and brilliant failures, Lawrence Vail. After she and her husband moved to the South of France they set up a ménage which was wittily described as being devoted to free love and free for all except Peggy, who paid for everything.

But she was not a fool. She herself said that she was not a 'real Guggenheim' because she did not have the requisite millions, her father's wayward career having largely disinherited him. But she had a keen eye for both people and the work they produced and her friendship with Marcel Duchamp – who introduced her to many of the surrealists, including her lover at the time she met Beckett, Hans Arp – was based on her human qualities and perceptions as much as on Duchamp's attraction to rich, not necessarily beautiful women. Learning the value of the Guggenheim name, she had cleverly obtained a reputation for being far wealthier than was really the case; and she had a deep passion for genuine creativity and originality. She described herself, with truth, as an uninhibited woman and as someone whose burning ambition was to find a place in the history of modern art.

After leaving Fouquet's on the night after Christmas the whole party adjourned to Helen and Giorgio Joyce's for the sort of relaxed and tuneful end to an evening that Joyce, who was wearing his grandfather's waistcoat, enjoyed. Beckett, silent as usual, had not appeared to be paying much attention to Guggenheim but, as she was leaving, he asked if he could walk her back to where she was staying, a borrowed apartment in the rue de Lille, a considerable distance away. As he loped beside her he took her arm and when they arrived at the rue de Lille he naturally came in.

Two editions of Peggy's memoirs have been published – one, rather heavily edited, in her lifetime, and a franker version after her death. In the first version she does not identify Beckett by name, but in both versions when they went into the apartment, 'he did not make his intentions clear but in an awkward way asked me to lie down on the sofa next to him. We soon found ourselves in bed, where we remained until the next evening at dinnertime.' At one point during this sojourn,

Peggy happened to mention champagne. Beckett rushed out and bought several bottles. As he was leaving, he politely said, as if they were never going to meet again, 'Thank you. It was nice while it lasted.'

Guggenheim was promiscuous; in fact since the death of her second husband, John Holmes, she had availed herself, as she put it, of 'every consolation which crossed my path', so a new affair was not a matter of great note. That news of this one soon spread was probably due to what people saw as a certain oddity about the pairing. Beckett had made an impression on her but, according to her account, she did not see him again for some time. Then, one night, they found themselves together on a traffic island in the boulevard du Montparnasse. Peggy had at this point moved to the house of a friend, Mary Reynolds, which she was renting while Mary was in hospital. They went back there and, according to her account, Beckett stayed on and off for twelve days, coming and going at odd hours of the day and night, mostly drunk and frequently bearing champagne. It was, Guggenheim wrote, the happiest time of their relationship, a time she remembered afterwards 'with great emotion'. She was in love with him and 'to begin with he was in love with me as well and we were both excited intellectually'. Certainly, in love or not, Beckett seems to have talked a lot; about his admiration for Joyce, Céline, Jack Yeats and the Dutch painter, Geer van Velde.

Unfortunately this account does not square with reality. Whatever length of time they were more or less together, it was not twelve days. In fact, assuming that the 'some time' which elapsed between the night after Christmas and the meeting on the traffic island was about a week, Beckett's comings and goings while Peggy was staying in Mary Reynolds's house can scarcely have lasted more than four or five days. There is also some reason to doubt that she brought the idyll to an end when she discovered that he had been unfaithful to her with 'a friend from Dublin', Adrienne Bethell, whom he had charitably put up in his room at the Liberia but allowed to 'creep into his bed' the night before. He told her that he had not enjoyed the encounter, that making love without being in love was like 'coffee without brandy'. This is an uncharacteristic statement, for in other company Beckett would say that casual sex without any emotional involvement was much more satisfactory; but Peggy interpreted the remark in the way she was probably meant to, as a flattering contrast between his feeling for her and his indifference to Mrs Bethell, a young married woman who was well-known in his circle for her libertarian attitude to sex.

Whatever length of time this part of their relationship may have lasted, Beckett was certainly back at the Hôtel Liberia in the early evening of 6 January, where he found a note from Alan and Belinda Duncan suggesting that they all three go to the cinema that night. The Duncans were on the fringes of the Joyce circle and were friends of MacGreevy's whom Beckett had known since his first arrival in Paris. They had a wide acquaintance in literary circles in both Dublin and Paris and Alan, who had been a major in the British army, had at one stage known W. B. Yeats quite well. They both liked to drink, Belinda especially.

After seeing a film Beckett and the Duncans went on to a café on the Left Bank, where they stayed until it closed in the early hours of the morning. As they walked home along the avenue d'Orléans, Beckett was accosted by a young man who spoke to him at some length and persisted even when he shrugged off his importunities and made to walk on. Beckett seemed to know him and from the facts as they were subsequently disclosed it would seem likely that he may have been offering the services of a girl and perhaps asking for money as well, for Robert Jules Prudent was a pimp and procurer who was well known in the quarter and Beckett had possibly had dealings with him in the past. It is unclear from the official account of the matter whether Beckett pushed him aside or not; contemporary newspaper reports said that after Prudent had accosted him 'an argument followed'. He was certainly impatient to rejoin his friends and at one point even threatened to call the police if Prudent did not leave him alone. Suddenly, however, the young man pulled out a knife and stabbed Beckett in the chest before fleeing from the scene. Though somewhat impeded by the overcoat he was wearing, the blade penetrated the layer of tissue surrounding the pleura. As he collapsed to the ground, the assailant ran off and the Duncans began to shout for help.

There were few passers-by at that hour, but the Duncans called a taxi and took him to their flat; the police and an ambulance eventually arrived and Beckett was taken to the Hôpital Broussais, where it was recognized that his condition was serious. The immediate problem was to deal with the wound and the loss of blood, but it was not until he was X-rayed a few days later that it was fully understood how close he had been to death.

Early in the morning the Duncans contacted a number of Sam's friends, including the Joyces. According to Joyce himself, on the day after the stabbing his house was 'like the stock exchange, telephone calls from everywhere'. One of these calls was from Peggy Guggenheim. Before going out to the pictures, Beckett had phoned Peggy, but according to her account she was still so angry that she refused to speak to him. In the morning, however, she had been better disposed and called at his hotel, where the proprietor knew what had happened because he had already had a visit from the police. Since he did not know to which hospital Beckett had been taken, Peggy rushed to all the hospitals in Paris and, failing to find him, telephoned Nora Joyce, who told her he was in the Broussais. She went there at once and left some flowers and a note to say how much she loved him and that she forgave him everything. When she went again next day, her arrival coincided with that of James Joyce, who was accompanied by his secretary, Paul Léon. While they waited to be conducted to the right ward, she rushed ahead and found Sam, weak but, as she put it, 'very happy'. Beckett's nature and outlook had their drawbacks. It may be regarded nevertheless as a blessing that he could be so stoical about calamity when it came. Whatever pain and discomfort he felt, he made much less of it than he did of the minor and sometimes imaginary ailments with which he was frequently afflicted.

An unexpected visitor was an acquaintance with whom he had played tennis as far back as 1929 when he was last in Paris and he and Alfred Péron used to take a train from Saint-Lazare to a club in the suburbs of which Péron was a member. Suzanne Deschevaux-Dumesnil had read of the stabbing in the papers and decided to go and see if he was being properly looked after and if there was anything he wanted. She was a thin, smartly dressed, attractive woman with chestnut hair, which she wore fashionably short, and remarkable grey eyes. Beckett remembered her and seemed glad to see her, so she came again. Her decision to visit him was to turn out to be fateful for both of them.

Joyce came more than once. On the second occasion he was accompanied by his friend and Beckett's, the writer Nino Frank, who has described Beckett lying in bed in the grey, cheerless hospital ward, 'his glasses sparkling, his lips tight-pressed, a wry smile on his face'. Frank thought that Joyce rather relished the atmosphere.

His eyes raised to the indefinite light of the window, he sat placidly facing Beckett, as the two of them marinated in intolerable silence. From time to time they exchanged a few words, a short laugh. I have never felt so close to Ireland, to its sentimental isolation, to the very air of *Ulysses*, as I did that day, sitting between those two brothers, in their shape and their keenness like twin knife-blades. Joyce confirmed my feeling, when, having had enough of silences, we left. 'He is truly Irish,' he told me. 'He doesn't hold it against the tramp at all, but do you know what he is mad about? The knife made a hole in his overcoat. He wants the judge to make it up to him and buy him another one.'

In fact, Beckett had refused to file a complaint, but was somewhat annoyed because the police had told him that they were going to prosecute just the same and he would have to testify in court. He had had enough of courts for the time being. After running away from the scene of the attack, Prudent, who lived on the avenue de Maine and was a familiar figure in Montparnasse, had gone into hiding. He realized that, if Beckett identified him, a fairly extensive manhunt would be organized, so in the early hours of the morning of the stabbing he moved into one of the cheap hotels which his girls used and stayed indoors for three days. On Friday the ninth he was arrested there by four policemen and charged with the stabbing. The police let it be known that they were acting on a tip-off; and it is possible that, when the heat came on, either the hotel proprietor or one of Prudent's own *poules* decided to give him up.

The news reached Ireland quickly, the *Irish Times* of 8 January carrying an account of the stabbing under the heading, IRISH POET STABBED IN PARIS, EARLY MORNING ATTACK. May and Frank immediately flew to Paris and stayed some days, during which May kept vigil for long hours by Sam's bedside. It was, in a sense, a providential way for mother and son to meet again, smoothing all awkwardnesses between them and bringing them closer together. It was also, for her, an introduction to his Paris friends, whom she saw in the best light possible, as kind and caring people. No doubt she was also impressed by the legendary James Joyce, a half-blind figure of great importance, carrying his stick and tended by his evidently distinguished and capable male secretary. On the third day after the attack Beckett was pronounced out of danger by Joyce's own doctor, Dr Fauvert, who had replaced a Dr Fontaine. Everybody was much reassured and cheered by Dr Fauvert's assurance that the wound would not leave any harmful after-effects. Beckett, as Joyce said, had had a lucky escape.

Lucia Joyce was still in the *maison de santé* at Ivry under the care of Dr Achille Delmas. Writing to Giorgio and Helen in Italian on 12 January, Joyce concluded by saying, 'So, as for these two unbetrothed lovers, both, to wit Enzo and Lucia are now in the hospital.' Lucia and Edgardo, for whom Enzo is probably a slip of the pen, are both characters in the opera *Lucia di Lammermoor* and are indeed unbetrothed lovers, since the plot of the opera hinges around Lucia being forced to marry somebody else. Lucia goes insane and Edgardo then dies of a stab wound. The reference seems to suggest Joyce believed that Sam and Lucia had at some stage been lovers. But, as always with Joyce, the allusion is richer and more complex that it appears at first sight. Since in the opera Lucia is forcibly betrothed to somebody else, as, one might almost say, Lucia Joyce had been betrothed to Ponisovsky, there may therefore be a suggestion of guilt, or perhaps culpability, on Joyce's part about the situation.

By this time Joyce had had Beckett moved to a private room and in the succeeding days gave further proofs of kindness in the shape of a reading-lamp and a custard pudding, which was one of Nora's specialities. Frank and his mother stayed in Paris, at the Trianon, for about a week, or until he was out of danger. His mother came to see him nearly every day and Sam was surprised by the 'gusts of affection and esteem and compassion' that he often felt for her while she sat at his bedside. 'What a relationship,' he commented to MacGreevy. He thought, however, that she looked poorly.

The wound continued to cause him pain, especially when breathing, which he continued to bear surprisingly lightly. To breathe, he had to lie on his back. The lights went out in the ward at nine o'clock and he found the night especially long and lonely. He was glad to observe the returning daylight and the signs of activity at around five in the morning.

Alan and Belinda Duncan were dealing with the police and, while Beckett was still in hospital, they appeared before the Juge d'Instruction, where they were confronted with Prudent. He seemed more cretinous than criminal, they observed to Beckett.

While still in hospital, he received the proofs of *Murphy* and, before he left, had nearly finished correcting them. He found himself making more changes than he had intended, chiefly out of boredom and the want of other activity. Reading it in proof, he was unenthusiastic about the book, finding it very dull, though painstaking and creditable. He was, however, infuriated by the blurb and by the manner in which it had been incorporated in the body of the book, being printed on a flyleaf – a common practice at the time and employed by Routledge with first novels. He wrote to Reavey refusing permission for it to appear between the boards of the book but was unable to have it removed altogether. Most authors are offended by the blurbs of their books and he was no exception. Viewed in retrospect, however, this one for *Murphy* does not seem too bad.

On 23 January, after two weeks in hospital, he was discharged. On his return to the Liberia he was greeted by an immense bunch of Parma violets sent by the Joyces. He had been warned that his wound would act as a barometer for bad weather in the future and also that he must take it easy. This he did, going out only once or twice a day, usually for a meal in a restaurant near by, retiring to bed between nine and ten, and lying in until at least midday.

In spite of this regimen the pain was still considerable – indeed he was inclined to think it was considerably worse. He now saw Dr Fontaine again, having rung her up to make an appointment, and she told him that the pain was only to be expected and that he must resign himself to it for some time to come. She prescribed a pain-reliever and then sent him on again to her assistant, Fauvert, who was equally reassuring. But it is notable that since the danger of his brush with death had receded Beckett was pitying himself more.

After the violets came Joyce himself, who sat in the lobby of the Liberia talking about Lucia, about whom he had had bad news from the clinic. Beckett had suggested that Geoffrey Thompson should be asked for an opinion and the possibility of Thompson coming to France and seeing Lucia was discussed. Joyce told him that he was going away to Zurich for a rest after the usual birthday celebrations in Paris and seemed on the point of asking Beckett to go with him, a prospect which rather pleased him, though he did not think he would be well enough. In his first days, or rather nights, in the Liberia a poem 'dictated itself' to him:

> they come
> different and the same
> with each it is different and the same
> with each the absence of love is different
> with each the absence of love is the same

This is the version which Beckett eventually published in English and it is the same as the French version, published in 1946. Peggy Guggenheim interpreted the poem as referring to their conversation about love-making without love and reprinted it in the first edition of her autobiography. There the last line read 'with each the absence of life is the same', though whether this was in a version which he may have given her is not clear. Peggy seemed to be under the impression that their affair, though interrupted by the stabbing, was, in some manner at least, continuing; and to be near her beloved she now moved into the Liberia. But this infuriated Beckett, and he objected so forthrightly that she had no alternative but to move out again to her sister's flat on the Ile Saint-Louis.

As he lay in bed in the Liberia he had moments of rare and complete happiness, such as are common with convalescents and those who have had a narrow escape from death. But he had also escaped from Ireland and this was part of his feeling, a patch of sunlight on the section of wall visible from his room seeming to him to be brighter than the whole of Ireland's summer.

For the compilation of a dossier it was necessary for him to have his injury assessed by a police pathologist, a Dr Paul. He was summoned to this appointment at 9.15 a.m. one morning but refused to go on the grounds that his condition did not allow him to get up before midday. Even though there was a possibility that he might receive compensation from the Ville de Paris he did not look forward with any enthusiasm to the legal proceedings against Prudent and co-operated only reluctantly with the prosecution. There were all kinds of reasons why he should play his part as a good citizen 'and all kinds of reasons for not doing so', he wrote to MacGreevy, possibly having in mind a previous acquaintance with Prudent and the disclosures that might result. 'But whatever I do and however it goes there are going to be plenty of unpleasantnesses before it can be called an *affaire classée*.'

On 2 February he was well enough to go to Joyce's birthday party at the Jolases', agreeing to go with Peggy, whom he was still seeing, even though Suzanne Deschevaux-Dumesnil, his former tennis partner, was now a frequent visitor at the Liberia. Before the party they went together to buy a blackthorn stick, the present which he suggested she should give Joyce. He himself wanted to give some bottles of the Swiss wine Fendant de Sion, which Joyce called the Archduchess because, while he thought it looked like piss when held up to the light, it indubitably looked like aristocratic piss – celebrated in *Work in Progress* as 'a rhubarbarous moundarin yellagreen funkleblue windigut diodying applejack'. This proved difficult to find until Guggenheim recollected a Swiss restaurant in the rue Sainte-Anne, where she had been some years before with Joyce and her then husband, John Holmes. When told that the wine was for M. James Joyce the restaurant was happy to sell them some.

At the dinner party Joyce offered 100f. to anyone who could guess the title of the book which up to then had been known only as *Work in Progress*. Guggenheim's later recollection was that Beckett had guessed it, but in fact the title remained a secret until Eugene Jolas hit on it later that summer. There were fifteen present at this dinner party, all adherents or disciples. Among them was Nino Frank who, on discovering that Beckett had an interest in film, offered to put him in touch with some French film-makers. The table was decorated with a model of Dublin, through which flowed the Liffey, represented by a green ribbon. What Beckett, so recently released from the place and very glad indeed to be out of it, felt about this we do not know; but when towards the end of the evening the tenor John Sullivan and Madame Jolas sang together, the performance was by no means to his taste. The proceedings came to an end with Joyce dancing his slow saraband, which Guggenheim seems to have thought was an Irish jig.

In the week after the birthday party Beckett at last saw the police surgeon. He still had to do breathing exercises but he felt considerably better and found he could now breathe with a great deal less pain. The police had impounded the overcoat, jacket and shirt he had been wearing on the night of the stabbing as part of the evidence against Prudent. The preliminary hearing had been fixed for the fourteenth, and though his apprehensions were still great he thought he might at least get his clothes back. This hope proved ill-founded; but at least these preliminary proceedings ended without unpleasantness. On Monday the fourteenth he went to the Palais de Justice. In the ante-room, to which he was directed first, he found Prudent sitting on a bench. In his account to MacGreevy he said that they 'exchanged amiabilities'.

Later he would tell people how, after some sort of greetings had been exchanged, he decided to ask his assailant why he had stabbed him. The reply, according to these later stories, was 'I don't know,' a rejoinder on which a great deal of criticism about the theatre of the absurd and the meaninglessness of all action has been founded.

A week later the trial itself took place. Still reluctant to press charges, Beckett attended with Alan Duncan. Prudent, who was said to be twenty-five, was described as an engraver's assistant; no evidence was offered of his second profession, that of pimp. When it came to their turn Beckett and Duncan offered only the barest minimum of evidence. Prudent, on the other hand, was well defended and, in the absence of evidence to the contrary, his lawyer was able to claim that provocation by Beckett had been sufficient to justify leniency. The court also seemed more than willing to accept the excuse that the accused had been drinking all day and was 'irritable' as a result. In spite of a previous conviction, Prudent wound up with only two months' imprisonment, though he was bound over to keep the peace under penalty of a longer period of two years. He served his brief term in the Santé, where, Beckett told Arland Ussher, he became the most popular prisoner. 'His mail is enormous. His poules shower gifts upon him. The next time he stabs someone they will promote him to the Legion of Honour.'

Beckett did not even get his clothes back, being told that he would have to offer evidence of ownership before this could be done. He consoled himself with the thought that he would have had to have the coat cleaned of blood and the rent made by the stabbing mended. He was evidently reluctant to wear the overcoat Joyce had given him, but in any case the weather was now getting warmer.

He had been approached by Jack Kahane, a publisher of pornographic books in English – whose son, Maurice Girodias, was subsequently the owner of the Olympia Press imprint and therefore an important figure in Beckett's publishing history – to do an English translation of the Marquis de Sade's Les Cent-Vingt Jours de Sodom. Kahane was a well-known operator in the world of Left Bank pornographic publishing and, as Girodias would later, liked to keep his name sweet in intellectual circles. For Beckett, this was an opportunity to execute what, by French standards, was quite a reputable job. De Sade's literary reputation had been growing since Apollinaire had edited a selection of his work and declared him to be the 'freest mind the world had ever known'. He had been adopted by the surrealists; had figured, garbed as Christ, in Buñuel's film L'Age d'or, and was now in France not only respectable, but a writer for whom extravagant literary claims were being made.

Beckett had been interested in de Sade for some time and he considered this to be a serious proposition. The offer was that he should be paid 150f. per 1,000 words, half of it on signing, on the basis that the work was about 150,000 words in all. When he met Kahane he insisted that there should be a preface in which he would make his own judgement of de Sade's importance and his interest in doing the first English translation quite clear, and that there should be no time limit. His own view was that 'the obscenity of surface is indescribable', but that nothing could be less pornographic. Writing to MacGreevy, he said he thought that the construction of the work was extraordinary, 'as rigorous as Dante's', and that the book filled him 'with a kind of metaphysical ecstasy'. He was very much inclined towards doing it and, of course, would not do so unless he intended to sign his name to the translation; but he was worried about the practical effect on his own future freedom of literary action in England and the English-speaking world. Would he not be known thereafter as the translator of de Sade and would not this reputation overshadow whatever fame might come to him through his own work? The idea of the preface was important to him because in it he could make his attitude and his motive as well as his admiration for de Sade quite clear.

A number of commentators have suggested that there is a considerable amount of what is loosely called Sadism in Beckett's own work. There is indeed a good deal of physical cruelty and even a certain delight in the gougings and thumpings that are described, but this is rather a long way from the quite specifically sexual cruelty of de Sade. Anyway, whatever admiration he felt and whatever his affinities with the divine Marquis, he decided eventually that the reasons for not undertaking the task were greater than the pecuniary and other attractions it held for him.

Peggy Guggenheim was now staying in her sister Hazel's spacious flat on the Ile Saint-Louis. It was sparsely furnished, containing little but Hazel's big bed and a number of her 'primitive' paintings. Into this bed it was Peggy's primary object to entice Beckett, but he proved a very reluctant lover. He did stay overnight once or twice, always taking care that if they went to bed together there should be a sufficient supply of champagne, whether to fortify himself or to distract them both, but the usual pattern was different. They would meet for drinks or a meal, after which he would walk her home over the Pont Saint-Louis. Arriving at the outer door of the apartment, she would urge him to come in and, according to her, Beckett would 'go through the most terrible agonies trying to decide' whether he would or not. But 'he always ended up by pulling himself together and running away.' She would then spend the night lying awake in a misery of rejection and desire. According to her subsequently published account, 'Beckett had little vitality and always believed in following the path of least resistance,' but it would seem that, in this matter at least, he was certainly prepared to resist.

Perhaps this would have been the case anyway, but the truth is that he was now seeing Suzanne on a fairly regular basis. Since he enjoyed Guggenheim's company up to the point where she began to attempt to entice him into bed, however, he spent much time with her too and they discussed many topics together. She called him Oblomov after the hero of Goncharov's novel, to which she introduced him. He found it interesting but was not perhaps as bowled over as she had hoped by this portrait of an anti-hero consumed by indecision, who, like himself, was very reluctant to get out of his solitary bed in the mornings.

Guggenheim was one of the people to whom he described his ante-natal experiences, saying that he had retained an oppressive memory of life in his mother's womb, which produced periodic crises during which he felt he was suffocating.

A sexually emancipated woman who was accustomed to going to bed with people she liked, Guggenheim was left frustrated and miserable by her relationship with Beckett. She had met Brian Coffey, who found her, as others did, quite attractive. In her frustration she often teased Beckett about this and said she was willing to sleep with Coffey even though she privately thought of him as 'a sort of dried up intellectual'. After one scene between them Beckett told her flatly that he was never going to sleep with her again and went so far as to suggest that she could go to bed with Brian, who might be able to satisfy her demands – a suggestion which ties in with the attitude of Beckett's alter ego, Belacqua, in the story 'Walking Out'. Belacqua suggests to his fiancée Lucy that she should go to bed with others, perhaps while he watched. Though she believed afterwards that it was her own stupidity and insistence that had moved Beckett to suggest this, Peggy was so stung by the suggestion that the next time they were all three out together she invited Coffey back to the Ile Saint-Louis. She thought Beckett looked sad but dignified as he walked away, but then one can presume she was hoping that he would be sad.

When she and Sam met the next day he asked her if she liked Brian. When she said, 'No,' it brought them, she thought, closer together, so that Beckett actually returned with her to the flat that night. Two days later, however, they met Brian Coffey for lunch. Rather preening himself on what he deemed his success, Coffey asked Beckett in her presence if he would object to Peggy and himself continuing their affair together. To Guggenheim's surprise and horror, Beckett said he would not and rather formally handed her over. She was so upset by this that she rushed out of the café and walked around in despair for hours; and then went to the Liberia and made a scene. As a pair, which many people now considered them to be, Beckett and Guggenheim were becoming famous for their scenes; but his decision at this point to leave the Liberia and find an apartment for himself probably had more to do with Suzanne than with Peggy.

The reasons he gave his friends, however, were that the Liberia was too dark; that he was thoroughly tired of hotel life; and he wanted to find a place where he could have his books about him. He was also beginning to feel that he would never do any work until he had such a place.

He spent many discouraging hours in coming to the conclusion that the more fashionable parts of the Left Bank were now quite definitely beyond his means before starting systematically to comb the 15th arrondissement. The 15th was rougher and tougher than Montparnasse, a heterogeneous, largely working-class and partly immigrant district. But even here there was hardly anything to be had except a few studios, which were also too expensive. Confronted by the difficulty of finding a place even there, he thought seriously of living somewhere in Italy instead of in Paris, where even a single room without charges and furniture seemed to be more than he could afford. But Suzanne was now also engaged in the search and, largely through her efforts, towards the end of April he found a place he liked and thought he could afford.

Suzanne Deschevaux-Dumesnil was six years older than Samuel Beckett. Her parents lived in Troyes, a sizeable country town in Champagne, about 150 kilometres south-east of Paris; and her father was a commercial traveller. Her people were French bourgeois, but she had Algerian connections on her mother's side – a fact which, together with her rather sallow complexion, somehow gave rise to an impression among some of Beckett's acquaintances that she herself had a *pied-noir* background and even led Peggy Guggenheim to refer to her as 'a simple native girl' when describing Beckett's new relationship. The Deschevaux-Dumesnils were cultured and fairly sophisticated people, and they had encouraged Suzanne to have a musical career as a pianist. She had not distinguished herself sufficiently to pursue this on the level of performance and when she met Beckett she was earning her living in Paris as a piano teacher; but she had a deep understanding and love of music and her tastes were wide enough to include the avant-garde experiments of composers such as Schoenberg and Berg. Her taste in literature was also well developed, and when Beckett began later on to do the difficult and experimental work on which his fame is based, she was among its first admirers and supporters. She would see some of his plays many times over and, on those occasions when she attended rehearsal, her comments were always keen and perceptive. Unlike Beckett, she had decided political views, left wing and, on most questions, consistent. An accomplished needlewoman and a good household manager, she was, however, uninterested in cooking, a lack of interest compounded by an enthusiastic belief in eccentric dietary theories.

The apartment Beckett had now found through his efforts was in the rue des Favorites, which is off the rue de Vaugirard in the 15th arrondissement, with the advantage of not being very far from Montparnasse, his favourite area.

The rue des Favorites, mostly composed of rather ugly and utilitarian modern buildings, some of which are factories and workshops, was

A portrait of the artist as
a somewhat neurotic
young man.

Willie Beckett, Samuel Beckett's sporting and gregarious father (right), sitting on the steps of the stand at Leopardstown Racecourse with the Clerk of the Course.

Samuel Beckett's mother in later years. She was a keen gardener.

Cooldrinagh, the 'Tudor-style family residence', in which, in the first-floor room with the bay window, Samuel Beckett was born.

ABOVE Barrington's Tower on Leopardstown Road, Samuel Beckett's childhood bolt-hole.

LEFT A lonely tree on Glencree in the Dublin Mountains. Perhaps this, or one like it, remembered afterwards, was the original of the tree in *Waiting for Godot*.

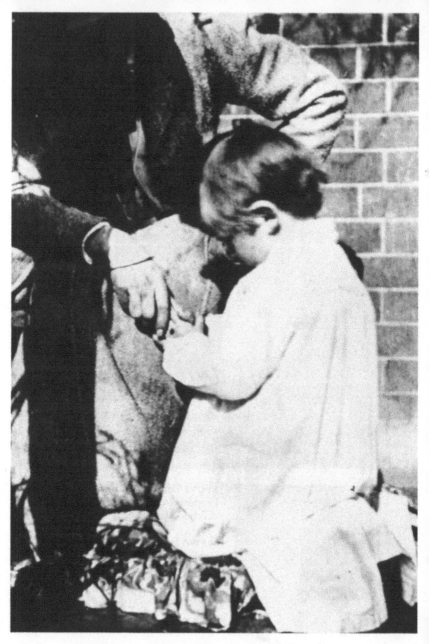

Samuel Beckett apparently saying his prayers at his mother's knee,
actually a staged photograph.

LEFT Sam (right) with his brother Frank at an early age.

BELOW May Beckett (left) in her donkey trap with Sheila Page.

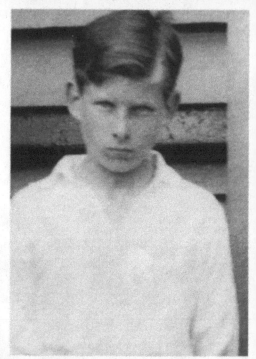

Samuel Beckett as a
14-year-old schoolboy,
regarding the world without
too much favour.

The Trinity Cricket XI.
Samuel Beckett, glowering, is
second from the right in
the back row.

Samuel Beckett with one of his mother's Kerry Blues, though not the one remembered in *Krapp's Last Tape*, whose death so grieved him.

Samuel Beckett during the stay at Yew Tree Cottage with (left to right) Peggy's daughter Pegeen, George Reavey, Geer van Velde, Gwynedd Reavey and Lisl van Velde.

ABOVE Peggy
Guggenheim, who
pursued him with
little enough success.

ABOVE RIGHT
Samuel Beckett,
Frank and Suzanne at
Ussy in 1952.

RIGHT The
unaesthetic Rue des
Favorites, where
Beckett lived between
1938 and 1961.

Meditating at
rehearsals for the
German premiere of
Godot, Berlin 1953.

OPPOSITE PAGE Samuel Beckett with Giacometti and the famous tree which was altered so often during late-night sessions in the sculptor's studio.

TOP Buster Keaton regarding Beckett, and vice-versa.

ABOVE Samuel Beckett with his 'Assistant Director', Rick Cluchy, in Berlin in 1972.

RIGHT The 'incomparable Billie.' Billie Whitelaw, Beckett's favourite female interpreter.

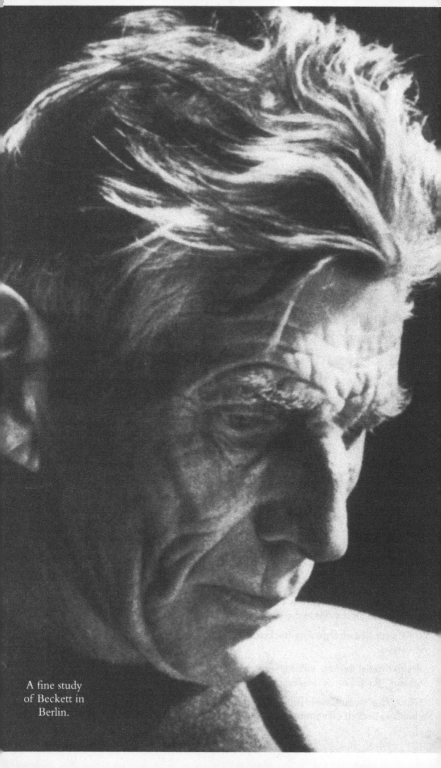

A fine study
of Beckett in
Berlin.

anything but a traditional Paris street, and a certain suspension of the aesthetic faculty would be necessary to like either the street generally or Number 6, the modern grey concrete block in which the apartment was situated. But Beckett was always ready for such suspensions and seems even to have rejoiced in the necessity for them. What he had found was on the seventh floor, but this was no great inconvenience since there was a lift. Compared to the Liberia, where the gloom and airlessness of his room had been oppressive, the apartment seemed to him to be bright and airy. It consisted of a studio room, bedroom, kitchen and bath. There was no telephone and the bedroom and kitchen were both quite small but at least he was finished with hotel life and, in a way, he was glad to have removed himself to a quarter into which people with artistic pretensions did not commonly enter. He was pleased by the fact that the rue des Favorites had been formerly called the impasse des Favorites and that it was not far from the impasse de l'Enfant Jésus. Although it was let as furnished, it had no furniture and so he had to search around for a bed, finding one eventually in a nearby emporium and making do on the floor until it came. Some of the furniture consisted of pieces that the Joyces had had in store and now contributed. They were, as one might expect, rather heavy and old fashioned.

While Beckett was still in the Liberia *Murphy* had appeared; and once again the reviews were on the whole favourable. In the *Spectator* the well-known Irish novelist Kate O'Brien had been more than enthusiastic. Reviewing five new novels together, she said that as far she was concerned 'Murphy swept all before him'. Rarely, she averred, had she been so entertained by a book, 'so tempted to superlatives and perhaps hyperboles of praise. It truly is magnificent and a treasure if you like it. Quite useless to you, quite idiotic if you don't.' Describing it as 'allusive, brilliant, impudent and rude', she compared it to Rabelais, Sterne and Joyce, 'the last above all', though she emphasized that Beckett was 'not like Joyce evocative of tragedy or Hell'. Praising the 'bright, clear lyricism' displayed in the description of Mr Kelly's kite-flying and in the creation of the 'lovely classic figure' of Celia, she declared that for the right reader it would be 'a book in a hundred thousand' and wound up by saying that her own great pleasure in it was 'not least in the certainty that I shall read it again and again before I die'.

If this was gratifying – and it was, for Beckett at this period was not indifferent to reviews – the twenty-two-year-old Dylan Thomas, writing in the *New English Weekly*, was both less enthusiastic and in some respects more perceptive. Though praising the obvious qualities of 'energy, hilarity, irony and comic invention' that the book displayed, he also invoked Joyce and demurred at the 'influence of those writers who have made *transition* their permanent resting place'. He found the book verbose and sometimes loosely written, and thought it failed principally 'because the story never quite knew whether it was being told objectively from the inside of its characters or subjectively from the outside', and 'because the minds and bodies of those characters' were 'almost utterly without relations to each other'. Nor was Thomas greatly taken by the book's humour, which he characterized in a phrase soon to obtain wide currency in Ireland, but nearly always attributed to various local wits, as 'Sodom and Begorrah'. He noted the pedanticism and unlikelihood of much of the dialogue and declared that while Mr Beckett seemed to suppose that he was writing about the lower strata of society, about the 'dispossessed and the regardless of possession', he in fact took a most romantic view of it.

An anonymous reviewer in *The TLS* made, in accordance with the custom of that journal, an attempt to summarize the plot:

> One might explain that the book's hero, Murphy, is an unemployed Irishman in London, who lives on the difference between the cost of his lodgings and the amount he claims for them from his guardian; that his mistress Celia is a prostitute who does her best to make him look for work; that his Dublin friend, Neary, loves a girl who loves Murphy, and they spend most of the book looking for him; and that finally . . . they track him to an asylum, where he has found a congenial job as an attendant.

Besides this brave attempt at a summary the reviewer confined himself to saying that one might suggest 'Murphy's attitude towards life by citing his admiration of the lunatics whom he has to attend,' and in furtherance of this end it instances his complaint about 'the complacent scientific conceptualism that made contact with outer reality the index of mental well being'.

In fact, though, Murphy's attitude to the psychotics he deals with in the Magdalen Mental Mercyseat is one of the weaknesses of the book. Murphy wants to retreat into what he thinks would be the peace and inviolability of his own mind. He imagines that psychotics have already achieved that blessed state; but the sad truth is that they have not. The psychotic state is one of continuous agitation and anxiety, persecution and other fears. No one is so unhappy or so beset by terrors as what might be called your average psychotic; and there is a curious lack of compassion as well as of reality – a relevant fact since Murphy is on the realistic plane throughout – in the way Beckett presents such persons in the novel and in the contrasts he makes between their mental state and that of the people in the outside world.

In spite of the presence of Suzanne in his life, the affair with Guggenheim still dragged on, after its fashion. To replace their sex-life, as she put it, they used to drink wildly and walk for hours around Paris, he apparently giving her the impression that some day they would be happy together. She asked Helen Joyce's advice and was told that she should rape Beckett. As, according to her accounts, they had already been to bed together, it was difficult to see how that would make any difference, but one night when they were in a restaurant and getting on very well together – to such an extent that he seemed, as she put it afterwards, 'to be yearning towards me and floating into me in spite of himself' – she told him that she was coming home to the rue des Favorites with him. As they left the restaurant, he took her arm and she had the illusion of everything being settled, but when they got to the flat and she announced that she was staying the night, he reacted with apparent terror and rushed off, leaving her in the flat by herself. At a loss what to do, she eventually went to bed but, of course unable to sleep, got up at four in the morning to write a poem which she left for him. It ended:

> Every step I take she battles
> Every inch the death-knell rattles.
> Shall I kill her holy passion?
> Destroying life, not taking action.

Beckett had already persuaded her to give a show in her new London gallery to his new-found friend Geer van Velde. Geer, or Gerardus, was one of two Dutch brothers, both painters, whose work Beckett admired and with whom he was to become very friendly. Geer was the younger, more outgoing and more ebullient of the two, and to begin with Beckett's interest seems to have been primarily in him rather than in his brother Bram. He was eight years older than Beckett and had originally been apprenticed to a house painter in his native town of Lisse near Leiden. In 1925, after an exhibition at The Hague, he had come to Paris where he had exhibited with the Independents. He was good looking in a rather soft sort of way and had a pretty young wife called Lisl, but Guggenheim certainly believed that the attraction he and Sam had for each other was primarily homosexual.

Having brought it about and having written a catalogue note, Beckett felt he had to go to London for the show, which opened in the first week in May 1938 to a generally expressed belief among such critics as deigned to notice it that van Velde's work was very much derived from Picasso's. Guggenheim thought so too, but privately, and to please Beckett she bought many of the paintings herself under different names, giving some of them to her friends. 'This was all because I loved Beckett so much and he loved van Velde,' she said later.

At the weekend she took the van Veldes down to her place near Petersfield and invited Beckett and the Reaveys. After much deliberation Beckett decided to accept the invitation. So, after giving up her own bedroom to the Reaveys, Peggy slept on a made-up bed in the dining room. After everybody had retired on the first night Beckett came to her there. He showed no inclination to get into the made-up bed with her but seemed very anxious to talk. After some hesitation he told her that he had a mistress and asked her if she minded. It was natural enough, perhaps, that he should tell her there was another woman in his life, but the uneasily apologetic note in his account of Suzanne made her seem, Guggenheim thought, more like a mother than a mistress. She had helped him to find his new flat, he said, and she was now making curtains for it. Peggy remembered meeting Suzanne once when she called at the Liberia and deciding that she need not be jealous because Suzanne was not attractive enough. He was not in love with her, Sam told Peggy, but Suzanne never made scenes as she did. 'She made curtains while I made scenes,' Guggenheim would say to people later.

Peggy now decided once again that she was not jealous but at the same time she began to feel that the affair with Beckett was more hopeless than ever. So she resolved that when she got back to London she would turn her attention to E. L. T. Mesens, a surrealist poet who was also the director of the gallery next door to hers in Cork Street. He had already made advances to her but she had temporized, even though she found him quite attractive.

The house in which they were staying, known as Yew Tree Cottage, was not far from Bognor Regis and on Sunday afternoon they all got into Peggy's Delage and went to the beach there. It was cold and wet with a wind blowing off the sea, and as they were walking along the beach Beckett suddenly stripped off all his clothes and plunged into the waves. He swam so far out in the choppy water that his head became invisible in the troughs and everybody began to be quite frightened, thinking that he intended to continue swimming out to sea. Then, much to his companions' relief, he was seen to have turned back. As he dried himself with his pullover and resumed his clothes there was an embarrassed silence which continued while they walked back to the car.

Was this episode, as Peggy thought, an abandoned suicide attempt? Or a suicide signal? And if so, had it to do with Peggy, or with Suzanne, or even perhaps with Geer, or simply life in general? But Beckett was a far stronger swimmer than any of the company knew; and it may have just been an impulse of the moment, the dangers of which they all exaggerated.

When they got back to London after the weekend Peggy took care to be flirtatious with Mesens in front of Beckett, hoping to arouse his jealousy. The little gallery director was an entirely opposite type to the Irishman, being short, cheerful and extremely talkative where Beckett was long, gloomy and frequently silent. Finally Peggy let Beckett know that she was making an assignation to go away for the weekend with Mesens and derived what she later called a 'diabolical pleasure' from doing so. However, the affair did not last long after she told Mesens how much she really loved Beckett, and she promptly took off for Paris again to be with the loved one.

She took her car with her and wired him that she was coming in the hope that he might meet her at Calais. He was not there, however, so she drove on to Paris and went to the apartment where, as she was to put it, he 'waited for me listlessly and seemed bored with me'. One afternoon at the flat when the van Veldes were there, Sam and Geer began to exchange clothes. This struck Peggy as a rather homosexual performance and she told Beckett so, accusing him of being in love with van Velde. He took this in silence but when they all went to a dancing place there was an awful scene, which ended with Beckett walking out into the night and leaving her alone with Geer and Lisl.

The estrangement which followed was a serious one and was only brought to an end when Peggy wrote him a letter telling him that she would like to say goodbye. There followed a meeting at which they became friendly enough for a tentative suggestion that they might both drive the van Veldes to their new domicile in the Midi to be discussed. The arrangement was made for the following weekend but Beckett temporized because he was now spending the weekends with Suzanne and it was not easy for him to go off with someone else. As Peggy later put it, 'he managed to get off at two o'clock on Sunday', however, and they headed for Marseilles in Peggy's Delage with Beckett driving. More or less cordial relations had now been resumed, which surprised the van Veldes, although, like a lot of other people in Paris, they had seen Beckett and Guggenheim fight so often that they were beginning to accept reconciliations as a matter of course. The piquancy of the situation was further enhanced by the fact that van Velde had been quarrelling with Peggy about the proceeds of the London show and, according to Beckett, who was forced to listen to both sides of the story, Geer and Lisl now revealed themselves in the process as 'unaccountably avaricious', with a pettiness that did not at all fit in with the rest of what he knew of van Velde.

After leaving the painter and his wife in Marseilles Peggy and Beckett started north again. She was in a hurry as she had to drive Yves Tanguy to London for an exhibition he was having at her gallery, but nevertheless on the way back they stopped at Dijon, a gloomy town, of which Peggy had many sad memories. Although Dijon is famous for its cuisine they succeeded in picking the wrong restaurant, where they had a dreary dinner in depressing surroundings. According to her, Beckett had no flair for picking the right places and she thought that in any case his Protestant upbringing made him automatically shun the pleasant things in life.

When they found a hotel he went to the desk and took a double room. This surprised Peggy, who had not expected that he would want to sleep with her, but she always found his behaviour unpredictable and made no remark. There were two beds in the room and initially they got into separate ones. Then, after lying awake in the dark for a while, Peggy crept into his bed, whereupon he got out and promptly got into the bed she had just left, saying that she was cheating. When she asked why he had taken a double room if he had not intended that they should sleep together, he replied that a double was thirty francs cheaper than two singles, a reason which, from her point of view as a rich woman, seemed both extraordinary and insufficient.

The next morning they walked about Dijon together, seeing such sights as there were, including the local art museum. When there were no differences about love-making or other matters Peggy always deeply enjoyed his company and on this occasion he seemed particularly charming. She told him that on the whole she thought he was a much nicer person now than he had been, an impression which was beginning to be shared by other people. It did not seem to strike her that this may have been due to Suzanne's humanizing, calming and reassuring effect. He replied that he thought that she too was nicer and that she was always nice when she did not make scenes. Because of her arrangement with Tanguy they drove back to Paris that day and, according to her, parted with sorrow, Beckett as usual regretting that he was relinquishing her.

Whatever his regrets, he was showing a much greater degree of commitment to Suzanne, who was now spending more and more time at the rue des Favorites, than he was to Guggenheim. Of course, Suzanne probably managed him better and was less demanding sexually.

Guggenheim's belief that he always followed the line of least resistance was probably a key to the situation. The line of least resistance with Suzanne was an easier and smoother path than it was with Peggy, who spelled trouble. If Peggy was all mistress or would-be mistress, Suzanne was certainly in large part mother, but mother without the self-centred gloom, the career demands and the arguments. There was even the discernible physical resemblance to May Beckett; and for someone with a growing desire for domesticity, as Beckett had at this point, her ability and willingness to look after the domestic side of things was a great advantage. The famous curtains were now up. The books he had thirsted for in the bareness and gloom of the Hôtel Liberia had now arrived from Ireland; and in some moods he declared that he never wanted to see other people again.

Later in life Beckett would occasionally speak of Suzanne as if she had rescued him from a long period of aimless and disorganized bohemian existence in hotel bedrooms. In fact he had been in the Liberia for only a little over two months before the stabbing; and though the life he was leading was disorganized and idle enough, on the surface he seems to be exaggerating the need for a rescue mission and the effect it had on him. It is true that he had been drinking heavily, rising late and doing no work; perhaps he also meant that he was slipping into other ways from which he thought it desirable to free himself – ways which are suggested by his seeming acquaintance with pimps and late-night promenades with Giacometti.

Another effect of his new quasi-domestic life was that he was beginning to free himself to a certain extent from the Joyces and their circle, which was, after all, something of an Irish ghetto, though he would be drawn back into it by various obligations. With a sort of obstinate goodness which had an element of guilt and perhaps also an element of masochism in it, he went regularly to see Lucia at Ivry and would continue to do so for some time to come.

Towards the end of May 1938 Beckett went to a party given by Helen and Giorgio at the Villa Schaeffer. After noting that all the usual crowd were present, he described to MacGreevy the evening as ending with 'all of the old songs and the old stupors' and remarked, 'quel ennui'.

And, under Suzanne's influence, he was not only becoming more of a Frenchman than an Irishman in exile; he seemed at this point to be almost about to become a French writer. As he settled into the new apartment he also began to write; and, possibly under the influence of Alfred Péron, he was beginning to write in French. He had always been drawn to Péron and was now more so than ever. They lunched together once a week at one of the little restaurants on the Left Bank and there was a suggestion that Péron might translate *Murphy* into French, which pleased Beckett very much. The poems in French he was now beginning to write can be seen in retrospect as an important departure. He would write twelve in all before the war, hoping at first to make a book of them. However, they would eventually appear for the first time in *Les Temps modernes* in November 1946. The first, beginning, 'à l'elle l'acte calme', seems to be about making love; but Lawrence Harvey suggests with some justice that the 'calm act' seems to be one 'apparently inspired by something resembling friendship more than love. The uncertainty of the narrator in "Echo's Bones" and in "Cascando" about his capacity for love is decided, in this poem, in the negative. She is "vide d'amour" . . . but he, "pur/d'amour", does not know love and possibly never has, though he gives himself to her as best he can.'

In spite of his growing relationship with Suzanne, Beckett fulfilled a promise to his mother by crossing to Ireland in August, hating the prospect but going all the same. Frank had written in May to say she had had an accident. She liked to read in bed by candlelight and had fallen asleep one night only to waken in a panic when she found that the candle had toppled over and set fire to the sheets. She had burned her hands while putting out the flames. When Beckett read this he found himself moved to tears. Since Suzanne, like many French people, liked to leave Paris in August and usually went to her parents in Troyes, the way was now clear for a visit.

He wrote to MacGreevy asking him to enquire about a room in Cheyne Walk, so that he might stay a night or two in London on the way, but in fact he turned up at Peggy Guggenheim's apartment, apparently hoping to stay there. Peggy, with her usual necessity to have a new – and, it must be said, distinguished man in attendance – had now begun an affair with the Yves Tanguy, who had succeeded Hans Arp and, of course, Marcel Duchamp, and would be succeeded by Max Ernst. Tanguy's interest in painting was said to have begun when he saw a Chirico in a dealer's window, a sight which so moved him that he jumped from the open platform of the bus he was travelling on at some risk to life and limb. His work had been regularly reproduced in *La Révolution surréaliste* and he was now a well-known member of the group, one of those surrealist painters who did not scruple to employ academic illusionism to achieve their shock effect. According to Peggy, when Beckett saw a picture of Tanguy and herself on the mantelpiece of the apartment, he grew very curious, and although she was leaving for Paris – and Tanguy – and offered him the flat, asked her to remain with him. She refused, telling him that she was forced to have other lovers in order to keep him as a friend, and blaming him for what she called her double life.

In Ireland Beckett found his mother nervous and depressed, but comforted himself with the philosophical reflection that her depression was no worse than it had often been in the past. He was not at ease in the house, where his cousin Sheila Page was staying with two noisy children, or in Dublin, where he said he felt like an amphibian on dry land. To take his mother out he went with her to the Horse Show and to the Abbey; but the only old acquaintance he saw apart from Jack Yeats was Mary Manning, whom he thought 'grown tedious and precious'. Finally he escaped to Donegal, where Frank and his wife had taken a house for the summer months.

He had his bicycle with him and he intended to go back by way of Brittany, calling on Péron, who was staying the summer there, and cycling on to Paris. Instead he went back to the French capital by boat and train. When he met Peggy there she told him that she and Tanguy were having trouble with Mrs Tanguy, who, though used to his affairs, was making a fuss about this one; whereupon, rather to her surprise, Beckett offered to lend her his apartment in return for her car, in which he would drive with Suzanne to Brittany. When he and Suzanne departed in the car, Peggy and Tanguy moved into the rue des Favorites, where they hoped to remain undisturbed since nobody knew they were there. But on the first night Tanguy disappeared, seemingly going to a surrealist party where a painter friend, Victor Brauner, had an eye put out by a bottle, an event which delayed him for several days. She says that she was still 'terribly in love' with Beckett and that one day Tanguy said to her, 'You don't come to Paris to see me. You come to see Beckett.' Years later she would tell Gore Vidal how deeply in love with Beckett she had been at this time, but how she had discovered him to be homosexual and had known then that it was hopeless.

Meanwhile the much-loved object of her affections was spending some pleasant and rewarding days in the village of Saint-Breyvien-l'Ausient in Brittany, where Péron continued to encourage him to write in French, the result of such encouragement being the unpublished essay 'Les deux Besoins'. In this, with the aid of a diagram, Beckett explores what was to become a dominant theme: art as an imperative for the artist. There is said to be an inner voice, which the artist must obey, but to which he can never abandon himself completely. The essay also reflects the interest in the pre-Socratic philosophers which is evident in *Murphy*.

September 1938 was the month of the Munich crisis, an event which convinced many people, including the apolitical Beckett, that war was now inevitable. He heard Hitler speak on the wireless and thought it was like hearing the air escape as from a slow puncture. But no matter how things went, he told Reavey, he would stay on in the rue des Favorites, like an ostrich, 'with my own handful of sand', adding that all he had to lose was 'legs, arms, balls, etc. and I owe them no particular debt of gratitude so far as I know'. But he thought that the crisis had made people more tender towards each other, 'even in the commerce of the rue de Vaugirard'.

And he was at last becoming more fully aware of the virulence and horror of Nazi anti-semitism. He had turned the pages of *Mein Kampf* and been shocked by Hitler's attitudes and stated ambitions. In early November Nazi gangs rampaged through the streets of German cities, beating up Jews, smashing up Jewish-owned businesses and burning down synagogues and houses. At least forty Jews were murdered on the famous *Kristallnacht*, and concentration camps were opened for another 25,000 who were arrested. Every day brought more refugees to France and sections of Paris like the Marais were becoming Jewish quarters. But when Beckett spoke to James Joyce about this, Joyce, who was cultivating disengagement and, as part of the exercise, annoying people by praising German efficiency and Hitler for 'getting a whole people behind him', told him that there had been similar persecutions before. Characteristically, however, at the same time Joyce had begun to help Jewish people such as Hermann Broch to get out of Germany.

The low-key nature of Beckett's relationship with Suzanne can be judged by the fact that in early November 1938 he went back to Ireland, where he stayed through Christmas and into the New Year. He went to see Joyce before he left and was handed the final ten pages of the great book, which had taken so long to write and in the composition of which he and many others had been involved. 'I done me best when I was let. Thinking always if I go all goes. A hundred cares, a tithe of troubles and is there one who understands me?' Beckett read them on the Métro and when he got to the Gare du Nord he was so moved that he telephoned Joyce to say how great they were.

As a sort of trial run for leaving Cooldrinagh May had taken a house in Greystones. It had, Sam remarked, a view of the graveyard, telling Ussher that 'from the window she can see the cemetery where my father is "at rest"'. It was, he thought, a cold, desolate spot in the November weather, calling it the 'côte de misère' and saying 'what does not face north faces east'. But he stuck it out, sitting by the fire, talking banalities to visitors, going for long walks by the sea in the bitter east wind, rarely venturing into Dublin, where those who saw him were surprised and where a rumour began to spread among his acquaintance that he had, after all, returned to live in Ireland.

In honour of Joyce he made a trip to Chapelizod, the little Liffeyside township named after Iseult of Brittany where Earwicker of *Finnegans Wake* has his pub, and collected a flat piece of limestone from the banks of the Liffey which he brought back to Paris early in the New Year, 1939, asking Stanley William Hayter to engrave it for presentation when the book was published. At Joyce's birthday party on 2 February there was again a horrendous-sounding centrepiece, with Paris represented by the Eiffel Tower and a windmill, Dublin by a church, and a bottle standing in for Nelson's Pillar and a cake with iced simulacra of Joyce's seven books in their original bindings. A bound copy was ready in time for the party, from which Helen Joyce read the closing section. Shortly Helen would begin to show symptoms of the depression which eventually developed into a nervous breakdown. Not only was she never again to be at a family gathering; there was never to be another. Joyce had hoped that the book might appear in February. When he had to wait until May he was fretful that the rising tide of war would sweep such matters as the publication of books aside.

In April 1939 Sam was thirty-three, as his beloved Dante had put it, 'Nel mezzo del cammin', and in a mood to reflect on the future. Writing to MacGreevy, he wondered 'if the second half of the bottle' would be any better than the first half and concluded that if it was, it would only be because he had 'got used to the taste'. Telling his friend about Suzanne he referred to her as a 'French girl', whom he was 'fond of', using a curious word – 'dispassionately' – to describe the degree of fondness and saying she was very good to him. 'The hand will not be overbid,' he adds; but the thought of her seems to cheer him up a bit none the less and provokes the reflection that maybe the wine remaining in the bottle of life was not so bad as he had feared.

French people, like others, were now viewing the future with some trepidation. On 16 March 1939 Hitler dismembered what was left of a defenceless Czechoslovakia and German troops entered Prague a few days later. A Nazi press and radio campaign against Poland made it clear that he was still not satisfied. Beckett, like many others, began to think that war was inevitable. 'If there is a war, as I am sure there must be soon, I shall place myself at the disposition of this country,' he wrote to MacGreevy. Even in far-off Ireland his brother Frank was also, like many Irish Protestants, wondering what he would do if Britain, the old focus of loyalty, once again became involved in a major war.

Every Tuesday Sam had lunch with Péron. In theory they were supposed to discuss a translation of *Murphy*, but they usually wound up by playing tennis and, apart from the poems in French he was writing, Beckett had no work to show. He had himself translated the story 'Love and Lethe', which had been in *More Pricks Than Kicks*, and given it to a French writer who was almost the same age as himself and who had connections with the *Nouvelle Revue française*. This was Jean-Paul Sartre, who duly passed the story on to Jean Paulhan, the editor, but Beckett heard nothing more. Apart from the few poems in French he had been writing, he had nothing else to show for his time in Paris. As he told Reavey, he was doing nothing there that he could not do just as well in the pubs of Chelsea.

He had read Sartre's novel *La Nausée* with enthusiasm, finding it 'extraordinarily good'. But his main literary enthusiasm at this time was for the work of an older writer, Louis-Ferdinand Céline, author of *Voyage au bout de la nuit*. Born twelve years before Beckett, Céline was a writer who ultimately pleased no party and was faithful to no cause. His most famous book, written when he was a dispensary doctor at Clichy, well placed to see the underside of French life, was a colloquial, slangy first-person narrative with a vast range, using his own experiences in the war, in Africa, in America and in Paris. On the surface Céline's two great pre-war novels are an attack on all levels of French society, the rich and the poor, and on French life in the twentieth century; but like all the writers whom Beckett deeply admired, his pessimism extends beyond this circumstance or that, this civilization or another. In his vision at its starkest, nothing really changes. As one cause succeeds another, calling forth meaningless loyalties and betrayals, we get deeper into the mire. 'We belong to suffering,' says Bardamu at the beginning of *Journey to the End of Night*. 'We have its fingers always round our throats.'

Beckett's École pupil and Trinity friend, Georges Pelorson, was now editing a magazine, *Volontés*, but Sam's dislike of the general orientation of this had caused a cooling off. He considered the fifth number, now just out, to be especially 'ignominious'. Pelorson was a conservative, concerned, like many others, about what was happening to France, and inclined to blame the Third Republic for whatever it was. There were many such. They did not all share the attitudes of *L'Action française*, the newspaper of the right, in which Maurras's young men wrote stinging front-page pieces about the corruption and venality of politicians and Jewish financiers; but when war and defeat came to France their traditionalist conservatism would lead them into strange ways. Beckett had also had a difference – not of course ideological, but the sort of trivial misunderstanding which arises between friends – with Tom MacGreevy. On the way back through London to Paris in January he had made a remark about the people in Cheyne Walk which Tom took as a reflection on himself. After Tom had failed to write for some time Beckett took the initiative and wrote asking if there was anything wrong.

307

I know on my side there is indolence and despondency . . . and there are things with which it is difficult for grown-up people to have patience. I know also that you have enough troubles without that of calling on your reserves of indulgence. But if my friendship means as much to you as it always has done to me, even when I may have appeared to neglect it, you will agree with me that it would be a great pity for perhaps a small thing to interrupt it. So let us clear it up, if you will, whatever it is.

When MacGreevy responded to this Beckett hastened to reassure him that a remark made in London had not been intended to apply to him so much as to other people in Cheyne Walk. 'I did not feel at all that way where you were concerned.'

The Joyces were in the throes of yet another move, staying a few days at a hotel while a new, smaller flat in the rue des Vignes was made ready. Joyce was visiting Lucia on Sundays but refused to talk about her condition. Beckett went once every week also, observing that her condition was worse. Helen's condition had now deteriorated, but this time her illness manifested itself through mania rather than depression, so both Giorgio and his father required help and sympathy of the Irish kind – that is, in the shape of a sympathetic drinking companion. *Finnegans Wake* appeared in early May, and perhaps because of the feeling of anti-climax that often accompanies the publication of a book, Joyce was feeling especially sorry for himself. The reviews were not on the whole favourable and he imagined that his financial situation was deteriorating so rapidly as to make desperate measures necessary, so he consulted Beckett about the possibility of a job. Since Beckett had little option but to take this seriously, he responded by telling him about the job in Cape Town as lecturer in Italian, thus adding the master to the list of people to whom he had passed news of jobs which had originally been given to himself. It would seem that Joyce affected to consider the chance of going to Cape Town seriously until reminded that the Cape was a place of frequent thunderstorms, a natural phenomenon of which he had a great fear. Beckett continued his visits to Lucia at Ivry, but he now thought her condition was getting considerably worse. His own physical condition was not good and he had a herpes sore or eczema on his face that was obstinately refusing to go. Nor was it a good time to be in Paris. Most people were unsettled by the prospect of war, and as the heat of summer grew he experienced a more than usual lassitude, getting up very late and finding it difficult to do anything whatever. He had promised his mother that he would be back again in July.

Suzanne was once more leaving Paris for the hotter months and the promise of some rest and recuperation in the sea air now seemed more attractive, so at the beginning of August 1939 he crossed over to Ireland yet again. He found his mother still very depressed and nervous. Cooldrinagh had now been sold and May had bought a plot of land across the road from it where work had been commenced on a new bungalow which had an even better and uninterrupted view of the mountains across some fields. He found the atmosphere strange. No one seemed to be concerned about the prospect of war and his mother seemed to have a vested interest in asserting that there was not going to be one, perhaps because it had nothing to do with Bill or his memory. For once he was anxious to talk politics, albeit international politics of a life and death nature, but after a while the prospect of war came to be regarded as a dangerous subject which was best avoided. He enjoyed seeing Frank, who lived near by with his wife and small child, and he did not object to his Uncle Gerald's children, John and Anne, who were in and out of the house a lot. Like Joyce, Beckett had strong family instincts and loyalties; but whereas Joyce's were largely part of his *amour propre* – if he was important, his ancestors were important – and did not extend beyond his immediate line to his collaterals, Beckett's did; and he would always be very obliging to his nephews and nieces. Later on, perhaps, this was partly an effect of his own childlessness, but even at this period it was beginning to be evident. John Beckett, who was young enough to be a nephew, but was in fact a first cousin, had inherited the musical talents which had come down through his Beckett grandmother and would eventually turn out to be a harpsichordist and composer of distinction. He was a dark, brooding boy, who even at the age of thirteen displayed some of the characteristics of the artist to be and Sam got on well with him. It was only when Sheila Page and Mollie Roe arrived with Sheila's two children and the house began to fill up that Sam began to find the small inanities of everyday life and gossip oppressive.

His impatience was alleviated as usual by long solitary walks, by bathing expeditions with John and Anne and by gatherings round the piano in the evenings. But the general oppressiveness of trivial talk about such Dublin concerns as the Horse Show was sharpened by the thought of Paris, where the international situation had had a reality for everybody, including Beckett, which was not felt in an Ireland already settling into the mindset of neutrality, and where people like the Becketts tended to think of the invulnerability and sheltering power of Britain and its empire when they thought of war. He began to long to get away, but he naturally found departure difficult. The more he insisted there was going to be a war, the less his mother could understand his anxiety to get back to France – an uncertain sort of place which would, after all, be in the front line. If people in Ireland knew anything about the situation, they knew that the war would be fought in France, as the last one had been. And so he stayed longer than he had intended, until the fateful first weekend in September in fact, when the German army invaded Poland and he came back from a walk along the cliffs with Mollie Roe and the dogs to hear Neville Chamberlain on the wireless saying that the ultimatum had now expired and Britain was at war with Germany. Sheila Page, whose husband expected to be called up, had already left for London. Mollie Roe too wanted to get back to England. When Sam went to his room and began to pack his mother once again questioned the necessity of going, to which he replied that he had promised a lot of people he would be back in the event of war. He may have been thinking of Joyce; he may have been thinking of what he had said to MacGreevy about placing himself at the service of his adopted country; but it is probable that, in the scales of his conscious motives at least, equal weight belonged to a simple desire to get back to the city he preferred and the people he felt were his fellows. Later he would say, 'I preferred France at war to Ireland at peace', but of course neither he nor anybody else knew what course the war would take, how complete would be the collapse of France and how brutal German domination.

That evening Frank drove Mollie Roe and Sam to Dun Laoghaire, which people of the Becketts' class still tended to call Kingstown, and they crossed to Holyhead on a mailboat which was already being hurriedly and inefficiently blacked out. In London, which had had an air-raid alarm on Sunday morning just as Chamberlain was finishing his speech, black-out regulations were also in force. He stayed just long enough to see MacGreevy and Reavey and then took the train from Victoria. At Newhaven he found that special permits were needed to board the boat; but when he argued that he was an Irish citizen, they let him through. It would be five and a half years before he saw Ireland or England again.

On arrival he found that many of his friends were already leaving Paris, some for the South, some for their countries of origin. Some, such as Péron and Pelorson, were being called up or standing by to join their regiments. Some of the Americans were going home. Others, the swashbuckling McAlmon among them, had decided to stay. Peggy Guggenheim was in two minds about it. Her family was in the South and she felt she should get them home to America, but she also wanted a visa to England in order to wind up the affairs of her gallery, so she sat in the Dôme drinking Pernod. In accordance with the sentiments expressed in his letter to MacGreevy, Beckett now applied to the French authorities, offering his services to his adopted country in any capacity they might think useful. Perhaps not surprisingly, his application was ignored. The Joyces were in La Baule in Brittany, to which Lucia and other patients of Dr Delmas's sanatorium were supposed to be evacuated. One night Joyce went to the local restaurant with a psychiatrist, Daniel O'Brien, who was also awaiting the arrival of the patients. The place was crowded with French soldiers. Someone began to sing the Marseillaise and Joyce stood up to join in. Hearing his splendid tenor, the soldiers hoisted him on to a table and insisted he should sing it again from start to finish. Then they all sang it a third time with Joyce leading.

Meantime Helen was running round Paris with two blue Persian kittens and a house painter boyfriend. Her brother had arrived to take her back to America and Giorgio was desperately anxious that she should go, but Helen resisted and, unfortunately for their relationship with Joyce, the Léons sided with her.

Giacometti was one of the few people Beckett knew who was still to be found in Montparnasse. On 18 October of the previous year he too had had a brush with death. As he stood on the pavement in front of the Joan of Arc statue in the place des Pyramides a car driven by a drunken American woman came speeding along the rue de Rivoli, swerved on to the pavement and knocked him down, after which it crashed through a shop window. When he recovered consciousness, it was to find that the metatarsal arch of his foot had been crushed. The injury left him permanently lame. They sat in the Dôme together in a silence which was broken only occasionally, and, in the early part of the night, sometimes not broken at all. Then they would walk through the blacked-out streets, Giacometti limping but no longer in pain.

On some nights in the Dôme or the Closerie des Lilas he would find himself in Guggenheim's company. She was staying in Mary Reynolds's house and one night Beckett came back there with Giorgio Joyce. He got very drunk and stayed the night, though of course nothing happened sexually. As they were preparing to go out in the morning – she was going to take him to meet Kandinsky – she missed her footing and fell down the stairs, landing on her right knee and putting it out of its socket. Beckett stood helplessly by while she squealed with pain, but Mary Reynolds's maid, Nellie, helped her to a sofa and sent for a doctor. She was taken to the American hospital, but was on crutches afterwards. It seemed as if everybody he knew was limping.

The Joyces arrived back in Paris on 15 October and Beckett was almost immediately involved in their lives, more especially and urgently because Joyce had suddenly decided to sever all relations with Léon, who for ten years had been his selfless helper and friend. And besides Léon, Joyce had lost Jolas, who had gone back to the United States, and Madame Jolas, who had transferred her language school to Saint-Gerand-le-Puy near Vichy.

Helen, whose behaviour was becoming more and more erratic, was running up mountainous debts. More to the point perhaps, she even denounced her father-in-law as a foreign agent to whoever would listen, including the French police. Giorgio now left her, moving for a while into the hotel where Guggenheim stayed after leaving the hospital. He was determined to get Helen to America and Guggenheim was involved in the negotiations with her brother Robert Kastor, but meantime Beckett and the disconsolate Giorgio had to do a good deal of drinking together – the Irish way of handling such a problem. The final dénouement was described by Joyce in a letter to Jacques Mercanton, a Zurich businessman whom he had known during the previous war. 'My daughter-in-law, in her turn gone completely insane, has been interned. The poor thing, after having ruined her family and home and run up debts of half a million francs is raving mad most of the time. My son, entirely disoriented, is now in Paris with friends.' In fact Helen was confined by her husband, brother and in-laws in the *maison de santé* at Suresnes near Paris. Robert Kastor flew backwards and forwards by the new Clipper planes which flew into Shannon in Ireland and then on to Nantes. Finally, just before France fell in the summer of 1940, with the assistance of two nurses, a doctor and drugs she was taken to Genoa and put on board ship for New York. At this point Joyce, who had always liked Helen better than Nora, would get quite sentimental about her, thinking of the elaborate preparations she had made to celebrate his birthday in 1939 and writing to Constantine Curran about the gold chain she had given him, which at that moment tinkled at the end of his fountain pen.

Now under the stress of all this plus his worries about Lucia (and still suffering perhaps from the aftermath of publication of a book that had taken seventeen years to write and was getting few favourable reviews), Joyce himself suffered a sort of breakdown, drinking and spending wildly and becoming, like his son, more and more dependent on Beckett. One day they went around to the rue des Vignes together and Joyce sat down at the piano. With shaking hands and quavering voice he played and sang for half an hour. Then he suddenly asked his younger companion: 'What is the use of this war?' There followed a difference of opinion. Beckett now thought the war was necessary to stop Hitler. Joyce was inclined to see it as an unwarrantable and unnecessary complication of people's private affairs and in part a plot to prevent his book from getting the attention it needed.

In spite of his general condition, however, he tried to cling to his rule about not drinking before six o'clock. Often when Beckett was with him in the late afternoon, he would notice his preoccupation with the time and his tendency to hasten his step to get to a favourite café just as the hour was striking. Once they got there too soon and Joyce broke his rule. 'We're going downhill fast,' he said to Beckett with morose satisfaction.

The autumn and winter of what the French called the '*drôle de guerre*' and the British the 'phoney war' were exceptionally cold. Beckett was one of the lucky ones who still had heating in their apartments, but others were not so fortunate. Nora Joyce decided that she could no longer make adequate housekeeping arrangements in the rue des Vignes and so she and her husband moved into the Hôtel Lutetia in the boulevard Raspail, Beckett going with Joyce to the flat to remove some books and personal effects. Those of Beckett's acquaintance who remained in Paris took refuge in the cafés for warmth, and although there was little to report from the fronts the war provided an endless topic of discussion, as did people's plans and movements. The possession of appropriate documents had become more important than ever, and Beckett was told that he should arm himself with a document from the Irish government saying that he was a citizen of a neutral country who had reason to reside in Paris. He went to the Irish embassy to make application for this, but had not brought his passport with him. Told to come back with it, he foolishly let the matter lapse, deciding that after all his passport would be sufficient for any contingency that might arise. Of course, nobody foresaw the imminent arrival of the Germans. Alfred Péron had been called up in September and Georges Pelorson shortly afterwards. Peggy Guggenheim had come back to Paris from the South and was going about as best she could on her crutches in the slippery weather, buying a picture a day with the funds of the Guggenheim Museum. Beckett was trying now to interest her in the work of Geer van Velde's brother, Bram, and a day was appointed to go to his studio, but at the last minute she decided she was not interested and made an excuse. He remarked sarcastically that she was more interested in needy painters such as Bracque and Dali. She had rented an apartment on the quai d'Orléans which had a big studio overlooking the river. Here she gave a series of dinner parties, to some of which Beckett came. Although it was early in his relationship with Suzanne they had already established a pattern of leading separate lives and accepting invitations separately. In December George Reavey arrived in Paris on his way to Madrid to take up a post at the British Legation, and he and Beckett and Joyce dined together at one of Joyce's favourite restaurants, but Joyce was gloomy and tired and the evening was not a success.

Just before Christmas Beckett put James and Nora on the train to Saint-Gerand-le-Puy, where they intended to spend Christmas with Madame Jolas and their little grandson Stephen, who had lately been the subject of a tug of war between his parents. He had now settled down in earnest to the translation of *Murphy*, which was finished in the New Year, but although he continued to write poems in French he was unable to start another prose work. He was not going out much, but Suzanne still visited the apartment on a regular basis and was nearly always there at weekends.

In February 1940 he went with Giorgio to visit the Joyces, now staying at the Hôtel de la Paix in Saint-Gerand-le-Puy. He found Joyce still sad about Helen, apprehensive about the future and disappointed about the reception of his book, which he seemed to think had been conspired against. The Russians had invaded Finland, which, under Marshal Mannerheim, was putting up a heroic resistance; and in the absence of any other major fighting, the eyes of the world were on the 'winter war' amid the lakes and marshes of the little northern country. Joyce had now chosen to think it 'odd' that Finland, 'until now terra incognita', was so much in the news. His book, after all, had been about the original Finn, a legendary Celto-Nordic hero. Also, a Finnish writer had now got the Nobel Prize. Of course, Finland had been attacked by the Soviet Union, but was it not something of a coincidence? he asked a silent but apparently receptive Beckett. In his black hat, long dark overcoat and dark glasses, Joyce was a conspicuous figure in the village. He stayed in bed in the mornings – 'What is there to get up for?' – and hated the numerous dogs of the place, against whom he went constantly armed with a pocketful of stones and a stick. In April Beckett returned and this time they all stayed at the Jolases' château, La Chapelle. Joyce wanted Beckett to stay for Easter and spoke of going to Moulins with him for the Holy Week services, but the trip never took place. While he was there the '*drôle de guerre*' suddenly came to an end, the Germans launching their blitzkrieg on Denmark and Norway. Just before Easter Georges Pelorson arrived from Paris on the way to join his regiment and they all stayed up late, Joyce talking again about Easter and its liturgy. Pelorson had written a poem about the Gracchi, but when he and Joyce began to discuss it at some length

Beckett went somewhat huffily to bed. Madame Jolas's school reopened after Easter and the Joyces moved to the Hôtel Beaujolais in nearby Vichy. Beckett returned to Paris and Pelorson went on his way. He would soon enough be back in Vichy for other reasons, attached to the Youth Ministry of the Petain-Laval government.

On the boulevards apprehension was growing. In the second week in May Holland, Belgium and Luxemburg were invaded, and on the following day the British government fell and Churchill took over. Still having heard nothing of his previous application to put himself at French disposal, Beckett now volunteered to drive an ambulance. Many writers from other countries had driven ambulances, the non-combatant's usual halfway house, in the First World War and, more recently, in Spain. To be ready, he now took out a heavy vehicle licence. It would come in useful eventually, but not till after the war, for this time also he heard nothing further. Perhaps because he expected to be taken soon, however, he still neglected to get the papers he needed and instead fulfilled an intention of going to Brittany, where Alfred Péron had been posted as a liaison officer with a British ambulance unit. This was partly to put the finishing touches to *Murphy*, partly simply to see his friend. Péron, a serving soldier, was worried about his family in the case of an invasion of France, and Beckett renewed his promise to do what he could for Marie and the children. Back in Paris he found that panic was growing and there was a new urgency in discussions of what one should do. He had heard nothing of his application to drive an ambulance; but time was running out for everybody. The Germans were now on French soil and the names of the battlefields of the First World War were briefly in the news, but the speed of their advance took everyone by surprise. This time they did not get pinned down in trench warfare, and before the month was out the British army was being withdrawn through Dunkirk and the French were in full retreat. In early June the roads began to fill up with crowds of civilians and the Germans increased the general panic by dive-bombing the refugees. It seemed as if Paris was about to fall. Nobody seemed to know what would happen to citizens of a neutral country when the Germans came. The one certainty among the Irish was that they would be cut off from their homeland and any possibility of returning there.

Communications were breaking down and Beckett now found that he could no longer draw money from his Irish account at the bank, an operation which required confirmation from the Irish side that the money was there; nor could he get the papers he wanted. In spite of the famous stand on the Marne which had saved Paris in the First World War, the general opinion in the cafés was that the nation had been defeated, undermined, some said, by pacifists, Jews and decadents. In any case Paris would fall. On 10 June the French government moved to Touraine. Beckett discussed the situation with Suzanne and it was decided that she should go to be with her parents at Troyes, while he went south. He decided to go as far as Vichy, where he knew people and at least had a hope of borrowing money. At the last moment Suzanne decided to come too. Nobody knew then that Vichy would shortly become the headquarters of Marshal Pétain's government and the capital of unoccupied France. It was simply a place to go, where he knew people and might be able to cash a cheque. They went to the Gare de Lyon where they caught a long, slow and crowded train. It was 12 June, and two days later Paris fell, cutting off their retreat. Vichy, which had plenty of hotels because it was a spa, was crowded with refugees. On 1 July it would become the headquarters of the French government. Georges Pelorson was there as well as the Léons, now reconciled to Joyce, who was deeply concerned about Lucia. Nobody could cash a cheque for Beckett, nor would the hotel do anything. Then Joyce thought of Valery Larbaud, the writer whose work had supposedly given him the inspiration for the interior monologue. Larbaud, totally paralysed and virtually unable to speak, lived near by with Nora Joyce's friend Maria Nebbia, and Joyce had already visited him. Now he wrote a letter on Beckett's behalf, heading it with the words, 'Samedi soir après le turbin . . .' – a quotation from a popular song, 'Viens, Poupoule', which they were both fond of. Larbaud obligingly cashed the cheque but Beckett now decided to go on southward. The hotels in Vichy were being requisitioned for government ministries evacuated from Paris and in fact two days later the Joyces had to leave the Hôtel Beaujolais and move back to Saint-Gerand-le-Puy. There seemed to be no more trains going south so they set out on foot after saying goodbye to Joyce. Of course, neither Joyce nor Beckett and Suzanne were aware that it was to be their last farewell.

At a little station on the way they happened on a train which took them as far as Toulouse. It was dreadfully crowded and very slow but at least they were moving, even if not at the same rate as the Germans, who were now rumoured to be advancing very fast southwards. He was worried about not having the proper papers, and as the train stood still outside the city, he and Suzanne decided to leave. This was probably wise as the French Gendarmerie were examining papers at the station and detaining aliens who were then interned in a refugee camp. They were now fairly near to the Spanish border and it occurred to Sam that they might make their way there and cross over. Although there seemed to be no way of leaving the city, certainly no form of public transport available, he wrote a postcard to Reavey, who had volunteered to work for the British Council and was now at the British Legation in Madrid, warning him that he might turn up. After two or three nights sleeping out on park benches and in doorways they at last found a bus. It was going in the wrong direction, to Cahors, which was northward, but they were fed up with Toulouse and again it was movement. At Cahors everybody was told to get off the bus. It was going no further. Beckett and Suzanne had no place to go but they wandered round and finally, with some other displaced wanderers, found refuge in a shop which dealt in religious articles. There, among the rosary beads and the statues of the Blessed Virgin, they spent the night on the floor. They were both filthy and exhausted but the following day they had a stroke of luck. From a sign Sam discovered that Arcachon, where Guggenheim's and Joyce's friend Mary Reynolds had a house, was not too far away.

A friendly lorry-driver gave them a lift in the back of his lorry and when they got there they found it easy enough to locate the house of the American woman, where, hollow-eyed and unshaven, Sam received a welcome which exceeded his best expectations. Mary Reynolds had a full house but she helped them to find a room in a pension, the Villa Saint-George on the boulevard de la Plage, and she also loaned Sam money. Mary's guests included Marcel Duchamp and he and Sam, both fanatical chess players, spent much time playing that game in a cafe on the seafront. Duchamp was a noted player, who would subsequently play chess on a professional level in New York, but unfortunately there is no evidence as to which, if either, usually had the advantage.

Ensconced here in relative comfort Sam now wrote to Reavey again, giving his new address, saying he had no money and repeating his wish to return to Ireland. In Madrid Reavey called on the Irish embassy and, as the diplomatic wheels began to turn, Beckett's address, request for money and wish to return to Ireland via Portugal were conveyed to Frank, who had been making enquiries in Dublin, and to the Irish Minister Plenipotentiary, who had now been appointed to Vichy. What Suzanne's plans may have been is not clear. As a French citizen she could neither have entered Spain nor gone to Ireland. It was the second week in August when Frank was contacted with a request to lodge money, but there followed a long silence from Sam which lasted until late September, when he turned up at the Irish embassy in Paris and asked that his brother should be contacted and his monthly allowance resumed. This was of course a great relief to Frank, who immediately deposited £15, or 2,640f., in Dublin for dispatch to Sam. What had happened was that while waiting for news in Arcachon Sam and Suzanne had heard that life in Paris was now returning to normal. The armistice which had been signed on 22 June meant that France was no longer at war. The city was peaceful and many of its inhabitants were apparently enjoying life as much as circumstance permitted. The prospect of returning to the rue des Favorites began to seem much more attractive to Sam than a prolonged stay in Ireland. Many others were returning and he and Suzanne had decided to return as well.

In June an armistice had been signed by which France was divided into occupied and unoccupied zones. The new government, headed by the aged Marshal Pétain, the hero of Verdun, had its capital at Vichy. At that stage most French people believed that the war was over and that Britain too would capitulate within a few weeks, after which the Germans would exact an indemnity, annex Alsace and Lorraine and go home, as the French had gone home in their turn in 1918 and the Germans in 1870. The majority tacitly approved of the Vichy regime and many assented to the penitential principles on which it claimed to base its government, the condemnations of Jews and freemasons it issued, its calls for an end to class battles, demagogic politics and the cult of pleasure, a return to the traditional values of *patrie*, *famille* and the Catholic religion. Those who did not fully assent were nevertheless chastened and humbled, anxious for a quiet life and rather glad to accept the Vichy version of events, which was that France had been undermined by decadent intellectuals, Jews and freemasons, though in fact the defeat had been principally caused by long-term military stupidity, to which Marshal Pétain had contributed his share.

In October 1940, the month in which Beckett returned to Paris, the first anti-semitic decrees were published by the Vichy government. But those writers, including Bair, who have Beckett as a member of the French resistance immediately after his return to Paris in October 1940 are wrong. In Paris as elsewhere there was as yet hardly any resistance to join.

The winter of 1940–41 was marked by severe food and fuel shortages, which caused long queues outside boulangeries and other food shops and reduced the more defenceless to desperation. To the effects of a bad harvest in 1940 were added the chaos of the invasion summer and the claims on the economy by the Germans as well as a British blockade of the Atlantic and Channel ports. But if the coffee was ersatz and meat so scarce that even the traditional *andouillettes* were absent from the menu at the Closerie des Lilas, the cafés of Montparnasse were open and life went on.

In this first, more relaxed stage of the occupation German soldiers were now fraternizing with a populace which was not displaying any great enmity to them; and after the British attack on the French fleet at Mers-el-Kebir in 1940 there was, if anything, more anti-English than anti-German feeling in Paris. In spite of continued British resistance and the check to the Germans represented by their losses in the air during the battle of Britain, the illusion that the war was not going to be of long duration persisted. On 24 October Pétain had a famous meeting with Hitler in a railway carriage at Montoire, after which he came to the microphone and proclaimed a policy of sincere collaboration with Germany. For most French people in both parts of France that was good enough. Doubts about an ultimate German victory still hardly existed.

In the rue des Favorites, where there was still by some miracle heating and hot water, Suzanne, like most people, did her best to keep some sort of food on the table; and as in most homes, a semblance of normality prevailed. Beckett rose late and spent much of his time queuing. In the evenings he would mooch down to Montparnasse, sometimes on his bicycle. He was, naturally enough, doing no work and, though others were now returning to Paris as he had done, there were few enough people that he had any desire to consort with. Giacometti was about, having returned to the city a few days after leaving it in June. Although he had been only a week absent he had seen more of the war than Beckett, returning after he and his brother had been overtaken by the rapidly advancing Germans along roads whose sides were littered with human corpses and the carcasses of dead horses, the combined stench of which he never afterwards forgot. Paul Léon had returned in September to his apartment in the rue Casimir. He too had turned up at Saint-Gerand-le-Puy, riding in a donkey cart in a state of utter exhaustion, and had been reconciled to Joyce, 'insofar', Joyce's biographer drily remarks, 'as reconciliation was possible for Joyce'. He had resigned his post as Secretary of the Archives of Philosophy and was now devoting himself to the rescue of Joyce's papers from the rue des Vignes, sorting and stuffing them into large brown envelopes before handing them over to the Irish Ambassador, Count O'Kelly. Alfred Péron had also returned, having been demobilized from the French army. He had resumed his teaching post in the Lycée Buffon, which was quite near the rue des Favorites, so that he and Beckett were able to meet frequently at lunchtime. Beckett had always been fond of him. Now they were closer than ever. But most of the Joyce circle was scattered when the news came through to Paul and Lucie Léon in January 1941 that the revered master had died in Zurich, the city in which he had also been a refugee from the First World War.

The Joyces had hung on in Saint-Gerand-le-Puy until December, partly because they were divided amongst themselves about where to go. They had turned down what might have been a life-saving offer from Robert Murphy, American chargé-d'affaires at Vichy, to get them to the United States *en famille* because neither the thought of flying there nor the United States itself appealed to Joyce; and had finally opted for Switzerland again when arrangements seemed to have been made to get Lucia out of Saint-Nazaire, which was in German-occupied France, and into a Swiss sanatorium. The Germans actually issued a *permis de sortie* for her without much fuss, but the arrangements fell through nevertheless. There were also problems about Giorgio, who had no exit visa and was a British alien of military age. And as the delays mounted there were finally problems from the Swiss end about Joyce himself, who was believed by the authorities there to be a Jew. (They had a policy of not admitting Jews.) His remark on hearing this was '*C'est le bouquet, vraiment*'; but his friends in Switzerland finally deposed that he was not a Jew, or, as he put it, '*que je ne suis pas juif de Judée mais aryen d'Érin*'. Events were catching up with even the most detached, apolitical and *déraciné* of people. The last snag was permission to leave unoccupied France, so he recruited the aid of his friend Armand Petitjean, now, like Beckett's friend Georges Pelorson, an official of the Vichy government.

When the family, minus Lucia, at length arrived in Switzerland on 14 December 1940, Joyce was utterly exhausted, though still capable of perking up a little in the right company and even singing a song or two. He died on 13 January of a perforated ulcer. Towards the end he was much cheered by a review of *Finnegans Wake* in the Vatican newspaper *Osservatore Romano*, which declared that the book was an assertion of spiritual values as against the materialism of the nineteenth century.

The news of his death naturally devastated what little was left of the Joyce circle in Paris, and Beckett was of course deeply affected. Lucie Léon decided that a few of them should assemble as usual on 2 February, Joyce's birthday; and it was with mixed feelings that Beckett made his way down a blacked-out rue de Rennes to the rue Casimir. Besides himself the only other guests were Léon-Paul Fargue, the translator of *Ulysses*, and his friend Madame Chériane. The talk, of course, was mostly of Joyce, and at a certain point of the evening the two recordings Joyce had made of extracts from *Finnegans Wake* were played. 'It seemed to me that there was no war that night in Paris,' wrote Lucie Léon afterwards; 'no German occupation, and that Joyce's great spirit was there in the room where he had spent so many hours talking with Paul.' But what was missing was the progression towards abandonment and gaiety, song and dance, which had marked the gatherings over which Joyce himself presided; and it was with a certain amount of sadness that the little party finally broke up.

Throughout 1940 and the earlier part of 1941 active French resistance to the Germans was practically non-existent. There were rudimentary resistance groups, sometimes set up with the assistance of British Special Operations agents, mostly in unoccupied France. But resistance movements and resistance activities are of their nature shadowy and hard to define. In Beckett's own country many people had engaged in 'resistance' during the years 1918 to 1922 who never heard a shot fired or ever handled a gun or an explosive device. That he was well aware of this analogy is shown by his use of phrases familiar in Ireland on the rare occasions when he discussed aspects of his experiences. 'On the run', he would say, or, 'on the trot', in describing his period in hiding in 1942. There were groups in 1940 and later who felt that they served – and did in fact serve – a useful purpose by listening to the British radio and circulating what they heard in mimeographed form to as many people as possible. Their task was to remind as many French people as they could reach that the war continued and that in spite of official propaganda the Germans were not having it all their own way, as well as to emphasize the sinister nature of German ambitions for the new Europe. Alfred Péron was a member of one such, the Musée de l'Homme group, consisting mostly of academics and intellectuals, which in mid-December 1940 had begun to publish a clandestine sheet, *Résistance*. Naturally Beckett, who also listened to the British radio both in French and in English, passed on much of what he heard during their frequent lunches together.

Gradually, however, some of these groups extended their activities, often after contact with the SOE. Among Péron's acquaintance was Jeannine Picabia, the daughter of the painter Francis Picabia. She had been a driver for a French medical group whose task was to distribute some of the supplies needed by the vast numbers of French prisoners of war which remained in German hands even after the armistice. The wide travelling this involved gave her ample opportunity to observe German activities of one kind or another. A reference to this during a visit to her father in Cannes led to an approach by a British agent, as a result of which she formed a *réseau* or cell, which would systematize such observations as far as possible and pass the results on.

This group or *réseau*, now called 'Gloria', was engaged like others in the collection and collation of information about German troop movements and German dispositions within France. The British were interested in even the tiniest indications which could be picked up by people who did no more than lounge in cafés or observe what could be seen in the streets and at railway stations, because up to the time of the attack on the Soviet Union an invasion of England had seemed almost certain. Even afterwards it could not be ruled out, while besides Britain itself there was Spain and the Mediterranean theatre generally to be worried about. A movement of troops towards the Spanish frontier would also be a concern of British intelligence, as were the movements of German vessels in and out of occupied ports. And after the possibility of further offensive actions by the Germans in the West began to seem unlikely, there was of course the question of an allied invasion of Europe. This kind of information–gathering, much of it possibly irrelevant or of doubtful value, was the kind of activity that Beckett, through Alfred Péron, initially began to engage in.

To begin with he agreed that the apartment in the rue des Favorites could be used as a '*boîte aux lettres*', or a 'drop' for the information collected by others. They came in uncomfortably large numbers to the rue des Favorites with their rudimentary, cryptic notes of this and that and he kept them until they were collected by somebody else. After a certain time it was agreed that he should collate the information, typing it out and translating it into English as briefly and concisely as possible, trying hard to get it all on one sheet and of course making editorial decisions about what should be included while he did this. He then took his typed sheet to an address in the avenue René Coty near the Cité Université. This is in the 14th arrondissement and not too far from the rue des Favorites, but his journey there was possibly the most hazardous part of the whole operation, for when resistance and repression began to become serious matters street searches became not uncommon and Beckett was a rather striking figure who could not hope to escape notice. At this second drop a member of the group known as 'the Greek' would put the sheet which Beckett had brought on to microfilm for its transfer to the unoccupied zone and from thence to England. Microfilm had the advantage in that it could be hidden in different objects about the person, such as the false tray of a box of matches, the barrel of a fountain pen, behind the mirror of a powder compact or in the sole of a shoe.

The person to whom the little pieces of film were now entrusted

323

was Jeannine's mother, Gabrielle Buffet, first wife of the stormy petrel of Dadaism, Francis Picabia. Gabrielle was a respectable-seeming, grey-haired, middle-aged woman and therefore judged to be an ideal courier, able to pass back and forth into the unoccupied zone without fear of being stopped. One of the ironies of the situation was that before the war was over Picabia himself would proclaim his German sympathies to such an extent that he would probably have been arrested and tried for collaboration had he not suffered a stroke in 1945.

The group with which Beckett was associated was, to begin with informally organized, though it had the code name Gloria. It contained others who, like himself, were citizens of neutral countries and one of these was Mary Reynolds, in whose house in the South he had stayed during his wanderings. The large house she owned in Paris, in which he had spent one or two nights with Peggy Guggenheim, now proved invaluable as a rendezvous. It had a walled garden which could be used as a hiding place for letters and documents and other things, such as an official stamp which was necessary for forged travel permits. To the members of this group Beckett was known as 'l'Irlandais', the thinness of the disguise being evidence of the somewhat amateurish nature of the proceedings during the early days, when the necessity for effective code-names and full anonymity were not thoroughly understood, and arrests were rare and torture rarer still. But though the Germans had not yet begun to react in the way that they later would; and no member of the group had as yet been arrested or tortured, narrow escapes were not infrequent. One such occurred when Jeannine Picabia befriended a kitten with a broken leg. Because she had to go on a journey she asked Suzanne to take it home to the rue des Favorites and look after it. A few days later Suzanne was detailed to deliver a message to the flat of two elderly sisters. When she arrived there she found it full of Germans, who had searched it, found nothing, but remained on to interrogate whoever else might call there. Asked the reason for her visit Suzanne said she had a message for the two women from her husband. Their kitten had had its leg set and was thriving. Perhaps suspicious of this explanation, the Germans asked to see the kitten and two of their number accompanied Suzanne back to the rue des Favorites. The kitten was indeed there; the Germans were somewhat mollified by its presence, but they decided nevertheless to have a look round the apartment. Legend, aided by biography, has it that when they found a copy of Mein Kampf among Beckett's books they saluted and departed happily. In fact though he had glanced through it he never possessed a copy of Mein Kampf, but he had a number of other German books and their presence on his shelves convinced the Germans that anybody so fond of the *Kultur* of the fatherland was not likely to be an enemy – of course, an entirely false assumption on their part, the converse being equally likely to be true.

The reason for Beckett's resistance activities is not something that can be arrived at easily. It was not necessarily out of any sense of identification with France, for France was a deeply divided country and, even in the early part of 1941, there were certainly as many French people who were for Vichy and collaboration as those who had other views. As Germaine Bree puts it:

The choice between collaboration or resistance, between Vichy and De Gaulle – and the ultimate consequences that choices implied – were matters of endless debate. This debate was pursued in relation to a multitude of individual viewpoints and loyalties, prejudices and bias too subtle and diverse to be accommodated by the oversimplified 'collaboration/resistance' option. When it came to staking their future the French had a wide gamut to choose from, ranging from total identification with the destinies of Germany, to total identification with those of Russia and Britain.

The divisions in French society were deep and bitter and have lasted to this day. Most of those who called themselves 'patriots' were actually at this juncture for Vichy and even for 'the new order'. Nor was it true – or certainly not true until later – that all who were 'civilized' or even 'progressive' were on one side and everybody else on the other: in other words, it was not a question of intellectuals against everybody else. What did, however, even in these early days, divide one sort of person from another was the treatment of the Jews; and there is no doubt that this influenced Beckett considerably. 'I was so outraged by the Nazis, particularly in their treatment of the Jews,' he told John Kobler, 'that I could not remain inactive.' And to the Dublin theatre director Alan Simpson he would later say, 'I was fighting against the Germans, who were making life hell for my friends, and not for the French nation.' In fact none of Beckett's close friends had at this stage been arrested or deported. Anti-semitism was rife in French life, thought and politics, and in the early days the Vichy government actually took the lead from the Germans in persecution, the infamous Statut des Juïfs of October 1940 leading to widespread dismissals from jobs and seizure of property. But German policy was to drain France economically, while avoiding policies that might lead to large-scale strikes or stoppages, as they had in Holland, where arrests of Jews caused a national strike in February 1941.

It may also occasion some surprise that so evidently neurotic and self-absorbed a young man as Beckett should suddenly be found to be active in a movement such as the French resistance. This is to leave out the often-observed and well-attested fact that the behaviour of neurotics often becomes active and even extrovert when their situation changes and definitive action is clearly called for. They have been, in a sense, awaiting their hour. Of course, Beckett was awaiting his hour in a more important, artistic sense, but the man of action, or at least the man capable of action in him, had been awaiting something too and had not found an ambience and a *métier* up to this point either. There is also the influence of Suzanne. All Beckett's later statements about the changes produced in him by co-habitation with Suzanne were on the lines of 'she made a man of me', or 'she rescued me'. He spoke of his neurotic state, his utter inactivity, his habit of staying in bed half the day and – even after their relationship had declined largely into habit and acquiescence – of Suzanne as having changed all that. And Suzanne was a Frenchwoman of decided views about many things. She was left wing and, particularly after the summer of 1941, would have been in no doubt about where she stood in relation to defeat and the occupation. There is little doubt that her influence in this regard was all important. Though she was not herself a member of Gloria, she had been in the Communist movement and had many friends there.

And of course there was the influence of Alfred Péron, Beckett's oldest friend among French people, a daily associate at this time even apart from their joint resistance activities, and a Jew who saw – and had reason to see – more clearly than most French people what the Nazis were up to, what their intentions regarding the new order really were and what future awaited the Jews should they triumph.

The stages by which resistance became a more serious matter are clearly marked, and by far the most important of them was the German attack on the Soviet Union in June of 1941. Taking its line from the Soviet Union, the French Communist Party had originally denounced the war as a capitalist-imperialist struggle; and after the armistice it had been foremost in advocating collaboration with the Germans and approving of fraternization with the occupying troops, even though these friendly advances had been ill-received by the Germans themselves. Now, suddenly, all was changed; and the communists brought their energy, discipline and fierce sense of dedication to the creation of an effective underground resistance movement. At the same time belief in the might and revolutionary destiny of the Soviet Union combined with continued British resistance and German setbacks in North Africa and the Mediterranean to create, for the first time, doubts about an eventual Nazi victory and a Nazi new order in Europe. Armed resistance groups began to come into existence and to carry out assassinations as well as acts of sabotage.

The first high-ranking German officer was killed in October 1941 and a few days later a German major was shot. The Nazi response was to take hostages and to kill them in batches of up to fifty at a time, Jews being favoured in the taking as well as in the subsequent killing. This taking of hostages, Jewish and otherwise, naturally had the effect of exacerbating resistance, while the propaganda insistence that all acts of resistance were the work of Jews and communists resulted in the Communist Party retrieving with interest a reputation for patriotism which had been lost by its previous policy of collaboration. And, as Alfred Cobban puts it, 'The campaign of the occupying forces against the Resistance was partly misdirected because of the lunatic Nazi pre-occupation with the Jews, who were singled out for seizure as hostages, and for execution, whether they were implicated in the Resistance or not.' To the arrest and execution of Jews for acts of sabotage was soon added their arrest and deportation merely for being Jews. There had been police sweeps as early as May 1941, in which Polish, Czech and Austrian Jews were arrested, but these were mainly French affairs. Now the Germans themselves began to arrest Jews and deport them on a large scale and transfer them to the German-controlled concentration camps at Compiègne, north of Paris, where the armistice had been signed. Yet another stage in the growth of resistance would be encouraged by the conscription and deportation of hundreds of thousands of French men and women to work in the German war industry. For those who were listed to go, those who were likely to go, and those who feared they might be suddenly sent, resistance was not only a patriotic but even a sensible option.

It was during this grimmer time that Paul Léon, who was still in Paris, was arrested. One day Beckett met him in the street and expressed alarm and misgiving at his continued presence. Léon explained that, yes, it was his intention to go south but that he had to wait until the following day when his son was taking his *bachot*. The following day he was arrested and taken to the concentration camp which the Germans had established at Drancy. In the spring of 1942 he was moved to Silesia and on 4 April he perished in a camp there.

Chapter Twenty-One

One of the largest resistance groups was that known as 'Prosper', which had a certain amount of overlap with Gloria, to which Beckett properly belonged. As things hotted up the German methods of counter-intelligence and their ways of infiltrating resistance groups naturally improved. By the summer of 1942 Prosper, which, like Gloria, consisted largely of intellectuals, was well and truly infiltrated; and arrests, followed by torture, were becoming frequent. In June Armel Guerne, the second-in-command of Prosper, was arrested and the activity of the whole group declined. M. R. D. Foot in his history of the Special Operations Executive lists Beckett among the intellectuals of Guerne's group who now decided to lie low. This is incorrect, but Gloria too was being infiltrated and many of its members had decided to scale down their activities as much as possible, among them Beckett, particularly after its infiltration by the Abbé Robert Alesch, a priest whose activities caused havoc and led to many arrests.

In mid-August 1942 Péron was arrested. He was making his way south to Lyons at the time, apparently to meet Jeannine Picabia there, but whether he was arrested as a member of Gloria or merely as a Jew was at first not clear. In any case Marie Péron now sent a telegram to Sam and Suzanne which read: ALFRED ARRESTED BY THE GESTAPO. PLEASE DO WHAT IS NECESSARY TO CORRECT THE ERROR. This seems to have been a warning, not a genuine request for assistance. On the same day Suzanne narrowly escaped arrest when she rang the bell of the apartment of another Gloria member. The Germans were already there and, aided by the silence of the person whose flat was being raided, she pretended to have rung the wrong bell. When she arrived back at the rue des Favorites it was to find the telegram delivered and Beckett anxiously awaiting her return. They hurriedly destroyed anything that might be found incriminating and left immediately, simply closing the door behind them as if they were going out to the shops or to a café. There followed a period in which they were both in hiding but still in Paris, sometimes together, sometimes not. During part of this time Beckett also slept out in the Bois in the company of another member of Gloria. When at one stage the Germans decided to round up all those who were sleeping rough, they searched the wood with dogs and the two hid in a tree while the searchers passed below. A day or two after they had found temporary shelter in a small hotel, this companion, also Jewish, committed suicide by jumping from a window. When the police were called Beckett had to leave the hotel and was shelterless once more.

While on the run, he and Suzanne also stayed for a while in Nathalie Sarraute's house, which was in the country outside Paris. There were others hiding in the house as well and there were personality clashes, not infrequent even in situations when everybody is in danger. Arrangements were primitive; the toilet was outdoors and everybody used chamber pots at night. But Sam was a late riser and at mid-day when the others were sitting down to déjeuner, he would carry his chamber pot through the kitchen. This was regarded as very bad behaviour by some members of the household.

Gloria, though now in disarray, was still able to arrange escape lines through into the unoccupied zone and procure the documents which would be necessary. In late September Beckett and Suzanne were able to get such documents and were put in touch with a passeur, who escorted them across the dividing line almost in view of the German sentries. They took a train from Chalon-sur-Saône, at the edge of the unoccupied zone, eventually winding up in Lyon. Suzanne had a friend who lived for some of the time in Roussillon, a village in the Vaucluse, so they decided to make their way there, moving as far as possible by night and sleeping in hay barns and in the woods by day. When they got there they found her friend was not in residence, but his sister was there with her family and so they decided to stay.

Roussillon is a place of red sandstone cliffs, red soil and red sunsets. To the craggy cliffs, with their protruding masses of rock, cling pine trees, and on the slopes and level stretches of countryside are vineyards, for the grape thrives here in the burning sun. The village is typical of the area, perched on its eminence, surmounted by its church, viewing the world around it with suspicion and some hostility. The people are dour, hard-working and self-contained. They do not impart any more information than is necessary to anybody, even each other, but especially to strangers. An American professor, Laurence Wylie, came here in

1950 to write a sociological study of life in a Provençal village. He spent a year in Roussillon, observing the people and talking to them. A year later, in the library of his American university, he idly picked up a blue book, *En attendant Godot*. It was a play and in it he found a reference to Roussillon and to a farmer whom he knew there named Bonnelly: '*nous avons fait les vendages, tiens, avec un nommé Bonnelly, à Roussillon*'. In his year in Roussillon he had never heard any mention of Beckett or his stay there.

When the footsore Sam and Suzanne arrived in October 1942 they decided to conceal neither their presence nor their identities. Roussillon was in the unoccupied zone. They were far from Paris; so far as they knew, they were not specifically wanted; and Sam was, after all, a citizen of a neutral country. The village boasted a hotel, the Hôtel Escoffier, though the promise of culinary delights was misleading, for Madame Escoffier who owned and ran it had no connection with the great eighteenth-century chef. They were, however, lucky to get a room because the place was already crowded with refugees from Paris, including a number of Jews. 'Here are two more who have come from Paris,' cried Madame Escoffier as she introduced them. Advised that they should go to the Mairie, or Town Hall, and register openly there Sam did this the following day, a card being filled out which stated that he had come from Paris, that he had no profession and that his nationality was Irish. His birthplace, though, was given by mistake or misunderstanding as Dublin, England. Because of his biblical name, Samuel, it was widely assumed in Roussillon that he and Suzanne were Jewish.

In November 1942, shortly after Sam and Suzanne's arrival in Roussillon, the Americans invaded French North Africa and the Germans instantly put into effect a long-prepared plan for the occupation of the whole country. The Vichy government had already become little more than an agency to facilitate the effective exploitation of France, which was becoming more rapacious every day. Now, with Pierre Laval at its head, the Vichy government would pretend to be the national representative of the French people, but its remaining power, already reduced to a derisory level, would decrease every day. The occupation left Roussillon comparatively untouched. Every so often, but usually at intervals of several months, a German patrol would come out from Avignon or up from the neighbouring village of Apt, but the village of Roussillon itself was never occupied. There were several Jews there, including a nephew of the one-time socialist Prime Minister, Léon Blum, who worked for a neighbouring farmer, but none of them were betrayed. For Sam and Suzanne, as for others, there was never a complete absence of risk, but the tensions which they had lived under for the past two years made their situation now seem relatively peaceful.

The village was not all of one mind about the Germans or about the Vichy regime, the division running more or less on the lines of right and left, the right being in favour of some measure of collaboration and the left believing that the war should be continued on the side of the allies. This division was a deep and bitter, if silent one and it would leave its mark for years to come. To say the inhabitants did not trust each other might be misleading. Like most peasant communities the people of Roussillon had never trusted each other in one matter or another; distrust was part of their nature, but so was ambiguity. In a way everybody was glad that both sides were represented. The village had a mayor, appointed by the Vichy government, who was in a position to obtain favours and ensure that it had a quiet life provided it did not call attention to itself by violent or dramatic activity. After the war Aimé Bonhomme, a communist and the leader of the local resistance, became mayor. In the meantime there was a sort of stand-off. There were no Germans, no overt collaboration and no unnecessary denunciations of Jews. But resistance was pretty carefully gauged too. Roussillon did not call attention to itself unnecessarily.

Another reason why strangers, including Jews, were comparatively safe there was that Roussillon – because of its location and picturesque beauty – had attractions as a tourist resort. It was accustomed to visitors from the coast and even foreigners. They contributed to its economy and there had, on and off, always been strangers resident in the area. This lessened the suspicion with which the present influx might have been viewed.

One foreigner already resident there was Noelle Beamish, a writer of detective stories whom the villagers believed to be English and who lived in a house on the outskirts with her younger companion and two somewhat fierce Airedale dogs plus various cats. Miss Beamish made no secret of the fact that she was a cousin of Winston Churchill's, proclaiming it fairly frequently to anybody likely to be impressed. Certainly the village was impressed and it was one of the first things strangers were told about Miss Beamish. About sixty, with short grey hair, and attired in the winter time in tweeds and boots, Miss Beamish was in fact Irish and, to Beckett's ears, she had an Irish accent. She was also lesbian, and her companion was said to be Italian. Sometimes Miss Beamish's long, utilitarian drawers would hang out on the clothes line together with her younger companion's frilly, feminine knickers, a sight which caused much amusement to the villagers.

Shortly before Christmas two more refugees arrived. These were

Henri and Josette Hayden who had also come from Paris. Before Madame Escoffier introduced them both couples were sitting in the tiny bar of the hotel. Henri and Josette were discussing the situation and she gave it as her opinion that they would be there at least another year when Beckett suddenly interrupted. This was an intolerable thought, he said, and it surely would not be the case – both the optimism and the interruption being uncharacteristic of the peacetime Beckett. As it was they were to be there for almost two years. Yet Beckett did not seem to Josette to be especially unhappy. He frequently whistled as he went about and, though often silent, was always cordial. He and Henri Hayden – a painter – soon became firm friends, drawn together by the chess games they played in the evening and the walks they took together. As Beckett became aware of Hayden's work and his dedication to it their friendship was cemented, for he was always drawn to fellow artists, particularly painters, whose work he could at all admire. Henri Hayden was at this time in his sixties. He had had exhibitions at the Zborowsky gallery in Paris and was reasonably well known. Josette, who was much younger than her husband, was a pretty, dark-haired woman of forthright opinions. Every day Henri took his easel out to the countryside and there are many paintings by him of the village and its surroundings. Some time after their arrival the Haydens moved into a little house. The winters in the Vaucluse were very cold and life in the hotel, which only had an outside closet, was not very luxurious. Everybody was now in any case short of money and, though the food supply was uncertain, they preferred to take their chance. The two couples and Miss Beamish and her companion spent that Christmas together in the Haydens' little house, Beckett and Miss Beamish singing the songs of their native land together. They were both drinkers and this was a bond between them since Suzanne did not drink and Henri Hayden consumed alcohol only in moderation. Josette thought Suzanne very attractive, even beautiful, with her chestnut-brown hair and clear eyes. She also noticed that in spite of the privations of life in Roussillon she always appeared to be smartly and tastefully turned out.

After Christmas Sam and Suzanne also moved into a house which was on the road to Apt, about 300 metres beyond the outskirts of the village and near where Aimé Bonhomme, the resistance leader, lived. T-shaped and in part two-storey, it was an eccentric house for the locality, being much more Norman than Provençal in design. It was partly surrounded by a wall and in part by a thick hedge, and at its back was the hill behind the village with its outcrops of red rock and its precariously overhanging pine trees. After the move they continued to eat one meal a day in the hotel, and every day at noon Sam would go down there religiously to hear the news on the wireless. For the satisfaction of their other needs, though, they had to depend on the local farmers, who supplied them with vegetables and wine in return for money or, when money was scarce, for work. The necessity to scrounge irked Beckett, as did the manual labour, which sometimes tasked his strength, but it was he and not Suzanne who did it all the same, chopping wood in the winter and helping with the grape and the cherry harvest. Contrary to the suggestion in *Waiting for Godot*, however, he never worked for the Bonnellys, though he used to visit and purchase eggs and wine there.

Before the Haydens arrived he had already made friends with a family named Aude, and after they had moved out of the hotel he and Suzanne would go to the Audes on at least one evening a week and eat an evening meal there, staying afterwards to converse with the family. The Audes were more open and friendly than the general run of the inhabitants of Roussillon and Aude himself had a genial and sunny outlook. Their family culture was somehow more gentle than the perhaps affected but seemingly aggressive roughness which was the norm. Beckett did occasional work for Aude and preferred this to the work he did for others. One of the principal consolations of life in the new house was that he had procured a piano and had had it moved in, recruiting the help on another sympathetic local, M. Blanc. Élie Blanc, who after the war founded a literary and artistic centre in Roussillon, was more literate than most of the village's inhabitants. He understood who and what Beckett was and often supplied him with cigarettes, for which he was very grateful. Another acquaintance he made was that of his new neighbour Aimé Bonhomme, the strongest character among those who favoured resistance.

In the early months of 1943 resistance activity in such places as Roussillon was beginning to be organized. There was a Conseil Général which directed operations for the whole area, the local leaders frequently making night journeys by back roads to Apt, where the resistance post office was a well-known brothel, for instructions. The President of the Conseil Général sometimes came and stayed at Bonhomme's farm, where allied airmen and others on the run were also occasionally sheltered. Grenades, rifles, radio components and supplies of cigarettes were dropped by allied planes on a plateau in the mountains beyond Apt. Two lorries were used to meet drops, the first one returning empty in the hope that this would fool the Germans if they should be stopped. Supplies had to be gathered quickly and there would follow a risky journey back over bumpy roads with perhaps sacks of hand grenades in the back of the truck.

It is evidence of the village's ambiguity that Bonhomme, who was a miller as well as a farmer, had an influence on the issue of ration cards and saw to it that those engaged in the resistance got cards whether they belonged to the locality or not. He knew Beckett and liked him and though, like everybody else in the village, he assumed he was in hiding, he knew nothing about what he had been up to in Paris. However, after a while he began to employ him to deliver coded messages when, for instance, a squad had to be mustered to meet a drop. From this it was but a short step to actually taking Beckett with him on night missions. On one occasion when they had to traverse a wood he thought that Beckett was visibly nervous, and all in all he decided that he was 'not the most courageous' of beings; that 'he wanted to protect a little his skin'. Nor had he a high opinion of his physical capacity when it came to carrying loads across rough terrain or even marching twenty kilometres in the dark. Nevertheless, according to Bonhomme, Beckett never refused to do as he was asked and the local man had a high opinion of one of his qualities and its value in this sort of activity at least: 'If he had a fault, it was certainly not that he talked too much,' he observed. The problem of storage was in part solved by using the caves among the rocks above the Bonhomme farm and Beckett was also employed on more than one occasion in this kind of ware-housing.

When Bonhomme was told much later that Beckett had been awarded the Croix de Guerre for his part in the resistance, he exhibited some amusement, remarking first that 'anyone could get it', and then conceding that lots of people had been given it 'who did less than he did . . . He at least made the effort to come with us.' Some of this mild denigration has the ring of the countryman judging the effete townie, particularly as regards Beckett's bravery and physical performance, and perhaps of the local hero who has received less than his due from Paris officialdom. Beckett was associating as a stranger in an unfamiliar countryside with men whose knowledge of the terrain was minute and whose familiarity with it had been lifelong. It may also be observed that not much use seems to have been made of the rifles and grenades that were dropped and collected, so it is possible that Beckett came to a conclusion early about the usefulness of Roussillon's contribution to the winning of the war. Nothing could be more calculated to reduce one's enthusiasm for a twenty-kilometre slog across rough country in the dark than a conviction that such a slog was not to much purpose. Though it could have been argued that the arms were being kept for a final uprising, their number seemed to be out of proportion to the capacity of the potential users. In spite of the copious supplies of rifles, explosives – and cigarettes – dropped, even on the eve of the landings in France, the Allied Supreme Command had little faith in the military value of the resistance groups – known by the Corsican word Maquis from the time of their first triumph, the liberation of Corsica in September 1943. Even Bonhomme admitted that they had too many arms and that there was a consequent difficulty in storing them.

It is also probably the case that Beckett was physically run down and mentally not in very good shape after his experiences of the past year. M. Blanc remarked that he did not seem to be physically up to the country tasks such as the chopping of wood and he pitied him for having to do them. Other perhaps apocryphal stories about Beckett and the resistance involved him storing dynamite on one occasion in his house – or rather, since he was apparently unwilling to have it in the house, on the terrace; and, which is perhaps the same story under a different guise, leaving a bag of grenades on the terrace in full view of passers-by.

There were Germans in Apt, a few kilometres away, and of course in the capital of the Vaucluse, Avignon, but they seldom came to Roussillon or even passed through it. On one occasion, however, some German soldiers did arrive in the village and ordered drinks in the bistro. Henri Hayden, who was having a drink there, immediately decided to leave. Since at this stage he looked like any other countryman, his presence and his departure passed without notice, but as the evening wore on and the Germans showed no signs of going anxiety began to mount among those who knew that a drop was expected that night. A lorry would have to be taken out to collect it, which would attract the attention of the Germans. Fortunately they left before the appointed time, so the arrangements could be carried out as planned.

These excitements apart, however, life in Roussillon was much the same as it had always been. The main enemy, for Beckett anyway, was boredom, somewhat alleviated by the presence of Miss Beamish, to whose house he would frequently go in the evenings to join in the fairly heavy drinking with which she was accustomed to pass the time, and by his rather more fully sympathetic – if often silent – communings with the Haydens. It was perhaps the sheer boredom and monotony of life that drove him back to literary composition, and after some time in the new house he began to write what was to become the novel *Watt*, eventually, after many rejections, to be published by the Olympia Press in 1953.

It had now been over five years since Beckett had written anything apart from a few critical pieces and a handful of short poems in French. *Watt* was a book which broke a silence, but it was more than that. It was a book about something which was central to Beckett and it could not have been written by anybody else. It not only signalled its author's break with the traditional novel, but it marked also his rejection of all assumed modes and accepted subjects. Gone, or almost gone – for it does mar the opening pages a little – was the form of satirical Irish whimsy – buttonholing, personal, would-be shocking and would-be charming at the same time – which had been the mode of other Irish novels of the 20s and 30s and was also the mode of *More Pricks Than Kicks* and *Murphy*.

Watt is an extraordinary book by any standards; but it is not perhaps such an extraordinary one for Beckett to have written in his current circumstances as would at first appear, for it is, in fact, about losing one's grip on reality. To account for it, some biographical commentators have suggested that Beckett had a breakdown in Roussillon. Those who were there knew nothing of this. But he had been through some very strange experiences, ending in this marooning in the depths of the French provinces while Europe went through one of its periodical fits of self-destruction. It was natural enough that he should ponder the nature of reality and feel how easy it might be to lose one's grip on it, such as it was.

On the surface, the events of the book seem comprehensible enough. Watt goes to work in the house of a Mr Knott, where he is responsible, in a subordinate capacity, for some of the household arrangements. He never encounters Mr Knott, nor do we; but he lays out the bowl from which his master eats in the dining room, feeds the dog, empties the slops, etc. On the surface these are simple arrangements; but it is their very simplicity that upsets Watt, causing him to look for another meaning in everything. As in the incident of the blind piano tuner, who, accompanied by his son, calls to tune Mr Knott's piano, it was the 'fragility of the outer meaning' that 'had a bad effect on Watt, for it caused him to seek for another, for some meaning of what had passed, in the image of how it had passed'. It is only towards the end of his stay in the house that Watt learns 'to accept that a nothing had happened, learned to bear it and even, in a shy way, to like it. But by then it was too late.' Too late because Watt, deranged – for once the word seems to have an exact meaning – is on his way to the lunatic asylum.

It is this wavering between an actual apparent incident and an elusive something beyond it, the fact that a pot, for example, has somehow ceased to be a pot for Watt and yet he can find no other name for it, his constant failure to form satisfying correlations between such things as the hour of Mr Knott's rising and the hour of his retiring, or to fix Mr Knott himself in any constant physical appearance, this reduction and, at the same time, dreamlike extension of experience and image, which gives the book a strangely disturbing, ghostly, and unsettling power.

Nothing is certain in *Watt*. Even in the comic prologue we are in the midst of uncertainties. When the tram stops opposite the bench on which a Mr Hackett and his wife are sitting and then moves on, it leaves behind on the pavement 'a solitary figure, lit less and less by the receding lights, until it was scarcely to be distinguished from the dim wall behind it'. Mrs Hackett is not sure whether it is a man or woman. 'Mr Hackett was not sure that it was not a parcel, a carpet for example, or a roll of tarpaulin, wrapped up in dark paper and tied about the middle with a cord.' To illustrate the fallibility of all knowledge, Arsène, the departing butler, tells a story about asking the time from a passing acquaintance one cold wet evening on Westminster Bridge.

> Securing me with one hand, he removed from the other with his mouth two pairs of leather gauntlets, unwound his heavy woollen muffler, unbuttoned successively and flung aside his greatcoat, jerkin, coat, two waistcoats, shirt, outer and inner vests, coaxed from a washleather fob hanging in company with a crucifix I imagine from his neck a gunmetal half hunter, sprung open its case, held it to his eyes (night was falling) recovered in a series of converse operations his original form, said, Seventeen minutes past five exactly, as God is my witness, remember me to your wife (I never had one), let go my arm, raised his hat and hastened away. A moment later Big Ben (is that the name?) struck six. This in my opinion is the type of all information whatsoever, be it voluntary or solicited.

In *Watt* for the first time Beckett achieved his characteristic style, a syntax full of reservations and uncertainties, denials and admissions that something else might be the case, with a superb use of the comma.

> Mr Knott was a good master, in a way.
>
> Watt had no direct dealings with Mr Knott, at this period. Not that Watt was ever to have any direct dealings with Mr Knott, for he was not. But he thought, at this period, that the time would come when he would have direct dealings with Mr Knott, on the first floor. Yes, he thought the time would come for him, as he had thought it had ended for Arsène, and for Erskine just begun.

Watt concludes with some addenda, characterized as 'precious and illuminating', the incorporation of which into the main body of the text was prevented only by 'fatigue and disgust'. They conclude with the stern, now famous injunction, 'no symbols where none intended'; and among these disjointed statements and phrases are 'never been properly born' and 'for all the good that frequent departures out of Ireland had done him, he might just as well have stayed at home'.

The umbilical cord to Ireland had certainly not been cut. Sometime after arriving in Roussillon Beckett had once again succeeded in contacting his mother and brother, and through the Irish Minister in Vichy was now in receipt of money. This alleviated the problem of existence somewhat, but the amount which Frank was now allowed to send had been reduced to a mere £3 per month, which barely paid the rent on the house. In response to demands, Sam was also able to send an occasional reassuring message, such as ALL WELL. HEALTH EXCELLENT. LOVE, sent in April 1943 in response to a message PLEASE CABLE HOW IS BECKETT'S HEALTH, received at Vichy; or ALL WELL LETTER DATED 5 APRIL RECEIVED LOVE SAM, received in Dublin on 5 June. Though such communications were necessarily brief and inexpressive, he may have felt glad that he was able to reassure his mother without undue loquacity, explanation or protestation; and he would only bother the Vichy legation to transmit a message after they had passed on the anxieties of home. For these cables, Frank paid the Irish Ministry of External Affairs at the other end, being charged 7s. 10d. for this one.

Life in Roussillon had always been frugal and its economy virtually a subsistence one except for petrol, tobacco, coffee and other imported items. The general shortages of war were not felt very acutely there and a little cash went a long way, but to repay the Audes for the vegetables, wine and other provisions which he and Suzanne received Sam continued to help them out with whatever tasks they gave him. It was rumoured in the village that he was in hiding, but to the normal taciturnity and secrecy of its inhabitants had been added distrust of each other due to the divisions of the war. Nobody wanted their neighbours to know too much about their activities. Some were in the black market. Some might be in contact with the Germans. Others were in the resistance; but of course as the tide of war swung against the Germans those who had been in favour of Vichy and collaboration more than ever kept a surly silence and the refugees and resistance activists were that much safer. The second front was opened in June 1944 and on 15 August the allies landed in southern France, taking Marseilles at once.

Concurrent with the allied landings, messages were broadcast ordering the resistance to start their guerrilla activities and sabotage. In some parts of France there were premature uprisings as a result, which caused much loss of life and a good deal of recrimination. Everywhere the blowing up of bridges, trenching of roads and uprooting of railway tracks hampered German troop movements; and in Roussillon too the resistance was mustered for these final activities. Bonhomme decided to lay an ambush beside the Route Nationale for a German column which was expected to pass that way and he brought Beckett along. It might have been a serious business. In the afternoon trenches were dug and there followed a long night wait with each man in his hiding place. Beckett was given a gun and waited with the others until dawn. But the Germans never came and the column returned home without a shot being fired. But they returned as heroes.

A previous biographer has painted a picture of Beckett disappearing with the Maquis for several days of sabotage after the American landings and then bringing up the rear in a glum and dejected fashion as they marched through the village after the German departure. In fact he was allowed to march at the head of the column, jubilantly carrying a French tricolore and singing the Marseillaise as lustily as the rest. Some bystanders, including Élie Blanc, who thought he knew Beckett fairly well by now but seemed to be unaware that he had engaged in resistance activities, interpreted this as simply a desire on the part of a foreigner to show that he was sympathetic to the French cause. It is at least evidence of a general feeling of goodwill towards him among the members of the resistance, who thought that he should be allowed to display his comradeship and solidarity with them. Though Bonhomme might not have considered him an especially heroic fighter, he certainly both liked and trusted Samuel Beckett.

When liberation came it was in the form of an American NCO driving a jeep who, after spending a little time in the village café, agreed to give one of the Jewish refugees who had been sheltering there, the Russian painter Eugène Fidler, a lift to Apt. Fidler remembered marvelling that Apt, which had seemed such a weary distance away over hilly winding roads when one had to go there on foot, could be reached in a few minutes in the soldier's vehicle.

After liberation, the refugees now began their departure from the village, but Henri Hayden had fallen ill and there was no question of he and Josette leaving for the moment. Beckett went into Avignon to see what the situation was and was told that, while there were trains to Paris, it was difficult to get a place on them. Another difficulty was that he did not have the money for his fare, which cost more than he had expected. Eventually, though, with money borrowed from the Audes, he and Suzanne got a place on a crowded train, on which some passengers had to stand or squat in the corridor for the twenty-four hours it took to return to Paris. Arriving at the rue des Favorites, they found that during their two-year absence the apartment had been broken into and furniture, cooking utensils and other things taken. On the little money they had, a hotel had to be chosen carefully, so Sam decided to return to the Liberia, Suzanne going to Troyes to be with her mother.

Back in Paris in November 1944 he was able to send a telegram which announced his arrival and continued, ALL WELL. IMPOSSIBLE TO MOVE AT PRESENT. LOVE TO YOU ALL. SAM, sending it to Frank at Clare Street rather than to his mother in Foxrock. A reply was immediately dispatched saying that his family would bear any expense incurred if it was possible for him to 'move' and ending, ALL WELL LOVE MOTHER AND FRANK. It was in fact necessary to make an application to an Anglo-French priorities board even to get to England, and an Irish neutral who simply wanted to go home to see his mother would scarcely have been looked upon with any great favour. Anyway, Beckett was not only reluctant to go, he found great difficulty in even writing. In response to a further query as to his address in Paris he replied through the embassy that it was 6 rue des Favorites, even though he was still in the Liberia. As Christmas passed without any further news of him, May insisted that Frank should take some action, so another message was despatched in January 1945: BROTHER ANXIOUS BECKETT'S WELFARE PLEASE WIRE REPLY AND ASK BECKETT TO WRITE.

Beckett had in fact sent a couple of postcards over the Christmas period – postcards being all that one was allowed to send – but they were apparently not received and he was somewhat peeved that the embassy should rather coldly remind him that, as he might not be aware, it was now possible to send plain postcards to Ireland from any post office in France. He also expressed himself as anxious to go home, but said that he was trying to obtain the necessary French and Irish return visas. When the embassy informed him that he did not need an Irish visa and offered, in response to further pressure from Frank, to include his name among others on a list they were submitting to the board, he agreed that this should be done without raising again the question of his return to France, but, as things were to turn out, he was right to be anxious about it. This was in February and he had now missed inclusion in the first group to be repatriated. It was to be April before all the formalities were completed for the departure of a second group.

All European countries, neutrals included, still had severe currency restrictions and, though the maximum sum which Frank could send had now been raised to £15 a month, this was still slightly less than Sam's normal allowance of £200 per year. It was arranged that the legation would send him cheques on the Guaranty Trust Company of New York in the place de la Concorde, but in order to effect these transactions Frank had to fill out a form known as E1A. On this he stated that Beckett's source of income in normal times was 'a gratuitous allowance from his mother which was lodged to his credit in the Royal Bank, Dublin, and by an arrangement with the Bank's Paris agents he was enabled to draw on a Paris branch'. Under the heading of 'Purpose for which he/she went to reside abroad', Frank stated that

> Sam first went to Paris as exchange student to École Normale. Found that Paris suited his literary studies and research and settled down there. Returned to Paris after outbreak of war to resume these studies and look after his possessions. He normally resided in France except for short periods when he was resident in London and during periodic visits to this country. He had been on one of these visits when war broke out and returned to his residence in Paris shortly afterwards.

The amount of his allowance was said to be £200 per annum. Frank also declared that this was his brother's only source of income.

Shortly before he left for Ireland, on 30 March, Samuel Beckett was awarded the Croix de Guerre with gold star. The citation was made by order of General Juin, the Chief of Staff, and signed by General de Gaulle as President of the Provisional French Republic. The suggestion that the award should be made had come from a prominent member of the Gloria group, so the citation said nothing about his activities in Roussillon, of which the authorities were probably unaware. Misspelling his name, it said that Becket, Sam, was: 'A man of great courage, who during the course of two years, displayed his qualities as an informant in an important intelligence network. He carried out this work with extreme bravery. Wanted by the Germans, from 1943 he was forced to endure a hard and clandestine life.'

When Beckett went home to Ireland he said nothing to anyone about this award. When his resistance activities were mentioned in after years, he dismissed them with the remark, 'It was boy scout stuff.'

Beckett left with his group on Sunday 8 April. On his way through London he had time to enquire from Routledge as to sales of *Murphy* during the war years, about which he knew nothing. He was desperately short of money and hoped that a royalty cheque of however small a sum would await him. He was told that of the 1,500 copies originally printed, almost 800 were still warehoused in 1943, when a decision had been taken to remainder them. When he queried how this could be done without his permission, he was reminded that he had disappeared from view in 1940 and that, far from being able to communicate with him about the matter, the publishers did not even know whether he was alive or dead. This explanation by no means contented him but he was obliged to accept it because of the possibility that the same publishers might be interested in *Watt*. Herbert Read, who had been enthusiastic about *Murphy*, was now expressing interest in the new novel and so, with some degree of hope, Beckett left it with him.

Wars quickly establish their own normality. For almost six years Beckett's normality had been that of a country where almost everybody's life was torn or disrupted by the exigencies of conflict. The virtually peacetime normality of Ireland was a shock. The experience was in its way like that of a soldier returning from the front to find that home was a strange place with strange and trivial preoccupations. His mother's new house had been given the Shakespearean name of New Place, though it is not clear who chose this. New Place was a considerably smaller, bungalow-style house, but his mother was still in very comfortable circumstances and employed a maid and a gardener. Like many people, perhaps particularly of their class and background, who are closer together than they would often wish to be, Sam and his mother were nevertheless repressed and undemonstrative with each other, so that their meeting after so many years in which Sam had been in danger and difficulties was not as overtly ebullient or joyful as it might have been between people whose bonds were looser and easier to throw off. As he had already learned from Geoffrey Thompson's doctor brother, Alan, May had begun to exhibit the more obvious symptoms of Parkinson's disease, including an occasional tremor in the hands. She seemed to him much frailer than she had been before the war.

Still living in Foxrock as she was, she continued to have friends who visited, and in spite of her illness she kept herself busy in the garden, as well as with the rituals of lunch and afternoon tea. She also had grandchildren in the shape of Frank's children – Caroline, now six, and Edward, who had been born during the war. The donkey was still occasionally harnessed to his little trap when they were brought over to visit and she also kept a noisy Pomeranian, which was feared and unloved by everybody except herself. Like most people in Ireland she had found the war an occasional inconvenience but little more, and of course Beckett's presumptuous and no doubt self-induced feeling that she needed his presence had proved ill-founded. She may have missed him sometimes, but she had on the whole fared as well in his absence as she would have done had he been present; and probably, on the whole, worried less about him than she would have been inclined to do had he been with her in the house during the intervening years and behaving as he had before he left.

His brother Frank had settled down into his role as a business and family man without any apparent regrets for India or anywhere else. Despite shortages building had continued during the war years and Beckett and Medcalf had thrived. As his father had been before him, Frank was a member of various clubs, a great walker and golfer, a deacon of his parish church and a good-natured philistine; but unlike his father he had not widened in girth and was almost as tall and angular as Sam. But their faces now clearly showed their differences of temperament and preoccupation. While Frank's moustache, expression and general bearing were those of a comfortable Dublin businessman of his time and background, Sam's appearance was drawn and ascetic. As is often the case, although Sam's experiences had been more intense and, superficially at least, less pleasant, he was much more youthful-seeming than his brother. Before the war he had sensed that a gap would open between him and the companion of his childhood and adolescence. Now he knew it was unbridgeable.

And of course he found he had even less in common with his other relatives, though he went through an unending ritual of visits and teas with as good a grace as he could muster. Imported commodities like tea and sugar were still scarce, as was white bread, but food was disgracefully plentiful and seemed never to have been otherwise. Confronted by such masses of good things as other people had yearned for during the war, Beckett seemed to make rather a point of not eating very much. On one or two occasions he made remarks suggesting a level of indulgence in Ireland which was unseemly by contrast with conditions in France. 'My friends eat sawdust and turnips while all of Ireland safely gorges,' he said, slightly exaggerating the privations suffered in Roussillon or even in Paris at the time.

And however trying he found family gatherings, with the exception of MacGreevy he did not seem to have a lot of time for friends either, making no effort to see anybody, bumping into the poet friends Brian Coffey and Denis Devlin only by accident and giving them fairly short shrift when he did. When he met Coffey in the Gresham Hotel he refused the normal offer of a drink and stopped to talk only for a few moments. He did, however, tell Coffey that he had finished a new novel and left him with the impression that he was quite pleased with it. To the fact that Devlin, now rapidly becoming a senior figure in the Irish diplomatic service and currently posted to Washington, was home to supervise the publication of a book of poems, his most sizeable collection to date – and in fact, as it turned out, his only sizeable collection – *Lough Derg and Other Poems*, Beckett gave only the minimum attention necessary, though he did again advert to the existence of a new novel and expressed the hope that he would be lucky enough to find a publisher.

MacGreevy, who felt he had seen enough of war as a serving soldier in the earlier conflict, had returned to Ireland in early 1941, but he had had by no means a comfortable time of it there, principally because he found it so difficult to earn a living. Largely in default of other outlets he had begun to write for a journal called the *Father Matthew Record*, which was edited by a Capuchin priest, Father Senan, who had literary and artistic interests, and published by the Capuchin Order. His own devout Catholicism combined with a certain natural affinity with clerics to make this connection more palatable than might otherwise have been the case, but of course there was an element of necessity in it too. Senan, who was in his own way an ambitious patron, had also prevailed upon the Capuchins to turn their annual into a literary production, to which during those years many Irish writers contributed and in which were published articles about painting as well as poems and short stories. From being a mere contributor to both publications, MacGreevy had progressed rapidly to editorial status.

However, the favour he enjoyed with Father Senan was combined with a level of piety which Dublin found rather surprising in a returned Parisian exile who had been one of the atheist James Joyce's inner circle. Not many in Dublin knew or remembered that Tom had always been religious, and what was regarded as his new-found piety allowed Dublin wits, then a flourishing and noxious band, to charge him with toadying and provided a subject on which they could exercise their talents. The publication of Yeats's *Oxford Book of Modern Verse*, in which two very beautiful poems of his had been included, had done something for his reputation as a poet. But as there was no sign of further poems his inclusion there could easily be dismissed as just another Yeatsian eccentricity; and most chose to regard him, with his bow tie and his exquisite continental courtesy, as something of a figure of fun. What had not crossed the Irish Sea, however, was any suggestion that he might be homosexual. Though the state was not unknown in Ireland, celibacy of one kind or another, official and unofficial, clerical and lay, was looked upon as a much more usual one and was not enquired into or examined too closely. With his evidently devout religious beliefs, his daily mass-going and his clerical connections, MacGreevy was regarded, like thousands of others, as one of nature's celibates.

There is no evidence that Beckett, who understood something of the exigencies of the literary life, and understood MacGreevy a great deal better than most, disapproved of his friend's relationship with Father Senan and the Capuchins. If anything he probably sympathized. In any case he now agreed to review his friend's book on Jack Yeats for the *Irish Times*, the editorial request that he should review it having been arranged by MacGreevy himself.

Yeats was enjoying a period of unwonted popularity in Ireland as well as an unaccustomed affluence, both of which had come about through the agency of Victor Waddington, a young English dealer. Waddington had opened a gallery in Duke Street, and by adroit press relations had succeeded in convincing a section of the Dublin business and professional class both that Yeats was a painter of some international importance and that buying paintings was a good way to invest money which, in war and post-war conditions, had very few investment outlets and could only be spent on an extremely limited range of commodities. Waddington had also published MacGreevy's little book which, as Beckett pointed out, had been written in London in 1938 and in the intervening seven years had been rejected by a dozen London publishers. Its merits, he said, would not surprise those who had read MacGreevy's essay on Mr Eliot, his admirable translation of Valéry's *Introduction à la Méthode de Léonard de Vinci*, 'nor those who follow, in the *Record*, his articles on writers and artists little known, as yet, in the Republic'. Two little oddities may be noted about the phraseology here: first that he could not bring himself to call the publication with which MacGreevy was connected the *Father Matthew Record*, and secondly that he refers to the southern Irish state as the Republic. This was not to be a term in common usage, official or unofficial, until the passing of the Republic of Ireland Act in 1948, the usual, and indeed the legal names employed before then being Ireland or Eire. The review is generally favourable, but, as has already been said, it demurs at the nationalist emphasis of MacGreevy's work, declaring that 'to some' it may seem that Mr Yeats's importance is to be sought elsewhere than in a sympathetic treatment 'of the local accident of a local substance. He is with the great of our time, Kandinsky, Klee, Ballmer and Bram van Velde, Rouault and Braque, because he brings light as only the great dare to bring light, to the issueless predicament of existence.'

While delivering this review, which was probably too elliptical for the *Irish Times*'s taste, Beckett took the opportunity of giving the editor, R. M. Smyllie, a four-line poem, 'Dieppe', for the Saturday book page. Poems were published in a box on the page, one every Saturday, and a poem there in those days commanded a certain amount of attention.

He also accepted the editor's invitation to go across to the Palace Bar, the principal meeting place of male literary Dublin. Those who saw him there noticed how withdrawn and pale he seemed. Everyone who met him at this time had also noticed how thin and generally worn out he was and they attributed his appearance to the effects of privations suffered during the war.

While in Ireland he heard the news of the death of Alfred Péron. Apart from the grief it caused him, the news emphasized the difference between the complacencies of Foxrock and the life-and-death nature of the war years for so many people in Europe. Arrested in August 1942, Péron had been taken in February of the following year to Mauthausen, a camp where many who had been involved in resistance activities were kept. He had managed to survive there until liberation and had then been transferred to a Red Cross camp in Switzerland. Here, as a result of his ill treatment in Mauthausen, he died in June 1945.

In spite of the comfort of his mother's house, Sam was having difficulty in sleeping; and now that the privations and hardships of the war were in the past, his mysterious ailments seemed to be returning, for he had begun to suffer from stomach pains. He asked Alan Thompson to examine him but Thompson found nothing wrong except for what he considered to be the appalling state of his teeth. He suggested that Beckett should go to their mutual acquaintance Andrew Ganly, by now a fashionable Dublin dentist and an Abbey playwright, whose wife Bridget O'Brien was a well-known painter. Ganly was shocked by the state of Beckett's mouth. He extracted some teeth which he thought too decayed to be saved and filled some others. This resulted in no noticeable improvement in Beckett's general health, but he was now very anxious to get back to France and began to make arrangements. However, two shocks were in store for him. He had hoped to collect some of the considerable balance of money due under the terms of his allowance which could not be transmitted to him during the war, and to bring it back to France. His brother's bank informed him that this could not be done due to the currency restrictions still in force. At the Department of External Affairs they not only confirmed this but also told him that it would be impossible for him to return to France in the foreseeable future. By agreement with the French government a visa would be required which would simply not be issued except in extraordinary circumstances, and in any case the French government were not issuing residence permits to aliens, whether with or without employment. This news placed Beckett in a quandary. Was he to be stuck in Ireland indefinitely? And, if not, how could he get out? A few days later Alan Thompson happened to mention a way by which he might at least return to French soil, if not to Paris and some semblance of normal life there.

Chapter Twenty-Two

❦

After the allied invasion of France in June 1944 the Irish Red Cross had offered assistance to the French; and after a tour of the areas in Normandy which had been devastated, some of them by allied bombing, the town of Saint-Lô in Normandy had been selected as being in need of a hospital and a medical presence. Before the invasion Saint-Lô had had a population of 13,000; but in one allied raid alone 8,000 people had been killed and thousands more injured. The dead included many of the inmates of the mental hospital, and in a large local jail ninety-nine prisoners who had been locked in had been burned to death. After the initial big allied air attack there were further smaller ones plus bombardments by heavy artillery. The town had been almost completely destroyed; a year later bodies were still being recovered and the survivors, a mere 3,000 men, women and children, were living a grim existence among the ruins. As Raymond Lelièvre put it in *Terre normande*, 'the city was like an upturned dustbin', but

in spite of the deadening atmosphere, in spite of the dust, the impassable streets, the darkness, the lack of water and of hygiene, the lack of everything; the winter with its succession of ills, in spite of the cold which never leaves you, the lack of heating and of shelter in this windy area; in spite of the mud which is everywhere, impregnating those few clothes you have left, life took up again, but it was exhausting, because the battle against the ruins became also a battle against sickness and death, but to be ill in Saint-Lô was unimaginable, though sadly inevitable for many. There was no longer a hospital and the overworked doctors who returned had no place to work and perhaps save lives.

The Irish had offered a hospital unit of 100 beds but there was no longer a building in Saint-Lô which could be converted to use, and so it was decided to erect wooden huts on the outskirts of the ruined town. The hospital thus created was to be staffed and maintained by the Irish Red Cross for as long as was necessary. Alan Thompson now suggested to Beckett that he might join the advance party, which was to leave almost immediately, as 'Quartermaster-Interpreter'. This job was to mean that as well as interpreting he would take charge of the considerable stores of the hospital. In fact his work began before they left Dublin. Given a blue uniform and various check-lists, he was detailed to help assemble and check all the supplies which were going and now lay on the dockside in Dublin. There were 174 tonnes of assorted equipment and supplies, plus six ambulances, a lorry and a station wagon. He later described the task of listing the supplies to Lawrence Harvey as 'a frightful job'. Leaving the supplies to go by sea to Cherbourg, he went ahead to Dieppe, where the Irish Red Cross had established temporary headquarters at the Hôtel des Arcades.

However much they had known about the devastation of Saint-Lô, new arrivals were still shocked by the state of the town, whose streets had to be traversed by stepping from one pile of loose bricks to another or from a rusty girder onto the end of a buried bedstead, and whose inhabitants lived in boarded-up cellars, sleeping on sometimes wet mattresses.

At the new hospital the eight members of the advance team occupied huts already erected by the French. Conditions to begin with were primitive. There was no electricity, no running water and no sanitation, the toilet for the advance party being a hole in the ground surrounded by sacking. When finally the first lavatory was installed, the housekeeper of the hospital, Madame Pilorgat, used to bring visitors to the hospital to see this proof of progress, which she would describe as '*magnifique*'. Meanwhile hot running water was only available at an American base about thirty miles away, where the luxury of a shower could also be had. The padre of the American camp was a Father Bardick, and shortly after their arrival he took the entire Irish party to a nearby château which was a convent. The Reverend Mother of the convent provided a surprisingly excellent seven-course dinner and afterwards they all got around the piano in the drawing room, which Beckett played while people sang individually and in chorus. They left with many blessings, expressions of goodwill and promises to return, the Protestant Alan Thompson remarking, to Beckett's murmured agreement, that he hadn't thought that convents could be such nice places. Neither of them had ever been in such an establishment before.

Meanwhile the stores were arriving and being hauled to the large granary of a stud farm near the hospital. Here everything had to be taken upstairs. As Alan Thompson was to report shortly after arrival, 'Some packets weighed over 2 cwts. Stores came in for days and days. All are stored safely now. It was a considerable task . . . The Storekeeper (Samuel Beckett) and assistant are making out stock cards for all material.' It was necessary to find labour and eventually the French authorities, who were on the whole extremely co-operative, assigned some German prisoners of war to help. Beckett's linguistic abilities came in handy in this respect also, for he could of course communicate with the Germans. This was the second time that the war had involved him in the problems of storage, the first being in the case of the arms cache in the caves of Roussillon. Wars usually do impose boring and unromantic tasks of one kind or another on the participants; but during the first weeks he also drove regularly to Dieppe and Cherbourg to collect supplies and the still arriving personnel of the hospital. One of the arriving doctors recorded how at Dieppe they were 'met by Colonel McKinney and Sam Beckett (Storekeeper). They had the huge big Ford V8 Utility Wagon with them and after going through Customs we got in. We weren't hungry as we had had an excellent four course lunch on board and later tea; but nevertheless Sam brought us 3 huge bags of pears, grapes and plums.' The new arrivals also remarked on the Storekeeper's driving. 'It was novel being driven on the right hand side of the road and Sam believes in getting the 150 miles done as quickly as possible.'

There were also trips to Paris over roads which had suffered from the depredations of tank tracks, shelling and mining. During these trips Beckett would try to visit the rue des Favorites, where Suzanne was now installed. He made no attempt to take such members of the Irish team as he might have driven to Paris there, or to introduce them to any of his friends; but characteristically he felt he should take them sightseeing all the same. There was very little traffic on these roads and most of it consisted of pre-war vehicles, which were not being pushed too hard by their drivers. In a letter to his sister, one of the Irish party recorded a trip to Paris 'in lovely weather during which Sam kept up an average of about 35 to 40 miles an hour': and told her that on

Saturday morning we did some of my business till lunch time and also Sam took me into Notre Dame which was magnificent. Sam has an assistant storekeeper here named Tommy Dunne, a very decent little Dublin chap. Sam is a T.C.D. graduate, interested in writing and in letters generally; he has lived in Paris the last 6 years or 7. He is a most valuable asset to the unit, terribly conscientious about his work and enthusiastic about the future of the hospital, likes a game of bridge and in every way a most likeable chap, aged about 38-40, no religious persuasion; I should say a free thinker but he pounced on a little rosary beads which was on a stall in Notre Dame to bring back as a little present to Tommy D. It was very thoughtful of him.

Of course, these trips to Paris had an element of frustration in them for Beckett. With one part of his nature he longed to be able to get on with his own life; but equally with another he enjoyed the set tasks, the discipline, the limited objectives and the suspension of larger problems. He does not seem to have set much store by the humanitarian aspect of what they were doing; but if the unending cheerfulness of the nurses and medical staff sometimes irked him, he had the consolations of routine and even of comradeship and the wearing of uniform. Something of the part he himself played in the matter of liaising with the French comes through in the talk he was subsequently to give on Radio Éireann about the hospital, in which he said that there doubtless were those

who would prefer the history of our difficulties with an unfamiliar pharmacopoeia and system of mensuration to the story of our dealings with the rare and famous ways of spirit that are the French ways. And yet the whole enterprise turned from the beginning on the establishing of a relation in the light of which the therapeutic relation faded to the merest of pretexts.

And some inkling of the difficulties that he certainly encountered as the go-between who was trying to accommodate two differing cultures and outlooks on life to each other is to be somewhat cryptically glimpsed when he says,

It would not be seemly, in a retiring and indeed retired storekeeper, to describe the obstacles encountered in this connection, and the forms, often grotesque, devised for them by the combined energies of the home and visiting temperaments. It must be supposed that they were not insurmountable, since they had long ceased to be of much account. When I reflect now on the recurrent problems of what, with all proper modesty, might be called the heroic period, on one in particular so arduous and elusive that it literally ceased to be formulable, I suspect that our pains were those inherent in the simple and necessary and yet so unattainable proposition that their way of being we, was not our way and that our way of being they, was not their way. It is only fair to say that many of us had never been abroad before.

There is a deliberate note of disregard and even disdain for the more obviously humanitarian aspects of the enterprise in this radio talk when he says that

> What was important was not our having penicillin when they had none . . . but the occasional glimpse obtained by us in them and, who knows, by them in us (for they are an imaginative people), of that smile at the human condition as little to be extinguished by bombs as to be broadened by the elixirs of Burroughs and Wellcome – the smile deriding, among other things, the having and the not having, the giving and the taking, sickness and health.

This is certainly a very extreme expression of disdain for the importance of such modifications of 'the human condition' as can be brought about by medical care. It is also notable as Beckett's first use of the phrase 'the human condition', a locution which, already used by Malraux as the title for his famous novel, was to become so fashionable in the post-war years. Malraux's novel had provided Beckett with an epigraph for Chapter Nine of *Murphy*.

When fully functioning the hospital had accommodation for almost 100 patients and an additional 200 out-patients were seen almost every day. Many of them were injury cases, including injured children. Mines lay everywhere and masonry was still falling at unexpected moments. There was also a maternity unit. It consisted in the end of twenty-five quite substantial wooden huts, which Beckett made a point of saying in his talk were lined with glass-wool and panelled in isorel, 'a strange substance of which only very limited supplies are available'. He also made a point of saying that 'the walls and ceiling of the operating theatre are sheeted in aluminium of aeronautic origin, a decorative and practical solution of an old problem and a pleasant variation of the sword and ploughshare metamorphosis'. All this was staffed in the end by ten doctors, each a specialist, thirty-one state registered nurses, many of them with specialist training, a pathologist, pharmacist and administrative staff. This may seem a little excessive for the size of the operation but no doubt many were glad to get out of claustrophobic Ireland and play some sort of useful part in things in a country so lately at war.

Beckett's particular friend among the doctors was Arthur Darley, who was a violinist and often played for the staff in the evenings. Darley was tall, good-looking, quiet and well-mannered. He was fluent in French and, also like Beckett, he loved music. Among the staff he was known as particularly hard-working and dedicated, being in charge of the tuberculosis hut, where he held an out-patients clinic that began at 9 a.m. and continued throughout the day. He received many presents from his patients, including innumerable bottles of Calvados, and he eventually had a large stock into which he and Beckett made frequent inroads.

Many of the patients in the tuberculosis section had contracted the disease in concentration camps. A little while after leaving Saint-Lô Arthur Darley himself was found to have it and many of his colleagues believed that he had contracted it there. When he died in 1947 Beckett's grief resulted in the poem 'Mort de A. D.', which appeared in the French *Cahiers des Saisons* in October 1955. Darley had certain spiritual yearnings which resulted in a habitual reading of Alban Butler's *The Lives of the Saints*. These are reflected in the poem

> la vie des saints une vie par jour de vie
> revivant dans la nuit ses noirs péchés

But whatever his pleasure in Darley's company and his initial enthusiasm for the work of the hospital, as the weeks passed Beckett became more and more impatient at being detained in Saint-Lô. Josette Hayden remembered him coming to Paris for a few days' leave in his blue uniform after some weeks there. He said he was fed-up and was actively negotiating with the embassy to prevail on the French to allow him to stay. There were no clothes in the flat and he appeared to have brought none from Ireland except the uniform he wore, so Henri Hayden lent him a suit for the last two days of his stay.

As things settled down in Saint-Lô and the need for his presence there became less, the artificial aspects of the situation were borne in upon him and he began to spend less and less time at Saint-Lô and more and more in Paris. One of his last acts was to meet a new matron, Mary Crowley, off the boat. Matron Crowley arrived at Dieppe on a night of snow and storm and was met by Beckett, who had driven through the appalling weather from Paris to take her on to Saint-Lô on Christmas Eve. The roads were perilous because of frost and ice and there were frequent detours where bridges had been blown up. Beckett drove on dourly and for the most part silently, stopping once at a village, where he disappeared and reappeared with a bowl of coffee which he told Miss Crowley was laced with cognac. When she refused it, though hungry, thirsty and almost frozen with cold, on the grounds that she never drank spirits, Beckett told her brusquely that she had better take it because there was no food available and would not be until they arrived at Saint-Lô; and so in spite of her scruples, she did drink it off. When they arrived in Saint-Lô midnight mass was just starting in the bombed-out cathedral. She said she wanted to go and, though not, it would seem, very forthcoming or enthusiastic about the idea, Beckett acquiesced. The roof of the Gothic building was missing, there was snow underfoot on the floor of the nave, and there were mines piled against the walls. Sufficient candles had been found, though, to light the altar of the Virgin Mary and music was provided by violins, while the congregation sang hymns beneath a clear sky which was full of frosty stars. Through all this Beckett stood in the shadows, and Miss Crowley felt that he was uncomfortable. When they got to the hospital her quarters were waiting for her, but Beckett was so rarely there by then that a bed had to be found for him. When she looked for him next morning Miss Crowley discovered that such was his impatience to be gone that he had slept for only a few hours before leaving for Paris at dawn.

His final service to the hospital was in connection with the rats which roamed the bombed-out town in large numbers and were an especial nuisance in the children's and maternity wards. In mid-January 1946, after Beckett had spent a couple of weeks in Paris, he received a phone call asking if could take advice on the problem and if possible secure some efficient rat poison, a commodity for which there was great need everywhere in war-torn Europe, and which was in short supply. He consulted some people he had known through Alfred Péron who worked at the Curie Hospital, and the Department of Pathology gave him a mixture which could be used to coat stale bread and other food oddments and left lying around for the rats to eat. Together with Miss Crowley he did this, but such was his impatience to get back to Paris again that he did not even wait to see the results.

That January Beckett formally resigned from the hospital staff. Restrictions on travel in and out of France were already being considerably relaxed and he had been informally advised by a member of the embassy staff that he should have no difficulty going, returning and even remaining for long periods in France since he had already had a residence there. Some writers have interpreted Beckett's Saint-Lô experience as a profoundly emotional one, but this does not seem to have been the case. He had, apart from anything else, little direct contact with the patients. It had some importance for him as part of his general experience of war, the part which best illustrated for him the ruin, devastation and sheer suffering that war could cause in more metropolitan centres than Roussillon and less untouched ones than Paris. But in the talk he was to give on Radio Éireann, after declaring that the hospital was not less important because it was a temporary structure – '"Provisional" is not the term it was, in this universe become pro-visional' – he went on to speak of it as if the major illumination among the ruins for him had been a symbolic one. The ruins of Saint-Lô had laid bare the basic conditions of human life; and, for those who took part in it, the Saint-Lô experience would give 'vision and sense of a time-honoured conception of humanity in ruins, and perhaps even an inkling of the terms in which our condition is to be thought again'. Shortly after severing his connection with the hospital he wrote a poem about it, published in the *Irish Times* in June 1946. The poem suggests a contrast between the mechanical rebuilding of civilizations and the effect on the human mind of their destruction. The Vire is the river which flows through the town.

> Vire will wind in other shadows
> unborn through the bright ways tremble
> and the old mind ghost-forsaken
> sink into its havoc

Chapter Twenty-Three

While Beckett was still in Saint-Lô the Paris review *Cahiers d'art* had agreed to a suggestion that he should write an essay on the painting of his friends Bram and Geer van Velde. This was eventually called 'La Peinture des van Velde, ou Le Monde et le pantalon'. The title relates to a joke he would use again in *Endgame* about the tailor who unfavourably compares the botched job God made of the world in seven days with the perfect pair of trousers he crafted in a somewhat longer period. He would return to the painting of the van Veldes in a piece called 'Peintres de l'empêchement', written in 1948. Lawrence Harvey, who discussed such things with Beckett, regarded the first essay as the most important of Beckett's essays on the van Velde brothers, but Beckett told John Fletcher that he thought this first essay was best forgotten. His other major excursions into art criticism were the well-known 'Three Dialogues with Georges Duthuit', which were written and published in 1949; a further discussion of Yeats, 'Hommage à Jack B. Yeats', written and published in 1954; and 'Henri Hayden', a tribute to his friend and companion of Roussillon days. Though Beckett was on fairly intimate terms with three other painters – Stanley William Hayter, whom he met through MacGreevy in his early days in Paris; Giacometti, with whom he was friendly during the 1930s and afterwards; and Avigdor Arikha, whose graphics would be used to illustrate his fiction – except for a brief catalogue note about Arikha, he never wrote about any of them. His views on the German expressionists of the decades after the war have already been discussed with reference to his sojourn in their country in 1936-7. He names quite a number of painters in the compositions listed above, but those who are praised most often and with least reservations are Wassily Kandinsky, Paul Klee, Georges Rouault, Georges Braque, Edvard Munch, Pierre Bonnard, Jacques Villon, James

356

Ensor, and those giants of our time, Henri Matisse and Pablo Picasso. Picasso's greatness is by no means regarded as unquestionable, however, or as being evident in every work.

With the exception of Kandinsky, none of these painters can be described as abstract, nor can any of those about whom he wrote directly, though much of the later work of Bram van Velde certainly verges on that state and has been described as abstract expressionist. In 'Peintres de l'empêchement' Beckett in fact says that 'it seems absurd to talk as Kandinsky used to of a painting freed from its object. What painting had freed itself from is the illusion that there exists more than one object of representation, perhaps even from the illusion that this unique object allows itself to be represented.' It will be remembered that in the essay on Irish poetry published in 1934, Beckett had spoken of the new thing that had happened, or the old thing that had happened again, namely the 'breakdown of the object'. In the first essay on the van Veldes Bram had been praised for his 'prodigious objectivity'; but the second essay was concerned with the idea of 'mourning for the object'. In other words, there is or was an object but there has been a breakdown in the subject/object relationship and this breakdown had perforce introduced a new kind of objectivity and a new kind of subjectivity. Although 'the history of painting is the history of its relationship with its object', there had always been hindrances to objectivity, 'the object-hindrance and the eye-hindrance'. But there was a sort of conspiracy to ignore them. 'People took account of these hindrances. An accommodation was reached. They did not form part of the representation or scarcely so.' Now, however, in the work of the van Veldes 'they do form part. The greater part one might say. What hinders from painting is painted.' The relevance of this view to his own work is clear. What hinders from writing would be written about. There would be an attempt to express the difficulty of expression, of any expression at all.

The first van Velde essay also contains one of Beckett's rare statements of political attitude. After flatly saying that the van Velde brothers 'are at bottom uninterested in painting; what interests them is the human condition', he goes on to wonder what will become of 'this solitary painting' in a time of hope for a collective salvation, of 'processions towards a happiness like that of a sacrificial sheep'; and he concludes that it will probably be 'pelted with stones'.

At the time when this, his first published piece of prose in the French language was written, he must have felt that this pelting was, if it even got that far, the likely fate of his own work; for not only had Routledge, advised by Read, rejected *Watt* but several other publishers had also done so. George Reavey, whose post-Madrid adventures during the war had included being torpedoed in freezing waters on the highly dangerous convoy route to Murmansk, was now back in London; and after Herbert Read's rejection of it, he had agreed to handle the novel. As Beckett passed through again on his way to Ireland in April 1946 they met, but Reavey, who had been on the receiving end with several rejections, was not hopeful. He himself was shortly to leave for the United States but there were now several copies of the novel in circulation and various people were sending it out, including, for a while, apparently, Curtis Brown and A. P. Watt, none of them with any success.

It was the usual story. All the publishers who read it recognized that the author had talent and that he could write – at times splendidly. None of them saw any prospect of commercial success. None of them was interested in actually publishing it. A typical letter of rejection from Fred Warburg said that as it showed an 'immense mental vitality, an outrageous metaphysical skill, and a very fine talent for writing, it may be that in turning this book down we are turning down a potential James Joyce. What is it that this Dublin air does to these writers? However what appears to us as his perversity is so considerable that we find ourselves unable to make an offer.'

Beckett's greatest creative period was 1946–50. It began with a revelation, occurring on this 1946 trip to Ireland, which transformed his whole outlook on writing. Because it is described as doing so in *Krapp's Last Tape* it is usually assumed that this famous revelation took place on Dun Laoghaire pier, a splendid Victorian granite structure which juts into the Irish Sea, forming one arm of Dun Laoghaire harbour. According to Krapp's broken and fragmentary account of the matter, it was on a night of storm, with huge waves breaking over the granite rocks which form a breakwater for the pier, and the anemometer or wind gauge at the end of it spinning like a propeller, that he suddenly saw the truth: 'What I suddenly saw then was this, that the belief I had been going on all my life, namely ... great granite rocks, the foam flying up in the light of the lighthouse and the wind gauge spinning like a propeller, clear to me at last that the dark I had struggled to keep under was in reality my most ...' Here Krapp impatiently switches off the tape-recording he has been listening to. But according to what Beckett himself told Eoin O'Brien, the Dublin medical consultant and author of *The Beckett Country*, all this – Krapp on the pier in the storm, the flying foam, the wind gauge – was invention. There had been a revelation and it had changed his attitude to his own writing, but it had taken place further down the coast on a much more humble edifice than Dun Laoghaire pier: the little jetty that juts into the sea at Killiney Harbour, which is overlooked by his brother's house.

Up to this point he had, like others, struggled to be knowing; indeed the 'knowingness' of his early writings is one of their most obvious characteristics. And besides this attempt at knowledgeability, there had been the struggle to do what the novelist is expected to do, to describe a world which would be a realistic simulacrum of the world about him. In other words, he had attempted to be creative in the wrong sense. But according to the revelation that he now had, instead of writing about that exterior world he should have written about the inner world, with its darkness, its ignorance, its uncertainty. The omnipotent, sly and sophisticated narrator of *Murphy* and *More Pricks than Kicks*, even *Dream of Fair to Middling Women*, was pretending to knowledge, experiences and abilities that inside himself Beckett knew he did not have. Their creator had tried to conceal that inner ignorance and darkness which could in reality have been his greatest strength. Now he resolved that he would let it prevail. From this point on there would an entire abandonment of pretence of any kind, including the ordinary fictive pretences of plot; a total renunciation of all certainties, including philosophic certainties of any kind; and there would instead be a reiteration of ignorance, a restitution to their rightful place in his work of the uncertainties and confusions of which life was made up.

The mode for such a reiteration and restitution would be the only possible one: first person monologue. All others would involve degrees of omnipotence, of pretence, of knowledgeability, as there had been even in *Watt*, however undercut and undermined. It would still be a fictive monologue, for reasons which will be discussed later. But in so far as fictions can be disowned by their creator and any knowledge of the purpose of their creation disclaimed, there would be a disowning and a repeated disclaimer.

In the history of Beckett's writing 1946 is also notable for a growing commitment to French. Apart from the abandoned philosophical essay 'Les deux Besoins', written during his visit to Alfred Péron in the summer of 1938, his first writings in French had been the twelve poems that he wrote in 1938 and '39 and that were eventually published in *Les Temps modernes* in November of 1946. But the transition from one language to another had been interrupted during the war by a return to English in *Watt* – and what English it is! *Watt* was followed by three *Nouvelles* in French, and would be Beckett's last writings in English until three little poems, apparently composed in both languages, in 1948 and '49. Then there would be no other writings in English until the radio play *All That Fall* in 1956; though there would be an abandoned work, 'A Prose Fragment', which was finally published in the *Evergreen Review* in 1957. Thus the departure from English into French was begun just before the war and resumed immediately afterwards, though a major work in English, *Watt*, written in Roussillon during the war, had interrupted it.

His decision to abandon one language for the other has been much debated, but it was not really a decision arrived at in the sense of saying to himself 'from tomorrow I will write only in French'. For some time he had to feel his way, through the poems, through the philosophical fragment, through two commissioned essays on the paintings of his friends Geer van Velde and his brother Bram. Initially he did not have the confidence that he felt was necessary to write in French, and it was only gradually that he gained it. Strangely, for this was not the case with other important turning points in his life, Beckett gave surprisingly consistent explanations over the years for this decision. In *Dream of Fair to Middling Women* he had said, 'Perhaps only the French language can give you the thing you want.' About ten years after he had begun to write in French he told Nicholas Gessner that 'it was easier to write without style in French', the implication being that he did not want to be a stylist, or at least to write in a particular identifiable style. About three years after that, in 1960, he explained to Herbert Brow that French appealed to him because it had 'the right weakening effect'; and about the same time he insisted to Richard Coe that he was afraid of English because 'you couldn't help writing poetry in it'. A couple of years later still, he discussed the question with Lawrence Harvey and said that

for him, an Irishman, French represented a form of weakness by comparison with his mother tongue. Besides English because of its very richness holds out the temptation to rhetoric and virtuosity, words mirroring themselves complacently, Narcissus-like. The relative asceticism seemed more appropriate to the expression of being, undeveloped, unsupported, somewhere in the depths of the microcosm.

And of course he was right. By comparison with French, English has a tendency to lure its user into the more obvious forms of poetry, of sentiment, of feeling, and of easy rhetoric. For an Irishman the temptations are even stronger. The greatest Irish poet of the century, Yeats, had begun by echoing Verlaine's famous injunction, 'We must take rhetoric and wring its neck'; but had ended up as a master rhetorician, saved only by his irony, his sudden colloquialisms and the 'quarrels with oneself' that turn rhetoric into poetry. And James Joyce too had a marked tendency to 'fine writing', which was always there when he spoke *in propria persona* and was only kept under by using the language of meaner mortals, or, finally, by the invention of his own language.

While in Germany in 1937, Beckett had written to Axel Kaun, an acquaintance he had made on his travels: 'And more and more my own language appears to me like a veil that must be torn apart in order to get at the things (or the Nothingness) behind it. Grammar and Style. To me they seem to have become as irrelevant as a Victorian bathing suit or the imperturbability of a true Gentleman. A mask.' But in fact the change of language did not allow him to get rid of Grammar and Style. What it seems to have done was to allow him to adopt a mock style, a sort of synthetically elaborate joke style, which is a totally successful ironic joke at the expense of all attempts at style and all styles whatsoever. This ongoing joke, which has its own elaborate beauties, provides one of the great pleasures of reading the later Beckett.

It would be much more dramatic and nicer all round if we could associate this change of language on Beckett's part with the revelation which he seems to have had at Killiney Harbour in 1946. But while we might admit the four shorter prose pieces, 'The Expelled', 'The Calmative', 'First Love' and 'The End', indeed written in French after his return to Paris in 1946, into the canon of 'true Beckett', the clean and satisfying pattern we might establish of an immediate change of mode as well as language, and of progress through the *Nouvelles* to the great trilogy of novels, *Molloy*, *Malone Dies* and *The Unnamable*, is marred somewhat by the composition of another novel, *Mercier et Camier*, in French in 1946.

Mercier et Camier was almost certainly begun shortly before and was undoubtedly finished fairly shortly after his trip to Ireland in April 1946, the trip during which the illumination on the jetty took place. For the biographer who rejoices in clear and definite patterns, transformations and creative crises it is a bit of a problem. Although written in French and more or less coincident with the first monologues, 'First Love' and 'The Expelled', it is what might be called an old-fashioned Beckett novel, set in an all too real and circumstantially described Ireland and, worse still, written in the old clever-clever, knowing and literary Irish mode. 'The journey of Mercier and Camier, is one I can tell, if I will, for I was with them all the time', it coyly begins; and though this flimsy justification of omnipotence is soon abandoned, the omnipotent tone of voice is not. 'What stink of artifice', the narrator remarks on the third page, and such a breaking-off to comment foreshadows the sardonic comments on the exigencies of a fictional construct which the narrators of the later novels will make. But unfortunately artifice is not abandoned and the artificial all-seeingness of the narrator is maintained to the end, as well as a highly artificial and whimsical tone of voice. Of course, being by Beckett, the book is not negligible. It has some good jokes and is at least notable in the Beckett canon for having an overtly homosexual couple as its central figures. It is in its way a better examination of what Joyce had identified as the principal Irish disease – paralysis – than *Murphy*; and it quite strongly conveys the indecision and unease which were characteristic of Beckett's periods in Ireland. But, like all the early work, it is none the less marred by condescension and the desire to show off and be appreciated for it. The voice of the author is too clearly heard and it is too concerned with the effect it is having on the reader.

It has been suggested also that the move into French was influenced by the fact that Beckett could no longer get published in English; but while it is true that hanging over the composition of *Mercier et Camier* was that worst of authorial impediments – the shadow of a finished work which nobody wanted to publish – his earliest publishing experiences in French were not of the happiest either. One of Beckett's seemingly numerous agents at this time, amateur and professional, was the novelist Toni Clerkxs, who was a sister of the van Veldes. She had now arranged in France for the publication of the French translation of *Murphy* and this would appear a year later, in 1947. The publishers were a small firm called Bordas, and at first the relationship seemed promising, but it would not survive the truly abysmal sales of *Murphy*, which sold less than a dozen copies in the first year, and five years later had not even reached 100. Bordas blamed Beckett's obstinacy about seeking publicity and co-operating with his publisher's efforts to arrange interviews for this débâcle. He therefore turned down *Mercier et Camier* as well as a book called *Quatres Nouvelles*, which was to consist of 'The Expelled', 'The Calmative', 'First Love' and 'The End' ('L'Expulse', 'Le Calmant', 'Premier Amour' and 'La Fin') – or in other words, the stories he had by then written in French.

Bordas's rejection of *Mercier and Camier* was in 1947. It was more or less in the middle of the writing of it that Beckett went back to Ireland in April 1946. The reasons for going back there were two. One was that, for good or ill, he was returning to the pattern of once-yearly returns to see his mother that he had begun to establish before the war. The other was that he was seriously short of money. Although the currency allowance had been increased at the end of 1944 – to such an extent that Frank, if he wished, would have been able to send more than Sam's monthly allowance – the value of this had been decreased by the introduction of the New Franc, which gave a much more unfavourable rate of exchange to Irish pounds. In these circumstances Sam had been reduced to translating articles from the English for the French edition of the *Reader's Digest*, and Suzanne had been exercising her talents as a seamstress in order to earn something. But the money that had accumulated for him during the war was still in Ireland, and in the more relaxed atmosphere of 1946 he hoped simply to be able to take it back with him in his pocket, which is what he eventually did.

Although to be with her had been one of the reasons for his journey to Ireland, by a not uncommon paradox or perversity in the human psychology, his mother got so little joy or comfort from his presence that it might have been better if he had stayed in France. He seemed to spend most of his time drinking with various acquaintances, arriving back at New Place at all hours and sometimes staying out all night, turning up the following afternoon with a hangover. During the war years Dublin had been a comfortable bolt-hole and there had been a number of artistic movements whose membership was largely composed of English refugees from the conflict. One of these called itself the White Stag Group. It was still around and its participants included Ralph Cusack, who was later to marry Nancy Sinclair. Cusack was an intelligent, ebullient, tortured and sometimes violent man who, like Beckett, had lived in France and spoke excellent French. At this time he was a painter but he would eventually write a book, *Cadenza*, which has a minor cult status. Like Beckett he loved music, but he differed from his more famous fellow countryman in that his indeterminacy of purpose was lifelong. Beckett spent a great deal of time in his company and that of the other members of the White Stag Group. There was no shortage of heavy drinkers among them and there was also a good deal of sexual promiscuity of a bohemian, inconsequential and probably not very enjoyable kind.

Of course, beings like Beckett are not easily able to lose themselves in this kind of thing. They remain apart from it, even while supposedly enjoying it. The sensation of being 'apart' while in company is not confined to literary artists, or even to the type which W. H. Auden characterized as 'the self-conscious man'; but it is commoner among such artists than among others; and Beckett, who sometimes had doubts about his own literal existence except perhaps as a consciousness, had it more strongly than most.

As has been said, for the biographer, with his or her assumption of omnipotence and desire to impose a tidy pattern on the subject's life, it would be a great deal simpler if, when Beckett returned to his Paris apartment in May 1946, he had plunged straight away into the composition of the trilogy of novels, *Molloy*, *Malone Dies* and *The Unnamable*, which are its greatest expression. A wonderfully satisfying pattern would then be established with the decision to write in French, the revelation on the jetty – largely, as he said, a revelation of his previous stupidity – and the adoption of the monologue as a mode all coinciding together and having such an immediate result. But things rarely happen quite like this in life, and they did not happen quite like that in this case either. *Mercier et Camier*, a novel in the old mode – older even than the mode of *Watt* although it was in French – had still to be finished; and even before the revelation he had already begun to experiment in the new mode of uncertain monologue in 'First Love' and 'The Expelled'.

Nevertheless, he was beginning now to realize what he could do and what he had to do. It was a matter of less rather than more, of dispensing with almost everything except a few fundamentals. Later on, he would describe the period 1946-50 as the 'siege in the room'. The phrase suggests a reduction to bare necessities, being driven back on oneself, being stripped of all resources save the ultimate ones of desperation and self-reliance. It also implies having to make do with little instead of much, with those things which are really necessary rather than with those which were once – before the siege – thought to be essential but are now recognized as luxuries, however desirable. And so, in effect, it was to be with him now. He would have to dispense with his native tongue and all its rich associations, Irish and English. He would have to dispense with the insidious cadences that seem to come so easily to Irish writers, as well as that tendency towards the baroque which the native genius encouraged. And besides those rich associations and insidious cadences, he would give up all the props and stays, the elaborate mechanisms and descriptions of the ordinary novel, the apparently life-giving fabrications, the buttressing descriptions. In terms of what he had to do, locale in the ordinary sense, story-telling as it was commonly practised, character as it was usually delineated, were a distraction and a nuisance.

But something else would now be banished besides plot and description – something that might be called the hope of salvation. Even *Ulysses* recruits the ordinary human interest in development and outcome, the reader's curiosity about what will happen to the characters, the hope of change. Nothing much may happen during the day on which Joyce's great novel takes place, but many critics have speculated about what will happen to the Blooms and Stephen afterwards, and have sought for clues in the novel – William Empson, for example, imagining that Stephen comes to stay with the Blooms and goes to bed with Molly, the result being the restoration of conjugal relations between the Blooms and in the blossoming of Stephen's genius. But in the novels and plays Beckett was to write there would be neither the hope nor the fear of any outcome. The soul would not be purified or purged, the character tested by crisis or event. Nobody would be found wanting because all Beckett's characters have already been found wanting. There is no hope for them. We do not tremble for the outcome. We already know what it is. Life, one could say, has already settled their hash.

For the work he was about to do *Mercier et Camier* can scarcely be called a trial run – rather something to be finished and cleared out of the way before he could properly begin it. But this does not apply to the four *Nouvelles*, or, as he called them, long short stories ('First Love', 'The End', 'The Calmative' and 'The Expelled'), written in French and finished before the end of the year. These monologues are truly *sui generis* with the great work he was about to do. All four have solitary heroes in a state of greater or lesser dereliction and they are much closer to the works to come than they are to the earlier books or even to *Mercier et Camier*. If they lack anything it is perhaps that strange élan in the midst of despair, that fierce joy in knowing and saying the worst that can be said about human existence that is the real taste of Beckett and that he was first to achieve in *Molloy*. One of these *Nouvelles* was accepted by *Les ·Temps modernes* in July 1946, shortly after his return from Ireland. This was 'La Fin', or 'The End', which at this time was called 'Suite'. Jean-Paul Sartre had founded *Les Temps modernes* in 1945, its first editorial being a manifesto in which he propounded certain of the ideas which would be expanded on in 'What is Literature?' in 1947. In this editorial he claimed that the French writers of his generation – a category into which he co-opted Malraux and Saint-Exupéry – had necessarily to produce 'a literature of extreme situations' which would 'restore to the event its brutal freshness, its ambiguity, its unfore-seeability' as well as restoring 'to time its actual course, to the world its rich and threatening opacity, and to man his long patience'.

This is brave talk, but it is still a long way from the place where Beckett was now arriving. Beckett and Sartre had been contemporaries at the École Normale, but they seem never to have met there. Nor did Beckett now form part of the existentialist circle, socially or intellec-tually. Even in his café life he did not mingle with the adherents of the movement which was attracting admirers and tourists to the boulevard Saint-Germain, since he preferred the traditional haunts of writers and artists in the boulevard du Montparnasse. But *Les Temps modernes* was a fashionable place to be, so the probability is that he sent his story to a magazine rather than to a man. In any case, after it had been accepted in May, half of it apparently became detached from the rest in the office of the magazine and so only half appeared in the July issue. The editors did not seem to notice that the story as published was incomplete, which, given the unusual nature of the writing, was perhaps pardonable, so no explanation was forthcoming. Beckett, appalled at the error, sent the second half of the story again; but now another editor, Simone de Beauvoir, chose to believe that the magazine had been hoaxed and refused indignantly to publish the supplementary half.

Beckett was naturally very annoyed about the affair, but was eventually prepared to overlook the matter, admitting that with writing of this density such a mistake could be made. De Beauvoir, however, now changed her ground. The magazine, she said, should not publish the second half because it was not in keeping with its general outlook or policy. This was a nasty impasse, but fortunately the working editor of the magazine, Paule Allard – whose mistake the whole thing may have been – wanted to patch things up and asked Beckett for a contribution to a later issue. The result was that the twelve poems written in French in the rue des Favorites before the war appeared in the November issue.

One further work belongs to this preliminary stage, his first play since the abandoned *Vanity of Human Wishes* about Samuel Johnson. This was *Eleutheria*, a strange and not altogether unworthy companion piece to *Godot*. It is a drama in three acts which belongs to the early part of 1947. By comparison with the later plays it is almost a conventional drawing-room drama about a middle-class couple called Krap who have a rebellious son, Victor. *Eleutheria* is a full-length piece with seventeen characters. By means of a divided stage the family way of life – conventional, comfortable and upper middle class – is contrasted with that of Victor, an angry and unhappy young man who lives in a hotel bedroom. Since in the first two acts we see the over-furnished living room of the family on one side of the stage and the squalid hotel bedroom on the other, the contrast is insistently brought to our attention. In the third act, however, Victor's bedroom occupies the entire stage. A weakness of the play as straightforward drama is that Victor seems unable to state the reasons for his rebellion against his parental background: indeed, as far we can discern he has no reasons except disgust. Like Beckett for much of his life, he is not a dedicated artist. He is simply somebody who has cut adrift and knows what he does not want rather than what he does.

In this the play bears some resemblance to the English drama of rebellion such as John Osborne's *Look Back in Anger*, which was to become fashionable a few years later. But, as the glazier who comes to mend his window tells Victor in one hilarious scene, his trouble is that nobody knows why he will not take his place in the ordinary world.

If they could only find a reason, he would be seen as a martyr and all would be well.

At that point you begin to live. Nobody is any longer of a mind to lynch you. You are the poor young man, the heroic young man. You are seen croaking like a dog at thirty, at thirty three, drained by your labours, by your discoveries, eaten away by radium, laid low by the sleepless nights, by the privations, died on mission, shot by Franco, shot by Stalin. Your praises are sung. Mother is dying of grief, girl-friend as well, it doesn't matter, men like you are needed, men with an ideal . . .

But as it is, Victor is simply seen as a disgrace to the family, though at a certain point in the play he rather limply says in his own defence, 'Perhaps it is time that someone were simply nothing.' M. and Mme. Krap's understanding of their son is naturally limited and neither of them are very sympathetic characters, though, as in Beckett's own case, the father seems to have more sympathy with the son's state of inaction than the mother has. 'Eleutheria' is the Greek word for freedom and this is what the play is about. Victor claims the freedom to do nothing, to disappoint everybody's expectations. In spite of the number of characters and the difficulty of the sets, *Eleutheria* could be seen as produceable, but it never achieved a production although it was some-times a near miss.

In the early part of 1947 Toni Clerkx had begun to send it out and reactions were not always unfavourable. Some producers thought it was too long, others that it had too many characters and would be too expensive to mount. One of those who thought it too long was Jean Vilar of the Théâtre Nationale Populaire, but he thought he might produce it if Beckett would rewrite it so that it had fewer characters. This Beckett refused to do, a brave act at the time; but in retrospect his obstinacy seems fortunate. If *Eleutheria* had been produced before *Godot*, it would almost certainly have been a flop. *Godot* came like a revelation. *Eleutheria*, with its multiplicity of characters and its conven-tional setting, would have seemed like an ordinary play gone wrong; and if it had been produced it might have prejudiced the chances as well as clouding the strangeness of *Godot*.

In May 1947, when *Eleutheria* was still going the rounds, Beckett summed up his position in a letter to his English agent George Reavey. *Murphy*, he told him, was due out shortly from Bordas, 'my own translation and not a very good one'. *Mercier et Camier* had also been accepted by Bordas and was expected out in the autumn. *Quatre*

367

Nouvelles was ready but had not yet found a publisher. Two of these stories had appeared separately – 'Suite' in the July number of *Les Temps modernes* (he did not say only in part), and 'The Expelled' in the review, *Fontaine*. Thirteen poems in French, dating from the late thirties, had also appeared in *Les Temps modernes*. And there was the play, which Toni Clerkx was handling.

During the period of intense activity which had now begun he worked every day. In the evenings he still would go to the cafés and bars, seeing some of the Montparnasse acquaintances he had had before the war, including Giacometti. He would also sometimes see Irish friends such as Desmond Ryan, an Irish land-owner who earned part of his living as a journalist in Paris, writing a Paris letter for the *Irish Times*. One night when Ralph Cusack was in Paris the three of them went drinking, winding up in the early hours of the morning on the terrace in front of the Basilica of Sacré-Coeur, that symbol of Catholic triumphalism which looks over Paris from high up on Montmartre. Huge numbers of chairs to provide extra seating for services are sometimes stacked here; and at one point Cusack, on a sudden impulse, hurled a chair down the several flights of steps which descend from the cathedral; then all three free-thinking Irish Protestants joined in the game, hurling chair after chair down the long flights with whoops of glee.

Suzanne never accompanied him to the bars, except occasionally to the Dôme at lunchtime. She did not drink: the very smell of wine made her ill, and her attitude to his drinking varied from a silent tolerance, to concern, to stern and outright disapproval – sometimes, though not always, expressed in words. The intimately shared life of the war years was now past. Though she sometimes felt, according to Josette Hayden, that she would like to have a child, he was adamant that this was out of the question. Socially and in other ways they were drifting apart. They were, she told Josette, a 'ménage à celibataire'. They would go to concerts together and occasionally to small dinner parties, but that was all. When his friends called to the rue des Favorites she would bring tea or coffee, exchange politenesses and then disappear again. It was understood that Suzanne did not speak English and Sam's English-speaking friends did not address her in that language, though some were inclined to suspect, both then and later, that she had rather more of it than she admitted to. When he rose, which was usually about noon, he would make his own breakfast, often scrambling a couple of eggs. On the nights when he went to the bars she would have been long asleep when he returned. From now on their social lives would differ considerably, Suzanne having her own circle of friends and he his. As time went by and some people who began as his friends were drawn into her orbit, it became a commonplace to advert to the fact that you could not be a friend of both parties. At a certain point you virtually had to choose between Suzanne and Sam.

But theirs remained a very close relationship on a certain level, if, as one might expect from two such individualistic people, an extraordinary one. Suzanne understood his writings very well. A short story of hers, 'F—', which Beckett translated and which was published in *transition* in January 1949, is so astonishingly Beckett-like that he is suspected of having written it. But a subterfuge like this would not be characteristic of either of them.

He was, as usual, complaining of various ailments: prolonged colds, referred to as flus, with pains in the joints of arms and legs, persistent headaches, neuralgia and sleeplessness; and as usual it is difficult to say how much of this was real and how much imagined, let alone whether the causes of what was real were physical or psychological. Drink may have been a factor, though drinking bouts such as the one with Cusack and Ryan were rare when his work was going well, and heavy drinking was usually a sign that he was stuck. At all events, a doctor he consulted could find no physical causes. Suzanne was probably aware by now of his tendency to suffer from ailments which were psychosomatic in origin, but she was herself a health faddist, so in a way, illness was grist to her mill. The meals she cooked when they ate together, which was by no means every night, were vegetarian and she was given to following the latest fashion in health diets and specifics. Newly enthusiastic about homeopathy, she consulted a homeopath about Sam's health; he suggested that he might need a change of surroundings and way of life. Some of Beckett's acquaintances were inclined to blame Suzanne's own ministrations and notions of nutrition for his general state of health, but they did not know that he had always been plagued by psychosomatic illnesses.

As a result of the homeopath's advice they took a cottage in the country for the months of October and November 1946, in Avondant in the *département* of Eure-et-Loir. Once installed there, Beckett found his condition improving – little wonder since his way of life also improved. He drank far less, walked more and in more pleasant surroundings, got up earlier and ate at home in the evenings. Altogether it was a soothing and revivifying regime which they were able to share and enjoy together in a way that they had largely ceased to do in Paris. The frenetic pace with which he had been working also slowed down in the country, where he had to give more time to Suzanne and this probably also produced an improvement, for Beckett was now descending deep into himself for his writing and the strain was considerable. If he had never been a neutral, superficial or impersonal writer, describing external circumstances and events which had little to do with his own history or psychology, he was even less so now. The kind of writing on which he was currently engaged involved a profound exploration of the self and a prolonged living on the nerves while the exploration was being conducted.

Back in Paris for the winter, however, his ailments returned and, with them, money worries. His way of life was always pretty spartan and this suited Suzanne more than it might have another kind of person; but their lifestyle was undeniably frugal and would remain so until he began to earn some money from his writings. Suzanne was a talented dressmaker and her work, making baby clothes, produced a small income which subsidized the household budget. Although they were on short commons, they were at least provided with the basic necessities of life by his private income and this allowed him a freedom as a modernist, avant-garde writer which was a factor of enormous importance in his life. It also had an influence on his work, in which there is never a question of characters earning their living and in which almost all the more or less free-floating characters are supported or looked after by somebody, however meanly. Molloy's communication with his senile mother is limited to knocking on her forehead: one for yes, two for no, three I don't know, four money, five goodbye. 'Dish and pot, dish and pot, these are the poles,' says Malone, the dish being supplied and the pot removed and returned by 'the woman'.

With the completion of *Mercier et Camier*, *Eleutheria* and the four long stories or *Nouvelles*, the first burst of work was over. He had fallen back into the pattern, established before the war, of promising his mother as he returned to Paris after a visit to make another before the year was out. In May 1947, in fulfilment of this promise, he set off for Ireland again, stopping off in London, where he discussed the present situation of his writings with George Reavey. *Watt* was still unpublished and so far Reavey had had very little success with it. From Ireland Beckett wrote to Reavey that *Watt* was 'an unsatisfactory book, written in dribs and drabs, first on the run, then of an evening after the clod-hopping during the occupation'. But it would have its place in the series of novels that was still to come; and at this point Watt appears in *Mercier et Camier*, also unpublished, surprisingly drinking pints, though in *Watt* itself he disdains interest in alcohol. A further letter from Ireland tells Reavey not to bother any more about *Watt* 'unless you feel like it', perhaps because he had now begun a new work.

Maybe out of remorse for his conduct during his last visit, or because shortly after arrival he stumbled on the beginnings of this new novel which excited him very much, in Ireland he sat dutifully with his mother at constrained mealtimes, walked, and visited his relatives. May's condition was certainly worse, the shaking in the hands making it difficult for her to handle tea things or cutlery at the table. Beckett consulted Alan Thompson about her and was given little hope, since there was no successful treatment for Parkinson's, though one would be developed in the United States a few years later. Under Suzanne's influence he had brought a homeopathic remedy with him from Paris and asked Alan Thompson to administer it and to observe the results, which Thompson, with some reluctance, agreed to do. When he went to the National Gallery to see an exhibition of French painting Beckett was pleased to find his friend Geer van Velde was represented. The visit served to remind him of the attractions of the gallery and he went back again and again during his stay. Once more he was having trouble with his teeth and again he saw his dentist friend Andrew Ganly about them. Ganly did what he could but he felt that sooner or later they would all have to come out and be replaced by dentures. At the end of June Beckett went back to France.

Maurice Sinclair and his wife, Mimi, had rented a villa at Menton on the Riviera near the Italian border for the summer and they invited Sam and Suzanne to stay. Somewhat uncharacteristically they agreed to go – but when they got there rented another house themselves in the avenue Aristide Briand because Sam had brought with him the exciting new work he had begun in Ireland and was eager to get on with it. The weather was poor, with a lot of unseasonable rain, and he wrote every day. When the weather permitted they cooked their meals in the open air over a charcoal brazier and the rest of the time just sprawled around the big, sparsely furnished house and its garden.

At Menton he received a letter from Reavey saying that Hamish Hamilton, who had been at school with Ralph Cusack and had heard from him about Beckett, was interested in *Watt* and would like to know his future plans. This placed Beckett in a bit of a dilemma. He was writing in French and had no intention of doing otherwise, but he could hardly tell that to an English publisher who thought there might be other books to publish, justifying the loss he would almost certainly incur on *Watt*. In the end he wrote ambiguously to Reavey, saying he had nothing to offer in English at the moment except translations from his own work, but 'perhaps, to encourage him with *Watt*, I should say I expect soon to resume writing in English, than which until now few things are less likely.' Then he added a coda. 'I am getting on with another book in French, entitled probably *Molloy*,' he said. *Molloy* was the work he had begun in Ireland, brought with him to the South of France and now carried back with him to Paris where the real 'siege in the room' was just about to commence.

Afterwards Beckett would say that 'Molloy and what followed' became possible 'the day I became aware of my stupidity. Then I began to write the things I feel.' In Molloy and most of what followed he would employ a narrator, a fictitious creation whose life and circumstances were not Samuel Beckett's life and circumstances and who felt things Samuel Beckett did not necessarily feel. In other words, the three books beginning with Molloy and now generally known as the trilogy would be novels or, if one prefers, fictions or, to be more accurate, a series of fictions, for in these books there are fictions within fictions and fictions within them again. Some of Beckett's narrators make up stories in which there are characters who in turn make up stories, and so on.

But the distilled vision of life and the human situation would now at last be Samuel Beckett's, not something imposed on Samuel Beckett by existing literature, adopted by Samuel Beckett out of a desire to imitate, to please or to impress. And so these books would be 'true' in some sense or another: 'true' to Samuel Beckett's experience and vision in any case; true, as he put it, to 'the things I feel'.

But it is a strange paradox of literature and the literary art that it is almost easier to arrive at the truth by means of a fiction than in any other way. Herman Melville, a great creative literary artist, said that he would have preferred 'the still, rich utterance of a great soul in repose' to the plays of Shakespeare. In other words, whatever aesthetic satisfactions he may have derived from Shakespeare's plays, he regretted the need for such creativity, regretted that Shakespeare had had to invent all those dramatic fictions with their poisoned sword-tips and discovered letters when he might have expressed his thoughts and feelings directly and simply, without masks or personae.

But great souls in repose, such as Melville imagined Shakespeare might have been had he not been involved in the turmoil of dramatic creation, souls dispensing with illusionist structures, messages from ghosts or hidings behind the arras, are too often inclined to be silent, perhaps even tongue-tied souls. Literature has very little tolerance for abstraction, even abstract, distilled wisdom. Though there are a handful of essayists and aphorists who have tackled experience more directly than the novel seems to do, even in their case there has to be an embodiment of some kind, a 'showing forth' by way of illustration or anecdote. And where the more difficult, murky and not easily fetched-up 'truths' of art are concerned, the prose embodiment usually has to be a fiction. Whatever may be the case with poetry – and one could argue that this has the same problem – for extended prose works, one might almost say that artistic truth is only to be arrived at by the creation of fictions.

These fictions are often nowadays undercut from within. It has become a commonplace of the post-modernist age to declare that your fiction is merely a fiction while still engaged in the very act of creating it. It has become almost a commonplace to mock at it or jeer at it in the very moment of writing it, to declare your dislike of the necessity, or at least your dubiety about it.

Beckett was not only one of the first to do this – another was his great Irish contemporary Flann O'Brien – he remains the most thoroughgoing. In the work of no other author does hatred for the necessity of creating a fiction shine through so clearly or is detestation of that necessity expressed with so mordant a wit. The trilogy may be seen as a series of subsidences. The creation of a character is followed by the character's collapse or merging with another. The more successful the creation seems to be as an act of creation – in other words, the furthest from its author – the sooner disgust and boredom set in. Malone creates a character called Macmann, and for a while he is happy to amuse himself with this character's story. But soon Macmann becomes, all too obviously, Malone: 'What tedium. And I call that playing. I wonder if I am not talking yet again about myself. Shall I be incapable, to the end, of lying on any other subject?' In the last book, *The Unnamable*, there are successive collapses of pretence, peelings away, monologue within monologue; and it is here that the subsidences are explained. Molloy crawling through the forest, Malone lying on his bed, Mahood in his jar under the tarpaulin, Worm, the reduction of them all whom no threat or torment can goad to action or even, seemingly, bring to life, are all, we now learn, fictional creations of the Unnamable.

But he is not fitted for fictional creation. He cannot keep it up.

Pretence, or creation, is the converse of truth. For that reason, if for no other, it is synonymous with boredom, disgust and collapse. Yet there is an incomprehensible necessity for creation and pretence. All these masks or personae, these attempts to create and sustain fictional identities, surrogates for himself are, it seems, made at the bidding of an obscure *camarilla* of masters whose nature is never clear but whose rule must be obeyed. It is they who have ordered him to create these storytellers and their stories. It is they who compel him to engage in this weary frenzy of creativity.

But this is not all, for behind them there seem to be others, or at least an other. And this other, or others, does not want fictions, or wants them only as a means of arriving at the truth. The object of the fiction must be truth of some sort; but by definition it is necessarily a lie. And it is this contradiction, this involvement in seemingly contradictory necessities, which is his principal torment, 'a labyrinthine torment that cannot be grasped, or limited, or felt, or suffered, no, not even suffered'.

There is on the one hand the command to give utterance, to be verbal, even loquacious – in other words, creative. But this loquacity, this creativity, almost necessarily involves fictions – that is, lies. And at the same time there seems to be a command, or a necessity to utter something important, something which once uttered will give him quittance, will restore him to his primal self:

> It is difficult to speak, even any old rubbish, and at the same time focus one's attention on another point, where one's true interest lies, as fitfully defined by a feeble murmur seeming to apologise for not being dead. And what it seemed to me I heard then concerning what I should do and say, in order to have nothing further to do, nothing further to say, it seemed to me I only barely heard it, because of the noise I was engaged in making elsewhere, in obedience to the unintelligible terms of an incomprehensible damnation.

His obedience to these contradictory commands seems to be a question of expiation. There is some obscure sin, perhaps the sin of mere being, of ever having been, for which amends must be made. And for him at least this expiation 'boils down to a question of words'. It could have taken other forms and sometimes he wonders what the form of expiation that is required of him might have been if it had not been a question of words. A simple job perhaps, sorting and arranging things, or carrying water in a thimble from one set of containers to another, tanks perhaps secretly connected by underground pipes, so that his work could be undone as rapidly as he did it. But because of his disabilities, for him it has to be a question of words. After all, he has no hands – and hence cannot be required to keep up a continuous clapping noise, as if calling the waiter – or legs – and therefore can scarcely be expected to dance the Carmagnole.

This need for expiation, of course, relates to Beckett's feeling that mere being was in itself an offence, a feeling powerfully reinforced by Schopenhauer. No doubt his Low Church Irish Protestantism, so near to Calvinism, had something to do with this too. If you were not among the elect, being was an offence. And how could someone of Beckett's temperament feel himself to be among the elect?

But since for him it is a question of words, it would seem that this is one of the things the trilogy is 'about': the position of the artist – more specifically the literary artist – who has been commanded, or seems to have been commanded, by an obscure power or junta of powers to give utterance, to create, but from whom not mere creation alone but also some sort of ultimate truth is demanded. And from the trilogy on, most of Beckett's works would involve people who are commanded by voices they do not know anything about to give voice themselves, to tell their story – or perhaps just any story. And they cannot even be certain that this is the ultimate commandment, for beyond and above the powers that thus command them there would seem to be an other or others enjoining them to be silent, or to earn silence by speaking some sort of ultimate truth.

Beckett was a perfectionist. Most artists are; and indeed perfectionism, à la Eliot and Joyce, was a feature of modernism, which rejected both the journeyman's code of 'needs must' and the slapdash, hit-and-miss methods of the great Victorians. But he was a perfectionist to a degree which was unusual and obsessive even among the modernist masters. He yearned for silence, the blank white page, the most perfect thing of all. As an artist he had had more false starts and false beginnings than most. The principal failing of his earlier work, so knowing but also so self-revealing in all the wrong ways, is the failure to achieve a form and a tone of voice which would allow him to express his particular truths. Perhaps this repeated failure made him feel more acutely than most the torment of marred utterance, of false utterance, of would-be significant utterance; and to feel also more intensely than others that the object of true, achieved and necessary utterance is silence – in some sense or other, a permission to be silent, whether granted by one's daemon or by one's creator.

This is something that many artists feel but few have expressed with such intensity and such longing. As if to make that longing clearer, and as if the hope was growing in him that he was about to achieve its end, his works would after a certain point get shorter and shorter, the fact that he was composing work whose crown and summation would be silence becoming ever more manifest.

But in the meantime, utterance, for whatever reason, has to be given. Stories have to be started and kept going. Even Moran, Molloy's counterpart, a man of action – besotted, like so many of his kind, with the thought of his own capacity and asinine in the conviction that capacity and application somehow amount to a morality in themselves – has to write a report of his search.

I have spoken of a voice telling me things. I was getting to know it better now, to understand what it wanted. It did not use the words that Moran had been taught when he was little and that he in his turn had taught to his little one. So that at first I did not know what it wanted . . . I understood it, all wrong perhaps. That is not what matters. It told me to write the report. Does this mean I am freer now than I was? I do not know. I shall learn.

In most cases the commandment seems to be in some sense to create as well as to utter: to create characters and circumstances and stories which will put these characters in motion. It is only through such stories that utterances of any length can be achieved, and length certainly seems to be part of what is required. Another requirement seems to be some form of externalization. The characters must be real fictions, differing from their creator. 'We are getting on,' says Malone after he has succeeded in creating the pimpled adolescent called Saposcat. 'Nothing is less like me than this patient, reasonable child, struggling alone for years to shed a little light upon himself, avid of the last gleam, a stranger to the joys of darkness. Here truly is the air that I need, a lively tenuous air, far from the nourishing murk that is killing me.' Even this may have a double edge, however, for Saposcat here sounds a little like the pre-revelation Beckett, exploring the self as Beckett had done in *Dream of Fair to Middling Women* and *More Pricks Than Kicks*, and convinced that it is in self-knowledge that wisdom lies. But in any case, the imperatives from which the storyteller suffers and the obscure commandants he seems to have been given are here again contradictory. For somehow the release is in the discovery of the self, and the reward of silence is only in the end given to those who have discovered it and saved it from being lost amid all the confusions of the world. 'If this continues it is myself I shall lose and the thousand ways that lead there. And I shall resemble the wretches famed in fable, crushed beneath the weight of their wish come true.'

And here is another part of the primary paradox of the literary artist's position. Just as it seems that the artist who is concerned only with the truth can arrive at it only by means of a fiction, that it is lies which are seminal, even of the truth, so it is only by the adoption of a fictitious mechanism and the entry into a labyrinth which may never lead back to the self that the self can be revealed. It is only by the saying of words on other subjects, 'even any old thing', that the words will come which will discover the true self and release it from its torment. And the fictions thus created must have as much substantive reality as they can be given. Otherwise the contradictory commandment would not be fulfilled. Though the trilogy and the *Nouvelles* which led up to it are an abandonment of the tradition of the realistic novel, the whole exercise would be pointless if the characters were not in some sense 'real' or real-seeming, if their circumstances did not have a certain logic and coherence and their stories carry some sort of conviction.

Beckett was aware of the medieval distinction between the literal, the allegorical, the moral and the anagogical meaning of the text and he invokes it to good purpose in *Proust*. The Unnamable's position may have allegorical parallels with that of the creative literary artist. The trilogy may have a moral dimension in as much as there is a view of life expressed, and an anagogical one in as much the position of Beckett's characters relates to the ultimate destiny of humanity. But the novels, since they are novels, must function in a storyteller's way also. The illusion, such as it is, must have some illusory power. Molloy's relationship with his mother, the forest through which he crawls, the fact that Malone is 103 years old, lying in bed in a strange room and writing with the stub of a pencil, and all the rest of it, must carry sufficient conviction to enable us to suspend disbelief to the extent that is required. And it is indeed part of Beckett's triumph that his text functions so well on a literal, realistic or, perhaps one should say, illusionist level while the other dimensions are still present.

The difficulty here was an extreme one and it was partly to do with the extreme circumstances in which he places his characters. It is undoubtedly difficult for the reader to believe in any literal sense in these articulate fellows who are yet so reduced in circumstance as to be beyond the pale of ordinary humanity. Of course, we are not required to believe in them in the sense that we would have to if Beckett had been writing naturalistic stories about *clochards* or tramps and creating what Molloy calls 'a wealth of filthy circumstance'. We are not given any 'filthy circumstance' as to their finances, precedent situation and acquaintance, the manner in which they arrived at their present state, etc. But, reduced as his characters are to the extreme simplicities of need and satisfaction, indeed by virtue of the fact that they are so reduced, Beckett does succeed in laying bare much of the reality of the human situation as well as the grossness of its perhaps necessary illusions. Some part of his strength is realism, about the body as well as about the soul; and a great part his cathartic power. Molloy enjoying the hospitality of the kind lady called Lousse; Malone waiting for the skinny old hand to push in his food and remove his chamber pot, scribbling away meanwhile in his exercise book with the stub of a pencil (a French pencil, he is careful to assure us); Mahood in his jar on the pavement outside the café near the slaughterhouse – they are real enough for Beckett's purposes. And while it may not be exactly the realism of Balzac, Flaubert or even of James Joyce, whose work is in many ways the culmination of the realist tradition, it suffices for the sort of conviction as well as for the sort of shock Beckett wishes to produce. In fact by reducing the element of circumstantial realism about his octogenarians, their diet of soup and their various paralytic ailments to the level he does, Beckett succeeds more brilliantly in making them part of our world than would be the case if these were literal accounts of the existence of *clochards* or crippled old men. We all have to eat, to shelter, to sleep and to make love after our fashion, which is seldom the fashion of the macho heroes or passionate heroines of another kind of fiction. Because Beckett's characters are at some remove from the particularities of the sociological-realist novel, they are also free to attain a degree of universality which would otherwise be denied them. They are and are meant to be different from each other, but the success of many one-man stage performers in amalgamating them into a single character shows that there is a 'Beckett man', whose differing incarnations have only about as much relevance as have the differing film incarnations of Chaplin's tramp.

By the time the 'Beckett man' was created the anti-hero had become a familiar figure in European literature, through the civil servant hero of Dostoevsky's *Notes from Underground*, Flaubert's Bouvard and Pécuchet and Madame Bovary, Kafka's K and, of course, Joyce's Leopold Bloom. Of these anti-heroes Beckett's is the one from whom all vestiges of the heroic or the admirable have been most thoroughly eliminated. Thus while Kafka's K has a highly developed moral sense which makes him anxious to do right and to see that it is done; and while Leopold Bloom has a great deal of love and fellow feeling in his make-up; the Beckett man is a lone individual who regards others with fear, hatred, impatience or contempt. And while Kafka's K and Joyce's Bloom have a certain amount of civic responsibility, the Beckett man has none. He does not believe in the brotherhood of man; and questions of equality are disposed of by the eager admission that he is, in all respects, inferior. He lays no claim to any virtue that can be named, except to a rather dubious humility and a too eagerly embraced resignation. His principal emotion when confronted with humanity at large is fear.

> Morning is the time to hide. They wake up, hale and hearty, their tongues hanging out for order, beauty and justice, baying for their due. Yes, from eight or nine till noon is the dangerous time . . . It may begin again in the early afternoon, after the banquet, the celebrations, the congratulations, the ovations, but it's nothing compared to the morning mere fun. Coming up to four or five of course there is the night-shift, the watchmen, beginning to bestir themselves. But already the day is over, the shadows lengthen, the walls multiply, you hug the walls, bowed down like a good boy, oozing with obsequiousness, having nothing to hide, hiding from mere terror, looking neither right nor left, hiding but not provocatively, ready to come out, to smile, to listen, to crawl, nauseating but not pestilent, less rat than toad. Then the true night, perilous too, but sweet to him who knows it, who can open to it like the flower to the sun, who himself is night, day and night.

The Beckett man has usually no past except, since he has been born, a mother or mother memory. He belongs to no recognizable community. He has no employment or qualifications for employment. Nor has he any sources of income except charitable ones. Molloy, as has been said, used to communicate with his mother by knocking on her skull even though he found it hard to 'ram this code into her ruined and frantic understanding'. Malone is dependent on an elderly female whose skinny hand pushes in his food and takes away his pot while he lies in a room which may or may not be in any sense his. The Beckett man lays claim to few talents or skills except those forced on him by the exigencies of his probably ludicrous and undignified situation. But he has a certain low cunning and, if set a useless problem, such as how to ensure that not one of sixteen sucking stones is sucked more than once in any one cycle, he will set about solving it with dogged method.

His attitudes are almost a complete reversal of those we have all been led, by our antecedents or our educators, to adopt. There is no hitherto esteemed notion, whether of honour, dignity, trustworthiness, courage, energy, ambition or goodwill, which he has not learned – sometimes by bitter experience – to scorn or to disregard. All systems of government are as alien to him as he is to them. He believes that most of what we call learning, or science or philosophy is a fraud, to be described as time-wasting except that it was probably invented to pass the time – and so is time-filling rather than otherwise. 'Yes, I once took an interest in astronomy, I don't deny it. Then it was geology that killed a few years for me. The next pain in the balls was anthropology and the other disciplines, such as psychiatry . . .' Born into an unending stream of instruction and exhortation, often maddeningly imprecise, very likely inaccurate, seemingly useless, the Beckett man believes that the most shameful moments in his career were those in which he was deluded enough to try to learn; and he holds by his rejections with a fierce tenacity, often catching himself in the nick of time, when another than he might have admitted to an emotion or a thought that was, in one way or another, worth the having or, worse still, to a virtue, such as faith, or hope or charity, the most ludicrous of these being hope. (The ludicrousness of hope is something on which both Schopenhauer and Proust had agreed, as he might even have pointed out.) The Beckett man's experience of erotic love is limited, alike by his ignorance of what is expected of him in such a relationship and by his incapability. As Macmann grows more intimate with his Moll he acquires 'some insight into the meaning of the expression, Two is company. He then made unquestionable progress in the use of the spoken word and learned in a short while to let fall at the right time, the yesses, noes, mores and enoughs that keep love alive.' After long trial and effort he and Moll finally succeed, 'summoning to their aid all the resources of the skin, the mucous and the imagination, in striking from their dry and feeble clips a kind of sombre gratification'. There is a terrifying passage in which Molloy wonders whether he has ever experienced what the rest of the world calls true love and recounts the experience he thinks may have been it. It was with someone who went by the peaceful name of Ruth, he thinks, but can't say for certain, for perhaps the name was Edith.

The Beckett man is utterly devoid of ambition and though sometimes, enraged, he may strike or beat to pulp with his stick or crutches someone even more helpless than himself, he does not seek power over others, or riches, or fame. He does not smart under indignity; it is axiomatic with him that one is insulted and humiliated at every turn. He is used to physical suffering and the degradations of the body; indeed he is something of a connoisseur of these and they might almost be said to give him pleasure. When Moran feels the first onset of paralysis in one leg and discovers the joys of immobility, he remarks:

Such are the advantages of a local and painless paralysis. And it would not surprise me if the great classical paralyses were to offer analogous and perhaps even still more unspeakable satisfactions. To be literally incapable of motion at last, that must be something! My mind swoons when I think of it. And mute into the bargain! And perhaps as deaf as a post! And who knows as blind as a bat! And as likely as not your memory a blank! And just enough brain intact to allow you to exult! And to dread death like a regeneration.

The background of these characters is empty. We encounter them *in medias res*; and though we are given some clues, there is very little prior circumstance and still less detail. There is a level of sophistication and a level of culture which might, if we were dealing in class distinctions and sociology, surprise us. Molloy understands that one can keep oneself warm by wrapping a journal round oneself under breeches and shirt, as indeed tramps do wrap old magazines and newspapers, but the journal he recommends is the *Times Literary Supplement*, which he believes admirably suited to this purpose. There is a general background of Irishness or at least a knowledge of Irish affairs which again, if we were concerned with nationality, might be thought significant. Moran likes Irish stew and knows something about Irish law. Malone is, like Beckett himself, familiar with the phrase 'Up the Republic!' and aware that it took the Lord Mayor of Cork, Terence MacSwiney, forty days to die on hunger strike; Worm will need his single eye to weep when he 'goes back to Killarney'; Macmann reflects that it is love that will lead him and his Moll hand in hand to Glasnevin, Dublin's principal cemetery, burial place of innumerable heroes and patriots. But one assumption about them may be as false, or as accurate as another. After reading the English version of *Waiting for Godot*, whose characters are also tramps or down and outs, Vivian Mercier remarked to Beckett that he thought he had made Didi and Gogo sound as if they had got Ph.D.s. 'How do you know they hadn't?' was the succinct answer.

Mercier also noted that photographs of the first Paris staging of the play, with which Beckett was intimately concerned, showed that Vladimir and Estragon were 'more shabby genteel than ragged'. Vladimir actually wore a stiff collar and tie and both had bowler hats, which suggested remaining aspirations to gentility. The bowler hat was of course *de rigueur* for male persons in many social contexts when Beckett was growing up in Foxrock (when he first came back in his beret, it will be remembered, his mother suggested that he was letting the family down by not wearing a bowler), and Willie Beckett commonly wore one. Of course, the bowler hat may to some extent derive from

Chaplin's tramp, who, like Beckett's characters, is of unexplained social status, or even from Laurel and Hardy, whose social background is also unclear. Like some of these other comic archetypes, the Beckett man is often courteous and mannerly, especially to ladies, and seems to be largely attired in the relics of old decency. Chaplin's tramp is a great hat-raiser when he meets members of the opposite sex, and Molloy also seems to have a good deal of punctilio in that regard. When he is reduced to crawling on his back, plunging his crutches blindly behind him into the thickets of the forest on his way to his mother, he keeps losing his hat until in a fit of temper he bangs it down on his skull with such violence that he cannot get it off again. At this point he remarks with some concern, 'and if I had met any lady friends, if I had any lady friends, I would have been powerless to salute them correctly.' Greater punctilio than that surely no man could have. Malone too feels that there are obligations in the matter of dress which should not be ignored. Lying helpless on his bed, he is visited by a mysterious stranger whom he first takes to be an undertaker's assistant concerned to measure him for a coffin. At a certain point the man backs away from the bed; 'it was then I saw he was wearing brown boots, which gave me such a shock as no words can convey.' If Beckett's men are tramps or *clochards*, then they are tramps or *clochards* who have come down in the world; but we must beware of their tramp status too as something which might be discussed sociologically.

There remains the question of how far Beckett's people are valid exemplars of human experience. He is of course a comic artist and the basic technique of comedy is exaggeration, everything pushed to extremes, calamity leading to calamity, the dire situation becoming ever direr. Stephen Leacock, about whom he had been enthusiastic as a schoolboy, was a master of this wild progression; and so were almost all the other great comic artists of world literature. Much of the trilogy is truly hilarious comic writing. A vein of caustic humour runs through it all, even through the most clogged and complex passages in *The Unnamable*. Its very syntax is a comic device, an exposure of stylistic mastery, of fine writing, of supposed narrative exactitude.

But comic creation or otherwise, one could argue that the Beckett man, in all his abysmal aspects, is 'truer' to humanity's real lineaments than most of what has gone before. Most of the 'great' literature of the past dealt with the heights of human experience. It took the exalted moments of lyrical emotion – of love, or even perhaps of hate – and it sought to enshrine them in words which matched their intensity. It celebrated those virtues which were proper to heroic struggle or great achievement – courage and loyalty, will-power, ambition and resolution among them. Even when it dealt in crime or villainy, these were not absent. For 3,000 years the bias of literature had been tilted one way, towards the heroic and the lyrical-poetic. Now it has been tilted the other, a process with began with the appearance of the first modern anti-heroes and culminated in Beckett.

Deep in our collective soul is a continuous unease about the contrast between the traditional ecstasies, nobilities and romantic passions of literature and what most of us actually feel, the state of mind in which most of us actually live most of the time. In its exposure of this gap Beckett's work has a profoundly cathartic effect. It may not be all there is to be said, but it is a needed addition to the sum total of what has been said. His work is a liberation from certain pretences; and, partly because it is a liberation, the experience of reading *Molloy*, *Malone Dies* and *The Unnamable* is a joyous and revivifying one.

Molloy took most of 1947 to write, the title character's story turning into Moran's at half way. Moran is an oddity among Beckett's men, since he has some sort of profession, whether as secret agent or otherwise, but by Christmas he too had been reduced to a state of dereliction, and as soon as it was done Beckett plunged straight away into the composition of *Malone Dies*. At this point he thought there were going to be two books, not three, and in the early part of 1948 he wrote to MacGreevy saying that he saw 'a little clearly at last' what his writing was about and expressing a doubt as to whether he had enough courage and energy to get the job done. 'The feeling of getting oneself in perspective is a strange one after so many years of expression in blindness.'

Although at this time he envisaged only the two novels, since he had now well established the author behind the author idea in his own mind, he decided that they were part of a series which had begun with *Murphy*. When he wrote *Murphy* he had had no notion of this form of continuance. Now, however, he began to introduce references to *Murphy* into *Malone Dies*. *Murphy* too had become part of his ultimate narrator's creative past, one of his earliest attempts to externalize himself and give utterance through fiction. This of course makes it plainer than ever that the author who is attempting all these externalizations is Samuel Beckett, as indeed he was. Criticism has on the whole rightly decided to ignore these attempts to bring *Murphy* into the scheme, but they are important in establishing how much Beckett's previous failures at fictional externalization had to do with the trilogy's underlying theme. What is perhaps more interesting is that there appears to have been no concept of the Unnamable as the ultimate author until he was well into *Malone Dies* – certainly not during the composition of *Molloy*. The trilogy was by no means a planned and executed work from the beginning. To MacGreevy he described *Molloy* as 'a long book, the second last of the series begun with *Murphy*, if it can be said to be a series. The last (*Malone Dies*) is begun and then I hope I'll hear no more of him', meaning the central character. The first French edition of *Molloy*, though published after *The Unnamable* had been written, took no notice of the later book as part of the scheme. 'This time, then once more I think, then I think it will be over and with the world too' stood in the French edition, which was published in 1951, and was not altered to take account of *The Unnamable* until the English edition in 1955, when it read: 'This time, then once more I think, then perhaps a last time, then I think it'll be over, with that world too.'

At the beginning of 1948 he looked back on what he had done, noting that in under two years he had written four long short stories, two novels and a play, but of all this work only two short stories had been published; and since he also had the unpublished *Watt* still on his hands it was obvious that a serious gap was opening up between his inner and his outer existence. Yet he pressed on: the completion of *Malone Dies* would mean that he now had four unpublished novels, an unperformed play and two unpublished *Nouvelles* in the apartment in the rue des Favourites. No wonder he was concerned with the incomprehensible imperative to create, for it was beginning to look as if he would never have any other reason for writing than a dumb obedience to it. His works, as he put it, went out into a void and he heard no more about them. In that obedience, though, he was surely now displaying literary heroism of the highest order. And it was not by any means all frenetic writing without any stoppages. Two months into *Malone Dies* he found himself completely stopped, filled with the desire to write but up against a complete inability to do so. He would sit at his table sometimes for two of three hours without being able to put down a word, mentally stuck in the room where he sat and unable to descend into the strange world where his creations had their being.

And there were circumstantial obstacles as well as mental blockages. Because of post-war currency difficulties which he did not understand, he could not get his money through from Ireland; and even when he did, his allowance was worth far less than it had been because of the revaluation of the franc. It was just about possible for a single person to survive on 40f. a week in Paris in the post-war years, but he was paying half this in rent which left him only 20f. Suzanne was still dressmaking and she gave occasional piano lessons, though blessedly not in the apartment. But it was not really possible for two people to live and pay the rent on what was coming in, so Beckett had to consider ways to supplement this income. Driven by necessity, he had recourse to James Joyce's old stand-by and informed those people who he thought might know of possible pupils that he was willing to give English lessons. His mother, it will be remembered, had once tried to prevail on him to give French lessons to the young ladies of Foxrock, something which he hated doing and eventually refused to do, but giving English lessons in Paris was certainly a more reputable expedient for a literary man. Through his friend Jean Thomas, who had risen high in UNESCO, he also tried to get translating work, but these efforts did not bear fruit immediately, though in late 1949 they did result in a book of translations from the Spanish. Nothing happened for some months, but Thomas had not forgotten and eventually he was commissioned to translate upwards of a hundred poems by thirty-five Mexican poets. Although struggling with *The Unnamable* at the time, he completed the task. In 1960 he wrote to Hugh Kenner, 'That lousy Mexican Anthology was undertaken to take the chill of the pot in the lean winter of 1949–50 and with scant Spanish. It had to be finished by April 1950 at the latest, immediate publication. Then nine years later when I thought all danger past and its merits discerned it comes rattling its doggerel at me.' What specially annoyed Beckett was being credited with the selection of poets and poems, which had in fact been made by Octavio Paz.

Some of the Joyce circle were now back in Paris and Madame Jolas suggested to him that he might help her to edit and put the final polish on her husband's translation of René Char's poetry, but he did not want to be associated in this manner with someone else's work. He had allowed most of his contacts with the Joyce circle to lapse. It was a part of his life that was over and, Joyce himself being dead, he had no great reason for associating with the little band whose principal claim to fame was that they had been the master's loyal admirers. Since his return to Paris at the end of the war he had, however, been visiting Lucia at Ivry. He naturally found these visits a strain and, whether because he wanted to or otherwise, he now convinced himself that they were a strain for her too and that they were impeding her possible recovery. Nor did he want to be living in the past. For the first time in his career he now felt himself to be his own man, functioning as a writer in a way that was completely his – not Joyce's or any other's – though of course as the results were not yet published nobody knew him as such. The Joyce circle would have tended to regard him as one of themselves, a mere servant of a reputation and a memory; and indeed when he went out to the cafés at night those young Americans who were now drinking in the Dôme or the Select would point him out to each other as someone who had once been the great man's secretary. Since they too were in search of the past, this tall, gaunt man, who frequently sat alone, staring moodily in front of him, was of interest, but they saw him as a mere survivor of something that was over, a beached and rather pathetic figure who was something of a joke.

Nor was he in any way part of the excitements of the post-war existentialist movement, or of its controversies. Camus's *La Peste* was hailed as the great post-war novel in this year, 1947, confirming the reputation established by *L'Étranger* during the war years. The latter book had begun with the famous and somewhat Beckett-like sentences: 'Mother died today. Or perhaps yesterday'; but its author had now moved from 'the cycle of the absurd' to 'the cycle of revolt', and the doctor who is revealed as the narrator of *La Peste* at the end of the book had assumed the responsibilities of action.

This was also the year of publication of Sartre's 'Qu'est-ce-que la Littérature?', a series of articles in which the editor of *Les Temps modernes* adumbrated a Marxist-existentialist theory of literature which would explain it as a function of the position of writers in the socio-economic systems of their time. The basic contention of these pieces was that changes in the economic and social position of the writer vis-à-vis his public could explain the prevailing conceptions of literature at any given point. Writing was a social act and the writer could choose for himself to go along with the social movements of his time or to oppose them. But if the work was to have any value, the writer's point of view must also coincide with the movement of history. In 1947 the French writer, almost by definition a bourgeois, had a choice. He could side with the emergent class, the proletariat, or retreat into his own original system of class values, thus missing the march of history and being ultimately left behind by it. It was of course a determinist position, for history had already determined certain matters without a shadow of doubt. In the great social struggle actually proceeding the proletariat would be victorious and its victory would usher in the future era, so that was that.

But this apparently determinist position was given an existentialist gloss by the introduction of the element of choice. For what it was worth, one had a choice: to fall into step or not to fall into step. How much or how little of this was understood by the young people who had been flocking since 1945 to Saint-Germain-des-Prés and indulging their existentialist anguish in the Flore or the Deux Magots is doubtful. Years later Sartre would describe them as a generation who had turned themselves into 'corpses'; yet although Guillaume Hanoteau, the historian of Saint-Germain-des-Prés, probably spoke the truth when he said they had become his followers 'under a misapprehension', they had nevertheless made Sartre the hero of the hour. Beckett, who was about the same age as Sartre and seven years older than Camus, was nobody's hero and he had no followers, or even readers. Whether he was retreating into a bourgeois system of class values or not – and after everything has been said about his protagonists' apparent freedom from the necessity to work and obliviousness to the march of history, it would certainly be straining definition to a risible extent to say that he was – he could only go on doing it. During these years he remarked more than once on how ill-at-ease he felt in the France of the immediate post-war era. It was not somehow the France he knew. 'It is hard sometimes to feel the France that one clung to,' he remarked to Mac-Greevy. And whatever sympathy he might, if he knew or cared, have felt for the earlier, more philosophical beginnings of existentialism, his own basic problem in regard to existence was an inclination to doubt its very nature.

In discussing this problem he sometimes referred to the idealist philosophy of the Irishman George Berkeley, who doubted if the existence of the physical, objective world could be proved, a suggestion which has always lapped round the edges of European philosophy, present in St Augustine, inherent in Descartes' *cogito*, discussed at length by Kant and being intrinsically present in Leibnitz's windowless monads. With Beckett it was more a feeling which haunted him than a philosophical idea; and it was allied with the sensation of never having been born but yet different from it. A phrase he used to describe it was 'existence by proxy', the inability to take a step without feeling that someone else was taking it. In most situations one went through the motions while having a feeling of 'being absent', a by no means uncommon feeling, if not to the degree that Beckett felt it. And along with this sense of existence by proxy went 'an unconquerable intuition' that there was another presence beside one's own, 'a presence, embryonic, undeveloped, of a self that might have been but never got born, an être manqué'. This feeling is of course specifically strong in the trilogy's attempts to call this other self or these other selves forth; and it is also part of Beckett's scepticism about the sort of richness and fullness of experience we find in so much literature, and undoubtedly more often in literature than in life. In most situations his narrators fail to feel what is expected of them. They are willing to pretend and anxious to please, but that is all. The attempt to find the true, lost self Beckett described in 'images of getting down, getting below the surface, concentrating, listening, getting your ear down so you can hear the infinitesimal murmur'. Paraphrasing what he said, Harvey described it as 'a grey struggle, a groping in the dark for a shadow'; and added 'on another occasion he said the encounter was like meeting oneself, like approaching home'.

In speaking thus to Harvey of striving to hear the infinitesimal murmur and groping in the dark for a shadow, Beckett was almost directly echoing the words of some of his narrators. And of course throughout his work there is perpetual going forth and coming back with an implication of meeting oneself on the way. *Molloy* opens with a haunting image of two figures passing and re-passing each other on a country road near a city, evidently in the foothills of the Dublin mountains, one sometimes going out, the other coming back, sometimes the other going out and the first one coming back. Once, on meeting, 'They turned towards the sea, which, far in the east, beyond the fields, loomed high in the waning sky, and exchanged a few words', but that is the extent of the communication between them. However one describes Beckett's doubts about the very nature of existence, his conviction of what he called 'the authentic weakness of being', there is no doubt that many people who have no philosophical terms for it share this feeling of being 'not there' when supposedly important events are happening, or feel at least the gap between expected emotion and what is actually felt – a gap widened, if anything, by literature, which has a vested interest in the intense and lyrical and has poured forth its descriptions of the supposedly proper feelings for various occasions with a profligacy which has probably on the whole weakened literate mankind's capacity to feel anything at all, or at least anything of the poetic kind.

But to explore one's own 'not thereness', one's sense of 'existence by proxy', to call into question – as, by implication at least, he was now doing – the whole corpus of lyric emotion, the pretence that we are all intensely present and intensely responsive in so many situations, that we experience richly and to the full the richness and fullness of the world – was a task likely to take a ferocious toll, if only because all these pretences appear to give the rest of humanity so much comfort. And in the case of a literary artist like Beckett it had to be done while obeying the command to create. 'What complicates it all is the need to make. Like a child in mud but no mud. And no child. Only need,' he told Lawrence Harvey.

Towards the end of the final revision in typescript of *Malone Dies* he grew pale and thin and Suzanne and such friends as he still saw were worried. Rather surprisingly, he decided to go to Ireland before it was quite finished and the decision shows that besides the imperatives of creation there were other imperatives still at work – primarily of course those lifelong ones involved in his relationship with his mother. 'All my life, I think, I had been bent on it,' says Molloy of the necessity to go to his mother and 'settle the matter' that was between them.

Yes, all my life, I think, I had been bent on it. Yes, so far as I was capable of being bent on anything at all a lifetime long, and what a lifetime, I had been bent on settling this matter between my mother and me, but had not succeeded. And while saying to myself that time was running out, and that soon it would be too late, was perhaps too late already, to settle the matter in question, I felt myself drifting towards other cares, other phantoms.

In Ireland, however, he finished *Malone Dies*, typing every day in his room in New Place with its view of the Dublin mountains, describing it to George Reavey as 'the last I hope of the series. *Murphy, Watt, Mercier et Camier, Molloy*, not to mention the four *Nouvelles* and *Eleutheria*.' Though he had begun *Molloy* there, it was the first time he had taken a major piece of work to Ireland with him. Relations with his mother were good throughout and although, as is the way when people get older, there was no longer such urgency or anxiety on her part about his future as there had been ten war-torn years before, the fact that he was busy typing in her house did no harm to the image of him as writer and translator which she now had.

And whatever the truth of that image, in fact Suzanne was now engaged in searching for a publisher for the books and a producer for the play. Earlier in the year he had hopes that someone would sooner or later be found who would be interested in the play at least. Now a young publisher had expressed an interest in the novels and Beckett told Reavey, 'I am preparing him for burial.' But it is syntactically unclear whether this refers to the putative publisher or to *Malone*. In any event there were the usual long delays inevitably attendant on alternative publishing and fringe publications, but the decisions were always finally negative. He returned to Paris still haunted by his creations and still enmeshed in his own rhythms. It was a beautiful autumn and he tried to relax but, perhaps because the catharsis of publication was lacking, he could not do so. Now as he struggled to get 'back into the light', he found there was another work waiting for him which he was later to say he began 'as a relaxation from the awful prose I was writing at that time' and in an attempt to escape from 'the wildness and ruleless-ness of the novels'. This was a play, *Waiting for Godot*, the writing of which he described as 'a marvellous, liberating diversion'. Years later he would say to the actor Jack MacGowran that it had 'saved his sanity'.

Waiting for Godot is still by far the most famous and the most popular of Samuel Beckett's works. Since for many people it is a first introduction to his peculiarly gloomy view of things, it may seem strange to describe it as having a holiday atmosphere, but so, in a way, it has. One of Beckett's most notable characteristics is his ability to make truly funny jokes about the genuinely worst aspects of human existence, and nowhere is this talent more evident than in *Godot*. One of its first admirers, the deservedly popular French playwright Jean Anouilh described it as 'a great music hall sketch of Pascal's Pensées as played by the Fratellini clowns'; and although, oddly enough, there is no record of Beckett admiring or even reading Pascal, one can see exactly what he meant. *Godot* contemplates the worst, but it does so with economy, logic and style. It has a peculiar buoyancy of humour which lifts it far above despond. Its roots are deep in the tragi-comedy of the variety stage, the circus and the silent film. Laurel and Hardy are not too far away. It exploits several archetypal forms and situations, all of which lend themselves to both comedy and pathos. Somewhere in the background are innumerable pairs of comedians, ever hopeful, ever failing, ever encouraging each other in the blackest of situations as well as recriminating and scoring off each other. There is the comedy of the married couple, deeply conscious of each other's weaknesses, but bound indissolubly by need, custom and, mysteriously, even love. There is the master and man situation which has a centuries-old history in European literature, from Roman times through Cervantes and Le Sage to Tolstoy and Goncharov.

And of course there are deep primal myths which are evoked, principally the Judaeo-Christian one of redemption and salvation. Beckett's Christian upbringing and his familiarity with the sacred texts of Judaism and Christianity is evident and is used in all his work, but nowhere more flagrantly but somehow fittingly than here. The tree, symbol of hope and resurrection as well as of crucifixion, is the salient object of the *mise en scène*, described simply as '*A country road. A tree.*' His characters have, as is usual with Beckett, no background. They have come down in the world – how, we do not know – but they are learned and discursive; and Estragon claims to be or to have been a poet, gesturing towards his rags to prove the point. As in Beckett's other works, the bowler hat is much in evidence, all four characters sometimes wearing one. There are Irish jokes and references, some of them quite arcane, as are those to the sort of clay pipe known as a 'dudeen' and the Dublin pipe and tobacco merchants, Kapp and Peterson. As in the trilogy, there is tremendous emphasis on making up stories, keeping things going; and as in Chaplin's early comedies, the music-hall and the circus, a lot of mileage is got out of the simplest props, including the characters' apparel – hats and boots, sticks, umbrellas, etc., all being made to play their part. Passing the time is acknowledged as a primary problem and whenever a little time is passed in story or game Vladimir and Estragon congratulate themselves.

> Estragon: That wasn't such a bad little canter.
> Vladimir: Yes, but now we'll have to find something else.

It is one of the deepest and most central paradoxes of the human condition that we so fear death while yet searching desperately for ways to shorten the time while waiting for it. In no other work of literature is this paradox so hilariously exploited.

When Vladimir suggests that Estragon should try on his boots, he insists that it would pass the time.

> Vladimir: It would pass the time. (*Estragon hesitates*) I assure you, it'd be an occupation.
> Estragon: A relaxation.
> Vladimir: A recreation.

The primary verbal device employed is this sort of antiphonal exchange.

> Estragon: And what did he reply?
> Vladimir: That he'd see.
> Estragon: That he couldn't promise anything.
> Vladimir: That he'd have to think it over.
> Estragon: In the quiet of his home.

Vladimir:	Consult his family.
Estragon:	His friends.
Vladimir:	His agents.
Estragon:	His correspondents.
Vladimir:	His books.
Estragon:	His bank account.

It chants itself; and, as Vivian Mercier has pointed out, these antiphonal exchanges, when Vladimir and Estragon swap brief phrases of similar length and often identical syntax, are reminiscent of the accented dimeters of Beckett's free verse.

Like many of the world's best plays *Godot* was written in an astonishingly short space of time, the child's exercise book in which it was quickly set down with hardly any emendations bearing the dates 9 octobre 1948 on the first page and 29 janvier 1949 on the last. The origins of the title have been much debated. There was a well-known French racing cyclist called Godeau in the early fifties (not, as has been said, in the days of the six-day bicycle races at the Vélodrome in the 1930s, though Beckett did, like many other intellectuals, attend these), and when Hugh Kenner asked him for information about this individual he rather flirtatiously provided it:

> Godeau (can't think of Christian name) is a veteran still going strong.
> He is a 'stayer' and I think specialist of 'poursuite' (derrière Derny).
> I don't think he has ever distinguished himself in the Tour de France,
> if he was ever selected for it, but he has had some success in town
> to town races, Bordeaux-Paris for example. I think he is an 8 jours
> man too. I have the feeling he is bald. I'll try to find out more about
> him for you . . .

But he did not say yea or nay to the implied question of whether this was the source. This was the explanation for the title he gave to Hugh Kenner in response to a direct query, but when, shortly after the play went into production, Roger Blin asked him point-blank where it came from, he said it was from the slang for boot in French, *godillot*; and to his friend Con Leventhal he related an incident which took place when he was standing at the corner of the rue Godot le Mauroy, where prostitutes plied their trade. When one of the girls unsuccessfully solicited his custom, she enquired sarcastically who then he was waiting for – Godot? As to the theological associations of the name Godot, he was certainly aware of them – it has also been derived from Nietzsche's *'Gott ist tod'* – which does not mean that they hold any clue to the ultimate meaning of the play; but whereas Americans are inclined to say God*ot*, with the emphasis on the second syllable, Beckett always insisted in putting it on the first.

Godot, as has been said, was finished in January 1949; and Beckett now had four unpublished novels and two unproduced plays to contemplate. Suzanne was still sending out *Eleutheria*, but with no success; and the novels in French were still, as he put it, 'dragging around somewhere', as indeed was *Watt* in English. He entered *Eleutheria* for the Prix Rivarol and seems, naturally enough, to have nourished some hopes until someone told him that all the plays and poems had been eliminated and only novels remained in the running. He also heard with some disgust that André Gide, whom he had suggested to Chatto before the war as a possible subject for a book, had agreed to be a judge only on condition that he did not have to read the entries. In the early part of the year he typed up *Godot* and Suzanne began trying to interest producers in the new work as well. Perhaps because of the simple nature of the text and the fact that there was clearly little necessity to read on to see what happened, it came back rather more quickly than some of the others, so that within a comparatively short time she had collected four refusals. Meanwhile the only form of publication he seemed destined to achieve was of translations of other people's works, among them 'Zone', the first poem in Guillaume Apollinaire's famous *Alcools*, which he thought contained admirable passages.

Most of these translations were done for the French journalist and critic, Georges Duthuit, who had been a contributor to *transition* and had had the idea of starting a new magazine which would make the best new French writing available in translation for readers of English, combining this function with art criticism which would bring the same readers up to date with the latest developments in French painting. Since paper was rationed, he found it impossible to do this until Eugene Jolas generously suggested that he could use the name of the pre-war magazine and try to convince the authorities that the new publication was in fact the pre-war one revived.

In fact the result, which was called *Transition Forty-Eight*, was a very different publication. Although Duthuit had in the old days been interested in verticalism and the unconscious, in line with the new mood of grim reality, there was less emphasis on these things now. And although Jolas's name appeared on the masthead as advisory editor for the purpose of satisfying the authorities and obtaining a paper ration, the old *transition* circle had little to do with the new magazine. With its eye on the large number of young literary anglophones now thronging the Left Bank, this provided translations of the work of many of the leading French writers of the day, including Jean-Paul Sartre and other existentialists. Naturally it had a voracious appetite for translation and Beckett, who knew Duthuit and needed the money, was soon roped in. From 1948 to 1953 he published something in almost every monthly issue, beginning with three short poems of his own which appeared both in French and in English in the first number. Many of his translations were mere hackwork, done mechanically and without a great deal of enthusiasm; and many were unsigned. In these cases he could not afterwards say with certainty what he had done and what not, but besides the small sums of money involved the magazine also provided him with a social focus which he was beginning to need and was particularly glad to have after the completion of *Godot*. In two years of dredging the depths of his own being, his social life had shrunk to practically nothing and he was becoming such a solitary that even Suzanne, who normally disapproved of his drinking companions, was worried.

Married to Henri Matisse's daughter Marguerite and himself an intelligent, gregarious and socially-minded man, Duthuit was admirably fitted to be the leader of a cenacle and had gathered a little circle round himself and the magazine. This had an office in the rue de l'Université, just round the corner from the École Normale, but like many magazine editors Duthuit spent a lot of time in a nearby café, where contributors and others foregathered for regular Tuesday lunches. Besides his writers and translators, a number of painters, including the French-Canadian-Indian Jean Riopelle, Nicholas de Staël, Zao Wou-Ki and Bram van Velde, would sometimes appear at lunchtime to join such as Pierre Schneider, the art critic, and other writers. They provided an undemanding, male atmosphere in which Beckett, who liked Duthuit and usually got on well with painters, found he could relax, occasionally playing the pin-ball machine. There was a good deal of light-hearted badinage on these occasions, though many of those who hung around the magazine and the café remembered him for his reserve. The conversation was sometimes about painting, a topic which Duthuit and Beckett also discussed together.

It was out of one of these discussions about painting that there grew the idea of Beckett's most famous and often quoted contribution to the magazine, but his original assent to the idea of the piece was largely because he wanted to give a lift to Geer van Velde's brother Bram, who was his closest friend at this time and who was ill. Bram van Velde was a painter almost as unsuccessful and devoid of support from the exhibiting establishment as Beckett himself was from the publishing. He was eleven years older than Beckett, but he somewhat resembled him in physique and in manner. Both were tall, thin and bony; and each was inclined to be silent when he had nothing to say. Both had presence, a quality immediately sensed by those who met them. Like Beckett too, van Velde had charm and could be cordial, sympathetic and even talkative in the right company. In spite of his poverty, he impressed people by his elegance, his physical grace and the expression of his sapphire-blue eyes. Some years later Patrick Waldberg would write in the *Mercure de France* of his expression, which was that of 'a lost and surprised child'. Waldberg also spoke of how from time to time 'a sudden quiet laugh' would 'break the melancholy of this face'.

In certain company van Velde, like Beckett, would relax and his grave demeanour would be transformed into that of an amusing and direct companion. Nevertheless it was probably his presence and his silence that appealed to Beckett as much as anything else. But he also thought of him as a soul-mate and companion in arms in the lonely struggle. 'Bram is my great familiar in work and in the impossibility of work,' he told Marthe Arnaud and there is no mistaking the especial warmth with which he regarded him. Though he hated writing art criticism or any other kind of criticism, he was a comparatively soft touch for a piece about van Velde, for a catalogue or otherwise. Knowing that Duthuit was going to New York in the autumn of 1948 to take part in a colloquy about modern painting, Beckett decided to interest him in the work of his friend, bringing him around to the rue des Favorites where two van Veldes were hanging. Duthuit was impressed and there followed a visit to van Velde's studio, after which he acquired a painting which was hung in his offices and consorted with rather well, everybody thought, with the examples of primitive art which already embellished it. To give a lift to van Velde, Beckett now suggested that he might write a piece about him, but after some discussion this idea was abandoned in favour of an exchange of views about painting between Duthuit and Beckett in which van Velde would figure largely. The result was the highly original 'Three Dialogues' with Georges Duthuit, which were published in *Transition 49* in December 1949. Although supposed to be dialogues, they were in fact written by Beckett, who uses Duthuit as a sort of straight man or Socratic interrogator.

Originally written in English, though subsequently translated into French by the author, they discuss the work of three painters: Pierre Tal Coat, André Masson and Bram van Velde. Beckett displays really little interest in the work of Tal Coat, an interesting one-time member of the so-called 'Young Painters in the French Tradition' group, whose exhibition at the Galérie Braun in 1941 had been intended to defy the

Nazis; but of the three he was perhaps least enthusiastic about Masson, at that stage and perhaps still by far the best known of the trio and a member of the original surrealist grouping who had, however, been expelled from Breton's entourage at the time of the Second Manifesto. In the dialogue Beckett says that 'Masson bears the scars of a confidence that must be painful to him'; and describes him as 'an artist who seems literally scared of the ferocious dilemma of expression', who yet 'continues to wriggle', though it is 'impossible that he should ever do anything different from that which the best, including himself, have done already'. Van Velde, on the other hand, is free from the assumption underlying all painting 'that the domain of the maker is the domain of the feasible'. Hitherto, he declares, 'the much to express, the little to express, the ability to express much, the ability to express little' have merged in 'the common anxiety to express as much as possible, or as truly as possible or as finely as possible, to the best of one's ability'. The history of painting up to that point is said to have been a history of its attempts to escape from the sense of failure 'by means of more authentic, more ample, less exclusive relations between representer and representee'; but van Velde has been 'the first to desist from this estheticized automatism, the first to admit that to be an artist is to fail as no other dare fail, that failure is his world and the shrink from it desertion, art and craft, good housekeeping, living.'

It will be seen that this – the necessity to paint one's failure to paint – is more or less the position adumbrated in the second of the two van Velde essays already written; but before he is finished Beckett seems curiously to have merged his own position with that of van Velde and, now speaking in the first person, he says,

I know that all that is required now, in order to bring even this horrible matter to an acceptable conclusion, is to make of this submission, this admission, this fidelity to failure, a new occasion, a new term of relation, and of the act which, unable to act, obliged to act, he makes, an expressive act, even if only of itself, of its impossibility, of its obligation.

What is significant here is the use of the words 'obliged' and 'obligation', which had also occurred in relation to van Velde in an earlier part of the dialogue. There, supposedly challenged by Duthuit to 'state again, as simply as possible', the situation in which he perceived van Velde to be, Beckett replies, 'The situation is that of him who is helpless, can not act, in the event can not paint, since he is obliged to paint. The act is of him who, helpless, unable to act, acts, in the event paints, since he is obliged to paint.' Here the word 'obliged' is repeated twice.

The 'Three Dialogues' are in fact a perfect example of the fact that a great artist rarely writes true criticisms of any art. What purports to be criticism in his or her case nearly always turns out to be a statement of his or her preoccupations and dilemmas and a clarification of his or her aims. The dialogues are the nearest Beckett ever came to writing a manifesto or a statement of what he felt to be his own position; and they are given special resonance by the fact that when they were written he had already embarked on the final part of what had now become a trilogy, the longest, most dense and difficult part, and one in which he was agonizedly concerned with what he saw as the artist's special burden and torment, the categorical imperative to create when combined with the impossibility of creation.

And categorical it has now become. There has been a shift from the 'need to express'; and even the 'compulsion' to express has now become an 'obligation' to express. A need is a lesser thing than a compulsion, but an obligation is an even more portentous matter, for it removes us from the sphere of psychology into that of morality, perhaps even into that of theology. Over and over again in the 'Three Dialogues', he seems in fact to be talking about the book he had already embarked on. In this he had more than ever been concerned with the categorical imperative to create, while at the same time rejecting all previous forms of literary creation. In the 'Dialogues' he objected to certain artists, including Matisse as well as Tal Coat, being regarded as revolutionaries, when in his view the only thing they had disturbed was 'a certain order on the plane of the feasible'. Challenged by his interlocutor to say what other plane there could be, he replied that logically there could be none, but that the true artist must now turn away from it in disgust, being 'weary of its puny exploits, weary of pretending to be able, of being able, of doing a little better the same old thing, of going a little further along a dreary road'. And asked what was then left to such an artist, he asserted, with a curious mixture of despair and pride, 'The expression that there is nothing to express, nothing with which to express, nothing from which to express, no power to express, no desire to express, together with the obligation to express.' It could be – it is – a description of *The Unnamable*.

In the immediate aftermath of finishing *Godot* in the spring of 1949, Beckett still had little notion of the book that was to come; indeed at this point he told MacGreevy that he foresaw no other writing for some time. The completion of *Godot*, merely adding to his total of seemingly unpublishable and unproduceable works, was followed by a let-down. He was exhausted; and Suzanne too was feeling the strain –

in fact she seemed even more worn out than he was. The flat in the rue des Favorites was noisy – one could hear the voices and the radios of neighbours; there were visitors to Paris whom he did not feel up to seeing or entertaining; and in these circumstances they decided to go away together for a while, to sleep and eat and do nothing. In 1949 the Haydens had bought an old farmhouse near the village of Reuil-en-Brie in the *département* of Seine-et-Marne. When he and Suzanne went there on a visit Beckett decided that he liked the countryside, which, strangely enough, was of a kind that had been in his thoughts. They were fortunate to find a room in a big dilapidated farmhouse in the village of Ussy-sur-Marne, one of the battlefields of the First World War, between Meaux and La-Ferté-sous-Jouarre, for a rent of only £6 a year. They liked the woman of the house, whose husband was out of work and who seemed willing to help Suzanne with cleaning and housework for a few francs. The room looked out on old fruit trees and lilac bushes, and the Marne itself, where one could bathe, was only 100 yards away. Beckett took immediately to the pastoral, rolling landscape, which he thought peaceful. Though the room had little furniture, they bought a few pieces and went there at Easter. It was the beginning of a long association with that corner of France.

Back in Paris, he was restless and dissatisfied. The idea for the last volume of the trilogy had come to him, but he was in some uncertainty about it. Originally it was to be called *Mahood*, a continuation of his line of heroes beginning with M; but soon *Mahood* had become *The Unnamable*. In July its composition was interrupted by his annual visit to Ireland. The symptoms of Parkinson's disease – the agitation, the shaking hands and nodding head – were now perfectly obvious in his mother but, perhaps because it was then incurable, she was obstinately refusing to accept the diagnosis. Sam saw no point in insisting, but he discussed the need for a permanent nurse with Frank, both agreeing that it had to come sooner or later, but hoping it would be later. This time round he showed no inclination to go drinking in Dublin with literary and other acquaintances, perhaps because the first question everybody asked him was what he was writing now and he found the evasions this forced on him – after all, he could hardly tell them about the four novels and two plays that were fruitlessly circulating and reproachfully returning to their author and his various agents – distasteful; but there was one painful occasion at the Ganlys' when the subject cropped up in the worst way possible.

Andrew Ganly, the dentist he had consulted about his teeth on an earlier occasion, was literary, belonging to what Brendan Behan once called that large and formidable body of people who had once had a play done at the Abbey. Bridget O'Brien, his wife, was a painter (in fact rather a good painter) and, being the daughter of the painter Dermod O'Brien, who was President of the Royal Hibernian Academy, she had a large acquaintance in Dublin artistic circles. When Beckett went to Ganly about the teeth which were still troubling him and was asked to come to a little gathering, he felt he could not refuse. At the party the conversation turned to his work. What was he writing now? Oh, nothing special, just a few bits and pieces. Somebody adverted to the titles of his works, *More Pricks Than Kicks* and *Whoroscope*, evidently, in this liberal company, relishing their obscene nature. Then people began to suggest equally indecent titles for other works by Samuel Beckett; and soon the game became one of suggesting the plots and nature of these projected works. This went on amid much laughter until somebody noticed Beckett; and then they all became aware that he was sitting hunched in his chair with his jaws clenched and his pale eyes reflecting his horror. The laughter dissolved and the embarrassment lasted until the gathering broke up. It was not the sort of scene that would have occurred a few years later; nor, a few years later, would Beckett have put up with it as he did on this occasion.

It was with some relief that he got back to Paris and the book which was to complete his trilogy of novels. Mahood, the original hero, had now become only another discarded fictional identity of the ultimate author. Others were someone called Basil and someone referred to only as Worm. The names may be different; the dilemma, however, is the same. 'All my life, since we must call it so, there were three things, the inability to speak, the inability to be silent, and solitude, that's what I've got to make the best of.' As Mahood, Worm and Basil are disposed of it begins to appear that their creator, the Unnamable, is also the source and fount of Beckett's previous fictional creations, 'all these Murphys, Molloys and Malones' who have made him waste his time, suffer for nothing, 'speak of them when, in order to stop speaking, I should have spoken of me and me alone'. When he thinks of these diversions (in both senses of the word: diversions of energy and resource, but also means of entertainment, passing the time) he gets quite angry. 'When I think of the time I've wasted with these, beginning with Murphy ... when I had me, on the premises, within easy reach, tottering under my own skin and bones, real ones, rotting with solitude and neglect till I doubted my own existence.' But of course, as he knows, without these fictions the dilemma of having 'nothing to express, nothing with which to express, nothing from which to express, no power to express, no desire to express, together with the obligation to express' would be even more acute. The book ends with some famous phrases: 'I don't know, I'll never know, in the silence you don't know, you must go on, I can't go on, I'll go on.'

These have been adopted by some Beckett enthusiasts more or less as a calendar motto, evidence that he was, after all, a bit Kiplingesque, even Churchillian. Like Molly Bloom's famous 'yes' they are seen as an affirmation of something or other. Since they come at the end of an immensely long sentence which closes a paragraph of over 100 pages this is, to say the least, to take them out of context. But nevertheless, perhaps the trilogy is a sort of epic, written in the only terms in which it may be possible to construct an epic today – one in which the heroic ideal is heroic only to the hero and the moral imperative is a categoric one, justifiable by no theology or system of ethics. The imperative is speech, a perhaps pointless utterance, an expression of the human state for its own sake, no more; but this utterance must continue and this record of the human state must be made.

> . . . you must go on, I can't go on, you must go on, I'll go on, you must say words, as long as there are any, until they find me, until they say me, strange pain, strange sin, you must go on, perhaps it's done already, perhaps they have said me already, perhaps they have carried me to the threshold of my story, before the door that opens on my story, that would surprise me, if it opens, it will be I, it will be the silence, where I am, I don't know, I'll never know, in the silence you don't know, you must go on, I can't go on, I'll go on.

Chapter Twenty-Six

❧

Almost immediately after finishing *The Unnamable* Beckett began writing the first of a series of thirteen short pieces which he called *Textes pour rien*, translated as *Texts for Nothing*. Their commencement was, he admitted, a failure to implement the last words of *The Unnamable*. But as the speaker in the first text grimly observes, there is 'nothing like breathing your last to put new life in you'; and so, although that book had, as he put it, 'finished me or expressed my finishedness', he went on. Still, the new pieces must be regarded as 'nothing more than the grisly after-birth of *L'Innommable*'.

The title, *Textes pour rien*, comes from the musical term 'mesure pour rien', which signifies a bar's rest. Perhaps the number was not an accident, in the sense that he may have decided when there were thirteen to stop there, or perhaps even arbitrarily aimed at this number to begin with, though he had now largely lost the somewhat precious belief in the arcane significances of numbers which he had adopted as a young man under the influence of James Joyce, along with tight shoes and a way of crossing his legs. The first text is notable, among other things, for its fully described location. We are on top of 'a mountain, no, a hill, but so wild, so wild', – in other words, on top of the Dublin mountains, falsely so-called, where the speaker, whoever he is, is lying in an old sheeptrack scooped deep by the rains. 'All is noise, unending suck of black sopping peat, surge of giant ferns, heathery gulfs of quiet where the wind drowns, my life and its old jingles.' In this first text there is a reference to his father telling him the story of Joe Breem, or Breen, the lighthouse-keeper; and some of the others are, by Beckett's standards, richly referential also. In the third text there is even a reference to a well-known Irish bookmaker, Dick Duggan, who would have bet at the Leopardstown track in Foxrock when Beckett was a boy, and whose nationally-known motto was: 'Duggie always pays.' In the seventh there is a rich evocation of the waiting room of the old Harcourt Street terminus of the South Eastern Railway, 'dim with the platform gloom as dispensed by the quarter glass self-closing door'.

Other references to events in Beckett's life are scattered through these obscure texts, including one to the recently completed play, *Godot*. 'Why did Pozzo leave home, he had a castle and retainers. Insidious question, to remind me I'm in the dark.' This is in the fifth text, where the narrator is 'judge and party, witness and advocate' and in which he sits, 'attentive, indifferent', at a trial in some 'obscure assize where to be is to be guilty'. It is in the third text that the reference already noted to Bibby, the nanny, occurs, while in the first the narrator's mother is evoked with 'Eye ravening patient in the haggard vulture face', along with an assertion which seems to fit Beckett's relationship with his own mother with tragic exactitude: 'we're of one mind ... always were, deep down, we're fond of one another, we're sorry for one another, but there it is, there's nothing we can do for one another.' These narrators, like those of the trilogy, hear voices and try to tell stories, but their efforts are flickering and desultory and the narrator of the fourth text says 'there is my life, why not, it is one, if you like, if you must, I don't say no, this evening. There has to be one, it seems, once there is speech, no need of a story, a story is not compulsory, just a life, that's the mistake I made, one of the mistakes, to have wanted a story for myself, whereas life alone is enough.' It will be noted that there is a curious inversion here: since there is speech there must be life, not the other way around.

With the writing of *Texts for Nothing* the siege in the room came to an end. It had lasted for four years and had been immensely productive. During this time had been written the three novels and the play that would establish him as one of the greatest of living writers – in some eyes, the greatest – plus another novel, several shorter prose pieces and another play. But all this work, and the previously written *Watt*, had brought him nothing in terms of the ordinary rewards of a writer's existence – not even publication, still less readership and some esteem, not to mention money or fame. The only offer of publication on the horizon for other than translations was of an almost derisorily peripheral nature. Those associated with an Irish literary monthly, *Envoy*, of whom the present writer was one, had been shown *Watt* by Con Leventhal and, immediately struck by the dissolving beauty of its prose, had offered to publish an extract.

Except to people like Leventhal who had been his contemporaries and boon companions in the thirties, and who now seemed to most of us to belong to a different era than ourselves, the name Beckett had only very vague connotations for the post-war generation in Ireland. It was known that he had published a comic novel of some sort in the thirties and that he lived in Paris, but he had very little identity otherwise. However, his composer cousin, John Beckett, belonged to the same age group as the editors of the magazine. When he produced a copy of *Murphy*, I had rather disliked its knowing tone and thought the first sentence falsely sophisticated and untrue. *Watt* suggested a different order of things. Perhaps Irish modernism had not after all come to an end with James Joyce and Flann O'Brien.

When contacted by the editors of *Envoy*, Beckett assented readily to the publication of an extract from *Watt*; but he had now given up practically all hope of publication or production of a more important kind and had not only washed his hands of Suzanne's efforts, but, worse still from her point of view, was now reacting with a mixture of irritation and disbelief when she made a new suggestion. His long immersion in the depths over, he seemed content to do absolutely nothing, sleeping late into the day and discouraging her efforts, as if waiting for the fates to somehow bring his work into the light. Among the trails which she was independently following, however, was one which had led her to Roger Blin, a director who had been influenced by Antonin Artaud's ideas and was at that time directing Strindberg's *Ghost Sonata* in the nearby Gaieté theatre in Montparnasse. When she mentioned Blin to Beckett he reacted as he did these days: What was the use? Blin might be an experimental director, working in very small theatres and open to new ideas, but it would only be the same old story. He would keep the plays for months and then return them with a discouraging letter.

He grudgingly admitted, however, to being slightly curious about how a disciple of Artaud's might deal with Strindberg and at length said that he would go and see. In the event he went not only once but twice, and so Suzanne felt emboldened to return to the attack. Finally he told her she could do what she liked. If she wanted to take the plays to Blin, well and good. In the normal course of events, many plays were left at the theatre, which the director glanced through with more or less impatience and deposited at the box office to be called for by their authors. The name Beckett rang a bell, however, because Blin had heard his poetry praised on the radio by the poet and critic Max-Pol Fouchet, who was a friend of his and had been an associate during the war, when Fouchet had acquired some celebrity as the founder and editor of the review *Fontaine*, a new journal intended to fill the gap left by the *Nouvelle Revue française* after it became collaborationist under its gifted but tragically-destined editor, Drieu la Rochelle.

Fouchet had happened to mention that Beckett was an Irishman living in France, and Blin had some sympathy with Irish dramatists because during the war he had discovered that though the Nazis did not allow productions of the work of English or American authors they had no objections to Irish ones, and he had consequently done several productions of Synge with great success. As it happened, the surrealist poet Tristan Tzara had also mentioned Beckett to him, and so when he took the plays home and sat down to read them he was in a slightly more receptive frame of mind than might otherwise have been the case. None of this would have mattered, of course, if he had not been impressed; but though he did not understand *Waiting for Godot*, he found himself quite enthusiastic about both plays. Beckett in his turn had liked Blin's production of the Strindberg play, which he thought true both to the letter and to the spirit of the author's work; and though he was well aware by now of the dangers of being hopeful, his hopes had in fact been raised when he found that even in the little Gaieté Blin was playing to half-empty houses. If the director and his players were prepared to go on playing to audiences of this size in a tiny theatre like the Gaieté-Montparnasse, then surely there was some hope for *Waiting for Godot*. In fact, however, it was *Eleutheria* which Blin first thought of producing. It reminded him of Pirandello and even the very difficulties of producing a drama with two sets on a divided stage might make it memorable theatrically if successfully surmounted. Besides that, there were no parts for women in *Godot*, something which was unusual at that time and which he found unsatisfactory. On consideration, however, he began to lean towards it rather than the more elaborate *Eleutheria*. *Eleutheria* had an enormous cast and there were the two sets and complicated lighting. He had no resources and simply no access to a theatre suitable for the kind of large production the play necessitated. In *Godot*, on the other hand, there were four parts and with the exception of Pozzo all the characters were down-and-outs. If it came to it they could wear their own old clothes and there was no set needed except a bare stage with some sort of a tree. Instead of the complicated lighting which *Eleutheria* demanded, a single spotlight would do. He decided for *Godot* and this decision he now communicated.

Beckett's response was to allow Suzanne to go along to the theatre to talk to Blin and at this meeting the director outlined the difficulties he would face. He was himself interested in the play but he doubted if any of those who had previously backed his productions would be. He had now done three productions at the Gaieté-Montparnasse, all of which had been box office failures, and the theatre was now in serious financial trouble. He had no theatre of his own, so both money and a venue might have to be found. There might be a long wait. (In fact the Gaieté-Montparnasse would close its doors the following year, leaving Blin himself liable for a large part of the debts incurred.) The impression Suzanne carried away from the meeting was not overly encouraging, but in view of the previous history of the two plays it undoubtedly represented progress.

Shortly afterwards Blin accepted an invitation to lunch at the rue des Favorites. He remembered arriving at a studio apartment which was extremely tidy, without any personal clutter, but also rather without character. Besides being shy, Beckett struck him as somewhat detached and rather cynical about the possibility of a production of the play; but they had a long talk which established what Blin would later call 'a solidarity' between them. It seemed that he and Beckett had much in common and were thinking along the same lines. Neither man was disdainful of the forms of popular theatre represented by circus clowns and the great comedians of the screen, such as Charlie Chaplin. Although the French had no tradition of pairs of cross-talk comedians, Blin had of course seen Laurel and Hardy as well as Chaplin and Buster Keaton. He envisaged a very simple production, faithful to the text and with no embellishments, and since Beckett already knew about his reputation for respecting authors' texts in the theatre he was quite ready to believe that he had found the right man. Blin's profound admiration for Synge also moved him, while his readiness to drink the Irish whiskey which Beckett offered did not displease his host either. They parted on good terms at the beginning of a long, affectionate and fruitful association.

As they sat and drank whiskey after lunch Blin had naturally repeated to Beckett everything he had said to Suzanne about the difficulties that had to be faced, but he now introduced what amounted to a positive gleam of hope. Since the play was an original one, in French, he would be able to apply for a grant from the government for its production, and the prospect of this would do something to convince a theatre management or a backer that the play was a viable proposition.

With these uncertain indications that at least one work of his might have some sort of a destiny, Beckett departed for what he thought was his usual annual visit to his mother in mid-June 1950, taking the manuscript of *The Unnamable* with him in the hope of typing and revising it in Foxrock. He had expected some deterioration in his mother's condition and both Frank and Susan Manning had written to him about her subsequent progress, which appeared to be good within limits. But there had been a sudden downturn and what he found now surprised and shocked him. Geoffrey Thompson advised him that she should be moved to the Merrion Nursing Home by the bank of the Grand Canal at 21 Herbert Street; and throughout July Sam spent long hours at her bedside, mostly sitting in an agonized silence. Afterwards he would go on immensely long walks with Geoffrey. At home, in a more than usually silent New Place, he got on with the typing of *The Unnamable* and with the translation of extracts from *Molloy* which Duthuit wanted to publish in *transition*.

By the third week of August it was clear that the end was very near and he spent three whole days and nights by his mother's bedside. Towards evening on 25 August, after spending most of the day there, he went for a long walk along the canal. When he came back he could not immediately face the prospect of going to sit once more in the silence which was broken only by her laborious breathing, so he sat on a bench by the canal, from which, as it so happened, he could see the window of her room. It was a cold late August in Ireland and there was an east wind blowing along the canal. He had sat there before, so often indeed that he had become a familiar sight to some of the regulars who used the canal bank and its towpath for a stroll, or just sat, as he did, on its benches. At a certain point he happened to look up. The blind of his mother's window, a dirty red-brown affair, was down. She was dead.

Eight years later he would imagine the writer Krapp sitting on that bench by the canal bank, waiting for his mother to die, 'wishing she were gone'. Krapp too would happen to look up and see the blind down. 'All over and done with at last.' Of course, it is never all over and done with; and it would certainly not be so for Beckett, who constantly, in an agony of remembrance, relived the things which caused him such depth of emotion that they even seemed to give meaning to meaningless existence.

His relationship with his mother had caused him over the years a great deal of intense anxiety and guilt. Of course, he owed her the gift of existence itself, but that was something which he professed not to value very greatly. And in latter years anyway, the relationship had brought him very little comfort or joy. Now that it was at an end, besides sorrow and remorse, he must of course have felt some sort of release. The funeral took place at the little parish church at Tullow, where the Protestants of Foxrock worshipped and which was within a few minutes' walk of May's home. Afterwards she was buried with very little ceremony in the old Redford Cemetery between the sea and the Wicklow hills at Greystones. From her room in the house that she had rented there, she could see this graveyard, and her younger son had imagined her longing to lie there beside Bill with the bleak east wind blowing off the Irish Sea. At her funeral he went through the obsequies without any display of emotion, his severe face revealing little of what he felt within.

In succeeding days at the house he destroyed all the papers he had kept there. Including as it did letters from Joyce and others, the burning of these was quite a holocaust. He was deliberately bringing other parts of his life to an end as well. In her will, of which Frank had been appointed executor, his mother left a personal estate of almost £2,000, which was now to be divided between her two sons, but her life interest in what her husband had left her, amounting in value to many thousands more, had now come to an end as well and was also to be divided between them. When Bill had died in the 1930s it had been valued at over £30,000 and much of it was in house property in Dublin which of course had more than maintained its value. For Sam the days of monetary hardship were over.

He was anxious to get back to Paris as soon as he decently could, but one little incident of his remaining days in Dublin was a visit to the *Envoy* office in Grafton Street. The extract from *Watt* had appeared in the January issue and Leventhal now suggested that he should meet the editors. He found the usual notice pinned to the door to say that these worthies might be found in McDaid's pub across the way but unfortunately, for once, nobody connected with the magazine was actually present and so Beckett never met Kavanagh or any of the younger contributors, and this small but potentially interesting inter-section of literary generations did not take place.

In Paris Suzanne was continuing her apparently fruitless search for a publisher and she now had yet another and, on the face of it, even less publishable work, *The Unnamable*, to add to the four novels and three plays she had been carrying round. In October she suggested that they should try a small firm called Les Éditions de Minuit, a suggestion which was met with the usual self-protective expressions of scepticism and indifference. Though it had not yet achieved the sort of reputation that was to come with the publication of the *nouveau roman* writers a few years later, it did seem to be open to avant-garde writing and it is both surprising and further evidence of how little interest Beckett him-self was taking in the question of publication, that it had not already been approached. Les Éditions de Minuit had been founded in 1942 by Pierre de Lescure to publish underground work by authors who rejected the drift towards collaboration with the Germans already reflected in the pages of such publications as the *Nouvelle Revue française*. One of its first publications had been the famous *Le Silence de la mer* by Vercors, which became a symbol of literary resistance to the Nazis and their ideology. After the war it had been one of the new publishing companies filling the vacuum left by older firms who had been compromised by their political affiliations; but it none the less had had a struggle for survival which at times had seemed very difficult.

Now, in late 1950, under an energetic new editor, Jérôme Lindon, it was still finding things difficult but it was at least acquiring a reputation for an openness to experimental work. The financial rewards of this were not very great, but it was giving the firm the kind of identity which Lindon had hoped for and which the arrival of the *nouvelle vague* novelists in the 1950s would confirm. At this time it had a reading committee whose secretary was Georges Lambrichs, who was to become very well known after he went to the larger firm of Gallimard, and it was to him that Suzanne now suggested they should take *Molloy*. Neither of them reposed a great deal of hope in the possibility that this time they might find a publisher. On the day in question they left the house together, Suzanne carrying the sort of manuscript parcel that now had a dreadful familiarity. Les Éditions de Minuit's office was then situated on the corner of the boulevard Saint-Germain and the boulevard Saint-Michel, where there is now a branch of the Crédit Lyonnais, and before Suzanne crossed the road they stopped at the Cluny, a café on the opposite corner. Here Beckett said he was getting fed up. This, as far as he was concerned, was the last try. If it failed there would be no more submissions to indifferent publishers. As Suzanne stood up to cross the street Sam asked her to get him a packet of cigarettes from the *tabac* which was also on the opposite corner. After leaving the manuscript she came back with them. Twenty years later she playfully reminded him that he had still not paid her for that packet of cigarettes.

There are two slightly differing versions of what happened next. One says that Georges Lambrichs read the manuscript and left it with a note for Jérôme Lindon saying, 'You must read this.' The other suggests that Lambrichs did nothing for the time being, but that after he left the office Lindon happened to see the manuscript on his desk and thought he recognized the name Beckett as having been mentioned to him by Robert Carlier, who had said that Beckett was an interesting writer who had been rejected by several publishers. One way or another Lindon began to read *Molloy* and, as he was to put it later, from the very first line the overwhelming beauty of the text assaulted him. It is an interesting comment. Beckett's writing *is* beautiful. It is the first thing to be said about it. 'I read *Molloy* in a few hours as I had never read a book before,' Lindon was to say. 'It was not a novel that had been published by one of my colleagues, a sacred masterpiece that I, as a publisher, could never have anything to do with: it was an unpublished manuscript, and not only unpublished, but one that been refused by several publishers. I could not believe it.'

He was also much taken with the humour of the work, something which, by and large, French critics would fail to be; indeed Lindon's high level of immediate perception was something that a great deal of criticism would fail to match, for in reading some of the hundreds of thousands of words that have been written about it since, it is not always apparent that Beckett's work is both beautiful and funny. At the office next morning Lindon decided to telephone the author to congratulate him and convey his enthusiasm, but there was no telephone number on the manuscript. On enquiry he found that it had been left by 'a young woman'. He then decided to send a *pneumatique*, a form of communication which had a guaranteed delivery within a few hours. The result was a visit from Suzanne on that very day. Lindon did not beat about the bush. He wanted to publish the work, he said. He had been running Les Éditions de Minuit for two years. *Molloy* was the most exciting work he had read in that time; and not only would he like to publish this: he knew already that he would like to publish the other two novels which had been left as well.

Suzanne was naturally overwhelmed at the news. After the series of rejections, often coming after a long delay and involving many unanswered enquiries, the speed of events was bewildering. She said she would have to consult Beckett and she would bring him in to see Lindon. When she got back she told Beckett what had happened and asked him to come with her to Les Éditions de Minuit. With what seems in retrospect to be either an extraordinary subconscious faith in his destiny or else an extraordinary indifference to the destiny of his work, he refused. When she went back again she took Bram van Velde's sister, Toni Clerkx – who had had experience of publishers and for a while had acted as a sort of agent for Beckett – with her.

Lindon made no bones about his position. The firm was not well off, there was not much money, but there would be great enthusiasm. He was prepared to publish all three works, but he would begin with *Molloy* and follow, at intervals, with *Malone Dies* and *The Unnamable*, bringing them out as soon as he could. He had had contracts prepared. Would Suzanne like to take them back to Samuel Beckett and return them to him signed, and would Beckett now like to come in and see his publisher for himself? All this took place on 15 November 1950. A few weeks later Beckett did come in.

Suzanne told Lindon afterwards that he returned home looking very unhappy. When she showed her surprise, asking if he was upset by the terms offered or had found his publisher a disagreeable person and was regretting the deal, he said no, he had found Lindon and everybody else at Les Éditions de Minuit very sympathetic, but he was saddened by the prospect of their bankruptcy, something which the publication of his books would inevitably lead to. Shortly afterwards, when *Molloy* was sent to the printers, it began to seem as if it might meet with further misadventures on the way to publication. The printer, a Catholic from Alsace, was offended on moral grounds by some of the text; he also feared that he might leave himself open to prosecution for a breach of public morality. Finally he went ahead, but he left his name off the printed text.

The book came out on 15 March 1951, Lindon paying 25,000 old francs (or around £16) in advance of royalties. Beckett was now a published author with a publisher prepared to advance him and look after his interests. It was not yet the era of the big hype, still less so in French publishing, nor were Les Éditions de Minuit the sort of people to provide it, but there was nevertheless a small quantitative leap from the world of the man alone, in whom nobody is interested, into the world of organized and organization-backed promotion which makes the individual a property to be expanded and developed. One of his publisher's first requests was for a photograph. Beckett did not have one, but he readily agreed to have one taken. The result was the photograph of him in profile by Giselle Freund – who had taken photographs of Joyce before the war and was the only photographer he knew – wearing a tweed jacket and a polo neck jersey, with his hair, in discernible waves, brushed up from his forehead, which was for several years the only one of the 'mystery man' known to exist. Beckett had also mentioned short stories and Lindon had thought he might be able to place at least one of them. And he had talked of his play, in which his publisher was also very interested. Shortly after the publication of *Molloy*, Beckett wrote to Lindon, thanking him.:

Your letter of yesterday to hand. Most cordial thanks for your generous advance.

I had the photograph taken this afternoon. It will be ready after tomorrow and I will send it you just as soon.

I know that Roger Blin wants to put on the play. He has requested a subsidy for it. I very much doubt that he will get it. Let us wait for Godot, but not for tomorrow.

The story, of which the first part under the 'La Fuite' has appeared in *Les Temps Modernes*, is at your disposal. Can it wait until my return? It is my first work in French (in prose). 'Le Calmant' that Madame

Dumesnil has given to Monsieur Lambrichs, would perhaps be more suitable. I leave the choice to you.

I am very pleased to hear that you want to bring out *L'Innommable* as rapidly as possible. As I told you, I prefer this last work, in spite of it having involved me in a mess of dirty sheets. I am trying to get out of them. But I'm not succeeding. I do not know if it will be able to make a book. It will perhaps be a time for nothing.

Allow me to say how much I am touched by the interest that you have shown in my work and by the difficulties that you are taking on yourself. And believe in my sincere and friendly regards.

Beckett was still on formal terms with his publisher but gradually their relationship would become more friendly and intimate. Finally, over the years, they would come to meet on a regular basis, with Lindon going round to Beckett's flat at least once a week, usually at six o'clock on Fridays. As time passed Beckett would also begin to take an active interest in the affairs of Les Éditions de Minuit and in its often difficult and uncertain continued existence. He would even support it financially. 'Though not a rich man he would have given me anything he had,' said Lindon. And in a tribute on Beckett's sixtieth birthday his publisher also said:

> I do not dare to express the enormous admiration and affection that I have for him. He would be embarrassed and on that account I should be so too. But I would like this to be known, and only this: that in all my life I have never met a man in whom co-exist together in such high degree, nobility and modesty, lucidity and goodness. I would never have believed that anyone could exist who is at the same time so real, so truly great, and so good.

Whatever anxieties its publisher felt were dispelled by the enthusiastic reception French critics gave to *Molloy*. Of course, there were the usual misconceptions. Joyce was invoked more than once and so was existentialism; but the lengthy considerations accorded to the book by Maurice Nadeau and Georges Bataille, who was something of a hero to the avant-garde, were remarkably perceptive. Whether through Lindon's efforts or otherwise, *Molloy* got into the right hands; and Nadeau at least, who had a reputation for discovering new authors, seemed to know a good deal about Beckett's personal history. Before *Malone Dies* appeared in the autumn *Les Temps modernes* printed a fairly lengthy extract which was headed 'Quel Malheur', and the novel itself appeared on 8 October 1951, again to widespread expressions of critical esteem.

The combined effect of the publication of both books in less than a year was to make Beckett a known author in France, mentioned and discussed in avant-garde circles in a way that was rather remarkable considering the obscurity in which he had previously lived; and to their publisher's relief the books even sold. Both had been published in editions of 3,000 copies and they went on selling throughout the first months of publication. In the course of time *Molloy* would sell 30,000 copies in this original Minuit edition and over 50,000 in a cheaper format. These figures may be compared with the thirty copies which *Murphy* had sold when in that year he bought the rights to the book from Bordas.

Lindon was also interested in bringing out *Waiting for Godot*, and the publication of *Murphy* and *Malone Dies* were undoubtedly of some assistance to Blin in his efforts to find backing and a theatre for the play, but these still hung fire. Nevertheless the tide was now running for Beckett and a broadcast of part of the play on the national radio in February was certainly a help, as was the grant of 750,000 old francs (or almost £500) which the Ministry of Culture offered for the production of first plays in the French language. Blin was acquainted with the Minister Georges Neveaux and had been much encouraged by a personal letter which assured him that Neveaux considered *Godot* 'an astonishing play' and was strongly in favour the grant. When this came through everybody's hopes were raised, but a theatre still had to be found. Blin first hoped that a production might be mounted in the little Théâtre des Noctambules, where Adamov's *La grande et la petite Manoeuvre* had been produced.

And as he was with Lindon, Beckett was slowly becoming more intimate with Roger Blin. 'Now that we are embarked on a dirty business together, I think we can say 'tu' to each other,' he wrote to the actor-director early in 1951. But negotiations for the Théâtre des Noctambules fell through. Then he had hopes of the tiny Théâtre de Poche in Montparnasse, where Chekhov's *Uncle Vanya* was having a long run. In August Blin went on holiday in Brittany and while he was there Beckett had news that the long run was at last coming to an end. He wrote at once to Blin advising him to 'swim, talk, void yourself – what awaits you is terrible'. But this optimism too proved ill-founded when the management of the Théâtre de Poche decided that the theatre would not after all be available for *Godot* until well into 1953. Then the even smaller Théâtre de Port Chasseur was considered, but an actual survey convinced them that the stage was too small even for a tree, and though Blin was pursued into the street by a voluble manager who kept insisting that she saw no reason why such an awkward prop as a tree should be necessary to any play, he turned the offer down.

Disappointing though all this was, Beckett was at least not dependent on a production of *Godot* for survival, for with the death of his mother his financial circumstances had improved considerably and there was now even a small trickle of income from his works. Gone were the days when the click of Suzanne's hand- and foot-operated sewing machine was heard daily in the rue des Favorites; when he had vaguely hoped that his books might make some difference to his circumstances and then gloomily reflected that a 15,000f. advance would last about a fortnight; when he had been forced to consider translation as a primary way of earning a living and even James Joyce's old recourse of English teaching as a grim possibility. Though Beckett was not one of those people whose way of life changes very much with the advent of money, he was now able to afford something which many Parisians desire – a retreat. When the money from his mother's estate came through one of the first things he and Suzanne did was to look for a small plot of land at Ussy-sur-Marne (where, since 1949, they had been renting the room in the farmhouse) on which they might build a little house of their own. The site he found was on an eminence with a good view of the green, gently undulating countryside of the Marne valley, a landscape of large fields, hedgerows and occasional coppices. Besides the landscape, he also liked the associations, for the great Irish missionary Saint Columbanus had passed through Ussy-sur-Marne at the beginning of the seventh century; another holy man, Agilbert, who had spent years in Ireland, was buried in the crypt of the church at Jouarre, as well as an Irish princess and saint he had never heard of, Ozanne.

When eventually the site was chosen he made, with his brother Frank's help, rough designs of the sort of house he wanted for a local builder. It was in some ways very like the plain, Irish slated cottages which were built around the turn of the century, with a long central room, a tiny kitchen and bathroom. In the course of time he built a wall around the small plot of land attached to the house, laid out a lawn and planted trees, bringing the grass seed back from Ireland with him after one of his visits. The Haydens lived not too far away and Henri helped him to plant the trees which, to the eye of Georges Pelorson (now more widely known by his *nom de plume*, Georges Belmont), seemed always to have a somewhat forlorn and bare appearance, like the tree in *Godot*. The long interior room was simply furnished with two single beds, some wicker-work chairs of the sort that can be bought cheaply and readily in Paris, a table and a desk under the window from which he could look out when he was working. A stove was

quarter, and then goes on to insist that Bram must not imagine that he was distancing himself from his friend. The truth was to the contrary. The more he shut himself up, the nearer to Bram he felt. Their aims and outlook were the same. If he could not any longer, as before, encourage him, that was only the effect of weakness and fatigue. It is evident from all this that the measure of success he was now enjoying had not produced much more elation or lightness of heart than failure had, but then both Schopenhauer and Beckett himself could have foretold that, believing as they did that the ground of misery and anxiety on which the patterns of success or failure are worked out remains the same. For a brief period after the appearance of *Malone Dies* in October 1951 Beckett had been oddly optimistic about his life and even about the outcome of Blin's negotiations for a production of *Godot*. At the beginning of 1952 he had even written a letter to George Reavey suggesting that it might not be such a bad year after all. Then this short period of comparative elation had apparently come to an end, his general outlook seeming to reinforce the belief that any alleviation of the human condition was fleeting and temporary, and the hope that any sort of success in any undertaking would change things an illusion. But a greater measure of success, which would test this belief further, was now at hand.

Godot appeared from Les Éditions de Minuit on 17 October 1952, a date which had been chosen some months before when it seemed as if the Théâtre de Poche had been secured and would be free by then. This hope had proved ill-founded but now suddenly things began to come right. Blin subsequently rejected with some indignation what he believed to be a biographer's suggestion that the long delay in achieving production was in any way his fault. On the contrary, he said, he fought against his natural laziness with violence, working for a production in a sort of dementia. Among those he had already approached was Jean-Marie Serreau, who controlled the little Théâtre de Babylone on the boulevard Raspail. He had liked the play when he saw it, but apart from anything else his theatre was not free. Unexpectedly it now became available for a production early in 1953, and he at once offered it to Blin. When he made the offer, Serreau had already decided to close the place and he honestly believed that *Godot* would hasten the process. 'I am going to close up shop and I may as well close up in beauty,' he told Blin. The Babylone was in fact a shop transformed into a theatre, with a stage which was four metres deep and six metres wide; but there were seats for 230 people in the auditorium. Serreau had been there since 1951 and had not sought box office success, though of course he was glad to have it when it came.

During the long period of fruitless negotiation Blin had approached several actors and had cast the play more than once. Now he had to cast it again but had the satisfaction of knowing that this time it was for real. This final casting, although some of it was accidental and there were even changes two weeks before the curtain rose, turned out to be inspired. Blin himself had at first wanted to play Lucky, and his desire to do so led to his first serious disagreement with Beckett as well as to his first acquaintance with Beckett's formidable ability to say no in matters connected with the production and performance of his work. Beckett wanted him to play Pozzo and, even though Blin insisted that in one of the many dreams he had about the play he clearly saw himself playing Lucky, so that he knew exactly what sort of performance he would give in that part, the author remained adamant. Ignoring Blin's protestations that he was quite the wrong build for that part, he told him he must play it; and so he did. Jean Martin, who eventually played Lucky and is now for ever associated with the part, has that rarest and most valuable combination of acting qualities, humility and intelligence. As Lucky, it fell to him to deliver the long stream of learned nonsense towards the close of the first act. Since this largely is nonsense it is the most dangerous moment of the play in terms of losing the attention of the audience. When Martin delivered it the effect was inexplicably electric. Lucien Raimbourg, who played Vladimir, was the oldest of the troupe. He had been around for a long time without any commercial success and was one of those thin, beak-nosed, sad-eyed actors who seem destined to be lugubrious man-servants or unfortunate husbands. Blin liked him because he thought he could play comedy tragically, as his own great idol, Buster Keaton, was able to do. Pierre Latour, who played Estragon, was a more business-like and worldly type. Blin had urged the play on him but he had agreed to take the part reluctantly and without, he said, any understanding of the text. Though he remained in it till the end of its run, he continued to claim that he never understood it and that he only did what Blin told him to do. As actors and producers got used to Beckett they would begin to realize that no overall understanding was required. The small boy in this first production was Serge Lecointe, who had also been found by Blin.

As soon as the play went into rehearsal Blin realized how much Beckett hated having to explain his writing; but fortunately he also believed that all the interpretations of the text had been placed there by Beckett himself, and that there was no theatre without ambiguity; so, like its author, he was content that the play should not be understood in any formal, paraphrasable sense. Beckett sat in throughout rehearsals but, given this initial identity of attitude, director and author got on well together, though of course there were disagreements from time to time. As an actor and director Blin was pleased that the play was subversive of the theatrical categories of the day, which he thought of as psychological, romantic and sub-Brechtian. He had recently seen Steinbeck's *Of Mice and Men* and he believed that something of what he felt about it had got into the production, the smaller actor, Raimbourg, being protective of the larger. The relationship between Vladimir and Estragon reminded him of an image he had recently seen: two monkeys in a miserable cage who laced themselves around each other 'with a kind of infinite distress'. He was the first to notice how 'married' they were: they had much need of each other, they could not be separated, but they could not bear each other either. Their need of each other was physical and when they came together and embraced they were diastole and systole of a single pulse, yet even on the physical level they rejected each other brutally. 'You stink of garlic,' says Estragon, pushing his friend away. As Blin put it, they are encrusted together and yet they tear apart with violence. But he rejected any homosexual element in their relationship, believing it more rewarding to see them simply as the eternal couple without reference to sex or youth or age. He believed also that Pozzo and Lucky had what he called the rapport of classes, of the domineering and the dominated, which is frequently, more or less unwittingly, portrayed by clowns in the circus. Though Blin thought that this aspect of *Godot* had a serious political import, Beckett resisted any attempts to give it an overt political meaning and did not like the contrast in costumes which Blin gave the master and the man, making Pozzo very elegant while Lucky was in rags. In fact, Beckett told him, in an earlier conception of the play he had decided that their costumes should be reversed in the second act. This would suggest some advertence to Yeats's idea that revolutions are simply a change of places between master and slave, the beggars changing places while 'the lash goes on'. He also told him – almost certainly thinking of the Harcourt Street line – that he had a station porter's costume in mind for Lucky, but Blin preferred that of a French doorman or hotel porter. However, Beckett seemed, when questioned, to have no clear idea of how the cast generally should be dressed, except, as he said, that they should all wear bowler hats, or *chapeaux melons*, as the French call them. Blin believed that it was wrong to attempt to play Beckett naturalistically. The lyricism of the text, its scansion, its 'breathing' all militated against what was called the style of the boulevard. What was needed was to play truly but not naturally. In his production he respected the punctuation of the text, believing that it was an essential ingredient of the poetry and that if the actors observed the punctuation the rhythms would be right; and in this he was absolutely correct. There is no writer in whom punctuation is as important as it is in Beckett and whose punctuation is so clearly a musical as well as a grammatical device. Blin also observed accurately that it was the punctuation that allowed the text to negate itself in the way Beckett's texts so frequently do, something that is also apparent in the novels, where one clause in a sentence is constantly undermined by another. He noted as well that all the characters had ailments or infirmities: Vladimir had prostate trouble and was constantly needing to piss; Estragon was always falling asleep; Pozzo had a heart condition; and he worked with the actors on these. Vladimir's restlessness he thought primarily due to this condition, but of course it gave movement to the play and he thought movement was important. He worked out the positioning of the actors as if he were working on a big stage instead of the tiny one of the Babylone and was glad to see, when finally the play was produced in a bigger theatre, how well these placings worked. There was a balletic, circulatory aspect to the play which he sought to emphasize. Like all those who have directed or played Beckett successfully, he refused to lumber himself with metaphysical preoccupations, working instead from the ordinary human meaning and import of the text.

As a director Blin knew himself to be lacking in method, someone who never prepared work for the day after or sought to plan out too rigorously what he was doing. Out of successive attempts and approximations the right thing would emerge, and he spent hours encouraging actors to do small things in different ways or with different expressions on their faces until they found one that seemed right. He was also a believer in having fun during rehearsals; but after a while he noticed that much of what was going on seemed aimless, casual or unnecessary to Beckett, who would sit out front, chain smoking and staring expressionlessly at the stage through his metal-rimmed glasses, not registering any amusement. As the days wore on Beckett's impatience with Blin's lack of organization became more and more apparent. He was particularly dismayed by the apparent inconsequentiality and impulsiveness with which the director would decide to rehearse one bit today and another tomorrow. Yet he would not impose himself. He was merely a silent presence, often apparently disapproving of what was going on; and this frequently unnerved the others. Since Beckett sat in throughout, he naturally had second thoughts about the text, as he always would have second thoughts about the text of his dramatic works whenever he was engaged in a production of them; and he made several changes, as he would still be doing twenty years later for a Schiller Theatre production in Berlin.

A serious disagreement between Beckett and Latour, who played Estragon, occurred towards the end. As word got round that there was a strange, funny play by the Irishman who lived in Paris and had written *Molloy* and other works in French in rehearsal at the Babylone people began to drop in to the theatre. A few days before opening night there was laughter at the point when, towards the end of the play, Estragon's trousers fall down. Mindful of his dignity, Latour told Blin afterwards that he was not going to let them fall again; he was not going to be a clown. Next day he was obdurate and an argument occurred. Only after Beckett and Blin had both spoken to him at some length did he reluctantly change his mind and let the trousers fall.

The play opened on 5 January 1953 before an evidently puzzled audience. There had been some laughter and there was a good deal of applause at the end, but opening night audiences, largely invitees, are never an indication of how a play has really gone, and a number of people had left the theatre halfway through. Suzanne came, but Beckett did not. He had retreated to Ussy and from there he wrote to Blin a couple of days later asking him not to hold the fact that he had stayed away against him. 'I couldn't take any more,' he said. But meantime the reviews had appeared and the play and its author had become famous. In *Libération* Sylvain Zegel referred to the audience's bewilderment but claimed that when the curtain fell it at least understood that a new playwright of an important sort had arrived, 'one of today's best playwrights'; and speaking of the two central characters, he made a good stab at saying what the play is 'about'. 'These two tramps, who represent all humanity, utter remarks that any one of us might utter. These two men are feeble and energetic, cowardly and courageous; they bicker, amuse themselves, are bored, speak to each other without understanding. They do all this to keep busy. To pass time. To live or to give themselves the illusion that they are living.' *En attendant Godot*, Zegel claimed, 'would be spoken of for a long time'. But it was Jean Anouilh, writing in *Arts-Spectacle*, who set the seal on the play's critical success. This was the most important notice of all, for Anouilh was not only one of the great figures of the French theatre, ranking with Montherlant, Cocteau and Giraudoux, but he was a popular playwright, 'the poor man's Giraudoux', who had drawn large numbers of people to the theatre himself. '*Godot*', he now said, 'is a masterpiece that will cause despair for men in general and for playwrights in particular.' He insisted that the opening night of Beckett's play at the Théâtre de Babylone was as important as the opening of Pirandello's *Six Characters in Search of an Author*, which had been presented by Pitoëff in Paris in 1923; and he added 'one can only raise one's hat – a bowler to be sure, as in the play and pray to heaven for a little talent. The greatness, the artful playing, a style – we are "somewhere" in the theatre.'

These notices were gratifying and one might have expected the principals to shower each other with congratulations, but when Beckett wrote to Blin he was still concerned with Estragon's trousers, about which Suzanne had reported to him.

> One thing troubles me, the pants of Estragon. I naturally asked Suzanne if it fell well, and she told me that he keeps them half on. He mustn't. He absolutely mustn't. It doesn't suit the circumstances ... He doesn't even realize that they're fallen. As for the laughter, which could greet their complete fall, there is nothing to object to in the great gift of a touching final tableau; it would be of the same order as the preceding scenes.

And he added something of immense importance – not only to the understanding of *Godot* but to the rest of his work as well: 'The spirit of the play to the extent to which it has one, is that nothing is more grotesque than the tragic. One must express it up to the end, and especially at the end.' And he finished by insisting quite firmly, 'I have a lot of other reasons why this action should not be tampered with but I will spare you them. Just be good enough to re-establish it as it is in the text and as we always foresaw it during rehearsals. And that the pants fall completely around the ankles. This might seem stupid to you but for me it's capital.'

Impressed by the reviews and by the now widespread notion that there was something going on which they should know about, people began to flock to the Théâtre de Babylone, which, to begin with at least, was now full for every performance. As Jean Martin was to put it, 'the snobs' began to come. Often they left after the first act, or left at the beginning of the second, when they saw the curtain rise again on the same old scene, so that sometimes the players, having begun with a full house, found themselves playing to an empty one before the performance was over. Responses were very mixed. Many people were bewildered; some were angered; others, a majority perhaps, were just bored. A great many came simply to be reassured that most avant-garde works were fakes and to mock those who had been taken in by it as well as the play itself. Others were convinced that *Godot* had a profound, allegorical significance which could be teased out and argued about all night, the most popular interpretation being that Godot was God and the play was about his disappearance from human affairs, although there were also those who believed that it was a morality play with a basically Christian message, Vladimir and Estragon being fallen and exiled mankind awaiting its salvation. With those who thought like this the tree was popular – and of course its blossoming. With others the word existentialist, still holding sway on the Left Bank, supplied a key to the play's meaning. Instead of accepting responsibility for their own situation Vladimir and Estragon had put their fate in the hands of some *deus* who never appeared *ex machina*, either because he didn't exist or because he didn't intervene in human affairs. Whereas the Christians saw Vladimir and Estragon as keeping faith to the end, those who believed in an existentialist interpretation saw their attitude as a typical example of what Sartre called 'bad faith'. All these discussions and arguments would be repeated as the play travelled to many other cities in the next few years. In the meantime the little Théâtre de Babylone was doing good business and in the end the play would run for a full year, before the end of which its author would be world famous.

A corrective to the allegorists and those who thought that *Godot* had a profound meaning which was just beyond reach came in February, in the shape of an extended essay by Alain Robbe-Grillet in the review *Critique*, which had been founded and was edited by Georges Bataille. This was entitled 'Samuel Beckett, auteur dramatique'. After summarizing the action, or non-action of the play, he said that although it was 'made out of nothingness' it nevertheless held together 'without an empty space'. That Beckett was already looked upon as part of a new trend which was becoming fashionable is evident from Robbe-Grillet's remarks on the changes in the intellectual and critical climate in France which made Beckett and Eugène Ionesco acceptable. That he had read *Molloy* and *Malone Dies* is clear from his remarks about the process of progressive dispossession and degradation suffered by Beckett's characters. 'What little had been given to us from the start . . . is soon corrupted before our eyes, degraded further, like Pozzo who returns deprived of sight, dragged on by Lucky deprived of speech.' The climax of this dispossession and disintegration is the point where the four characters fall to the ground together, creating a formless mass from which Didi's voice emerges, asserting 'We are men!' For Robbe-Grillet Beckett's play was 'the most banal, the least audacious expression imaginable' of the human state, the state of simply 'being there'.

Towards the end of the 1950s Robbe-Grillet became known as the chief theorist of a new wave of novelists – Nathalie Sarraute, Claude Simon, Robert Pinget, Marguerite Duras – whose work, like Beckett's, was mostly published by Les Éditions de Minuit. Like Beckett, they disdained traditional story-telling and plot construction. Like most new movements, the *nouveau roman* school sought to identify an ancestry for itself among its predecessors; and in 1961 Robbe-Grillet published a well-known essay, 'Nouveau Roman, homme nouveau', on this theme. Declaring that it was not their intention to make a *tabula rasa* of the past, he made a list of predecessors which he thought would find general agreement among the new novelists. They were Flaubert, Proust, Kafka, Joyce, Faulkner and Beckett. Beckett certainly did not disdain the overtures or acknowledgements of influence which emanated from this group. Partly to please Jérôme Lindon he sometimes went to their launchings and other public appearances. In 1959 he allowed himself to be photographed with a group which included Robbe-Grillet, Claude Simon, Robert Pinget and Nathalie Sarraute. He was particularly close to Robert Pinget, whom he recommended to his English publisher after he had acquired one himself. He made a translation of Pinget's play *La Manivelle*, to which he gave the Irish-sounding title, *The Old Tune*.

Besides the new developments in the novel – or anti-novel – with which Beckett was now being identified, there was also a theatrical movement that went by various names, including a-theatre, anti-theatre, theatre of the absurd, experimental theatre, method theatre or the theatre of ridicule. The other names principally associated with this alleged movement were those of Arthur Adamov, a Russian émigré who had been born two years later than Beckett and whose first play, *La grande et la petite Manoeuvre*, had been produced three years before *Godot* at the Théâtre des Noctambules; and Eugène Ionesco, a Romanian born in 1912 who had settled in France in 1938. Ionesco's first plays were also staged at the Théâtre des Noctambules, beginning with *La Cantatrice chauve* in 1952. While Blin was fruitlessly hunting for a theatre for *Godot*, Adamov's *La Parodie* and Ionesco's *Les Chaises* had been produced. The principal characteristics of the anti-theatre or theatre of the absurd movement were usually accounted to be a marked minimization of scenery and action and a generalization and depersonalization of characters, who were usually given little social identity and sometimes no name, as well as elements adopted from the music-hall or the circus, such as clowns or marionettes. The passive characters of *Godot*, as well as the music-hall and circus associations and the fact that there was no action in the ordinary sense of dramatic action, gave plenty of excuse to critics to make Beckett part of a movement. Though – perhaps for Jérôme Lindon's sake, perhaps for the sake of company in the struggle for expression – Beckett would assent to and even encourage the association of his name with the *nouvelle vague* in the novel and would become friendly with other members of the movement, the public association with Ionesco, Adamov and the 'theatre of the absurd' would always annoy him and he would discourage it in every way possible short of public dissociation.

Chapter Twenty-Seven

Beckett was now a well-known French author, standing indubitably, in terms of age, originality and dates of publication, at the head of one new movement – that in the novel – and widely touted as a principal figure in a second – that in the theatre. He was, however, still virtually unknown in the English-speaking world and he even had a novel in English, *Watt*, written during the war and rejected after it by a score of publishers, which was still unpublished. Few except publishers' readers knew of it, and few even of *Murphy* – indeed about this time even Bram van Velde was surprised to discover that his friend had ever written in English. Beckett himself was more or less prepared to let *Watt* and his pre-war writings rest in obscurity, and to go on writing in French for the rest of his days.

There was by now, however, a new generation of anglophone writers in Paris and there was even a new outcrop of little magazines in English. Associated with one of these was a young American, Richard Seaver, who had come to Paris with the usual dreams of what might happen in that city. He found quarters behind an antique shop in the rue du Sabot, just around the corner from the new offices of Les Éditions de Minuit at Number 7 in the rue du Dragon, a premises which had housed a bordello until the famous closing of such establishments in 1948. On his way to the cafés of Saint-Germain-des-Prés, Seaver passed through the rue du Dragon and one day in the publisher's window he noticed two books, the titles of which he found intriguing. They were *Molloy* and *Malone meurt* by Samuel Beckett. He had a vague notion that the author was Irish and had been associated with James Joyce, one of the mythical figures of the bygone era in Paris which, to many young Americans like Seaver, was more important than the present. He even remembered that a Samuel Beckett had contributed to the

Exagmination Round his Factification for *Work in Progress*, copies of which were still being passed from hand to hand.

Finally one day he went into Les Éditions du Minuit and bought the two books. He found them so 'stunning' that he went back to Minuit and asked if there was anything else by the same author in print. There was nothing, he was told, but a third novel, *L'Innommable*, would shortly appear. However, there was an earlier work called *Murphy* which Bordas had published and which Minuit thought might still be in print. Bicycling over to Bordas he found that not only was *Murphy* still in print but, by the look of the stock in the back of the shop, the original printing was virtually intact. So he bought that too and took it home to read. 'Not *Molloy* or *Malone*, I felt, not great, but only a notch below.'

As Seaver began to enthuse about his discovery to people a friend told him that the French radio was shortly scheduled to broadcast part of an unproduced play by Samuel Beckett called *En attendant Godot* and that it might be possible to get tickets for the recording, at which Beckett would hopefully be present. He went to the taping session and thought it also a beautiful work, but the author did not turn up. Shortly afterwards, however, he discovered in the French literary review *Fontaine* a story by Beckett, 'L'Expulse', which he thought was of the same calibre as *Molloy*.

All this might have remained a private discovery except that around this time Seaver became involved with a newly founded English language magazine called *Merlin*, whose editor was a Scots Italian named Alexander Trocchi. Moved by Seaver's enthusiasm, Trocchi agreed to publish an essay about Beckett in the autumn 1952 issue. It was called 'Samuel Beckett: An Introduction'. This set out the facts about Beckett's career, more or less accurately, saying that he had been associated with Joyce in the late twenties, had been recognized at that time as an astute critic, had undertaken the first translation of 'Anna Livia Plurabelle' into French, and had latterly forsaken English himself in order to write in that language. Whatever its merits – and it had many – Seaver's essay was the first extended discussion of any kind – apart, that is, from book reviews – that Beckett's work had received in the English language. For this alone it is notable. But its author's enthusiasm was to have further results.

After the issue of *Merlin* containing his essay was published the editors wrote to Beckett asking for a contribution. There was no reply and no way of knowing whether the letter had been received. However, Seaver had by now penetrated into the offices of *Molloy*'s publishers, Les Éditions de Minuit, and made the acquaintance of Jérôme Lindon, who in conversation let slip that Beckett had 'a final work in English' written during the war and still unpublished, more or less implying that the work was hidden away somewhere and might be difficult to extract from its author. The editors of *Merlin* now wrote to Beckett asking if they could publish an extract from *Watt* in their magazine, but again there was a prolonged silence. Then one rainy afternoon, when they had more or less given up hope of hearing anything from him, a knock came at the door of Seaver's pad behind the antique shop in the rue du Sabot, which also served as the headquarters of the magazine. When he opened it he saw a tall, thin, raincoated man with steel-rimmed glasses, who thrust a black binder containing a manuscript at him and disappeared into the rain. That night the editors of *Merlin* – Alexander Trocchi, Seaver himself, Jane Lougée, Christopher Logue and Patrick Bowles – sat up half the night reading the mysterious novel aloud to each other, taking turns until they finished it. Enclosed with the manuscript had been a note specifying precisely the sections from *Watt* they could use. These were Mr Knott's inventory of possible variations of attire – 'as for his feet, sometimes he wore on each a sock or on the one a sock and on the other a stocking, or a boot, or a shoe, or a slipper, or a sock and boot, or a sock and shoe . . .' – and so on; and the similar variations possible in the positions of the furniture in his room – 'thus it was not rare to find, on the Sunday, the tallboy on its feet by the fire, and the dressing table on its head by the bed . . .; and, on the Monday, the tallboy on its head by the bed . . .', etc. Thinking it over afterwards Seaver believed that Beckett was testing *Merlin*'s integrity by limiting them to these rather arid passages, which, unless they were truly enthusiastic and discerning, they would surely reject. Much later he confronted Beckett with that accusation and received in reply only 'a broad bad-boy grin'. In any case *Merlin* published the extract in its next issue and, though encountering a certain amount of reader opposition, said they would be publishing more of his work when and if it became available.

Beckett now had an enthusiastic band of anglophone followers in Paris, the 'juveniles', as he called them. He met them from time to time to deliver further extracts and he could sometimes be seen thrusting himself forward with his peculiar gait along the rue du Sabot while they trotted to keep up with him like enthusiastic puppies.

Of course, publication in *Merlin*, which had a very limited and haphazard Left Bank readership, would not have set the English-speaking world alight any more than publication in *Envoy* had, but the editors now made a further proposal. They were going to publish a series of books: they would like *Watt* to be the first, and they could offer Beckett an advance of 50,000 old francs or roughly $100 and bring out the book quite soon. Their further plans would include a translation of Jean Genet's *The Thief's Journal* and a work by the Marquis de Sade.

To this proposition Beckett agreed. After all, nobody else had offered to publish *Watt* in the eight years it had been doing the rounds. However, no sooner was the suggestion made and the contract signed than a hitch occurred. Foreigners conducting business in France were required by French law to have a French partner, and the editors and proprietors of *Merlin* did not have one. One of them was, however, acquainted with a certain Maurice Girodias who, though French in upbringing, manner and appearance, was a son of that Jack Kahane who had once asked Beckett to translate the works of the Marquis de Sade into English. Girodias had published books in French under the imprint of Les Éditions du Chêne, but he now intended to publish pornography in English for tourists who associated Paris with sexual licence. He wanted his pornography to be mildly classy and even para-literary, not outright sleaze, for the market he was aiming at would be aware of the works of Henry Miller, but it would nevertheless be pornographic in intention and, presumably, in effect.

He now suggested that the Olympia Press imprint which he intended to use for his pornography could have a subsidiary imprint, Collection Merlin, over which the *Merlin* juveniles would have editorial control. When asked what was in the deal for him, he replied that he too was interested in de Sade and Genet; and, besides, he was looking for writers: perhaps some of the editors or their friends might occasionally turn their literary talents to the production of works for the ordinary Olympia Press imprint.

Girodias was, in Seaver's words, 'elegant, suave, compelling'. He had a shiny black Citroën, which at the time of their first business conversation stood before the courtyard door in the rue du Sabot like a symbol of prosperity. They did not know that he was almost broke and that the Olympia Press venture represented a rather desperate last throw of the dice. So they brought him to meet Beckett. Girodias later recalled the meeting as an absurd one. Beckett was silent almost throughout, staring at his interlocutors with impenetrable blue eyes, but contributing nothing to the discussion. Trocchi later asserted that Girodias never read the book and published it only because he thought it might appeal to the sort of readership he had in mind.

In any case this modification of the original deal was agreed to by all parties and *Watt* – which appeared in 1953 after the success of *En attendant Godot* – became the first book published by Collection Merlin, the Olympia Press imprint. Two thousand copies were printed and over the years all were sold, though probably many of its purchasers were making the same mistake as its publisher and took it to be just another dirty book.

In September of the same year, 1953, another English language publisher entered Beckett's life – Barney Rosset, the ebullient and genial owner of the American Grove Press imprint, which was then about three years old. Barney had actually taken a course in the Paris avant-garde of the thirties and forties and had studied the magazines like *transition* with which they were associated. He was therefore aware of the name Samuel Beckett when *En attendant Godot* opened in Paris and became a rumoured new portent in London and New York; and soon after the Paris opening he had interested himself in acquiring the English language rights. When he arrived in Paris he had actually bought these from Les Éditions de Minuit and expressed an interest in doing the trilogy as well. To this end he had corresponded with Beckett, who had introduced a cautionary note:

> With regard to my work in general I hope you realise what you are letting yourself in for. I do not mean the heart of the matter, which is unlikely to disturb anybody, but certain obscenities of form which may not have struck you in French as they will in English, and which frankly (it is better you should know this before we get going) I am not at all disposed to mitigate.

Barney had replied that he felt they should not worry about such things until they got going. 'Sometimes things like that have a way of solving themselves'; but Beckett was still worried and returned to the subject, saying that 'a clear understanding on this matter' was essential for both parties before they set to work. In all this he may have been underestimating his man, for it would be Barney's glory to fight several battles with censorship in America, most famously on behalf of unexpurgated versions of *Lady Chatterley's Lover* and Henry Miller's *Tropic of Cancer*.

Rosset was on honeymoon and had his young second wife with him when he came to Paris in September 1953, rang Beckett and suggested a meeting. Somewhat reluctantly, Beckett agreed and an arrangement was made for six o'clock in the basement bar of the hotel in which the Rossets were staying, the Pont Royal near the bridge which gave it its name. Beckett arrived on time, wearing his raincoat and peering myopically round the bar for the Americans he was supposed to meet. When Rosset approached he seemed to be in a hurry and insisted that he could stay only a few minutes. About four o'clock the next morning they were still together and Beckett was buying the honeymoon couple

430

champagne. It was the beginning of another long and fruitful association.

Rosset's judgement was not infallible and much dross as well as some works of genius would appear over the years from the Grove Press imprint. But all the same, he had remarkable intuition and his firm would come to be associated with most of what was worthwhile in fifties modernism as well as in the counterculture of fifties and sixties America. As Stan Gontarski has put it, Grove 'had no particular program or ideology of its own, beyond a broad-based avant-gardeism, a general post-war dissatisfaction with the status quo, a militant anti-authoritarianism, and an unwavering commitment to absolute freedom of expression in speech, print and, finally, film'. It would eventually publish many of Les Éditions de Minuit's new novelists, as well as Ionesco and Genet. What was also important to Beckett, however, was that Barney Rosset had proved congenial. A man of great charm and openness as well as of a somewhat unconventional but none the less fierce integrity, he became Beckett's friend as well as his publisher, and shortly after their first meeting he acquired the rights to *Molloy* and *Malone Dies*. Barney's wife at the time of this first meeting was a German, Hannalore Eckert, known as Loly, and she also became friendly with Beckett, though the friendship was not of the same immediate and congenial nature as that with Barney. Loly had been in Paris during the war and though she had told Barney something of her history, it was of such an extraordinary nature that he did not quite believe it all. Her father, a First World War veteran, had been a major in German intelligence during the war and was assigned to the Paris garrison. Loly, then in her late teens, got a job on the war-time version of *L'Humanité*, which was produced under the occupier's supervision. She and her father were quite anti-Nazi and made no secret of the fact to the French who worked in and around *L'Humanité*, either as journalists or contributors. Some of these contributors had belonged to the Cocteau circle and had an ambiguous attitude to the Germans, though some also were in the resistance. Loly would share jokes about the Germans with those French people she met in journalistic and theatrical circles, where she became quite well known. Finally, in spite of her father, she was arrested and ended the war in a work camp near the Swiss border, afterwards succeeding in emigrating to America. When she and Barney met Beckett in the Pont Royal on that first occasion they all went to *Godot* together. Both Barney and Beckett were astonished when after the performance she was recognized and embraced by Jean Martin and others as the little German girl who had been so popular during the war.

But when he came to Paris at this time Barney also looked up his former wife, the painter Joan Mitchell; and it was through his new-found American publisher that Beckett met Joan, who was to become an important person in his life. He already knew the person she was living with, Jean Riopelle, also a painter, who had been around in the Duthuit days and sometimes at the lunchtime sessions in the café, though he was not one of Beckett's chosen band of significant painters. In fact in so far as Beckett and Riopelle were ever more than sometimes hostile acquaintances it was because of the attraction Sam felt for Mitchell. Beautiful in a bold, rather brash American way, she was also a forthright and formidable woman and she was as fond of drink as Sam was himself. John Ashbery has described her 'a tough lady, a hard drinker, hard on herself and harder on her friends', saying that nevertheless 'her thorniness made you want to hug her as one thinks about embracing a rose bush'. Once when asked by a French critic what he thought was her principal characteristic, Ashbery replied, 'She has a knack of putting you ill-at-ease at once.' Like other women friends of Beckett's, though, she was very talkative and outgoing and when she was in good humour could be very jolly. In New York in the 1940s she had been one of those, among them Franz Kline and William de Kooning, who forgathered at the Cedar bar and took part in the famous Ninth Street Exhibition, one of the events which marked the arrival of abstract expressionism in the United States.

She and Barney Rosset had had a romantic history. As high-school sweethearts in Chicago, he more in love with her than she was with him, they had decided to go to New York and for a while lived there together, Barney constantly urging her to marry him. In 1948 she agreed, saying she wanted to go to France to paint large pictures and needed someone to carry them for her. Though they were not formally divorced until 1952, they soon parted. When Beckett met her through Barney she was already living with Riopelle, a French-speaking Canadian-Indian whose native French, like his English, was a patois. By the late fifties he and Joan were sharing a studio over a garage in the rue Frémicourt, which is in the 15th arrondissement and not far from where Beckett and Suzanne lived in the rue des Favorites. Joan was by then doing her 'black' paintings, in fact sometimes dark green abstractions, which seemed to suggest foliage in an overgrown garden. Shown retrospectively in the Robert Miller Gallery in New York recently, these have achieved a considerable measure of critical acclaim.

That Beckett was fond of her there is no doubt, and many of their acquaintances were in two minds as to whether they were having an affair. As Barney puts it, 'there was a deep interest between the two of them, there was a wonderful feeling there between the two of them'.

And he thinks that if they did not have a sexual relationship it was not Joan's fault. Some of those who observed them together when they were more or less inebriated thought that Joan sometimes treated him rather roughly, with a sort of studied indifference to his feelings, and that Sam rather liked this. Whatever the relationship between them, certainly Riopelle was jealous and on more than one occasion he displayed his feelings. One night when they were all three in the Rosebud bar in Montparnasse together he became so insulting that Beckett had to leave the table and sit elsewhere. Joan told Riopelle to behave himself and eventually he quietened down; but whatever cause for jealousy he had, or may have thought he had, it is possible that one aspect of the relationship was summed up by a remark which Beckett was overheard to make on another occasion in the same bar, when he told Joan, 'Stick to Riopelle, he can fuck and I can't.'

Beckett often went back to the studio in the rue Frémicourt, however, with or without Riopelle, and on some occasions he stayed there very late. Entrance to the studio was across the forecourt of the garage, which had large steel gates, only closed late at night. Often on the stairs visitors encountered a pair of fierce Sealyham terriers, which, apart from whatever threat they may actually have posed, made a fearful din. One night Beckett left at about three o'clock in the morning and, when he came downstairs, found the gates closed. Perhaps because he did not want to wake the dogs, he decided not to go back upstairs for the key and drunkenly attempted to climb over them. During this operation he fell off, staggered backwards a few paces and fell into the oiling pit of the garage. Eventually he succeeded in climbing out, getting over the gates and going home, but when he awoke the next morning he was in pain. Two or three days later he mentioned the pain to Georges Belmont, giving him the impression that he had short-sightedly stepped backwards into an oiling pit in a garage when going in to enquire about his car. (He told other people that he had fallen through an improperly placed man-hole cover coming out of a station.) It was one of the occasions when Belmont felt that he was enjoying his pain, though he did not know how the accident had really come about; but he urged Beckett to go to a hospital for a check-up. When he did he found that he had broken a couple of ribs. By coincidence, he did shortsightedly fall into another garage oiling pit several years later, when his car was being checked out at La Ferté in the Marne valley. On that occasion he was laid up for several days at the Hayden's house in Reuil.

Joan Mitchell was the sort of woman that Samuel Beckett liked: intelligent, talkative, with freely expressed opinions, seemingly extrovert and convivial. Such women often tended to be, in John Calder's phrase, physically 'ample'. Even Peggy Sinclair, for all her apparent silliness, had conformed to the pattern of talkative, intelligent jolliness and prettiness, though the ampleness of her charms had been found somewhat dismaying by the young Beckett when revealed. Ethna McCarthy also conformed to type, being intelligent, talkative and seemingly outgoing. True, Suzanne, who was sharp and talkative but scarcely convivial, did not; but then Suzanne played a different role in his life from the others. Almost from the beginning, as Peggy Guggenheim had swiftly diagnosed, she had been transformed into a mother-figure. It was her function to help, to stabilize and, in many matters, to be a standing reproach.

The advent of English-language publication raised again a question which had first become insistent with the success of *Godot*: what was to be done about translations of the French works? Beckett had made a rough, preliminary translation of that work so that it could be offered for American or English production, but he felt considerable reluctance to embark on translations of the novels, so he willingly accepted the offers of help with translation that had come from two of the young enthusiastic editors at *Merlin*, Richard Seaver and Patrick Bowles. They have both testified to the illuminations and difficulties of translation with Beckett, and what emerges is that he was not content with mere translation for translation's sake but was keenly aware of how much a change of language may change not only the meaning but the quality of the statement.

When Seaver began work they would meet each day at La Coupole, one of the famous Montparnasse cafés which Beckett frequented. There they would work over what Seaver had done, weighing every word and phrase. Beckett always took care to express a measure of approval before introducing a reservation. He would often declare that a word could simply not be translated, then he would suggest an English equivalent. Bowles had the same experience. When discussing his draft, Beckett would emphasize that the French could not and should not be directly translated. They would have to write 'a new book in a new language'. And with the transposition of speech there would inevitably be a transposition of thought, 'and even at times, of action'. With Bowles he was supposed to work every morning at the rue des Favorites, but his translator's habits were somewhat irregular and he did not always turn up.

Though the Bowles–Beckett *Molloy* has admirable qualities, and is to many people in the English-speaking world the work itself, Rosset felt that Beckett was unhappy with Bowles as a translator and would perhaps be unhappy and irritated no matter who tried to do it. Finally he urged that Beckett should do the job himself, that he was the only

434

person who could; and though he had commissioned a translation of *Godot*, he abandoned this when he got Beckett's. *Malone Dies* was entirely Beckett's and from that point on no other translators were considered. He told Barney that doing *Malone* from scratch had been child's play compared to revising Bowles, with whom he had lost touch for long periods towards the end of their joint effort. Of course, since these first translations were prepared for an American publisher, there was always the question of transatlantic and cisatlantic idiom. Beckett worried about this but he knew very well that anything he did was 'bound to be quite unamerican in rhythm and atmosphere'. He did, however, initially agree that when Barney pointed out words incomprehensible to an American reader, he would consider alternatives. A reader may well feel that a more germane matter is the problem of Irishisms, in which Beckett's text is thankfully rich, and which could sometimes be regarded as having even less warranty in the French text than Englishisms.

Beckett felt in the autumn of 1953 that an endless prospect of translation was stretching out before him which would perhaps last into 1955. There was now a question of a German translation of *Godot* in which, with his knowledge of German, he felt compelled to collaborate, and he found this going back over old work 'with all the adventure gone' very dispiriting, particularly since it seemed to be driving out new inspiration.

And, given his temperament, into the future also was beginning to stretch an endless process of supervising theatrical productions. After its first successful run at the Babylone *Godot* was taken off, but revived again before the end of the year, this time without poor Raimbourg, who had left for a new play by Armand Salacrou, which was not a good move since his performance in it earned him no notices at all. Beckett again took an active part in rehearsals, which sometimes lasted until midnight, and found himself very tired and edgy as a result. He found Raimbourg's successor adequate but no more and hoped that the original Vladimir would be back and, as he put it, 'the old team whole again' for the tour of the provinces and Switzerland, Milan and Germany that was coming up.

Beckett's willingness to involve himself in productions of his work was if anything to increase as the years went by. Of course, a playwright may regard each new production as a work of art which must be brought as near to perfection as possible. But there is no doubt also that Beckett sought in the theatre a release from the ultimate loneliness of the writer's form of creation, and that to some extent he can be accused of yielding to the temptation which Yeats in similar circumstances found himself yielding to – that of seeking 'creation without toil', or at least toil in that ultimate uncollaborative sense in which it presents itself to the writer.

The tour was a shoestring affair for *Godot*, though internationally famous as the obscure, experimental, avant-garde play to end all avant-garde plays, was not yet big business, but at least it was easy for the company to take the set with them: there was only the dismantled tree to think about and it could be packed in a suitcase. When the company arrived in Berlin, Beckett was glad to take a week off and join them there for he was finding Paris distressful. In spite of frequent sojourns at Ussy, he was feeling chronically tired and stupid and even Ussy, where there were tasks to be done, seemed in his present mood to be more of a burden and an encumbrance than a refuge.

L'Innommable had appeared from Les Éditions de Minuit in July. The principal response had been an important essay by Maurice Blanchot in the *Nouvelle Revue française*. Blanchot was a novelist who was also an influential critic and his essay was a milestone in the progress of Beckett's reputation. Describing the trilogy as 'an experiment without results, continuing with increasing purity from book to book by rejecting the very resources, meagre as they are, that might permit it to continue', he went on to suggest that in the case of *L'Innommable* readers were perhaps 'not dealing with a book at all, but with something more than a book? Perhaps we are approaching that movement from which all books derive, that point of origin where, doubtless, the work is lost, the point which always ruins the work, the point of perpetual unworkableness with which the work must maintain an increasingly initial relation or risk becoming nothing at all'. This is very perceptive, but it will be seen that Blanchot had evidently read the remarks about van Velde in the 'Three Dialogues with Georges Duthuit'.

The reception of the trilogy made Beckett an important figure in the French literary consciousness. Reviewing *Malone Dies* in *Mercure de France* in the previous year, 1952, Maurice Nadeau had said that *Molloy* 'gave the impression that it was impossible to go farther in the conquest of Nothingness. *Malone Dies* pushes back the boundaries of the undertaking which is pursued in another work still to be published. After which it is difficult to imagine that there could be anything left for Beckett but silence.' Now Beckett felt these words were perhaps true. 'I feel more and more that I shall perhaps never be able to write anything else,' he wrote to MacGreevy. 'I can't go on and I can't go back. Perhaps another play some day.'

Whatever his reputation in France and in avant-garde circles in the United States, in England he was still no more than a distant rumour, someone who had had a sort of success in a little theatre with a 'joke' play to which audiences came out of curiosity and then left at the interval, completely baffled as to what was supposed to be going on. Those few who knew about his work hoped that this might change when a London West End production of *Godot*, the English version of which he had almost finished, was announced – to take place in the first six months of 1954. In his present mood Beckett felt otherwise. 'I don't think it can do any good in London,' he wrote, adding, 'but then I didn't think it could do any good here.'

As the months passed, however, this production – by Peter Glenville and Donald Albury – had continually to be put back. There were difficulties about a theatre, about casting, and endless and sometimes seemingly insoluble difficulties about the Lord Chamberlain's office, which had to license plays for production in the West End and would sometimes, for reasons that seem ludicrous now, refuse permission on the grounds of indecency. The most offensive passage was held to be the dialogue between Vladimir and Estragon on the effect of hanging in giving the victim an erection. But once again the seemingly eternal question of Estragon's trousers falling down was raised. Trousers had fallen often enough in pantomime, in the cinema and even in Whitehall farces without anybody being much the worse for watching them do so, but the Chamberlain's rules for serious intellectual works were much more rigorous. Intellectuals were dangerous chaps and must not be allowed to get away with anything. There were also minor but never-theless serious matters, worthy of the attention of a serious official – calling somebody Fartov, for example, or Mrs Pozzo being said to have had clap.

As correspondence with the Lord Chamberlain dragged on, Peter Glenville approached several well-known actors whom he hoped might play Vladimir and Estragon. John Gielgud dismissed the play as 'a load of old rubbish'. What happened with Ralph Richardson is best told in his own words. Beckett would visit London in 1954 partly to see the Jack Yeats exhibition there, and a meeting was arranged between him and the actor, who at this stage seemed quite interested but had drawn up a sort of 'laundry list' of things he didn't understand but would like to discuss. Beckett was invited to come backstage at the Haymarket, where Richardson was then doing a play.

Beckett came into my dressing-room – wearing a knapsack which was very mysterious – and I started to read through my list. You see, I like to know what I am being asked to do. March up that hill and charge that blockhouse! – fine – but I wasn't sure which was the hill

437

and where the blockhouse was ... But Beckett just looked at me and said, 'I'm awfully sorry but I can't answer any of your questions.' He wouldn't explain. Didn't lend me a hand. And then another job came up and I turned down the greatest play of my lifetime.

Beckett's account of his meeting with 'Ralph Richardson, Knight', whom he 'disliked intensely', closely agrees with this. 'He wanted the low-down on Pozzo, his home address, family background and curriculum vitae. As I know nothing of these matters we parted coldly,' he wrote to Mary Manning.

Finally, since the Lord Chamberlain had no powers over theatre societies whose audiences were members, the play would go on at the Arts Theatre Club and with a less distinguished cast. It would take over eighteen months for this apotheosis to come about. Since the era of *Hair* and *Oh! Calcutta* and multitudinous strip clubs was not far away, the objections to Estragon's pants seem even more curious now than they did at the time. But when Peter Glenville decided to transfer the play from the Arts Theatre Club to the Criterion Theatre they became a pressing matter again. Some compromises were then arranged. Fartov became Popov and the unfortunate Mrs Gozzo was made to suffer from warts instead of the clap. This is Beckett's only traceable decision to compromise in the latter part of his career, though when *Murphy* was receiving rejection after rejection, it will be remembered, he had declared himself ready and willing to make almost any changes that were thought desirable.

In the days when Beckett had been desperately casting around for a career other than in literature he had discussed with MacGreevy the possibility of becoming a recognized art expert – something more than a critic: in fact, a connoisseur. Not only had the trip to Germany had been presented to his mother as a step towards such a possible career, but his travels in that country had been quite genuinely motivated by this ambition. Of course unless one set up as a professional dealer or authenticator the possibilities inherent in connoisseurship were virtually limited to finding a job as a gallery curator or director. MacGreevy's steps towards such an outcome had been much more deliberate than Beckett's half-hearted ones and now, somewhat against the odds, he had actually landed a job as Director of the National Gallery of Ireland. He had been able to state on his c.v. that he had graduated from the École Normale Supérieure of the University of Paris, had worked on the Parisian art review, *Formes*, had served on the editorial staff of the *Connoisseur* magazine, had been a lecturer at the English National Gallery and, at the outbreak of war, had assisted at the evacuation of its pictures to safe havens elsewhere in the country. And of course, this was before the time when increasing academicization in every sphere of life would make it impossible to get such job without the requisite university degrees in art history, appreciation, museology and all the other things that academia delights in. A critic, or even a gentleman, could still land such a job in the early fifties, and perhaps Beckett's own toying with the idea had not been so ludicrous after all.

Dublin gossip, however, inclined to the view that MacGreevy had got his job mainly through clerical influence. Apart from Father Senan, the editor of the *Capuchin Annual* and the *Father Matthew Record*, a certain Monsignor Shine, a well-known collector and prominent figure in certain circles, had supported his application. In the event his record as a gallery director was not very inspiring, which is scarcely to be wondered at since the grant in aid of purchases from the government was a mere £2,000 a year (when the gallery became the beneficiary of one third of the income from Bernard Shaw's estate in 1958, the quality of the purchases recommended to the Board by the Director naturally improved), and all in all MacGreevy's record, his talents and his personality certainly entitled him to the post.

Beckett's response to his friend's appointment was one of whole-hearted congratulation, though he did not envy him what he called the 'nerve-wracking business' of decisions about the apportionment of limited funds. 'Scraps come through to me from my few contacts in Dublin of your great work for the Gallery. I hardly dare hope it is appreciated. Don't wear yourself away and the happiness of the work.' MacGreevy's health was not good and in a previous letter Beckett had declared flatly that he thought it 'impossible to have health in Dublin. Of any kind.' He had also told him he felt he was sure that 'the depressing feeling of always working for others and never for yourself does not make things easier for you'.

For whatever reason – success, removal from Ireland, the realities of the war years, a comparatively settled and domestic routine – the number and variety of Beckett's complaints had diminished. He would still get cysts from time to time, his teeth would give him trouble and so would his eyes, but the panic attacks which had impelled him into psychoanalysis were now a thing of the past.

His affection for MacGreevy was manifested towards the end of 1952 by the enormous trouble he took to entertain two porters from the National Gallery, Thomas and Vincent Hickey, father and son, who had decided to spend a holiday in Paris. MacGreevy had asked him to do something for them. His response was extraordinarily wholehearted and revealing of his own character. He met them at the airport and saw them every day for a week, proudly exhibiting all aspects of Paris to them, from what he called the cultural heights of the Louvre to the carnal depths of a Montparnasse brothel, where he seemed, they observed, to be well known and on very good terms with the management, though they did no more than have a drink there. He seemed to them to be a simple, unaffected, but extraordinarily caring companion and host, taking them to dinner at the Coupole and to a variety show at the Bobino as well as to the pictures. Through the fifties and sixties, virtually only two such music-halls survived – the Olympia on the Right Bank and the Bobino on the Left. Besides being a temple of 'Le Chanson', where singers like Georges Brassens and Juliette Greco attracted the sort of sophisticates who liked demotic culture, the Bobino also offered comedians and acrobats who appealed to less rehearsed tastes and were just as much an attraction as far as Beckett was concerned. He went there frequently, even in the afternoons, and nearly always by himself, though sometimes, as in the present instance, with visitors whom he was showing around Paris. He owed his erudite knowledge of Parisian argot, which far surpassed that of most respectable Parisians, as much to the singers and comedians of the Bobino as to any other milieu, though no doubt the denizens of the late-night establishments which he visited with Giacometti and others contributed their share. His spoken French, incidentally, impressed many people as virtually accentless; though some, such as Georges Belmont, who had a good deal of acquaintance with Irish people, thought that it betrayed recognizably Irish characteristics; and others such as Josette Hayden chose to be linguistically snobbish about it, Josette even going so far as to say it was deplorable.

However this may be, he exhibited to the Hickeys an extraordinary knowledge of Paris which he shared generously with them. He himself insisted to MacGreevy that it was no trouble to take them in hand and show them round and he seems genuinely to have enjoyed their company and their ordinary Dublin working-class responses to what they saw. From Paris, Hickey wrote profusely to MacGreevy, saying, 'Mr Beckett is an Angel. I shall never forget him . . . No father or mother could be kinder or more careful of us in every possible way. Our food, our health, even our sleep he enquires about. I will never, never be able to repay him for what he is doing to make our stay here happy and interesting.'

His showing of the city to the Hickeys was quite a bravura performance, but it would always be his disposition to be kind to Irish visitors provided they had the right credentials, even to the extent of allowing this to interfere with his work. Members of his family were always welcome, including, in the course of time, his nieces and nephews; and so, for the most part, were Trinity personages like H. O. White, who came around this time and was also taken to the Bobino as well as to some well-known Montparnasse cafés and to what were becoming Beckett's favourite restaurants, Chez Mario's and Les Iles Marquises. White came with Leventhal and thus was a special case, but others who had no such claim were given much the same treatment.

A visitor of a different kind was his old flame, Ethna McCarthy, who arrived towards the end of 1953. Ethna had changed direction in the 1940s. She had ceased to write poetry and ceased to hope for a career in the French department at Trinity. She qualified as a doctor, specializing in paediatrics. She had now applied for and, as she thought, had been given a job in the World Health Organisation and had severed herself from Dublin in order to take it up. On arriving in Geneva, however, she found she was confronted with the necessity of taking a severe medical test which she unfortunately failed. Back in Paris she spent her time going from office to office in a fruitless attempt to have this decision reversed or at least to be given an opportunity of taking another test.

Beckett helped her in any way he could, urging her to see Constantine Cremin, the Irish Ambassador, with whom he was acquainted, for he was on the list of Irish residents in Paris who were sedulously invited to the embassy party on St Patrick's Day and other functions. But there is little indication of tenderness in his comments on her visit, which is perhaps not surprising given the lapse of years. If anything, though indignant about the way he believed she had been treated, he was a little impatient with the necessity to remain in Paris until the thing was decided one way or another when he had intended to be at Ussy.

The interruptions, distractions and social necessities of the mid-fifties were less unwelcome to Beckett because in these years he was writing very little. Apart from *Texts for Nothing* he had written nothing since the great creative push that had resulted in the trilogy, the *Nouvelles* and *Godot*. 'I'm horribly tired and stupefied but not yet stupefied enough. To write is impossible but not yet impossible enough,' he told Barney, adding, 'That's how I cod myself these days.' Creatively, he had reached an impasse. He was groping for something and he even made a tentative draft of a mime which would eventually become *Endgame*, but it would not reveal itself as such for some time to come. And meantime, during 1954, there were plenty of distractions. There was a Yeats show at Wildenstein's Galérie des Beaux Arts, which he visited many times. This was supposed to represent a breakthrough for Yeats and to some extent it did, the Musée d'Art Moderne buying a picture, though the gallery was not well attended and the reaction of critics not so wide-spread or enthusiastic as he had hoped. But, as he wrote to MacGreevy, in his own small circle all those whose opinions he valued were enor-mously impressed, 'Bram van Velde, the most difficult and critical of painters' going to the show at least ten times.

In the April number of the *Lettres nouvelles*, those friends and associates who had rallied to Bram van Velde now rallied to Yeats, but Beckett felt guilty that he could not write at more length himself, managing in the end only a short *hommage*. 'I have dreadful difficulty with this form of writing, it is a real torture, and spent days before the blank sheet before I could do anything and the result is no more than the most clumsy of obsequies. But I hope the bowing down is what matters. To prove or defend or describe or situate seems to me all equally superfluous in the presence of such an achievement,' he told MacGreevy, asking him to tell Jack Yeats that he had 'lit a fire that will spread'. He thought they should all aim for a proper show in the Musée d'Art Moderne in the following year and that Wildenstein's had not been the right place.

A more minor distraction, but still a worry, were the approaches of a young American scholar, Richard Ellmann, who, with the approval of Giorgio Joyce and Joyce's executors, was writing a biography. Beckett co-operated with Ellmann, though he would later claim that he did so reluctantly. Perhaps later, as his disapproval of Ellmann grew, particularly after the publication of Joyce's intimate letters to Nora in the 1970s, he would come to exaggerate the degree of his reluctance and the inaccuracies of Ellmann's account of what passed between them, but a very young poet, Desmond O'Grady, was present at some of Beckett's meetings with Ellmann and, while his account does not throw a great deal of light on Beckett's attitudes to the American personally, it is of interest in so far as it suggests that he was well-disposed to the American's project at this stage. It also suggests a Beckett who in the early summer of 1954 had a fair amount of time on his hands, some of which he was prepared to give over to a casual association; and it is revealing in the glimpse it gives of Beckett's attitude to a rather naive young compatriot.

O'Grady had left a Cistercian school and come to Paris to write in accordance with the Joycean gospel of rejection of motherland and mother church. He had very little money and after a while, following Joyce's example, he got a job teaching English at the Berlitz School and rented a little attic room in the rue Mouffetard. One of his colleagues at the Berlitz was an elderly Dublin man named Fred Gallagher. Gallagher appeared to have no literary interests, but he too had left Ireland as a young man and had roamed around the world and then settled in Paris, where he lived with his Peruvian wife. One evening he asked O'Grady if he had heard of an American professor named Richard Ellmann. O'Grady said that he had and that he knew of Ellmann's book on Yeats. Gallagher explained that he had had a letter from Ellmann, who was coming to Paris and wanted to meet him. For a moment O'Grady thought that Gallagher was supposed to know something about Yeats, and it was not until his friend said that he had once known someone called James Joyce who had written books which he understood to be famous, though he had never read them, that O'Grady suddenly realized that he was talking to the prototype of Ignatius Gallagher in the story, 'A Little Cloud', in *Dubliners* and that Ellmann too knew this.

At Gallagher's request, he agreed to go with him to meet the American professor at the Deux Magots on the boulevard Saint-Germain. They found Ellmann sitting at a table on the terrace with a tall, thin, ascetic-looking man in his late forties who was introduced as Samuel Beckett. O'Grady had heard a lot about Beckett from George Whitman, then and since the proprietor of the Shakespeare and Co. bookshop, who was a fount of information about literary Paris and had a great deal to say about the Irishman who had known Joyce and written *Waiting for Godot*. The young poet had read *Murphy* and *Watt* and, at Whitman's urging, he had bought and was about to read *Molloy*, so he was naturally pleased to meet Beckett and contrived to sit beside him. After some time the author of the books he had read courteously enquired what he was doing in Paris, to which O'Grady replied that he had left Ireland for ever and come to the French capital to write poems. At this point Beckett 'almost smiled'. Then, O'Grady's account continues,

> very politely and warmly interested, he asked me if I had published anything yet. Was there a book? I said that, besides being published in my school magazine, I had published several poems in the 'Poets' Corner' of the *Limerick Weekly Echo*. Another almost smile. Now, I added, I had a book forming in my head. And what did I write about, he enquired apologetically, softly. Love, Ageing, Death. Another almost smile. Where was I living in Paris? Rue Mouffetard. He congratulated me on my choice of street . . .

For the rest of the evening Beckett was solicitous that this Irish youth's glass was never empty, that he always had a cigarette and that he did not have to pay for anything. He made very few contributions to the conversation between Ellmann and Gallagher and none at all on the subject of James Joyce, appearing more interested in Gallagher and the Dublin he remembered than in Ellmann's biographical researches.

As they parted he told O'Grady he was happy to meet him and that he looked forward to meeting him again. He then enjoined him to be sure to eat properly and to take care of his health. 'He was very gentle, even fatherly as we shook hands and said goodnight.' According to O'Grady's account, for the rest of Ellmann's stay in Paris the four of them met daily. While Beckett continued to be interested in Gallagher, he turned his attention more and more to the youthful poet exile from Limerick, asking him who the young people were reading in Ireland and if he had ever read Denis Devlin, Thomas MacGreevy or Brian Coffey. Did he and his generation read the Europeans? He also encouraged O'Grady in his resolve to survive outside Ireland and to write poems. Even after Ellmann left they met occasionally.

A more serious inroad into Beckett's time and of course a matter of deep emotional importance to him was the death in August of his brother Frank. He never avoided being present at the illnesses and deaths of his close relatives, and when he heard from Frank's wife Jean that her husband was seriously ill he left more or less immediately for Ireland. What had seemed to be flus and colds had been diagnosed as lung cancer and it was very far advanced. Beckett stayed at Frank's house Shottery in Killiney, sitting with his brother in the garden while they talked of their childhood, and busying himself with the creation of a lily pond which would be sizeable enough for a little boat. There is a photograph of the two brothers sitting by the newly created pond, Sam staring expressionlessly at the camera, Frank wearing a crumpled linen hat and a pullover under his sports coat on what looks like a warm summer's day, tea things on the table, the teapot under a cosy.

When, after a while, it was decided that Frank would have to be told the nature of his illness and that his case was hopeless, Beckett undertook to tell him and resolutely did so.

By mid-August Frank was so ill that he could not sit in the garden and, just as he had done with their mother, Sam kept vigil by his bedside. As he grew worse the stench of illness pervaded the room but grimly and tenderly he remained for long hours with his brother. 'It is sometimes difficult not to get up and walk out,' he told Mary Manning. However different the two brothers were, their bond had always been close and Sam had felt the intensities of the small family world very deeply indeed. He made very few trips into town but Geoffrey Thompson remembered walking back and forth through the courtyard of Trinity late one night when he spoke of his brother's sufferings and the cruelty of a god, if there was a god, who could preside over such a world.

Five weeks after having been told that he faced the end, Frank died.

At the burial in Deansgrange, Sam presented the same stony-faced visage to the world as he had at his mother's funeral. Partly to relieve himself of the burden of work when he originally began to feel unwell, Frank had sold a partnership in what was still a prosperous practice. It was expected that Jean would be comfortably off and would not want for money, but Sam insisted that she should call on him for anything that might be needed in the future, particularly where the children's education was concerned. Then he returned to France and went almost immediately, with some sense of relief, to Ussy.

He spent much of the autumn and winter of 1954 to 1955 there, in a bleak mood, sometimes trying, without much success, to expand on early sketches of what would eventually become what is by common consent his bleakest and most pessimistic play, *Endgame*. He was beginning to use the typewriter more, for correspondence and otherwise. Like many writers he never learned to type properly with all five fingers, still less to touch-type, so that he had to look at the keyboard; but, like most, he worked up a fair speed, banging away with his two index fingers. However, initial composition was still always undertaken in his almost illegible handwriting, which had become if anything more crabbed and difficult to decipher now than ever.

He was beginning to discover that there were people in the world on whom his writing would have a sort of obsessional effect, almost as if they were sacred texts which contained a key to some sort of salvation. In later years some of these people, often young and neurotic, not to say slightly deranged, would go to great lengths to meet him and would even camp outside his door. His first brush with such a person occurred at this time and was very disturbing in its effect. Some weeks before, he had received a letter from the warden of a jail in Germany, Lüttringhausen, where the prisoners wanted to stage a German version of *Waiting for Godot*. According to the Governor, a particular prisoner had been the inspirer and prime mover in this and his life and character had been transformed by contact with the play, which had changed him from a violent and troublesome inmate into a co-operative and gentle one. The Governor asked if Beckett would be willing to correspond with this inmate about the play. When Beckett replied in the affirmative, a correspondence followed, in the course of which Beckett declined an invitation to visit the prison but felt impelled by courtesy to say that if any of the prisoners were ever in Paris they could, of course, visit him.

One day in January Roger Blin received a visit at the theatre where he was working from the convict whose life and outlook had been so deeply affected by the play. He seemed half-starving and half-frozen so

Blin took him for a meal and then back to his apartment for the night. There he discovered that the man had broken his parole to come to Paris and was obsessed with the idea of meeting Beckett. There was still no telephone at the rue des Favorites so he left messages. At last, as was sometimes the outcome, Suzanne returned his calls. When Blin explained the situation she was adamant that they should have nothing to do with the convict, who should be sent back to where he came from. This was little use to Blin, however, for the convict was now wearing his clothes and had settled in the apartment. He had no identity papers and nowhere to go. Finally Blin made contact with Beckett himself but discovered that he was quite as determined as Suzanne to have nothing to do with the man or his situation. He suggested that Blin should tell him that he was out of Paris and would be for some time, and gave him a large sum of money – quite enough to enable the fellow to get back to Germany and to wherever he might intend go thereafter.

Armed with this money, Blin returned, explained that Beckett would be absent for quite a long time and handed over the money. But to his dismay, his guest refused to go, and worse, he seemed to be enjoying life in Paris, going out to the bars and returning late at night. The actor's fears for his own safety had now increased, more especially since he had learned that the original offence for which his guest had been imprisoned had been a violent knife attack on somebody. He was therefore immensely relieved when one day he came back to the apartment to find a courteous note from the convict thanking him for his hospitality and saying that he was returning to Germany. The note made no mention of Samuel Beckett.

Chapter Twenty-Eight

In the early summer of 1955 the deferred production of *Godot* in London
began to take shape. Beckett approached this, as he approached all
things English, with mixed feelings. London had been for many years
the focal point of his hopes for success as well as the place of his failure;
but when the difficulties with the Lord Chamberlain began he found
himself unable to care whether they were resolved or whether the play
ever saw the boards there or not. He had had little input into the
production and equally little confidence in its reception. Partly to protest
against the Lord Chamberlain's interference he decided not to attend
the opening. On the first night, 3 August 1955, the hostility of the
audience became apparent to the actors not long after the curtain had
risen. There were titters and laughter of the wrong sort and soon a
noisy exodus began. At the line 'I have been better entertained elsewhere', there was ironical laughter and cheering and when one of the
characters yawned, a wit in the stalls yawned even louder and longer
to much applause.

At the beginning of the second act there were few people in the
theatre and the curtain came down to not very enthusiastic applause
even from them. Hardly anybody came round afterwards and those
who did had evidently spent a good deal of time in the bar for, according
to Peter Bull who played Pozzo, 'most of them were in a high state of
intoxication and made even less sense than the play'. Nor were the
notices the following day of the sort to restore the flagging spirits of
the director and cast, most of them examples of good old English-
highbrow-bashing, not to say pretentious-French-highbrow-bashing,
an activity of which Milton Shulman's review in the *Evening Standard*
provided a good example. He saw *Godot* as just another of those plays
which hoped to impress people through obscurity and thought that,

though dependent on its symbolism, its symbols were 'no more profound or demanding than those in a nursery version of *Pilgrim's Progress*'. As Beckett was to put it, the 'shopkeepers' – that is – the English, had made mincemeat of his play. But he did not really care. He was tired of what he called the 'endless misunderstandings' of it. 'Why', he asked, 'do people have to complicate a thing so simple I can't make out.'

But at the weekend the tide dramatically turned. The two principal English opinion-makers of the day were Kenneth Tynan in the *Observer* and Harold Hobson in the *Sunday Times*. Both were courageous, enthusiastic and, on the whole, perceptive. Both urged people to go and see the play. Hobson said that though a playwright philosophizing might be questionable, it was essential that he should philosophize with swagger. 'Mr Beckett has any amount of swagger. A dusty, coarse, irreverent, pessimistic, violent swagger? Possibly. But the genuine thing, the real McCoy.' And he said of the exchanges between Vladimir and Estragon that they had 'the simplicity, in this case the delusive simplicity, of music-hall cross-talk, now and again pierced with a shaft that seems for a second or two to touch the edge of truth's garment. It is bewildering. It is exasperating. It is insidiously exciting.' Beckett found himself oddly moved by Hobson's notice, which he said was touching and courageous.

But though Hobson was to become legendary as a supporter of Beckett, at this distance of time Tynan seems the more wholehearted of the two. Claiming that *Waiting for Godot* 'asserts and proves' that 'a play . . . is basically a means of spending two hours in the dark without being bored', Tynan went on to say that the author of *Godot* was an Irishman now living in France, a fact which should have prepared the audience for 'the extra, oddly, serious joke he now plays on us. Passing the time in the dark, he suggests, is not only what drama is about but also what life is about.' He too stressed that debt to vaudeville, invoking Buster Keaton and Charlie Chaplin and declaring that Laurel and Hardy would have been the ideal casting for Pozzo and Lucky. 'The play', he said, 'sees the human condition in terms of baggy pants and red noses . . . it summoned the music-hall and the parable to present a view of life which banished the sentimentality of the music-hall and the parable's fulsome uplift. It forced me to re-examine the rules which have hitherto governed the drama and, having done so, to pronounce them not elastic enough.'

With these two notices *Godot* and its author were launched in England, more especially since Tynan had called attention, with some slight exaggeration, to the play's 'enormous success in Europe over the past three years'. The *Times* critic returned to the play during its second week to see what kind of audiences it was now attracting and reported that 'in the attentive silence one could almost hear the seeds of a cult growing'.

On 12 September the production transferred to the Criterion Theatre and it ran there for eight months to more or less full houses. Beckett himself agreed to come over to see the play in early December and he went every night for five nights in a row with Alan Schneider, who was to direct the American production in the New Year. Schneider remembered Beckett touching his arm every now and then and quite audibly whispering – at least in Schneider's version of his accent – 'It's ahl wrahng! He's doing it ahl wrahng', when various parts of the production displeased him. 'Every night also', Schneider said, 'we would carefully watch the audience, a portion of which always left during the show. I always felt Sam would have been disappointed if at least a few hadn't.'

For the moment, however, the degree to which Beckett was known in England was almost entirely contingent on *Godot*. He was the author of a strange controversial play which, as in Paris, had become a topic of conversation and had supplied, in its title, a phrase which rapidly attained almost proverbial status. The fact that it had been translated as *Waiting for Godot*, while the French *En attendant Godot* more exactly meant '*While* Waiting for Godot', gave it more universal applicability. It could be used in dozens of minor human situations as well as with reference to the fundamental, major one – and it was. Meanwhile, everywhere it went, controversy raged about the 'meaning' of it. A piece by G. S. Fraser in the *TLS* fuelled, in intellectual circles anyway, the debate about Beckett's optimism or pessimism, Christianity or atheism. Existentialism, and more recent currents of French thought were still a bit of a mystery to most people in England and even to English intellectuals, but Beckett's play and Beckett's attitudes could be handily labelled existentialist and discussed accordingly.

The fact that his celebrity as a controversial playwright far outshone whatever reputation he gained by his prose narratives in the English-speaking world irked Beckett and would continue to do so over the years. He was jealous on behalf of his other work of the fame of his plays, *Godot* in particular. It had been written quickly – as he sometimes pointed out, in three months. He had not 'expected it to go so far', and he had put far less sustained effort into it than he had into his comparatively unread novels. When the Olympia Press edition of *Molloy* was published later in 1955, some review copies were distributed in England and a little of the balance was redressed, a few of the reviewers – Vivian Mercier in the *New Statesman*, for example, and the present writer in *Time and Tide* – at least being aware that Beckett's prose writings were as important as his plays and, in certain circles, more highly esteemed.

The most adverse review was Philip Toynbee's in the *Observer*. Toynbee declared that Beckett belonged to 'the dwindling phalanx of the European avant-garde' and that 'his reputation is very high along the Boulevard St Germain'. Yet, he said, he was 'by no means a solitary or original figure for his play and his novels follow the current Paris fashion without demur'. Claiming that he belonged to a tradition in French nihilistic writing which included Jarry, Lautréamont and Sade as well as Sartre, Camus and Genet, Toynbee alleged that what Beckett had done was to carry 'his despair and disgust to ultimate limits of expression – indeed beyond them'. He went on to raise an objection to the very existence of the novel, which would remain currency in certain circles for some time, claiming that

> surely the whole point of the thesis that life is horrible and meaningless and nothing else must be that there is no more than this to be said about it . . . By continuing to live, and still more by continuing to write, the author refutes his own message and it is no use saying, in such a case, that we must not confuse the creator with the creature and so on. This book is a serious statement of a personal attitude or it is nothing. I am inclined to think that it is nothing.

The point is of course that even if existence is 'horrible and meaningless' it may be incumbent on the sufferer to soldier on. There may be moral reasons arising from a sense of obligation to give of one's best to one's fellow sufferers or to avoid giving pain to one's intimates. There may be the imperative or seeming imperative for certain natures of some sort of creative or artistic expression. There may indeed be an imperative so inexplicable and in itself apparently meaningless that it can only be described as categorical. Toynbee was on surer artistic ground, perhaps, when he called for a more inclusive vision, saying that *Molloy* expressed 'an attitude to life which cries out for at least some opposing one'.

Molloy was swiftly banned in Ireland, though scarcely for its lack of inclusiveness. It was one of the last serious works of literature to be prohibited because it contained passages which were 'obscene' in 'general tendency'. Since the Censorship Board had no apparatus of its own for the purchase of books, it depended largely on institutions and private busybodies who called its attention to what was considered obscene matter, or on the customs authorities who sometimes referred copies of imported books to it. It had been saved the trouble of banning *Ulysses* because the British customs had seized the book and no copies ever reached it.

In November *Nouvelles* and *Textes pour rien* were published in Paris by Les Éditions de Minuit. The review by René Lalou in *Nouvelles littéraires* summarized the growth of Beckett's French reputation to date and said that though *Murphy* had been largely ignored, he had 'conquered the audience of literary critics four years later when *Molloy* and *Malone Dies* appeared' and many had declared that 'this Irishman, with his complete mastery of our language, was one of the most original post-war writers'. It listed those critics – Jean Anouilh, Robert Kemp, and Gabriel Marcel – who had been among the first to hail *Waiting for Godot* and 'to proclaim the value of this tragedy of despair not even lit by a glimmer of consciousness'. Lalou thought that the work under review exhibited once again Beckett's 'constant use of monologue as an artistic technique, his implacably pessimistic vision and his insistence on the degrading functions of the human body', and he ended, somewhat plaintively, by saying: 'In searching for the man behind the writer, I would like to ask Samuel Beckett the following – is he wholeheartedly committed to the grievances that these beings . . . who have known only the baseness of life . . . seek to bring against the human condition?'

One cannot of course imagine Beckett answering such a question, except through his work; however, since he was not working on anything new, it was a time for café life, mostly still in the Dôme, the Select, the Coupole and the Closerie des Lilas – though he tended to go to the latter for solitude, or for prearranged meetings on a one-to-one basis. For late-night drinking he went to bars such as the Falstaff and the Rosebud, both small and crowded places in Montparnasse. Here he would sometimes meet Joan, an encounter which always pleased him, whether or not Riopelle came along as well; but a new late-night drinking companion was the illustrator of *Textes pour rien*, the Romanian-Jewish painter Avigdor Arikha. With Arikha he would drink whiskey into the small hours, moving from bar to bar and sometimes covering great distances as they crossed and re-crossed the Seine from Right Bank to Left. With his red curly hair, bull-like aspect and general ebullience and volatility, Arikha supplied some want that Beckett felt. His friends tended to fall into two categories: those with whom he felt sufficiently at ease to maintain a mutual and comprehending silence, and those who were talkative enough to fill up the silence and distract him – even, in some cases, with gossip and chatter. He was a good listener, seeming to give all his attention to what one was saying and inclining his head a little as he listened.

Arikha at this time was emerging from an abstract phase, partly influenced by Giacometti, who had urged him to become more figurative. In the sixties he would renounce abstraction altogether and at his first non-abstract exhibition Beckett positioned himself 'like a Swiss Guard' in front of the paintings, gazing in his sometimes disconcerting way at the critics circulating round the gallery as if challenging their response. He wrote a text for the catalogue in which he described the process Arikha was now engaged in as 'siege laid again to the impregnable . . . eye and hand fevering after the unself'.

Beckett did not have a telephone until later in the decade and those who wished to communicate with him urgently used the *pneumatique*, a form of letter service which virtually guaranteed an answer since the messenger would wait until the return form had been completed. Neither did he have a television set, but he would acquire both telephone and television towards the end of the 1950s, the television being a small black-and-white one which he got largely in order to watch sporting events, principally rugby and tennis. He still played tennis occasionally in the afternoons, sometimes with Suzanne, whom he had met originally on the tennis-court. In the evenings he would play billiards, another game at which he excelled, often at Les Trois Mousquetaires, a well-appointed billiard saloon in the rue de la Gaieté, a street between the avenue du Maine and the boulevard du Montparnasse. Here he played with a wide range of acquaintances, whose only bond was their fondness for the game. French billiards differs from English: the table has no pockets and scoring is a matter of contact between the balls; and Beckett's billiard career resembled his linguistic progress in that, though brought up to the English game, he became a master of the French.

Since he went out virtually every evening, either by prearrangement with somebody or without; and since a Frenchwoman of Suzanne's outlook and disposition had no intention of hanging round the billiard tables of Les Trois Mousquetaires or drinking whiskey in the Rosebud, they saw little enough of each other and he had very little home life. They still went to concerts together, sharing an enthusiasm for the works of the moderns – Bartók, Schoenberg and Berg (whose *Wozzeck* Beckett considered to be one of the great works of art of the century). There were also small dinner-parties, frequently with the Lindons, very much formal French occasions, with dinner at a polished wood table in the dining area of the Lindons' flat; but apart from these outings they were leading largely separate lives. However, though Suzanne could frequently be snappish and sarcastic, they were usually on friendly terms and the bonds of association were still strong, more especially since Beckett was exceptionally loyal, found the ties of old association hard to break and seldom in any case really wished to be free of them. Strong though this loyalty was, however, it did not rule out flirtations, casual sexual encounters or even affairs, including one in 1954 with a thirty-two year old visiting American, Pamela Mitchell, which he took immediate steps to terminate when she announced that she had decided to settle in Paris. The life he was leading at this time was an unsatisfactory one, as he knew, writing to MacGreevy to say 'it is time now I made big changes in my way of living, but I doubt if I have energy.'

The summer of 1955 was, however, a very beautiful one and, sometimes with Suzanne, sometimes alone, he spent much time at Ussy, where he read a lot and the days passed in a way that seemed less wasteful and more harmonious. He would sit for long periods watching the birds in nearby trees through a pair of binoculars.

This was also the year of Beckett's solitary excursion into mime, a form whose popularity in France had been renewed by the war-time success of *Les Enfants du Paradis* and the triumphs of such artists as Deryk Mendel and Marcel Marceau. Perhaps he hoped that a wordless mime might end his creative blockage, but his involvement came about partly through the fact that John Beckett spent part of 1955 in Paris. John was Sam's uncle James's son, so he and Sam were first cousins, though in terms of age they could well have been uncle and nephew. John was a composer, a pianist and a harpsichordist. Portly and dark, he was quite unlike Sam in appearance but, as an artist himself and with a knowledge of painting as well as books and music, he was closer to him in spirit than the rest of the family. In Dublin he belonged to a circle which included Brendan Behan and Patrick Swift as well as the present writer and he admired Sam and his work enormously. Beckett saw a great deal of him in Paris in 1955, taking him to his favourite bars and cafés and when he left he missed him a great deal. Together they collaborated on what became *Acte sans paroles*, a mime written in response to an invitation from Deryk Mendel. Although it was the last balletic mime that Beckett would write, he did compose what might be called schematic ones for German television many years later.

In 1960 he was asked for a second ballet by Patricia Ryan, wife of the editor of *Envoy*, who was pioneering the production of ballet in Dublin and commissioned scenarios from poets Patrick Kavanagh and Donagh MacDonagh. Writing to thank her for the request, Beckett said,

Perhaps some day I shall be able to offer you something of the kind you want, but for the moment I am afraid it is quite impossible. My friend Deryk Mendel, who played *Act without Words* in London and Paris, has already asked me for a ballet with words (argument and text). I feel quite unable to write such a work. If I ever do I must give it to Mendel. And then if I ever do a second, I shall be happy to give it to you.

Plans were now well under way for an American production of *Godot*, to be directed by an enthusiastic young American, Alan Schneider. Since he had an affluent American producer to back him Schneider was able to come to Paris and lodge at what he called a 'very plush' hotel near the Étoile. Within an hour of sending the *pneumatique* he had a phone call from Beckett saying he would meet him in the lobby but insisting, as he had done with Barney Rosset, that he only had half an hour or so to spare. Schneider had armed himself with a large bottle of Lacrimae Christi as a present and with this in hand he stationed himself in the 'rather overdone' lobby and waited for the unknown quantity who was Beckett. Promptly on time he strolled in, 'his tall athletic figure ensconced in worn short coat; bespectacled in old fashioned steel rims; his face as long and sensitive as a greyhound's'. After initial greetings it seemed to be decided between them that they would neither stay in the hotel lobby nor go to a bar and so they wandered out with a vague intention of finding somewhere where they could have a drink of the Lacrimae Christi together. Beckett brought the state of indecision to an end by suggesting that they take a taxi back to his apartment in the 6th and there, his appointment seemingly forgotten, they wound up drinking a large part of the bottle together at the kitchen table.

Schneider, who had been very excited by the commission to direct and had been poring over the play on the boat coming over, plied him with all his pent-up queries. Beckett, he thought, tried to answer directly and honestly; but as far as the larger questions were concerned provided little illumination. The first question was: 'Who or what does Godot mean?' to which the answer was immediately forthcoming: 'If I knew I would have said so in the play.' Beckett seemed perfectly willing to answer questions of specific meaning or reference but would not go into questions of larger or symbolic meanings, preferring his work to speak for itself and letting the supposed 'meanings fall where they may'.

Before Schneider departed they arranged to meet the following evening at what the American remembered as 'one of his favourite restaurants in Montparnasse', probably the Iles Marquises, which he was beginning to favour to the exclusion of all others except occasionally the bistro of the Closerie des Lilas. Under Suzanne's influence he had now virtually renounced meat, and though he ate it when he was a guest at somebody's dinner-table, left to himself he ate fish. After the meal Schneider persuaded him to come to a performance of *Anastasia* at the Théâtre Antoine. The American had directed the New York production and was interested in seeing how it would be done in Paris. It turned out to be 'very artificial and old fashioned, and Sam's suffering was acute'. Evidently Beckett did not go to any great lengths to disguise his suffering.

After the theatre Beckett took Schneider to Fouquet's, Joyce's old haunt on the Champs-Élysées, where, after drinking a good deal of wine, they separated just before dawn. Schneider was one of the many people who, on closer acquaintance, fell under Beckett's spell. 'I came', he said, 'with respect; I left with a greater measure of devotion than I have ever felt for a writer whose work I was engaged in translating to the stage.' The proposal with which he had come to Paris was that Bert Lahr and Tom Ewell should play the main parts in *Godot*. Lahr was a well-known comic actor who had had a huge popular success with his performance as the cowardly lion in *The Wizard of Oz*, and Ewell was also a familiar name.

Partly because of the image of these actors in the popular mind and partly because of the producer's advertising campaign which called the play 'the last sensation of two continents', the opening in Miami in January 1956 turned out to be a disaster. There was a celebrity audience which included Tennessee Williams, and at least two thirds of them left after the interval. In the days that followed the actors received several abusive letters, in one of which, addressed to Bert Lahr, the writer wanted to know 'how can a man, who has charmed the youth of America as the lion in *The Wizard of Oz*, appear in a play which is communistic, atheistic, and existential?' After two weeks the production was taken off. As John Lahr was to put it in his biography of his father, *Cowardly Lion*, the producer, Myerberg, 'had billed the production falsely, mounted it outrageously, and brought it to a town with no sympathetic audience to sustain an experimental play'. In the aftermath, Schneider was fired and Ewell left the cast.

A broken-hearted Schneider wrote Beckett a short letter in which he took the blame. Beckett's reply displays his magnanimity and sense of commitment to people to whom he had once given his loyalty – 'for the moment all I can say and all I want to say is that this Miami fiasco does not distress me in the smallest degree, or only insofar as it distresses you.' And the letter contained a statement of his attitude to such matters: 'Success and failure on the public level never mattered much to me, in fact I feel much more at home with the latter, having breathed deep of its vivifying air all my writing life up to the last couple of years.'

But for all his faults the producer, Michael Myerberg, was still courageously set on bringing the play to New York – and with Lahr. This time he was determined to ensure that there would be no advance misconceptions; and perhaps overdoing the honesty and forewarning bit, he placed an advertisement in the *New York Times*, warning away all but the 7,000 intellectuals whom he felt might be receptive to the play. The play, it said, would be offered for a limited engagement of only four weeks and respectfully suggested 'that those who come to the theatre for casual entertainment do not buy a ticket to this attraction'.

In fact the play, as in other cities, ran and ran, arousing conflicting emotions in its audiences and a disproportionate amount of discussion of its meaning. Symposia were organized by the producer at which these discussions took place and, still in search of meaning, many bought copies of the recently published Grove Press paperback offered for sale in the foyer. (Through the years this would sell well over a million copies.) As Lawrence Grover has written, 'in a few weeks, *Waiting for Godot* was being discussed less as new play than as a cultural episode.' There were essays by Norman Mailer and others which apprised people of the significance of Beckett's work; and this time the critics bore in mind the play's European reception, so that even those who were somewhat dismissive, such as Walter Kerr of the *Herald Tribune* and Brooks Atkinson of the *New York Times*, were cautious.

A week after the opening an ungrateful Beckett wrote to MacGreevy to tell him that *Godot* was playing on Broadway 'in what seems to be a dreadfully wrong and very vulgar production. On the whole the criticisms I have seen have not been too bad, but they don't tell me how the play is going and I am quite in the dark as to that.' It is possible hearsay had deceived him here, for some accounts of the production, which was directed by Herbert Berkhof, notably Eric Bentley's, suggest a fair degree of sensitivity. There is no doubt that productions everywhere benefited from the feeling that obscure works of art hold large meanings or messages which can be separated from their text and paraphrased; that they can, in other words, be 'interpreted' as allegory or otherwise. This feeling, a hangover from some notion about the greatness and wisdom of artists which grew up with the romantic movement, was certainly stronger forty years ago than it is now. There is of course no means of knowing what proportion of *Godot*'s early audiences were moved by those things in the play which exist irrespective of meaning in the larger sense: the bleakness, the little joys, the large hopes and the futile loyalties of the human state.

Later in 1956 Beckett discussed the worldwide success of the play with a Dublin acquaintance, Alec Reid, and told him that he thought it was based on basic misunderstandings. Both critics and public were eager to insist that the play could be interpreted in allegoric or symbolic terms when the reality was that it strove at all costs to avoid definition. 'The end', Beckett said, 'is to give artistic expression to something hitherto almost ignored – the irrational state of unknowingness where we exist, this mental weightlessness which is beyond reason.'

Writing in the *New Republic*, Eric Bentley was, incidentally, the first to note that there is a once well-known play of Balzac's in which the characters spend the whole evening waiting for a character called Godeau who never arrives, or at least never comes on stage, for his arrival is announced just before the final curtain falls. Beckett would deny to more than one interlocutor that he had any knowledge of Balzac's work before he wrote his own play.

There had now also been a Dublin production, directed by Alan Simpson, at a little theatre called the Pike. Beckett had originally been worried that Simpson might have to ask for alterations in the text for the Dublin stage because of certain crudities of language, if for no other reason; and of course he would not consent to this. However, Simpson courageously staged the play as it was, altering, unbeknownst to Beckett, only one phrase, though not for censorship reasons; indeed he restored the phrases that the Lord Chamberlain had insisted be removed for the London production, incurring some obloquy for doing so. The general reception of the play was, however, much the same as in the other capitals, though perhaps the audience's enjoyment of certain subtleties was keener.

457

Towards the end of 1955 Beckett finally overcame the creative blockage from which he had been suffering for almost four years with the beginning of serious work on a new play, which finally became *Fin de partie* or, in English, *Endgame*. It undoubtedly had a long gestation, perhaps going back to 1954; and to begin with it seems to have been merely another mime, 'Mime du Rêveur', or 'Dreamer's Mime'.

Of the various drafts, or seeming drafts of what was to become *Endgame* that survive, one concerns someone called Ernst who has a wife, Alice. Like Hamm in the finished play, Ernst sleeps with a handkerchief over his face. He spends his days lying on a cross, a mechanical device that has to be raised during the day and lowered during the night. The fragment is full of Christian imagery and references. Alice calls Ernst 'mon petit Jésu' and even washes his feet. But all this is of course too overt for Beckett. Somewhat closer to the finished play are two other fragments, one of which is actually called 'Avant fin du partie'. In the latter, which is in red ink, X and F are a typical Beckett couple, bound together by a mutual need that is indistinguishable from – or at least all that is left of – love. But they are also, like Hamm and Clov, master and servant. Like Hamm and Clov too, they are waiting for the end of things, but meantime they have to amuse each other. One of their means of doing this is by playing games. In one particular game X creates a situation in which his mother disappears. Some people say that she has gone off to Paris to try her luck as a whore, but he knows she will come back. When F, entering into the spirit of the game, enters dressed as the mother, X asks for a kiss, then wants to be taken for a walk. But when he asks her if she loves him she answers, 'No, *mon pigeon*, I do not want to lie to you.'

Some of this was possibly also too overt for Beckett but gradually, as he retreated to his 'hole in the Marne mud' and went on 'struggling with a play', what transpired became *Endgame*, at first in two acts, then in one but a longish one. *Endgame* again exploits the master and man situation which has been so fruitful in the world's literature. And again, like Didi and Gogo, Clov and Hamm are waiting, in this case perhaps for the end of all things. The running device which can be relied upon to keep the action going when all else fails is Clov's threat to leave Hamm, but they are still a couple, bound, like all Beckett's couples, by deep ties which do not preclude impatience, anger, even loathing of each other.

Much of the play had its genesis in the life and circumstances of Cissie Sinclair towards the close of her days. During these years she was living in a small house at Raheny, County Dublin, with her daughter Deirdre. Totally crippled by rheumatoid arthritis, Cissie was confined to a wheelchair, and when Sam came from France to see his mother he would visit his aunt. If he had a car he would take her for a drive; otherwise he would push her in her wheelchair along the sea road. When these visits were imminent she used to tell her friends with great pleasure, 'Sam is coming'; and afterwards she would advert for days to the fact that 'Sam was here'. The arthritis had made her body hard and stiff like marble. Once when Beatrice Glenavy called she found Cissie leaning sideways in her chair and was asked to 'straighten up the statue'. As Beatrice did what she was asked she saw tears of what she interpreted as laughter in Cissie's eyes.

Hamm's statue-like confinement to his chair comes from Cissie's situation, but so do other things. Cissie had an old telescope, given her by Beatrice Glenavy, through which she watched ships in Dublin Bay or seabirds feeding on the sands at Raheny when the tide was out. Like many Dublin strands the one at Raheny is a long foreshore, and when the tide is out the sea recedes a considerable distance. At most seasons of the year it has an extraordinarily bare, deserted appearance, the empty world effect increased by the enormous expanses of sea and sky. Clov looking out over the empty world through his telescope was inspired by Cissie's surveys of sea and strand at Raheny. But however much these details spring from her life, there is nothing of Cissie in the character of Hamm, though there may be a little of Beckett himself in Clov, patiently waiting on various invalids and pushing Cissie in her wheelchair.

Attempts have been made to interpret *Fin de partie*, or *Endgame*, in terms of the game of chess, for which Beckett had a lifelong passion. The endgame is the last of the three parts into which games of chess are said to fall – opening, middle and endgame – and is the part of the game when advantages are pressed home or desperate bids to nullify them are made. When it was produced in Berlin in 1967 Beckett told one of the actors, 'Hamm is a king in this chess game lost from the start . . . Now at the last he makes a few senseless moves as only a bad player would . . . He is only trying to delay the inevitable end . . . He's a bad player.'

Work on the new play was as usual interrupted by various distractions. In the spring of 1956 an old friend wrote to say that she would be passing through Paris on her way to the South of France where she had a place and asked him to have lunch with her. This was Nancy Cunard, and in view of their past association Beckett came up from Ussy to see her. Nancy was now sixty and mental deterioration and a rather sad and lonely – indeed Beckett-like – end were not many years away, but they were reasonably cheerful at lunch and Beckett was moved by the remembrance of her kindness to him almost thirty years before, of the money she had given him and the publication of *Whoroscope*. She told him that Henry Crowder was dead and gave him some news of Aldington, who was living in France near Montpellier. She was short of money and was selling copies of *Whoroscope* for £1 apiece. They gossiped about other old acquaintances, including Walter Lowenfels; but when she had gone he was impatient to get back to the country and away from Paris, which he found more and more tiring.

A more momentous visitor to Paris than some others who sought a meeting with him that summer, however, was John Morris, then Controller of Programmes for the BBC Third Programme. A noble experiment in minority programming, the Third was then at the height of its prestige and influence and had gathered round it an assorted bunch of highly talented and sometimes eccentric people. As well as being unafraid of broadcasting lengthy poetic dramas and features, the Third had come to have a frequently Irish flavour. This was partly because of the number of Irish people it employed, like the poets Louis Mac-Neice and W. R. Rodgers, or those who had intimate knowledge of Ireland, such as David Thompson, R. D. Smith and Francis Dillon. John Morris was an enthusiast who had seen *Waiting for Godot* and had been greatly delighted by its language, its mood and its humour. The possibility of doing it on the air had been discussed, but it had been decided on high that it would be impossible to do an uncut version while the Lord Chamberlain's office were still worrying about such matters as Estragon's trousers and Mrs Pozzo's clap. After some discussion with his colleagues Morris wrote to Beckett asking if he had considered the possibility of writing for radio and reassuring him about the receptivity of the medium. Beckett had not considered doing so, but he did not rebuff the advance; and a few days later he had the beginnings of an idea, writing to Nancy Cunard at the beginning of July to say that he had 'never thought about a radio play technique, but in the dead of t'other night got a nice gruesome idea full of cart-wheels and dragging feet and puffing and panting which may or may not lead to something'. What it led to eventually was *All That Fall*, but it was still a fairly inchoate idea when John Morris arrived in Paris later that month to seek to persuade and enthuse Beckett further.

They met in Montparnasse on the morning of 18 July to talk about it and Morris was sufficiently encouraged to write that evening to Val Gielgud, the head of radio drama, to say that Beckett was 'extremely keen to write an original work for the Third Programme', adding, 'I got the impression that he has a very sound idea of the problems of writing for radio and I expect something pretty good.' When he got started Beckett worked rapidly and Morris soon appointed a producer. This was Donald McWhinnie, a stage director who loved and under-stood radio and was then under contract to the Third. It was a lucky choice. McWhinnie had very original ideas about production, but he was at the same time deeply sensitive to Beckett's aims and had no intention whatever of interfering with his text. Although Val Gielgud, who was what used to be called a low brow, indeed something of a professional one, was never enthusiastic about Beckett, the word had now spread in the Third and there was a growing band of enthusiasts there which included – in addition to Morris and McWhinnie – Barbara Bray, the Third's chief script editor, whose relationship with Beckett was to become one of central importance in both their lives.

Barbara Bray was a highly attractive and vivacious young woman who had come down from Cambridge after the war with a good degree and, after lecturing for a period in Egypt, applied 'with some scepticism' for a job in the BBC. She had been married, but the tragic death of her husband, also a university lecturer, had left her a widow with two young children. Extremely intelligent and fiercely independent in her views and outlook, she was a notable addition to the Third. Outside broadcasting circles she was already acquiring a reputation as a reviewer and translator, mostly of books which had already had some impact in France, where there was a great deal going on in the 1950s which was known about in England but not very well understood there. In fact comparatively few people could discuss these matters at that time with such verve or authority. Gifted with a sense of humour, she was an eager and good talker with a wide range of interests in literature and public affairs. With her sceptical Jewish background, she was tolerant of differences in character and temperament while being rather fierce and implacable about standards of conduct in public affairs. She worked closely with Donald McWhinnie, and after John Morris's visit she corresponded with Beckett about the production of *All That Fall* and came to Paris with Donald a couple of times to continue the discussions. When, early in 1957, McWhinnie was seconded to television, most of the Third's dealings with Beckett, which now included a further drama project as well as readings from his prose, devolved on Barbara; and as the relationship between Beckett and the BBC grew and developed in the following months, so did his relationship with her, until finally it became a close personal one. One wonders whether he was at all struck by the coincidence of appellation between Barbara and Mrs Bray, his unforgotten nanny, whose name he had given to the figure in his early unpublished story, 'A Strange Case', who had been the hero's nanny and was the object of his earliest sexual desires.

Besides *Fin de partie*, his task for the summer was supposed to have been an English translation of *L'Innommable* for the Grove Press, but he found this job both onerous and difficult, requiring a courage and persistence that he sometimes felt he had not got, so he was glad of a respite. *All That Fall* was mostly written at Ussy between July and September 1956, its author finding it 'queer to be struggling with English again' after so much writing in French.

He had hoped to finish *Fin de partie* in time for Blin to do it at a festival in Marseilles during the summer but it was not finished until the end of June 1956. Having started as two acts, it wound up as one longish one which he thought would take about an hour and a quarter and which he described to Schneider as 'Rather difficult and elliptic, mostly depending on the power of the text to claw, more inhuman than *Godot* . . . I'm in a ditch somewhere near the last stretch and would like to crawl up on it.' Since the stakes are higher in the theatre than in publishing, rejections and refusals are common, even in the case of well-known playwrights at the height of their success; and in spite of its author's worldwide reputation the play met with refusals when Blin started to offer it to various managements. As Beckett put it to Mac-Greevy, theatre owners seem to expect an author to arrive with his text under one arm and millions of franc notes under the other. 'With *Godot* after all, we had a State grant of 750,000 and now nothing but a gloomy graceless act, a complicated mime, and nos beaux yeux.' In succession, the Théâtre des Champs-Élysées, the Marigny, the Théâtre des Oeuvres and others refused the play, sometimes quite cursorily, sometimes after a little toying with the idea.

It was a dark wet autumn and, with *Fin de partie* and *All That Fall* finished and no other wind of inspiration carrying him forward, Beckett was depressed. He carried his gloom with him even when he went into company and wondered how even old friends could put up with him. Like Murphy, who had dreamed of the reversal of the direction of the kick, he even had thoughts of revenge on the people who were denying him the opportunity of seeing his new play on the stage, something he desperately wanted to do in order to find out, as he put it to Schneider, whether he was 'on some kind of road, and can stumble on, or in a swamp'. It seemed extraordinary to Blin and others that in the midst of his supposed worldwide success Beckett could not get a new play produced.

As he brooded sullenly he was aware of his condition of alienation. 'It is not easy to get through the ages from self so estranged,' he wrote to MacGreevy. In Paris, when he socialized, there was the constant temptation to drink, but he could not do it without drink. Probably he drank too much, and he was certainly aware of the danger. 'The trouble with palliatives is that they become the whole of life,' he wrote to Barney, who was also depressed. At least in Ussy he was content with 'a few glasses of wine at dinner'. To console the unhappy Barney – whose difficulties were, as was often the case, domestic – he declared his belief that it was difficult to be anything other than unhappy for more than a few minutes at a time 'with the help of dope, or work, or music, or the other' – the other being, presumably, sex. One should, he said, 'stick it out for the sake of these'; but he did acknowledge the possibility of domestic happiness by saying, 'And if you have found someone you'll be all right.'

Music had been all his life one of his chief consolations, but he did not listen to it lightly and sometimes grew very angry when it was treated as mere background. When he listened to recordings of music he liked, he would do so very intently, sometimes leaning forward in his chair and seeming to follow every note.

For much of the autumn of 1956 he was at Ussy. There, too, it rained incessantly. He walked as much as he could in the Marne mud but the weather did nothing to cheer him up. In fact the silence and stillness of the countryside seemed to convey futility more than the bustle and activity of the city. 'Everything is drenched and dripping here, and without definite work on the stocks there does not seem much point in this deep silence and emptiness,' he wrote to MacGreevy. But he was reading a great deal. He had started Milton's *Paradise Lost* and the line, 'insuperable height of loftiest shade', kept running through his head. And he re-read Racine, who was a marker for him, for his first acquaintance with the great French playwright had been at Trinity. By reading him now he could measure the distance he had come, and perhaps he would even find out something about where he was going. He heard a production of *Bérénice* on the radio, Jean-Louis Barrault playing Antiochus, and marvelled again at the fact that nothing happens in a Racine play. The characters just talk, but what talk it was and how lovely to hear it spoken. So he read through *Andromaque* and *Phèdre*.

Over the New Year of 1957, still at Ussy, with the rain still pouring down, he heard his own play *All That Fall* broadcast from London but could not make much of it because the reception was so poor. Shortly afterwards, however, the BBC took the unusual step of sending him a tape-recorder as well as a tape of his play. In spite of his early interest in motor-bikes, Beckett was not very mechanical or very adept at such things and even though the BBC also sent him a manual, it took him some time to make the recorder work. When he finally heard the tape he knew that McWhinnie had been the right choice for producer. Sending his script to John Morris at the BBC in September, he had included a note about the special quality of bruitage that might be necessary. 'It is a text written to come out of the dark,' he told Alec Reid. Radio did come out of the dark in those days, when people frequently sat around a set with only the glow of the dial light to illuminate the room. The two ways of establishing décor were by evocation in language and by sound effects. This is done in *All That Fall* in a way that undercuts both, as Beckett always undercuts the illusionist devices of the medium. When Mrs Rooney mentions a cow it then moos. When she speaks of the birds they give a brief chirp. Yet the evocations are strangely very strong.

McWhinnie had seen from the start that a new kind of radio production could be brought into being with Beckett's text. He decided that it was necessary to get away from 'standard realism', as he wrote to Beckett, who was glad to hear this kind of thing, suggesting that the sound effects should aim to get away from the excessive naturalism and illusionism of other radio productions in order to recreate 'the enclosed, subjective universe' of *All That Fall*. Thus he had used humans imitating animal noises rather than the recordings of the animals themselves kept in the BBC effects library, and he had processed other sounds to give them a curious abstract but meaningful quality.

Largely due to the Third Programme, radio was fashionable in the late fifties. After the broadcast critics immediately recognized that Beckett and McWhinnie had done something new with the medium. Roy Walker in the *Listener* said that the play was a 'radio classic', while Christopher Logue in the *New Statesman* called it 'Radio triumphant'. There were dissenting voices. The poet Donald Davie, while being perceptive enough to note that the syntax of the play parodied itself,

464

thought nevertheless that there was too much 'derivative slapstick' and that, in the manner of old-fashioned short stories, the play depended on a trick ending.

All That Fall is in some ways the most human of Beckett's works. It is set in Foxrock and there is a very real nostalgia and tenderness in the sleepy village setting. Mrs Rooney, his first female central character, is an entirely successful creation – humorous, strong, intelligent, but at the same time 'destroyed with sorrow and pining and gentility and churchgoing'. Marriage is again one of the play's primary themes, Mr and Mrs Rooney being another indissolubly wedded Beckett couple, even though Mrs Rooney pines for more than her husband has to offer. 'Love, that is all I asked, a little love, daily, twice daily, fifty years of twice daily love like a Paris horse butcher's regular. What normal woman wants affection?' And there are jokes about other people's marriages. When they hear a cry from the houses as they pass, Mrs Rooney says compassionately: 'Mrs Tully I fancy. Her poor husband is in constant pain and beats her unmercifully.' But the principal Beckett theme, the balance of suffering against the supposed joys of existence, the aggravation of it by effort and ambition, is never far away, though more humorously and gently dealt with here than elsewhere. It is 'suicide to be abroad', Mrs Rooney reflects; but then asks herself, 'What is it at home? The gradual dissolution.' And when Mr Barrell the station-master congratulates her on being up and about again after her illness, she replies: 'Would I were still in bed, Mr Barrell. (*Pause.*) Would I were lying stretched out in my comfortable bed, Mr Barrell, just wasting slowly, painlessly away, keeping up my strength with arrow-root and calves-foot jelly . . . just drifting gently down into the higher life, and remembering, remembering . . . all the silly unhappiness . . . as though . . . it had never happened . . .'

More than six months had now passed since *Fin de partie* had been finished and there was still no Paris theatre available for its production, so early in 1957 an offer from the Royal Court in London to house it for a brief run was accepted, though ironically this little run had no sooner been arranged than the Studio des Champs-Élysées became available for a Paris production. Putting on a new French play in England was in any case a risky venture and the experiment was not a success. When it opened before an English audience on 3 April, Beckett described the event to Schneider as 'rather grim, like playing to mahogany, or rather teak'; and the English critics were, with the exception of Harold Hobson, either hostile or baffled. Kenneth Tynan in the *Observer* made the curious error of suggesting that Beckett was, like Joyce, a lapsed Catholic and that the play echoed the cry attributed by

George Orwell to Joyce: 'Here is life without God. Just look at it!' Beckett's new play made it, he said, 'clear that his purpose is neither to move nor to help us. For him, man is a pygmy who connives at his own inevitable degradation'; and he objected to this message of despair being forced down his throat.

Tynan admitted to being influenced by what he called the facile pessimism of *Acte sans paroles*, the mime which followed the play, and certainly at this distance of time it does seem to have been a mistake to put the mime on as well, for in its representation of a parched hero who is constantly foiled as he clambers around after a flask of water, which is lowered from above only to be whisked out of his reach, and who winds up by attempting unsuccessfully to hang himself, it strikes one as almost a parody of Beckett's fundamental attitudes. Tynan described the sort of pessimism displayed as 'not only the projection of personal sickness but a conclusion reached on inadequate evidence'. He was ready to believe, he said, 'that the world is a stifling, constricted place', but not if his informant was 'an Egyptian mummy'.

The same production opened at the Studio des Champs-Élysées a little over three weeks later. Beckett's relationship with his actors was becoming firmer and more assured, not to say authoritarian, and there were constant arguments during the rehearsals of *Fin de partie* at the Royal Court and again at the Studio des Champs-Élysées. Rehearsals had already been in progress in London for a couple of days when Beckett arrived, and he disagreed immediately with some of what he saw. It seemed to Blin that he conceived the play like a musical composition which had to be performed exactly in accordance with the score. Beckett insisted that on the several occasions when Hamm calls Clov, Blin's voice should always be on the same note and that the actor should pay no attention to the evolving relationship. Blin wanted to sound like the boss at the beginning and like a suppliant towards the end. But Beckett would not have this. He also reproved Jean Martin for not drawing out the phrase '*ça va finir*' enough. He wanted the word '*finir*' to be very long- drawn-out, and when Martin said that it simply could not be done in French he disagreed.

While respecting the text, Blin wanted to play Hamm with more feeling: he thought it was necessary for an actor to feel sympathy with the person he was playing. Beckett, on the other hand, wanted the words said simply in a neutral voice with an odd shout from time to time. 'There is no drama whatsoever in *Fin de partie*,' he told Blin. 'There is a heap of words, but there is no drama.' If any concept was guaranteed to make things difficult for actors, this was it. Yet he wanted them also to indicate with their voices the passage from anger to laughter at several points in the play in a way that they found literally impossible to do. Though Blin knew that the text was not to be altered, he did request a modification of a line to make it more sayable. In reply he received a shrug and a statement which suggested that everything was now immutable and unalterable. 'It is written like that, isn't it?'

All in all Blin had a hard time of it. Like all actors he found it almost impossible to convey blindness with his eyes; and when he took off the dark glasses – in his case snow goggles – which he wore at the indicated points of the text, he tried to get over this difficulty by turning his eyes up to show the whites. For the moment when he takes off his headgear, a toque, he wore a bald wig. One evening during the three-month run at the Studio des Champs-Élysées, the wig had been stuck down so firmly that his skull hurt and the curtain had to be rung down and the performance stopped while something was done about it. Blin had learned by now not to ask certain questions, so he never asked Beckett why Hamm's parents should be in rubbish bins, knowing that if he did the answer would be, 'Why not?' Although, since the practice of putting old people in institutions which are little more than refuse bins has grown to be almost the norm in our civilization, it seems odd now that he should ever have needed to ask. Disagreements, or what Blin called 'impassioned discussions', took place throughout the rehearsals and most of the time Beckett got his way.

Because of its apparently stark and unrelieved pessimism, *Fin de partie* or *Endgame* has always been the great divisive work among Beckett critics, and the severest test in the obstacle course which even his hardiest followers have had to run. As has been said, Beckett described the play to Alan Schneider as 'mostly depending on the power of the text to claw' and 'more inhuman' than *Waiting for Godot*. Speaking of this first French production, he said that in such a small theatre 'the hooks went in'; and told Nancy Cunard he preferred it to others because it was 'more like what I wanted, nastier'. When even such an enthusiastic early admirer as Vivian Mercier found himself confronted with this work, he was so appalled that he abandoned a book he had intended to write about Beckett and allowed contact with him to lapse for several years. On mature consideration many years later Mercier thought that *Endgame* was 'only one of a spate of works of art directly promoted by the existence of first the atomic and then the hydrogen bomb'; but that where it differed from other works dealing with that situation was that whereas they described 'efforts, however futile, to preserve human life', Hamm and Clov are 'dedicated to the philosophical position that the entire human experiment has been a failure and must not be repeated'. Admitting that other themes are dealt with in *Endgame*, he still thought that 'the underlying monstrous assumption is that the end of the world is at hand – and good riddance!' Although it was possible to concede artistic validity to the play because it could bring about the willing suspension of disbelief which for Coleridge constituted poetic faith, what he would deny it had was 'any philosophic validity, except in a dialectical process where such a village-atheist oversimplification is countered by an equally extreme antithetical statement in favour of the preservation and continuation of human existence on Earth'. But Mercier did add that in conversation Beckett himself denied the validity of an extreme pessimist attitude to life and recounted how he had once said to him: 'That would be to judge and we are not in a position to judge.'

In 1973 Harold Hobson, one of Beckett's most loyal critics, who had welcomed the play on its initial production at the Royal Court, felt moved to complain that in recent years there had been some danger of Beckett's work being sentimentalized and of audiences self-defensively persuading themselves that his plays were not really filled with terror and horror but were, at bottom, jolly good fun. 'Well, they are not jolly good fun. They are amongst the most frightening prophecies of, and longing for, doom ever written.'

At the time of the first French production, however, Marc Bernard in *Nouvelles littéraires* was cool and amused, saying that he constantly had the impression that he was listening to a medieval fantasy or comic poem in which allegorical characters, fake scholasticism and Aristotelian reasoning were made into a mixture in which metaphysics suddenly took on a farcical tone. On his dusty throne was seated 'the intellectual, paralysed, blind, as talkative as a fourteenth century doctor. He is waited upon by the Common Man, half way between man and beast' who, so that there should be no misunderstanding, 'has been given a simian appearance: long dangling arms, curved spine. The intellectual's father and mother are stuffed into two dustbins; from time to time a lid is lifted and one of the parents begins to talk.' Surrounding these four characters who symbolized in an abridged fashion all mankind, 'the universe of course continues to turn, aimless, gloomy, absurd, desolate'. Yet he thought that the play was 'resolutely reactionary', since it gave man 'no chance, no hope' and exhibited no faith in science, progress, man or God. Beckett, he thought, was 'intoxicated with this nothingness' and indulged in it 'with a masochistic voluptuousness'; from it he drew 'along with the bitterness, its inner sweetness'. Negation was his strong point and in it he found 'a kind of strange pleasure'.

Kenneth Tynan and Marc Bernard were not alone in relating Beckett's vision to his personality and outlook. Other critics noted from time to time that in his work there is a joy in the stripping bare of human experience and a masochistic pleasure in the exposure of our pitiful state. Some of them alleged that its validity was weakened by its relationship with his own individual psychology or pathology.

Certainly many of his friends and acquaintances noted that he seemed to find a kind of joy in gloom, even in physical misfortune and mishap. To adopt the beautiful phrase which Robert Welch has used about Edmund Spenser, he was 'a fabulist of his own distresses'. Georges Belmont – then known as Georges Pelorson – had had a sense of Beckett in some way relishing the extreme discomfort of an enormous cyst on his neck when they were in Trinity. When they went walking in the Dublin mountains together he had the impression that Beckett was enjoying the walk as much as he himself was until suddenly, still striding along, he would begin to complain of his sore feet or the tightness of his shoes. Others felt his zest in gloom and deteriorating situations of one kind or another. When he recalled Joyce saying 'we're going down hill fast' with relish, his own enjoyment of the remark was manifest.

There was sometimes an element of self-parody in his refusal to admit to enjoyment of certain circumstances; but the underlying attitude was very real. When in the mid-sixties he was staying with John Calder in London they went with some of his BBC friends to Lord's cricket ground to see England play Australia. In fine weather one could sit in the members' enclosure drinking beer and watching cricket, both of which Beckett appeared to enjoy. As they sat there someone remarked on the lovely summer weather; and everyone else, including Beckett, concurred. It was a truly beautiful English summer day – 'The sort of day that makes one glad to be alive,' remarked someone else. 'Oh I don't think I would go quite so far as to say that,' Beckett replied.

When challenged about his view of life he would sometimes refer to the objective evidence that was to be encountered everywhere. He agreed with Tom Driver that his plays dealt with human distress. 'Some people object to this in my writing,' he said.

At a party an English intellectual – so-called – asked me why I write always about distress. As if it were perverse to do so! ... I left the party as soon as possible and got into a taxi. On the glass partition between me and the driver were three signs. One asked for help for the blind, another help for orphans and the third help for the war refugees. One does not have to look for distress. It is screaming at you even in the taxis of London.

Beckett came to London with Martin and Blin for the final rehearsals and production of *Fin de partie*. He had business there with the BBC, and in any case London was becoming a more attractive place to him since he now had admirers and sympathizers there who would come to be his friends – Donald McWhinnie and, on another level, Barbara Bray. To begin with he stayed with the two Frenchmen in a digs in Ebury Street which had been recommended by the Royal Court, but on the third day Beckett decided that, as he told acquaintances, he could stand it no longer, so he moved into the Royal Court Hotel a few yards from the theatre in Sloane Square. He wanted to be on his own and to have a place where he could meet people and come and go as he chose.

He promised the BBC a new radio play, which, though he did not know it yet, was to be *Embers*. Later that year he approved of their plans for a reading from *Molloy* with music by his cousin John Beckett, which would be broadcast on the Third Programme in December; and they had also secured his permission for a reading of 'From an Abandoned Work'. Both of these were to be read by an Irish actor, Patrick Magee, for whose fitness as an interpreter of Beckett everybody in the BBC, including Barbara Bray, was enthusiastic.

Magee had a harsh, gravelly voice which had little superficial charm but had a hypnotic effect on the listener. With his voice and his light blue eyes went a faint air of menace (which would later secure him parts as the mysterious villain, sometimes a former Nazi, in more than one English movie). He was grey-haired but ageless and could combine debility with menace, as Beckett characters with their suppressed violence often do. Before his current career in the BBC and in London generally he had had a long theatrical history, going back to the small touring theatre companies of Ireland, in whose productions he had often played non-Irish roles – sometimes the aristocratic villains in English melodrama. For these he had developed a rather strange accent with only faint Irish overtones and prolonged vowel sounds. The general effect was strangely déclassé but still indubitably Irish and thus ideally fitted for the performance of Beckett. As a person, he was engaging but somewhat unpredictable, for the suggestion of barely suppressed violence was true also of the man, who could sometimes be quite truculent when drunk. In his outré way he was attractive to women, who seemed to fall into what one is tempted to call his clutches in large numbers. He was intelligent but non-intellectual and had a fund of macabre Irish stories, told while fixing the listener with his steely blue gaze, though his party piece was, quite incongruously, the ineffably sentimental Victorian ballad about the toy soldier who continues to stand sentry long after the little boy has died. As an actor he had the good sense to see that one played Beckett for the weight and mood of the words and the situation without bothering about the ultimate philosophical import. With his chequered and, up to this point, not highly successful career, there was a sense in which, as an actor, he had been waiting for Beckett just as Beckett had been waiting for him. When they met they got on famously. Beckett understood Pat's Irishness but was happy that it was no stereotype, and Pat in his turn treated Beckett with respect but without deference, as one Irishman to another. His voice and his basic persona as an actor, so near to Beckett's own conceptions, were an important source of inspiration to the playwright, who began early in February of the following year a work whose first draft is entitled 'Magee Monologue' but which became before long *Krapp's Last Tape*.

While *Fin de partie* was in rehearsal he heard the news of Jack Yeats's final illness and death. Yeats died on Thursday 28 March 1957 and when Beckett got back to his hotel from the theatre that evening and found a telegram from MacGreevy to say so, he made what he called 'a serious attempt' to go over to Ireland for the funeral. He found, however, that there was no seat on the Aer Lingus flight on the Friday afternoon and that he would have to go via train and boat from Euston. Although this would have got him there in time to the funeral in Mount Jerome, he found that he could get no assurance of a seat on the flight from Dublin back to London that afternoon. Unable to face the journey on the boat, he telegraphed MacGreevy asking him to arrange a wreath. It is hard to escape the impression that he was reluctant to cut short his stay in London to go back to Ireland for another funeral. He felt some remorse about this understandable delinquency, and went to some lengths to explain, ending his letter 'come back to me soon, dear Tom and let us hear and say what friends have as Love to say about him.'

Yeats's dealer, Victor Waddington, was in Paris shortly after the New Year in 1958 and there was talk of another retrospective, for which Beckett promised to give all the help he could, though he viewed the possibility of a likely posthumous increase in the painter's reputation with some scepticism. He was gratified, though, that Waddington bought a number of pictures from Henri Hayden, whose work he also intended to exhibit in London, either in a group or a one-man show. Beckett thought that Hayden had been doing some magnificent work and he noted with some irony that he was getting some appreciation at last at the age of seventy-four.

Meanwhile there was the question of an English translation of *Fin de partie*. He had signed a contract with the Royal Court Theatre and had also given Alan Schneider in the United States the rights to an off-Broadway production before an English translation existed. In April he wrote to Schneider saying that it seemed strange to be making plans for 'a text which does not yet exist and which, when it does, will inevitably be a poor substitute for the original (the loss will be much greater than from the French to the English *Godot*)'. Writing to Barney Rosset he even put a figure on this loss, saying the French was 20 per cent 'undecantable into English'.

Though he was not contracted to finish until August he actually finished in July, somewhat depressed by the thought that all the sharpness and the rhythms of the French had gone. When the English version was produced in London the following year, however, the critics' reservations were certainly not on linguistic grounds, the doubts expressed having more to do with the now familiar subject of Beckett's vision, his 'pessimism' and his 'despair', than they had with his use of language.

In this first French production, however, the play at least proved popular enough for a continuation of the run in another theatre when the booking of the Champs-Élysées came to an end. This all seemed to be arranged and rehearsals had even begun, but at the last minute the new theatre management backed out. Everybody, including Beckett, was very angry and there was talk of an action. But he was not keen on any kind of legal proceedings in which he himself would be involved. In some disgust he retreated to Ussy, where he had been unable to snatch a few days since the beginning of the year. He hoped to be able to stay for some months as he was very tired. Since the beginning of the year he had been translating or assisting with translations: *All That Fall* into French and German, *Fin de partie* into German and English, *L'Innommable* into English, *Malone Dies* into German and *Echo's Bones and Some Other Poems* into German. He told people how sick and tired of translation he was and what a losing battle he always felt it to be, adding, 'I wish I had the courage to wash my hands of it all, I mean leave it to others and try and get on with some work.'

Complain as he did about translation or about the necessity to become involved in theatrical productions of his works, however, the impression remains that his attitude was deeply ambiguous. So much of the time he felt unable to write any more prose narratives or poems and welcomed these other tasks as a distraction from this inability and a substitute for new creative writing. When, in 1956, Vivian Mercier saw him in Paris, he told him that he felt 'all dried up , with nothing left but self-translation'. And about the same time he told MacGreevy, 'I just don't happen to be a professional writer, so can't go on and on till the cows come home.' Though he complained about Paris as if his life there were to blame, sojourns at Ussy were also mostly given over to translation.

If the critics who complained of the limited, private and neurotic basis of Beckett's writings had known his history of psychosomatic disorder they would no doubt have thought it was confirmation of their judgement. But in fact he was no longer subject to so many neurotic and often imaginary ailments and panics as before. However, he was still plagued with cysts, including one on a lung in 1952, which was at first thought to be a malignant growth and had been subjected to radiation treatment. Now, his stay at Ussy after the termination of the French run of *Fin de partie* was brought to an end by a minor but tricky operation on a cyst in a very awkward place – above and behind the palate of his mouth. Even worse, its presence had called attention to the condition of his teeth, and he had been told very firmly that he must have them out. Wearing dentures changed Beckett's appearance and the difference was apparent to the old friends who saw him this summer, among them H. O. White and Con Leventhal, now married to Ethna McCarthy, who was back in Dublin and practising as a paediatrician.

Since his fame it was becoming something of a thing in Dublin, especially in Trinity academic circles, to seek Beckett out when in Paris, more especially when this became easier with the installation of a telephone in the rue des Favorites towards the end of the year. Soon he would institute a regime whereby he would answer the telephone's clamour only between the hours of eleven and twelve in the morning. At other times, when Suzanne answered, from those undaunted by her insistence on speaking French, she would take a message. He was not listed in the telephone directory and the number Blomet 09.11 was given to friends with the severe caution that they should not pass it on. This was no deterrent in Dublin, so visitors often arrived armed with the number, but in Paris few people gave it out. There was now also a telephone at Ussy, the number of which – Ussy-sur-Marne 18 – is even more deeply graven on many people's minds, for somehow it seemed a more serious matter to disturb his work or his solitude by ringing him there.

Whatever inescapable gloom there was in Beckett's outlook on life found some alleviation at Ussy, of which, since his first dissatisfactions with it, he had been growing more and more fond. At Ussy he seemed 'to recuperate something in the silence and solitude'. Although frequent trips to Paris were necessary he found them an irritant. Fortunately there was a good train up early in the morning and another back late in the evening, giving a full day in the city to attend to his affairs. The pull of the country was becoming very strong. He had, he told MacGreevy, a craving for its silence and solitude and he had given up resisting it. Far from declaring that, but for Suzanne's fondness for Ussy, he would give it up, he was now making the opposite assertion. If it were not that she preferred to be in Paris more often than he did, he said, 'I think I would give up the apartment and dig in here for the duration.' As he came to be increasingly alone there, he found to his surprise that he was able to bear this state with equanimity. 'It is lonely here, and very isolated, but not sinister, and I don't find it difficult, though I wouldn't have managed it in the old days, or indeed, five years ago,' he told his friend. 'What I find more and more difficult to cope with is Paris, and people, and speech. I can't do it without drink, and alone I am quite content with a few glasses of wine.'

A typical day alone at Ussy might consist of some writing or reading, often in the garden; getting out the two-stroke mower and cutting swathes through the long grass which always seemed to greet him on his return – molehills were a problem – followed by a simple dinner, the radio, the newspaper, a bath, and bed. Sleep, however, was less certain than rest for, even in the state of comparative relaxation that the country induced, he was often awake for hours on end and enjoyed only a troubled slumber when he did sleep. He had little desire to travel and asserted that all he asked was to be allowed to alternate between Ussy and the Left Bank 'for the few years that remain'. To Rosset he spoke half jocularly of spending his 'last years keeping bees and growing lavender'. But Beckett was complicated and contradictory and could not foresee the future. At the very time of these protestations, he was on the verge of new relationships, interests and excitements and also on the verge of a very great deal of foreign travel.

Chapter Thirty

Whatever Beckett might say about the desire to spend the rest of his life between Ussy and the Left Bank, 1958 was a year of travel and movement. He had given a half-promise to Bram van Velde that he would go to Berne in the spring for an important retrospective of the artist's work at the Kunsthalle. At the last moment he tried to back out of it but Bram insisted and he drove down with him at the end of April. His hostess was Ida Meyer, daughter of the painter Marc Chagall and wife of the director of the Kunsthalle, who had arranged a programme of receptions and visits to galleries and theatres which he found oppressive. The view of the Bernese Oberland pleased him but the Föhn was blowing and this he felt to be almost unbearable.

Among those he met was the doctor in whose arms Joyce had died and who told him that, but for tardy diagnosis of the ulcer which killed him, he could have been saved. Infirmity and death were much in Beckett's thoughts. In a private collection he saw one of the last of Cézanne's self-portraits, painted almost at the end of his life, and found it overwhelmingly sad, 'a blind old broken man' – though one could interpret it as Cézanne's continuing determination to (as he had written in a letter at the time it was painted) 'realize the dream of art that I have been pursuing all my life'.

Before leaving for Switzerland Beckett had written what he described in a letter to Barney Rosset as 'a brief stage monologue for Pat Magee, a rather sentimental affair, as my work in English tends always to be, but with I think an ingenious situation'. This was *Krapp's Last Tape*, which would be produced at the Royal Court before the end of the year. This burst of creativity did not alleviate the feeling of frustration caused by his translation work and involvement in productions. Later that year he wrote to Rosset that he felt he was 'getting more and more entangled in professionalism and self-exploitation and that it would be really better to stop altogether than to go on with that'. What he needed, he said, was 'to get back into the state of mind of 1945 when it was write or perish', but he supposed there was no chance of that. The memory of his great creative period after the war still haunted him as a possibility and a few days later he told Barney: 'The only chance for me now as a writer is to go into retreat and put a stop to all this fucking *élan acquis* and get back down to the bottom of all the hills again, grimmer hills than in '45 of cherished memory and far less than then to climb with . . . It's not going to be easy, but it's definitely the only last gasp worth trying to pant as far as I'm concerned'.

Meanwhile though, he was glad of any break that offered, glad, as he told MacGreevy, to get away from 'pen and paper'; however, he immediately exhibited his usual ambiguities by adding, 'they are a sorry pair, but finally I suppose about the best one has'. In June he surprisingly yielded to pressure from Suzanne, Blin and Jean Martin by consenting that all four should take a holiday together in Yugoslavia, where dinars which could not be spent outside the country were accumulating from a successful run of *Godot*. Apart from trips to Ussy, this holiday in Yugoslavia was the first that he and Suzanne had taken together for some time. It was undertaken, like many later trips, partly out of guilt and concern for her. She had not been well during 1957, when a trying visit from her mother and sister which lasted most of the summer had been succeeded in the winter by a mysterious tiredness with no obvious cause but which obstinately refused to go away.

They flew to Belgrade, then on to Zagreb and the coast near Fiume, though Beckett regretted not having seen Sarajevo and Dubrovnik. Making, as most people did, no distinction between Croat and Serb, he found the Yugoslavs to be a 'very sober, quiet, serious, likeable people, in great poverty'. At Lovran, where they stayed for most of the three-week holiday, they bathed a good deal; but though 'stupefied . . . by sun and sea', Beckett often seemed bored, especially with Suzanne, and, though the holiday had supposedly been partly undertaken for her sake, there was often friction between them because of his drinking and his frequent insistence on having another and then another in the evenings. When Jean Martin began to write a novel about his father in an exercise book, he and Sam embarked on a kind of game. As Martin wrote a page each day he left the opposite page blank, on which Beckett would write comments. As the comments became freer and more elaborate, it became a sort of collaboration, with Martin departing from his original intention of following Beckett's suggestions.

Godot had been published by Faber and Faber in 1956. It was followed from the same publishers by *All That Fall*, but Fabers had shown a reluctance to have anything to do with Beckett's novels. These were only just beginning to appear in England from a much smaller firm, that of John Calder.

John Calder, like Barney Rosset, could be described as a loner and an eccentric. Like Barney too he came from a moneyed background, and as a young man publishing had not been the first of his sometimes costly enthusiasms. By the time he came to it his capital was limited and he did virtually everything himself, even bringing review copies round to the literary offices to save packing and postage and acting as his own representative with the booksellers. He was heavily dependent on fairly cheaply bought translations, and Calder lists always contained more translations or American titles than they did original English works; but he had always been at home in France and with French culture and to some extent the list reflected this. His imprint was the first to introduce many of the *nouveaux romanciers*, including Nathalie Sarraute, Robert Pinget, Claude Simon and others, to the English consciousness. Contacts with Grove and with Olympia Press proved rewarding, giving both an avant-garde appearance and a risqué air to the list. Like Barney, Calder published titles which could be regarded by some as masterpieces of contemporary literature, by others as mere pornography, the emphasis varying according to your point of view.

When he saw the first production of *Godot* in Paris, Calder was rather baffled; but he had a friend who wanted to see it, so he agreed to go again. It was one of the most fateful decisions of his publishing life. This time he was deeply impressed and he wrote immediately to Les Éditions de Minuit asking about the English rights and saying he was prepared to make an offer. They had just changed address, so it took some time for his letter to be delivered. When it was, it was one post behind a letter from Faber and Faber; but in the meantime Calder had managed to get Beckett's address and telephone number. He and Beckett were immediately drawn to each other and, after some further meetings, Beckett wrote to him with a proposition. Fabers, he said, had turned down the novels on the grounds of their general difficulty and obscurity as well as the danger of prosecution for obscenity and he asked if Calder would like to publish them. To Calder's 'astonishment and delight', he became Beckett's principal English publisher and remained so for the rest of the author's lifetime.

Beckett was inclined to romanticize those publishers – Lindon, Rosset and Calder – who had approached him while he was in relative obscurity or who were with him more or less from the beginning of his post-war fame. He also made them his personal friends. Almost as a matter of principle, Beckett trusted his publishers as one would a friend. In an affectionate letter to Barney Rosset in 1957 he wrote:

> As you know, I am incapable of understanding contracts. My 'method' consists, when they are drawn up by those in whom I have confidence, in signing them without reading. Any contract drawn up by you, and involving me alone, I shall sign in this fashion, those involving Lindon and me I submit to him and sign when he tells me to. To the former category belong my translator's contracts with Grove and I leave it entirely to you to formulate them as you think fit. I know your friendship for me, and your probity, and the possibility of your deriving from them excessive benefits does not at any time cross my mind.

And afterwards, when a confusion had arisen between Faber, Curtis Brown, who were now his agents, and Barney about *All That Fall*, he wrote, 'Sorry again, Dear Barney, for all the mess. Put it down, or part of it to my stupidity. You know my feelings towards you of affection and indebtedness and I should hate you to think they were losing edge.' He often repeated these assurances and frequently stressed his limited understanding of business matters, which he described as 'null'.

In this year, 1958, Calder issued the first Beckett book to appear under his imprint, *Malone Dies*. Its dispatch to the literary journals was accompanied by a copy of the Olympia Press edition of *Watt*, over which Calder had pasted his own imprint as distributor. My own acquaintance with Beckett began shortly afterwards as a result of a review I wrote of these books in the *Times Literary Supplement*. In spite of the fame which their author was now enjoying from *Godot*, *The TLS* asked me for only 500 to 600 words. Since I had been familiar with *Watt* for some time and had read and reviewed *Molloy* when copies of the Olympia Press edition had been sent to English journals in 1956, I suggested what was called a middle – a long piece of 2,700 words. My piece appeared some time in the summer of 1958. Barbara Bray sent it to Beckett and when he came to London that autumn for the production of the English version of *Endgame* in a double bill with *Krapp's Last Tape* at the Royal Court, there was some discussion of it, with the result that Donald McWhinnie asked me if I would like to meet Samuel Beckett. It was arranged that I would meet Pat Magee in a BBC pub, the George in Great Portland Street, and we would travel down to Sloane Square together.

I had by now learned a good deal about Beckett from Dublin and other acquaintance, but my image of him was principally the result of the powerful impact of his work, which conveyed among other things an impression of rejection of the world's affairs and even of its comforts, a sardonic asceticism if not quite a saintly resignation. And there was a growing legend of an enigma, a solitary who despised or was indifferent to the joys, such as they were, of ordinary human association.

I met instead an agreeable, courteous, indeed almost affable man with an accent which I was immediately able to place as south Dublin Protestant, even to a slight lisp and a difficulty with 'r' sounds which was common in that class. We talked, amongst other things, about mutual acquaintances, and I discovered that he was by no means averse to a bit of news and gossip. Although I had the good sense to keep off the meaning of his work – a topic I would have been unlikely to broach with any fellow author – I did ask him about geographical and circumstantial detail; and he seemed pleased to confirm, for example, that the seat by the tram-stop at the opening of *Watt* was indeed the – at that stage still existing – seat by Portobello Bridge. I had somehow the notion that he would not be very interested in contemporary novelists; but when McWhinnie asked if I had read Malcolm Lowry's *Under the Volcano*, saying it was a great book, Beckett immediately endorsed the view – 'Yes, a very great book.' When the question of lunch arose, he dispelled any delusions about asceticism by suggested a place in the King's Road where he had been the day before, saying we could get a good bottle of Beaujolais there.

It seems to me now that the legend of Beckett's reclusiveness grew largely out of his refusal to give interviews about himself and his work and, above all, to discuss its meaning at a time when most authors were becoming quite anxious to do these things. We were even then moving rapidly into a time when the media would become ever more voracious in their demands and authors ever more willing to prove themselves compliant, discussing their books, their opinions and even their private lives without reservation. Most writers would prove glad to have any opportunity to speak at length about their latest opus, its origins in personal experience, its implications and intentions. In this climate Beckett's reticence would set him apart and, since the media are mystified by anybody's lack of enthusiasm for publicity, it was but a short step from this setting apart to the legend of the complete recluse, the enigma who had cut himself off from the world. Of course, it was easier for someone known to the general public only as a playwright and author of impenetrable novels to become an enigma than it would be for a novelist or poet whose work was overtly concerned with personal experience. The play is, on the surface at least, a more impersonal medium than the novel or the poem; Beckett's plays were more than usually enigmatic and the works by which he became known suggested, if they suggested anything, an outcast or a solitary like his central characters. Some of the powerful magic his invisibility behind these works had generated would in time be dissipated by the re-publication of *More Pricks Than Kicks* and the appearance of *Mercier and Camier*, which is a more personal and revealing book than the later novels. But on the other hand there would be his practice of not answering the telephone, his frequent disappearances to Ussy and, of course, his refusal to be impressed by the Nobel Prize and disdain for personal publicity in relation to it.

His sense of privacy was, however, acute and his resentment of what he considered intrusion strong. When he was in London Barbara Bray had prevailed on him to meet and talk with Rayner Heppenstall, a fellow BBC employee who also contributed to the *Observer* and wanted to write a profile for that paper. Beckett's misgivings were considerable and were not lessened by what he regarded as intrusive questions during the interview. When the piece appeared he was quite disturbed and Barbara was distressed; but even by the standards of that era it was innocuous enough and, considering the sort of self-revelation indulged in nowadays, at this distance of time looks completely harmless. On the question of his marital status, it said that Beckett 'seems a little in doubt whether he should be described as a married man, but is certain there are no children'. This seemed to give particular offence to Beckett and when I met him again, a day or two later, he spoke of it with indignation.

Jack MacGowran was playing in *Endgame* under George Devine's direction on the Royal Court bill, while Magee was directed by McWhinnie in *Krapp's Last Tape*. They were both, in their different ways, very Irish, and it was obvious that Beckett enjoyed their company as much for that reason as for others. Out of several meetings during the days that followed I remember in particular one evening during which Magee recounted the story of the Waterford postman who had disappeared without trace in the early 1930s, how he had been wrapped in a blanket and buried in a boghole and how a whole village had remained silent under police questioning for months afterwards. Beckett listened to Magee's version of this *cause célèbre* with great intentness.

One thing you soon discovered about Sam Beckett was that he was a good listener; but if he felt impatient he conveyed it, albeit almost invisibly, with a clouding of the eyes. And he was one of those rare people whose authority is not dependent on the adoption of a manner or a mask. He was – or seemed – completely natural and at ease, though the ease was probably the result of a severe discipline imposed on his natural shyness; and his presence was a powerful one – a field of force in which, even if there were no lightning bolts, you were conscious of the electric tensions. I never experienced the long, brooding silences which other people did in Beckett's company, but then these first meetings took place in the presence of other people with whom he was already at ease; and to some extent they were, because of Magee and MacGowran, Irish occasions. In a letter to Barney Rosset written at this time, he mentions having met the author of the *TLS* piece; and says that he has turned out to be 'a Wexfordman', which is true; but to find him mentioning it is rather surprising. Many Irish people are highly conscious of these geographical distinctions. I would not have thought he would be.

He did not like George Devine's direction of *Endgame* and seemed to pay more attention to McWhinnie and Magee in *Krapp*. Under Donald's direction, Magee's Krapp was far less senile and more agile and energetic than in other versions I have seen since. In accordance with what was becoming his usual custom Beckett did not attend the opening night of either play, though he stayed in London, moving at this point from the hotel into Magee's digs in West Cromwell Road. This, as digs went, was a fairly upmarket establishment. One ate breakfast at a large mahogany table before an open fire and the morning newspapers were placed on the sideboard alongside the breakfast dishes. The Scots lady who ran it was evidently very fond of Magee, but much concerned about his morals and the number of different women he brought back home at nights. On the Sunday night following the opening of *Krapp* at the Royal Court, a symposium about Beckett's work was held at the theatre. The participants included myself, Alan Brien, then drama critic of the *Spectator*, and a young English dramatist named Harold Pinter. In the *Spectator* Brien had called *Krapp's Last Tape* and *Endgame* 'exercises in peevish despair'. In *Endgame*, he had alleged, 'Beckett perversely clings to his reputation for obscurity, but any trained play burglar can unpick the lock in the first ten minutes. This is science fiction, filled out and shaded in with echoes of Noah's flood and Prospero's exile.' Both plays, Brien declared, had left him saddened,

> partly because, as Beckett intended, I became once again conscious of my own mortality. But mainly because I became even more conscious of Beckett's refusal of immortality as a dramatist. As the floodwaters rise he burns his bridges, scuttles his boats, punctures his water wings and tries to forget how to swim. Art is illusion and Beckett seeks to destroy even that by creating deliberately inartistic works of art.

At the symposium Brien attacked in these terms both Beckett's vision of things and his failure to embody it in dramatic form. I replied that where art was concerned, one truthfully expressed vision was as good as another; that this truth is seldom anything but partial except in the case of one or two very great, very inclusive artists, such as Shakespeare; but that even such a partial vision had immense value if its truth had never been encompassed before. This argument still seems to me to be central to a defence of Beckett, if defence is needed.

It is notable that even an intelligent critic such as Brien should still at that point have hankered after plausible situations with real-life characters – in other words, an illusionism which had long been abandoned in every art except literature. Brien also complained about the fact that even if we accept Beckett's situations as 'real', they are not pushed to their ultimate limits. The playwright himself did not attend the symposium, but he was always annoyed when anybody adverted to the fact that these situations are not fully and plausibly developed. When, during a rehearsal of *Happy Days* a couple of years later, the actress who was playing the part questioned him as to whether Winnie really was buried up to her neck in the earth or not and what this implied in the way of her day-to-day arrangements and survival, he was quite peeved.

Pinter's first play, *The Birthday Party*, had at this point just been produced in London. As a nineteen-year-old actor on tour in the Irish provinces, he had ten years earlier picked up a copy of *Irish Writing* and read the passage from *Watt* which began, 'It was about this time that Watt was transferred to another pavilion, leaving me behind in the old pavilion. We consequently met, and conversed, less than formerly. Not that at any time we had met, or conversed, very much, for we had not. For we seldom left our mansions.' Pinter read this passage over and over, his hair standing on end. He had never heard of Samuel Beckett and when he was back in London found that nobody else had heard of him either. 'Not even the Westminster Public Library knew the name'; but eventually that library discovered that there was a copy of *Murphy* in the Bermondsey Public Reserve Library. When this was procured it was clear from the absence of stamping that although the book had been there since 1938 nobody had ever taken it out. He therefore appropriated it, a criminal act which he has never regretted, and he is still in possession of the copy. When he began to write his own plays the influence of Beckett was clear. In about four years Beckett had moved from being completely unknown to having a serious and profound influence on the course of English drama.

English critics did not on the whole like *Endgame* at the time of its first production, the doubts expressed being mostly to do with the now familiar subject of Beckett's vision, his 'pessimism' and his 'despair'. Harold Hobson, the playwright's staunchest supporter among the London critics, was ill and his stand-in was not enthusiastic. In the *Listener*, T. C. Worsley, who enjoyed *Godot*, was less than enthusiastic about *Endgame*. 'In *Waiting for Godot*', he said, 'Mr Beckett's neurosis and mine were for quite long stretches on the same theme; in *Endgame* they never tangled. He has, in *Endgame*, . . . expanded not the public but the private images. He has concentrated not on what is common between his audiences and him but on what is private in himself.' The 'desolation' of the play was not 'the desolation of the world but of one individual mind'. It told one 'a good deal about Mr Beckett's solitary despairs' if anybody was interested in them. 'But who is?' he asked.

The general impression left by the critics was of a refusal to face the bleakness of Beckett's vision. *Godot* had had a certain amount of charm and humanity. As Beckett himself put it, '*Godot* is rollicking beside this awful thing, *Gott hilfe mir, ich kann nicht anders*'; or 'God help me, I can do no other.' In *Godot* there were far more laughs, and the dilemma of a pair of tramps was a more acceptable way of symbolizing the human condition than the situation of the blind but evidently powerful Hamm. Between Vladimir and Estragon there was warmth, an acknowledged mutual dependence, even a vestigial sort of love. Hamm is hostile or indifferent to all others, including his father and mother. Though there is a certain amount of mutual dependence between Hamm and Clov, there is little of the tenderness that exists between Vladimir and Estragon, and only Hamm's slobbering old parents Nag and Nell, the couple in the dustbins, have anything that could be described as love for each other. When Nell is woken by Nag, she speaks to him tenderly and asks, however grotesquely, if it is time for love. They even try to kiss each other and, having failed, gaze at each other fondly, if dimly. Nag saves Nell three quarters of his biscuit and when she says she doesn't want it he promises to keep it for her. They have evidently had a loving relationship in the past and she remembers being happy on her honeymoon, an admission of happiness rare among Beckett characters.

But Nag and Nell were not enough to convince critics that there was any sort of warmth in Beckett's vision. Their existence in the dustbins had rather the opposite effect. It confirmed both Hamm's unpleasantness and the author's willingness to entertain a very saturnine and cruel idea of human life. Though, as one critic said, making Hamm and Clov the central characters was a bit like elevating Pozzo and Lucky, the master and man of *Godot*, to that position, Clov has none of Lucky's pathos, being nearly as cold and selfish-seeming as his master. Though Hamm has no longer large estates to manage, nor so many dependent 'paupers', as he calls his former tenantry, he is still powerful enough to be alienating; and he misuses such power as he has. And while he and Clov are another of Beckett's apparently inseparable couples, their mutual need is not of the kind which Auden said amounted to love. It is merely circumstantial. Hamm knows the combination of the lock on the larder, so Clov is dependent on him; Hamm needs Clov to perform certain essential physical services. But we do not feel that there would be any heartbreak on either side if Clov really did leave as he constantly threatens to do. Oblivious as they are to the warmth of the relationship between the two old people in the dustbins, and unappreciative of what humour there is in the play (there is very little by Beckett's standards), the critics had nothing to distract them from the ultimate nature of Beckett's vision of things. To confirm the terrible coldness of that vision they instanced the moment when Clov finds a flea in his trousers and Hamm cries out 'catch him, for the love of God!' declaring that 'humanity might start from there all over again!'

Vivian Mercier, the young critic who had ceased to write about Beckett for more than ten years after *Endgame* appeared, contributed a piece entitled 'How to Read *Endgame*' to a magazine called the *Griffin* in June 1959. Oddly enough, the fact that his wife was diagnosed as having multiple sclerosis in that year, a fact which should, as he put it, have confirmed 'Beckett's gloomy estimate of the human condition', had instead the effect of turning Mercier against his work. Perhaps for those who really are old or ailing or abandoned or who have near at hand spectacles of debility and suffering, it is difficult to experience the joy in his extreme vision that Beckett conveys to some others.

In 1958 the English critics paid a good deal more attention to *Endgame* than they did to *Krapp's Last Tape*, although the latter had been written in English specifically for Magee and the former had been translated from the French. Reviewing *Krapp's Last Tape* in the *Observer* Kenneth Tynan wrote a would-be humorous parody about the agonies of a bored and bewildered critic, ending with a piece of dialogue which became mildly celebrated: '"Is that all the review he is getting?" "That's all the play he's written." "But a genius. Could you do as much?" "Not as much. But as little."' *Krapp*, which is a one-act play and therefore shorter, has, however, over the years received its due share of attention, much of it biographical. When he directed the American convict-actor Rick Cluchey in it in Berlin in 1977, Beckett summarized it for the American, telling him that Krapp was a sixty-nine-year-old writer, 'in no way senile', who has 'something frozen about him' and is 'filled up to his teeth with bitterness', in fact full of violence 'like a tiger in a cage'. He lives alone, talks to himself in monologues, and every year on his birthday mulls over the twelve months just gone by, making notes and recording his impressions and state of mind on a tape-recorder. On these melancholy occasions he sits 'like a badger in his hole', listening to a tape of a year gone by before recording a new one.

On this particular birthday he listens to a tape recorded after a year of critical decisions on his thirty-ninth birthday. It was in that year that his mother died; that he had a vision on a night of howling wind and storm at the end of a jetty concerning the direction his work should take. In that year too he had bidden 'farewell to love', agreeing with a girl with green eyes with whom he was drifting among reeds in a punt that it was 'hopeless and no good going on and she agreed'. Now, thirty years later, he sits past midnight, listening with absorption to this old tape and making a new one. On the new, he has little to record: his pleasure in the word 'spool', a couple of visits from Fanny, a 'bony old ghost of a whore' who comes in occasionally, a tearful re-reading of *Effi Briest* which he had first read with the girl in the punt, the sale of seventeen copies of the magnum opus which was apparently the result of his pseudo-vision on the jetty. Perhaps, he says, his best years are gone, 'when there was a chance of happiness'; but he wouldn't want them back, 'not with the fire in me now'.

There are biographical coincidences between Krapp's life and Beckett's: the nursing home on the canal where his mother dies, the fact that his father had died twelve years before her, the insight about his work on Dun Laoghaire pier, the emphasis, as always where a woman is concerned, on her eyes; and above all, perhaps, the reading of *Effi Briest* with a girl in Germany who holidays among the pines on the Baltic coast and says farewell to him on a railway station platform wearing a green coat, and seems also to be the girl in the punt.

One oddity which nobody noticed in 1958 is the dating. The tape-recorder had not been invented thirty years before that date, still less forty-two years before when Krapp's father died. Wire recorders first came into use during the war, wire being supplanted by tape some years later. Something else that escaped everybody's notice until much later is the use made of Manichaean ideas in the construction of the work. When Beckett directed the play in Berlin in 1971 he kept a notebook, which listed these correspondences at length. The most important are that Krapp is in violation of the three seals or prohibitions of Manichaeanism for the elect: the seal of the hands, forbidding engagement in a profession, the seal of the breast against sexual desire, and the seal of the mouth, which forbids the drinking of wine. The dichotomy of light and dark which is central to Manichaean doctrine is also carefully worked out and Beckett's notebook lists no less than twenty-seven points in the play at which the alternation of light and darkness is stressed.

All this dependence on a doctrinal system seems – and indeed is – terribly mechanical, though so well integrated into the play that critics did not notice it until the notebook entered the public domain. But the real question which arises is whether Beckett was truly influenced by these ideas or whether he was merely playing the game that James Joyce used to engage in, using them dispassionately as structural devices. Manichaeanism was basically a Gnostic heresy. Its adherents believed that the world was ruled by evil powers, against which the god of the whole of creation struggled as yet in vain. The French critic Guy Scarpetta, a member of the Tel Quel group and a follower of Philip Sollers, viewing Beckett's *oeuvre* as a whole has suggested that it does indeed exhibit Gnostic beliefs, that it expresses a sort of negative theology, god being most manifest through his absence. However convincing such a reading might be, the fact is that Beckett seems to have known no more about Manichaeanism than is contained in the article of the eleventh edition of the *Encyclopaedia Britannica*, which he possessed, and that he never bothered to read any further; so on balance we must, I think, conclude that the heavy overlay of Manichaeanism in *Krapp's Last Tape* is after all structural, that it is like a template laid on ordinary experience which allows some of it to show through in orderly fashion, but no more than that.

Another peculiarity of the play which is perhaps far stranger and makes it unique among Beckett's works is the impression that it is about a possibility of happiness which has been missed. Krapp certainly seems to think he missed such a chance, though the fact that he thinks it is of course no evidence that Beckett did. Once, breaking all the rules, I said to him, 'Krapp seems to think he missed something, if not the last chance of happiness then of some sort of romantic fulfilment.' What I really wanted to know was whether he believed there ever could be anything of the kind. 'Yes,' he replied, 'Krapp seems to think that perhaps he did.' Then, sensing perhaps that he had not answered my implied question, he added with a smile, 'But that's no proof of anything, is it?' Discussing the play with Pierre Chabert on another occasion, he said, 'I thought of writing a play on the opposite situation, with Mrs Krapp, the girl in the punt, nagging away behind him, in which case his failure and his solitude would be exactly the same.'

In the various drafts of the play, from the 'Magee Monologue' on,

Beckett steadily reduced the directly autobiographical references. Thus Miss McGlone was originally given the name of Beckett's Roussillon drinking companion, the cheerful Miss Beamish. And the woman with whom the younger Krapp lived in Kedar Street, a liaison he considers himself 'well out of' and calls a 'hopeless business', was first given the name Alba, which was the name he had given Ethna McCarthy, before becoming Celia, which was the name of Murphy's companion; then Furry, perhaps after Mrs Rudmose-Brown; but finally winding up as Bianca. Another curious change was in the references to Krapp's sex life. At the age of twenty-seven he had been making plans for a 'less ... (*hesitates*) engrossing sexual life', but in three earlier drafts he had prayed for a fuller sex life, pronouncing the word 'intercourse' with lip-smacking enthusiasm.

The spell with Magee in West Cromwell Road (not to mention sessions with others such as the present writer) involved a lot of drinking and Beckett arrived back in Paris in a state of collapse, still, he felt, coughing up what he had imbibed in the murk of London. He desperately wanted to retire to Ussy, but within a couple of weeks he had to go to Dublin on an errand of mercy. Ethna McCarthy was now very ill with cancer and both she and Con expressed a wish that he should come over to see her. Although he remarked ruefully to Mary Manning that the trip would just about finish him too, he went. He thought it odd, though, that when he had met Con in London he had found him his usual suave and silky self – a fact which now puzzled him, for it meant that either he had not known about Ethna's condition at that point or he was happy to forget about it. Charitably deciding that it was the former, Beckett went to Dublin in the first week of December.

It was a grim week. Apart from anything else, it was bitterly cold. He stayed with Frank's widow, Jean, and went every day to talk with Ethna in Baggot Street. In the flat, which he found very depressing, Ethna crouched all day over a smoky coal fire. They went out to Jammet's twice, but the effort was obviously too much for her and the excursions were anything but festive. Distressed though he was, he stayed on every night, drinking and talking till it was time for the late train back to Killiney from Westland Row. It was clear that there was no hope, but since Ethna had arranged to go London to see a Harley Street specialist as a last resort, he encouraged her to do so, if only so that she would get away from the depressing flat.

Beyond sadness for her, he felt little emotion. Except for MacGreevy, a visit to Beatrice Glenavy and an accidental bumping into the artist Norah McGuinness in Switzers book department, he saw hardly anybody he knew, but Jean took him for a drive along some of the roads around Foxrock and in the foothills of the mountains. The 'dear old' Ballyogan Road, which is directly on the other side of Croker's Acres from Cooldrinagh, he found very evocative and unspoiled save for Electricity Supply Board pylons along one side. It was still a place of high, unkempt hedges – known as ditches in Ireland – and untidy tinker encampments, as it had been during his own solitary ramblings as a boy.

MacGreevy came out to Killiney on the Sunday before Beckett left and they went for a walk together. It was not much of a meeting because Beckett was, as he said, 'perished with cold and sadness', but he was touched that his friend had made the gesture of coming all that way, more particularly because he himself had not been well since the beginning of the year. (He had suffered a heart attack in 1956.) MacGreevy's illness had led to an increase of tenderness between them. When he first heard of it, Beckett wrote to say, 'I won't try to tell what I feel, because you know, and between us the big words were never necessary. I know how difficult it is to be careful, for one who never spared himself. You owe yourself ten years of simple mental and physical rest, have some of it now unprotestingly. You're a man much loved and much needed, don't forget that.' Much loved he may have been by Beckett and one or two others, but despite his important position, the proudly-worn ribbon of the Chevalier of the Legion of Honour in his buttonhole on all occasions, and his appearance of being a central figure in the cliquish art circles of Dublin, he was in fact lonely. In a later letter Beckett spoke of Tom's *'besoin d'un frère'* as something 'very deep and early in you'. MacGreevy had come to Paris in September and they had made a point of going to the Cochon de Lait together, the restaurant near the Odéon where they used to go, as Beckett reminded his friend, thirty years ago, and where they had made friends with the waiters Mario and Angelo. 'Ah well!' he said in a letter referring to this sentimental return.

Beckett flew back to Paris direct from Dublin, arriving on 8 December 1958 in a bad fog. He did not feel well and was afflicted by a cough but he went almost immediately to Ussy, where he stayed alone until after Christmas, though he was back in Paris for the New Year. Like others whose creativity finds various forms, he was unhappy that he should neglect the one that seemed to him most important, the novel, though it certainly did not seem so to others, including most critics and members of the public; and he was now making an effort to return to prose composition. He started a new work which he found very difficult and which in the course of time would become *Comment c'est*, *How It Is*, but there were days when he felt it was unlikely to become anything. At Ussy he was spending long days wrestling with it and he returned there shortly after the New Year. By March 1959, after what he felt had been months of false starts and rewriting, he had only about ten pages and they were a mere approximation to what he wanted. But he was determined not to try anything else until it was clearly hopeless. No sooner had he begun to assuage the guilt of neglecting one form than he began to hanker again after another. He now felt the call of the theatre just as strongly, and described the urge to write another play instead as 'one of those queer compulsions'.

Ethna died in late May. Before she died he had sent her some violets, picked in the wood near Ussy, saying, 'This is just my heart to you and my hand in yours and a few wood violets I'd take from their haunt for no one else.'

In June, however, he was still resolutely struggling with the new prose work and finding it horribly difficult. Even though he could see clearly what he wanted to do, and that it should be only about 100 pages, he felt he was making very little progress, or only just enough to keep him from giving it up in disgust. 'I rely a lot on the demolishing process to come later and content myself more or less with getting down elements and rhythm to be knocked hell out of when I am ready ... It all takes place in the pitch dark and the mud, first part man alone, second with another, third alone again. All a problem of rhythm and syntax and weakening of form, nothing more difficult,' he told Barney Rosset. Yet, comically perhaps, he was once again hankering after other forms of composition – theatre or radio. 'I hear their siren voice and tell them to stick it up.'

The year had begun with the completion of a radio play, which was broadcast on the Third Programme towards the end of June. First called *Ebb*, it then became *Embers*, a hauntingly beautiful but obscure work whose narrative development is difficult enough to follow on the page and was, in spite of Donald McWhinnie's best efforts, still more difficult to follow as a broadcast. Evidently influenced by his stay in Killiney, whose beach – by contrast with most Irish beaches – is notoriously composed of shingle and pebble, *Embers* is set beside the sea there, and as it opens we hear the sound of the sea and Henry's boots on the shingle. As always, though, Beckett does not tie himself down to a specific geography. Although the island at the beginning, when all the shore is in shadow 'and the sea out as far as the island', may be Dalkey Island, there would be scarcely much difference in the evening light on either side of Killiney Bay; and his father's preference for 'the other side of the bay' because of the sun on the water at evening could only refer to Dublin Bay.

Like so many Beckett characters, Henry tells himself stories, including a halting and inconclusive story about an old man called Bolton who is in great trouble and calls out his doctor, Holloway, in the middle of a winter night. Like other Beckett works this too contains a couple who seem to be bound together only by memories and past associations and yet cannot separate. Henry finds Ada's chatter sometimes unbearable and believes 'that's what Hell will be like, small chat to the babbling of Lethe about the good old days when we wished we were dead'. Yet when, shortly after she joins him on the beach, she says she must 'be getting back', he begs her to stay. 'Not yet! you needn't speak. Just listen. Not even. Be with me (*pause*) Ada! (*Pause. Louder*). Ada!' Henry's father, a figure slightly reminiscent of Bill Beckett, dies of a heart attack while taking the evening bathe that he loves. He and Ada have a daughter, Addie, who is learning to ride and play the piano. 'It was not enough to drag her into the world, now she must play the piano.' In the BBC production this became 'to bring her into the world'. In 1959 the BBC was still censoring Beckett's work which, in view of the latitude permitted nowadays, may seem strange. Bolton 'trying to toast his arse' had to be cut and so had the expletive 'Christ!' Because it is written in English the play is replete with Irishisms. Thus in Bolton's story a beach is referred to as a strand, which is in more common use in Ireland; Bolton often hears a sound 'above in the house and walking the roads' and he tells us of his impatience with his little daughter's refusal to run away and play when he is 'mad to talk to himself'.

Embers met with a mixed reception for, in spite of his advocates, his following and his now worldwide fame, the general tone of English criticism was still somewhat hostile to Beckett. The two 'serious' journalistic critics of the day, Frank Kermode and Karl Miller, both disliked it. Miller, writing in *Encounter*, held it up as an example of the 'technical weaknesses, boring repetitiveness, obscurity', which he felt to be part of the 'pathetic and presumptuous' mood of Beckett's art, work which was lacking in the 'means of control which art is supposed to need'. While the playwright John Whiting praised 'the dramatic which has the precision of fine poetry', even some of Beckett's admirers had, and still have, their doubts about the work's obscurity. For anybody anxious to grasp in a literal way what is going on, it is indeed an obscure work to listen to. Since I had missed the broadcast McWhinnie arranged for me to hear it in Broadcasting House. Apart from finding it difficult to comprehend in any larger sense, I found it very difficult to follow, probably because Jack MacGowran, who spoke it brilliantly, was asked to do too much, playing both Henry and Henry's father in recounted scenes between father and son, and both Bolton and Holloway in the scenes which Henry imagines taking place between them. Nevertheless, like some obscure poems, the play had many beautiful and significant moments and was best allowed to do its mysterious work on the psyche without too much irritable reaching after fact and circumstance.

Beckett's adherents in the Third Programme were at this point in the ascendant, though questioning voices were being raised. In January Pat Magee had read some lengthy extracts from *The Unnamable*, to which I contributed a prefatory talk. For once Beckett was less than wholeheartedly enthusiastic about a Magee interpretation and also about the prefatory remarks. 'Cronin delivered his discourse before *The Unnamable* broadcast, I had the text, but God knows what I've done with it, it was all right, not very exciting. Barbara Bray would let you have a copy,' he wrote to Barney Rosset.

In early April a lengthy sojourn at Ussy and work on *How It Is* was interrupted by a trip to Paris to see the great French rugby team of the late fifties play Wales at Colombes. Beckett retained all his life an intense interest in sports, principally tennis, rugby and cricket, and whenever he could he watched them on television. No discussion of his attitude to the excitements and distractions of our mortal lot would be complete without reference to this absorption. He may have refused to get excited about public events whose outcome was allegedly more important than that of rugby or tennis matches and he seems not to have asked himself whether he could justify an interest in sporting events philosophically, but enjoy them he did. He certainly enjoyed this particular match, for which Claude Simon, *nouveau romancier* and author of *La Route des Flandres*, gave him tickets. This legendary French team suffered, Beckett thought, an unwarranted defeat by Ireland two weeks later at Lansdowne Road. Though, on occasion, his loyalties were divided when Ireland played France, the greatness of this French team excited him and he was disappointed when they were 'unjustly' beaten by his native country. Their victory over Wales on a very hot day when the temperature was over 75 degrees gave him great satisfaction.

In July work was interrupted by a trip to Dublin. Trinity had offered him an honorary doctorate and, after a good deal of indecision, he decided that he could not refuse and that he would go to Dublin to receive it in person. Beckett had what would in Ireland be called 'a soft spot' for Trinity and a continuing sense of guilt about it. He was also, for such a revolutionary writer, somewhat over-partial to the attentions of academe. In the post-war era it began to be clear that academic writers were to be the principal guardians and propagators of James Joyce's reputation – and also, perhaps, that whatever their excesses and distortions, they were doing Joyce more good than harm. Beckett would have been aware of this and of how academic criticism fed on the obscurities and supposed profundities of Joyce's work. As an obscure and supposedly profound writer, he himself had obvious attractions for expository criticism. It was at about this time that academics began to seek him out, and on the whole he responded readily to their blandishments if not their questions.

To H. O. White, who had written to him with Trinity's invitation to accept a doctorate, he replied to say he felt honoured and moved and would accept with gratitude. Even in such a dutiful letter, though, his misgivings were inclined to surface. He adverted to his preference for private obscurity and his dislike of public appearances and publicity, remarking that he felt both 'unworthy' and anything but 'doctoral and literary'. But he went to Dublin nevertheless, and seems even to have been eager to go, arriving a good ten days before the conferring and tolerantly fielding the questions put to him by reporters at the airport about Ireland and whether he would ever live there again. He stayed in Killiney with his sister-in-law and was happy to find that Jean had now fully sorted out the details of Frank's estate. In avuncular fashion he observed that Caroline seemed to have dozens of young men at her beck and call and that his nephew Edward, still at school, was a really promising flautist who should develop into a first-class player if he had the proper training.

Edward was taking private lessons from a French musician who lived in Dublin, but there was a family conflict because his mother wanted him to go to Trinity to study engineering in the expectation of following his father into the business in Clare Street. The compromise arrived at hinged on a special arrangement with Trinity. It was that he should study engineering for a year, then go to Paris in the hope of becoming a professional musician with a promise from Trinity that he could return to take up his engineering studies if it turned out that he was not talented enough. This arrangement was the result of Beckett's intervention with Leventhal, H. O. White and others, as Leventhal was now Registrar of the University.

Somewhat to everybody's surprise Suzanne had decided to go to Dublin too and arrived a few days after Sam accompanied by her friend Marthe Gautier. At the weekend Caroline celebrated her twenty-first birthday with a large gathering of young people at Shottery. It was no place for aunts and uncles, so everybody moved out for the night in question, Beckett spending much of that evening with his 'cher vieux', MacGreevy, the person in Dublin with whom he had most in common and felt most at ease.

By contrast with preceding visits – melancholy occasions when he had been quite abstemious – Beckett had this time round been making free with the whiskey when he met acquaintances; and Suzanne's arrival seemed, if anything, to trigger off greater indulgence. Later in the week, during a gathering at Arland Ussher's at which he was imbibing fairly heavily, she begged the somewhat cynical and ironic Ussher to stop Sam drinking, even, as he was to tell it in Dublin, going on her knees to him to do so. Whether this consumption of alcohol was evidence of his enjoyment at being in Ireland or the reverse is hard to say. Both probably, since his attitude was of course ambiguous.

He was currently very much out of humour with official attitudes there because of the stance the organizers of the Dublin International Theatre Festival had adopted to a production of Sean O'Casey's *The Drums of Father Ned* in the previous year. The Archbishop of Dublin had refused to say the Mass which was supposed to get the festival off to a holy start (the poet Patrick Kavanagh said you might as well have a Mass for the opening of the flat-racing season) until parts of O'Casey's play thought to be offensive were expunged. Rather than risk a row with the Archbishop the organizers had withdrawn the play. Beckett had consented to a performance of *Act Without Words* I and II at the festival, but when he heard of the row about the O'Casey play he withdrew permission. Indeed he withdrew all permissions for the performances of any of his plays in Ireland then or in the future. This was a blow to Alan Simpson of the Pike Theatre, who had bravely put on *Godot* with some success and was looking forward to doing *Krapp's Last Tape* and *Endgame*.

Simpson's feelings about the matter were not improved when he read in an evening newspaper that another theatre group, having succeeded in 'talking him round', had received permission from the author to do *Endgame* and *Krapp* in a Dublin theatre. As it turned out, Curtis Brown had 'gaily' given this permission without consulting the author. Simpson wrote to Beckett in terms which suggested that he was 'wounded to the heart (naturally)'. Beckett tried to stop the production and failed, but later in the year he was slightly mollified when he heard from various friends, including MacGreevy, that the production had been quite a good one though he was very upset and embarrassed by the affair, more particularly since he felt 'morally bound to the Pike Theatre'.

The withdrawal of the O'Casey play and the subsequent events, though almost the last example of such censorship in Ireland, would be instanced by him for many years as an example of Ireland's bog-trotting obscurantism, and he would continue to bring it up in conversation long after the situation in Ireland had changed. In the course of time he would add a ban on productions in South Africa to the one he had imposed on them in his native land; but like other authors and playwrights he perhaps somewhat illogically adopted a different policy in relation to Eastern Europe. There, on the presumption that only dissidents were interested, he was already allowing and even encouraging productions by waiving royalties. His vision of human existence seemed to strike a special chord in the hearts of Eastern Europeans and when approached by dissident producers or playwrights he would always respond. Some, most notably Václav Havel, felt that they had a special affinity with him; and Beckett returned the compliment by dedicating a short play, *Catastrophe*, to Havel in the early eighties. However, he naturally gave no encouragement to the Yugoslav playwright Miodrag Bulatovic, who wrote an unauthorized sequel to *Godot* called *Godot Came*; nor probably would he have been very pleased if Bertolt Brecht had fulfilled the ambition he had before his death in Eastern Germany in 1956, which was to write a counter-play to *Godot* in which the relationship of Pozzo and Lucky would have been worked out in accordance with the Marxist view of history.

In consenting to receive an honour from Trinity now, Beckett was evidently exempting the university from any implication in the goings-on of official Ireland. The Orator's address at the commencement ceremonies had been composed by Leventhal, but of course it was delivered in Latin. Rather surprisingly perhaps, it described Beckett among other things as a satirist. Although the justice of this description might have been wryly conceded by Austin Clarke, widely held in Dublin to be the original of the poet Ticklepenny in *Murphy*, who was in the audience, and by one or two others who had figured in Beckett's earlier works, it was still held by many to be odd. Rather inaccurately too, the Orator's address said that Beckett had been 'for a while one of our Lecturers in French, then a friend and helper of James Joyce in Paris'. Beckett had been Joyce's friend before he became a lecturer, but a little synthesis is perhaps allowable on these occasions. The address concluded by saying that since it was well known that Mr Beckett cared little for the outward emblems of fame, 'he does us all the more honour in coming to receive our honours today'. While the degree was formally conferred Suzanne sat quietly at the back, and though she attended the reception afterwards it escaped the notice of the press that she was in Dublin, on her first and last visit to Ireland.

After the ceremony both Sam and Suzanne stayed for a few days; but when she left with Marthe Gautier they flew straight back to Paris, while he went to England the following day to visit his cousin, Sheila Page, in Surrey and, as he rather vaguely put it to MacGreevy, 'to see to some things in London'. Beckett, with his strong sense of family and his deep reluctance to let go of the past, was certainly pleased to visit his cousin; and he had business to discuss with Faber and Faber, John Calder and with the BBC. But one of his reasons for the London stopover was more personal, albeit to meet a BBC employee, Barbara Bray. Though they saw each other only at intervals – when he came to London or she to Paris – his relationship with Barbara had by now become a very intimate and personal one.

On arrival back in France he went straight to Ussy. He told people that Paris was 'nightmarish', a place where quiet was almost impossible to obtain, but perhaps he wanted to be on his own to think things over. In the spring he had got out his old bicycle and now he went for long, fairly brisk rides along the country roads. When not cycling, or occupied in the space which he persisted in calling a garden – though it was anything but garden-like in appearance – he read. The Olympia Press had recently published the Diaries of the Irish patriot Roger Casement, documents which had, over the years, been the subject of controversy and were alleged by Irish patriots to have been a forgery by the British government as they contained frank accounts of Casement's homo-erotic experiences. Beckett read these with he called 'great absorption', and came to the conclusion that they were 'quite authentic'. Read as they were during the composition of *How It Is*, it is not fanciful to see the sedulous anatomical thumpings and pokings of Beckett's book as deriving to some extent from the Casement Diaries. The long process of goading the central character into utterance involves gougings in the armpits with the narrator's nails, thumps on the skull, bangings with a pestle on the kidney area and gougings on the buttocks with the blade of a tin opener. Likewise it is possible that the tin opener itself and the sack of tinned food derive from the author's domestic circumstances at Ussy, where he ate largely out of tins when he was on his own.

Besides the Casement Diaries he read the recently published third volume of Ernest Jones's monumental biography of Sigmund Freud, which also contains a lot of anatomical detail, in this case very grim stuff, for during the last twelve to fifteen years of his life Freud had undergone no less than thirty-three operations for cancer of the jaw. Though he found it all 'quite horrifying', he was amazed at Freud's 'unbelievable fortitude'. He also read Robert Pinget's new work *Le Fiston*, which appeared that year. Pinget, whose third novel this was, was now firmly identified with the school of the *nouveau roman*. His attitude, expressed in the declaration that 'anything that isn't writing is death', much resembled Beckett's own and Beckett continued to admire him, finding that he liked the new work very much. A less happy experience was his attempt on Aldous Huxley's *Point Counter Point*, which somebody had sent him in the Penguin edition. He found it no more rewarding than the other Huxley books he had attempted in the past.

At Ussy this time he was for once lonely and discontented, perhaps because he had so recently been in London and because his friendship with Barbara had now become a close one, but declared to Barney Rosset, 'I see nothing else for it' – meaning that for the sake of his work, specifically for the sake of finishing *How It Is*, he was determined to stick it out. As the leaves began to fall he was still struggling with it and still very dissatisfied. He knew how important it was not to hurry and he had now decided that it might take another year at least. Nevertheless, to force it along he was adopting a more professional approach and was trying to do at least half a page a day, though on some days he stopped working without achieving this. None of his extended prose works had given him so much trouble as this comparatively short last novel. His dominant feeling was that he was on the wrong track, but since there seemed to be no other that he could possibly take, he persisted.

His long sojourns at Ussy had to be interrupted occasionally for trips to Paris, however. Some were to see people who had come either on business or on a visit with hopes of seeing him, as MacGreevy did in September. He would usually agree to meet people who had been told to telephone or even call by somebody he trusted, perhaps more especially young people and even more especially if they came from Ireland, following in this the example of James Joyce who was always willing to meet people, especially young writers, from Ireland if they came with a recommendation from someone he trusted. With them, Beckett would often discuss Dublin and how or whether it had changed. Though he always asked about their work he rarely asked to see it and made few bones about the fact that if it was sent to him he was unlikely to read it. In the sixties also there was a succession of friends' and relatives' daughters and nieces who came to Paris and who were entertained and looked after. Mary Manning's daughter Fanny was met at the Gare du Nord. He arrived a little late and found her outside, looking, as he put it, 'like a frail blown leaf'. Elizabeth Ryan, Tom MacGreevy's niece, and Sean O'Casey's daughter Shivaun came, as did Niall Montgomery's daughter Rosemary and of course his own niece, Caroline. Depending on age and level of sophistication they were taken to restaurants, cinemas, music-halls and circuses. Some were just out of school and, by most of their standards, they felt themselves to be royally entertained, sometimes by Sam and Suzanne together.

One visitor to Paris during 1959 was Brendan Behan. He had first met Behan in 1952 when he had arrived unannounced at the rue des Favorites at 6.30 in the morning and made so much noise that at length Beckett got out of bed and admitted him. He was wearing an open-necked shirt and a crumpled and apparently blood-stained blue serge suit. He was a friend of John Beckett's but there was no question of him arriving humbly with a recommendation: he was there under his own steam. Within a few minutes of being seated at the kitchen table Brendan launched into a long string of anecdotes and reminiscences, many of which concerned people they both knew. Beckett had some whiskey himself from a bottle he produced for his guest and continued to be amused for an hour or so, after which the performance began to weary him. Besides, *Godot* was then in rehearsal at the Théâtre de Babylone and he had to be there by 9.30. When he announced this, his guest exhibited some disgruntlement, but he then had the bright idea of depositing Behan at the *Merlin* offices on his way to the theatre. When they got there they found Christopher Logue in charge and Beckett, feeling that Logue might have more time to spare and more stamina than himself, made his escape.

However, Behan had now identified his compatriot's apartment in the rue des Favorites and, ever one for rousting out possible drinking companions at odd hours of the day and night, began to turn up at such hours. Equally, Beckett and Suzanne learned to identify the manner of his arrival, which was usually noisy; and would keep mum and still until the storm blew over, which it usually did after twenty minutes or half an hour, though the nearby café might continue to hold terrors for longer. At least once, however, Beckett had responded to a message of distress from his fellow dramatist. According to Ulick O'Connor's biography of Behan, 'Beckett went to Brendan's lodgings in rue des Feuillantes and paid some rent due to a landlord who had been threatening Brendan with eviction'. A version of what may have been the same episode that Brendan, who was proud of the acquaintance, circulated around Dublin, though, was rather more dramatic. According to this he was arrested one night and charged with being drunk and disorderly. When, the following morning, he proved unable to pay the fine, he was allowed to make one telephone call and naturally telephoned one of the most eminent Parisians he knew, the other candidates for the honour being Camus and Sartre. In response to this telephone call Beckett came around to the police station. In Brendan's words, 'he paid them what I owed them, and he took me away, and he gave me 10,000 francs and a double brandy and a lecture on the evils of drinking.'

By 1959 Behan's fame, which had achieved phenomenal dimensions since 1956, was almost the equal of Beckett's; and whereas Beckett had only recently had difficulties in getting a play produced, Behan had two in production, one at the Théâtre des Nations and the other at the Oeuvre. Naturally this necessitated a visit to Paris and, naturally also, while there Brendan expected to see his fellow playwright, who was his companion in fame and was besides a fellow Dubliner, a matter of almost mystical importance to him. Beckett took care to avoid a meeting and special precautions were observed at rue des Favorites, but since he had to be in Paris and since Brendan covered enormous tracts of the city in his roamings there was some prospect that a meeting might come about accidentally. Beckett took a certain amount of avoiding action and was highly relieved when it did not. Though he had ample opportunity to do so, he did not go to either of Behan's plays.

Towards the end of the year Beckett acquired a motor-car, the first vehicle he had owned since his AJS motor-cycle. It was a Citroën 2cv, fondly known as a Deux Chevaux, a car popular among the French literary community and intellectuals generally, as well as hundreds of thousands of ordinary people. His was blue and he drove it with his usual mixture of abandon and apparent ineptitude. He used it to meet friends at the airport as well as for trips to Ussy, and the tales of those who regarded themselves with some justice as survivors of these perilous journeys are many. As the sixties progressed he began to suffer from cataracts and his worsening eyesight made his driving seem even more hazardous.

Someone who had an especially vivid recollection of his driving was Professor Arnauld of Trinity, who had been Professor of French at the time Beckett held his lectureship. As a Trinity man and closely associated with what guilt Beckett felt, Professor Arnauld was well looked after. They met at Chez Francis opposite the Pont de l'Alma, where Beckett had often gone with Joyce. They had a pleasant meal and drank some wine. Then Beckett offered to drive Professor Arnauld to his hotel, which was on the other side of Paris beside the Gare du Nord. This offer pleased the professor, but when they got into the Deux Chevaux Beckett, evidently determined to find the most expeditious route, produced a map of Paris which, to his passenger's disquiet, he propped behind the driving wheel. It was rush hour and the traffic was heavy. To consult the map it was necessary for Beckett to take off his glasses and lean forward short-sightedly over the wheel, and this he did as they approached intersections or whenever he was in doubt about the route. Near the Gare de l'Est the car came to a stop in the middle of the traffic. 'It has run out of petrol,' announced Beckett. The professor was deeply relieved, but Beckett had a can of petrol in the back and he replenished the tank while angrily hooting drivers took avoiding action. Then, to Arnauld's despair, he got back in and they drove on.

Chapter Thirty-One

❧

Towards the end of 1960 Barbara Bray resigned from the BBC with the intention of taking up residence in Paris. She had applied for the vacant job of Paris representative of the Corporation and, though offered instead the possibility of becoming a full-time producer, she decided that she would leave and go to Paris anyway, a move which had been discussed more than once with Sam. She was becoming known as a translator, particularly of the works of the new French novelists, one of which, Marguerite Duras's *The Square*, had been suggested to her by Beckett – though he turned down Calder's request that he might write a preface – and through translations, reviewing and freelance work for the BBC she felt she could support herself there. In fact she would have gone earlier than she did, but 1960 was an unsettled year in France. At its beginning, Algerian settlers, a group of army officers and some of de Gaulle's former adherents such as Bidault and Soustelle had attempted a coup against the Fifth Republic which had come into being with the Presidency of the General. The coup was a failure and rapidly degenerated into a terrorist campaign. Barbara, who was the mother of two children, decided against moving to Paris while there were still bombs going off and tanks in the streets, so, in the end, she did not make the move until June 1961.

By then, Sam's life had also changed. He had finished *Comment c'est*, his last extended prose work. He had bought and, with Suzanne, moved into a new, much larger apartment at 38 boulevard Saint-Jacques – a busy modern thoroughfare with a noisy flow of traffic at all hours in the 14th arrondissement, within walking distance of the boulevard du Montparnasse and thus in his favourite quarter of Paris. And he had got married.

The move into a new apartment, which, apart from anything else, would have more work space, was long overdue. In the rue des Favorites, with its single, all-purpose living room and its small bedroom, Sam and Suzanne were very much on top of each other. The new apartment was much larger and, to make it larger still, they took the adjoining studio apartment, which had its own entrance. Number 38 was a smart new apartment block with plate-glass windows, aesthetically neutral though less functional and plain than 6 rue des Favorites. There was an intercom system and after the outer door had been opened one went up in a lift.

His work space in the smaller apartment had a writing table and a chair, as well as a narrow studio couch which could be either sofa or bed. On the specially built shelves stood those books which he wanted to have near him when he was working. There were his own works and those of James Joyce, various early editions of the works of Samuel Johnson, French and English dictionaries, and works of reference including *The Oxford Companion to Music*, Bibles and Bible Concordances. On the remaining wall space hung the Yeats painting he had bought in Ireland in the thirties and his van Veldes. The fact that he had a smaller apartment for his exclusive use has been seen as a move towards separation, but neither Sam nor Suzanne presented it to their friends in this light. True, they now had separate bedrooms and separate hall doors, though they shared a kitchen and a living room where meals were eaten when they ate in; but this degree of separation is not exceptional among married couples and it was long overdue. 'We simply must have our rooms where we can shut ourselves up,' he had written to Mary Manning when the move was first mooted.

A drawback was that at the back the apartment overlooked the Santé prison. The grim, reproachful buildings of the prison with their many cell windows were visible from Sam's studio and he could hear the shouts – or howlings, as he once described them – of the prisoners at regular times of day. Later, when he mentioned this to visitors, he would sometimes point out that he could see the white dome of the Panthéon as well, as if this compensated for whatever distress he felt. In time he would establish a form of communication with one long-term prisoner, waving to him in salutation at certain hours of the day.

Suzanne began to lay carpets, hang curtains and move in furniture in August 1960, just as Sam had finished *Comment c'est* and, feeling 'pretty flattened', was looking forward to 'a month or so just idling out of doors at Ussy'. When he did move, in January 1961, he was 'troubled by the strangeness and unfamiliar noises' – including of course the cries from the Santé – and felt it would 'take some getting used to', but at the same time he was leaving rue des Favorites 'and the abominable Vaugirard without a pang – after 21 years'.

As Suzanne furnished it, the living room was a simple, rather functional room with a formica-topped table. Shortly after the move Sam's nephew Edward Beckett arrived in Paris to take up his music studies and every Friday he would come to tea there, arriving at seven sharp. Since there was as yet no fridge, the tea was what Edward called 'tea à toothpaste', served with concentrated milk from a tube. He thought that Sam was clearly anxious to create a family atmosphere; but Suzanne did most of the talking. As Edward soon realized, 'you had to mind your Ps and Qs when you were with her. She couldn't abide prattle,' and 'If you said anything silly she'd pounce on it.' But she was kind and forthcoming where he was concerned, taking, with Sam, an interest in his appearance, going with him to the Inno France department store to buy clothes more suitable for Paris than his Irish corduroy jacket, and sending him to the hairdresser for a cut more fashionable than his Irish quiff. When Edward's sister Caroline had come to Paris as a young au pair she too had been taken shopping, in her case for a bikini, Suzanne professing herself appalled that she was going to a seaside place in the South without one.

The move to the new apartment had its traumas for Beckett; and, although he was cool and matter of fact about it, so had the marriage. It was arranged rather hurriedly. The first step was a telephone call to John Calder, asking what were the formalities in England and if it could be arranged quietly and quickly there. Nobody was informed of the plan except MacGreevy and Calder – and Barbara, of whom he had been seeing a good deal. She had lately made a number of trips to Paris in connection with the production of *Embers* and the radio play *The Old Tune*, a translation Beckett made of Robert Pinget's *La Manivelle*, which Barbara produced. He explained to her that the marriage was for testamentary reasons; and no doubt these did play an important part in the decision. 'In any case I always urged the other party's cause,' Barbara says.

After Sam's telephone call Calder made enquiries and then advised him that marriage would require two weeks' residence in England, but that the bride-to-be need not fulfil this requirement. Early in March Sam drove alone to Calais, crossed with the little Deux Chevaux to Dover and then drove on to Folkestone, which had been decided on as a place for the ceremony. There he booked a hotel room, writing to MacGreevy from the Bristol Hotel, The Seas, Folkestone, that he was there 'to deal with the little matter I mentioned to you'. He found the Registrar, who bore the Irish name Cullen, sympathetic and helpful, though Beckett cautiously told him no more than he felt was necessary. Cullen informed him that it was not necessary to remain in Folkestone night and day for the next two weeks, but that 'whatever his movements he should keep on his hotel room for that length of time'. He was avoiding London and possible publicity but he decided that sometime during the following week he would go up to Ilford to see his cousin Sheila Page. For the moment, however, he was happy to be in a hotel room by himself, 'unhurried and undisturbable'.

He found the Bristol extremely comfortable and also very quiet. He had brought with him, as well as an accumulation of unanswered mail, what he hoped was a near-final draft of his new work in English, *Happy Days*, and he thought perhaps he might finish it and show it to a London producer before he went back to France. Dividing his time between it, his correspondence and occasional paddles in the sea in the bright, warm March weather, Beckett found that the first few days passed tolerably enough. Then he began to find it rather dreary hanging around Folkestone, doing his best to be invisible. After spending a few days with Sheila Page to relieve the monotony, he went on a little outing to Brighton. Suzanne arrived on Wednesday the twenty-second and two days later, on Friday 24 March, the ceremony took place before Mr Cullen and two witnesses, employees of the Registrar. Beckett gave his occupation as writer and his address as the Bristol Hotel. Suzanne gave no occupation. Questioned later as to whether they had had a few glasses of champagne, she answered drily, 'a wedding day is nothing to celebrate'. On the following day they drove to Dover, crossed on the ferry and arrived in Paris late on the Saturday evening. Sam's immediate reaction was: 'Thank God it's done at last.' For the moment they told no one but when, as it gradually did, the fact and something of the circumstances of their marriage became known among their friends, Beckett would intimate that it was, as Joyce's had been, for testamentary purposes, to ease Suzanne's inheritance of his copyrights and property under both French law and English.

Their marriage made no difference to their relationship, nor was it ever intended that it should. Like countless other couples they were achieving a degree of separation within marriage – not in any dramatic or, still less, tragic sort of way, but matter-of-factly and with accommodations on both sides. The impression among their friends was that he was bored in her company. The sexual bond, which had probably never amounted to very much, had long been dissolved and, except for music, there was little in the way of shared interests. Suzanne had believed with a certain amount of passion in him as a writer and had been determined that he should be published and recognized. She understood his work and, as a cultured French woman, its importance. But after he had been well and truly taken up by others, her role as promoter was gone, something which irked her and led her to make occasional sarcastic remarks about his relationship with his admirers. Many of their friends thought that she was now jealous of his fame and the adulation it brought him. Nor was she much interested in his leisure activities. She vehemently disapproved of the drinking which kept him out late at night, and of most of the friends with whom he drank. 'Sam chooses his friends like a dog chooses arse holes,' she told John Calder.

And in any case, by now most of their friends were observing one rather strange characteristic of the couple – that you could not be a friend of both. Sooner or later, you had to decide which of the two you were going to see most often and resign yourself to rarely seeing the other. This choice naturally caused distress, but oddly enough Suzanne sometimes won out in cases where one might expect the opposite result. A few people, like the actress Madeleine Renaud or the famous director and apostle of 'Total Theatre', Jean-Louis Barrault, who had a professional relationship with Sam, managed to remain to some extent in both camps. But others, such as the novelist and playwright Robert Pinget, found themselves cast, sometimes not altogether happily, as among Suzanne's friends – in Pinget's case at least, an odd result since he had a deep admiration for Sam and had, after all, been his discovery.

All things considered, it may seem surprising that Sam and Suzanne had remained together, but she was a very practical person and so in a way was he. Up to this point he had needed both to be alone and to have companionship of a sort. He could be alone with Suzanne in a way that he probably could not have been in a new relationship with Barbara. If Barbara now offered more than this, his refusal of it had a great deal to do with the inertia and abandonment of initiative which were deeply part of his nature. To an extraordinary extent he was a creature of habit; and in so far as Suzanne represented the disapproving mother, he was bound to her by the habit of guilt, a bond which can be as hard as any other to break. But there was more to it than that. He was deeply loyal, someone for whom old ties and associations had a strong – sometimes even, one might say, an overly sentimental importance – and he had also, as his relationship with his publishers demonstrates, a keen sense of gratitude. He owed, he felt, almost everything that had happened in the way of publication and success to Suzanne. The decision to marry her now clearly demonstrated where his loyalties lay. He was making the bond more difficult to break, not less.

As Barbara Bray came to understand him and to understand better both his psychology and his situation, she fully acquiesced in this attitude. 'One of Sam's main characteristics was loyalty and gratitude and I would never ask anybody whose main characteristic was loyalty and gratitude to be disloyal or ungrateful. On the contrary,' she has said. But she knew that he agonized over the choice he was making. And, though he never spoke to his friends about such matters, a few others were aware of it as well. When, over·a decade later, Harold Pinter's marriage to Vivian Merchant ended in a blaze of unwelcome publicity, Beckett, who knew Vivian Merchant very well and liked her, wrote Pinter a letter of understanding and goodwill. When they met subsequently he said that break-ups were always very painful but that to leave somebody was at least action; it was doing something about a situation instead of letting it drift. Throughout the conversation Pinter felt that what he was saying was that it would be very difficult to leave Suzanne and that he would if he could but he simply could not be responsible for hurting anybody to that extent.

But Barbara Bray was a very attractive and intelligent woman who understood Beckett's work extraordinarily well and was one of the first to fully understand his position in the scheme of things. She was one of the few with whom he was prepared to discuss his work, though usually textually, in terms of the search for a word or a phrase for a feeling or an idea. She also had a sense of humour which contributed both to her understanding of his work and to his pleasure in her company. She admired him profoundly both as a man and as a writer. There was a deep affection between them and a deep level of intellectual understanding. Unless Beckett was entirely devoid of belief in the possibility of a human fulfilment, he cannot have been oblivious of the choice he was making. Yet he made it and he grimly stuck to it, at whatever cost.

Barbara arrived in Paris in June 1961, and to begin with she leased an apartment in the fashionable 16th arrondissement, on the Right Bank, not far from the Étoile. After some time in this rather *haute bourgeois* area she moved to a contrasting one, the rue Séguier at the heart of the Latin Quarter, off the rue Saint-André-des-Arts. She needed a place big enough for her children, who were then aged twelve and thirteen, and where she could work. Sam lent a hand with these moves, but she always remained independent, travelling to England frequently to appear on a well-known radio programme of the time, *The Critics*, reviewing for the *Observer*, translating for the Oxford University Press and other English publishers. Almost every day Sam came round, often spending a good part of the day or a large part of the evening there and often bringing new work with him to discuss. When they went out in the evening, which was rarely as, apart from anything else, Barbara had the children to look after, they would go to one of his favourite restaurants, such as the Iles Marquises, or to the Falstaff bar in Montparnasse, which he began to frequent in the 1960s.

Beckett had always had a passion for secrecy. He imparted information about his private life to his friends only very selectively and resented any but the most minimal references to it when they appeared in print. He had gone to London in June before Barbara came over and while he was there he had lunch with John Calder and Harold Hobson. Hobson, the *Sunday Times* drama critic, was his most faithful English adherent, putting forward claims for his greatness as a dramatist from the very beginning, when such claims were rare. Yet when Hobson wrote a piece in *The Times* newspaper about this visit Beckett felt it to be too informative and was upset. When Hobson came to Paris in July, he asked him over lunch to be more discreet where possible talk about his private life was concerned. Hobson, though he did not believe that his piece had given anything away, accepted the reproof with a good grace.

Beckett's resentment of intrusion and personal comment was strong. Yet he was strangely not consistent. It was as if he felt the need to flirt with danger; as if there was also a streak of exhibitionism in him, or at least of confessionalism – as there is to varying degrees in most authors.

Towards the end of 1961 a young American academic, Lawrence Harvey, approached him with an introduction from a mutual friend. Harvey wanted to write a book about him and Beckett found him sympathetic. They met in the Closerie des Lilas and then Beckett took him to the Iles Marquises, where after the meal they had several *marcs*. After this meeting Beckett told the friend who had effected the introduction, Herbert Myron of Boston University, that he hoped to see more of Harvey and that he was willing to help him in any way he could, short of talking about his opus.

Harvey was staying in the suburbs and shortly afterwards, on a night of wind and heavy rain, his bell was rung. When he opened the door he found Beckett standing there in front of his Deux Chevaux. To his amazement Harvey found a bulky sheaf of papers thrust into his hand with a few hurried phrases to suggest that he might find them interesting. Pressed to come in, Beckett at first declined, but then stepped inside only long enough to exchange a few words before departing again into the rain. When Harvey opened the bundle he found that he had the manuscript of a rumoured but unknown early work, *Dream of Fair to Middling Women*, an unpublished short story entitled 'Echo's Bones', some miscellaneous critical writings and the typescript of a play which Beckett was known to have written but which nobody except the producers and the publishers who rejected it had since seen, *Eleutheria*.

These then unpublished works are far more revealing psychologically and circumstantially than Beckett's later writings; and though Harvey, who was very soon 'Dear Larry', was told that the author would be at his disposal at any time 'for talk I hope about any old thing under the sun but me', he surmised correctly that Beckett would have no objection to him adding a biographical dimension to whatever he intended to write. There followed in 1961-2 and 1964-5 what the American called 'a score of conversations ... that often lasted far into the night' and many 'faithful' replies to queries in letters. That he talked at length about himself to this comparative stranger is evident from the book Harvey wrote. And, somewhat to Tom MacGreevy's disquiet, he also directed Harvey to him. When MacGreevy expressed his surprise and unease about this Beckett replied, 'My feeling about the whole thing is that it will be done in any case, and the more accurate the better.' Harvey submitted certain early sections of the book to Beckett so that it was quite clear what he was doing. The long chapter about *Echo's Bones*, for example, even contains a section called quite plainly 'The Personal Sources and the Unity of Echo's Bones' – yet when it was sent to Beckett he approved, apparently wholeheartedly, declaring himself 'greatly touched by such scholarship and research and depth and patience of investigation'.

Harvey's first intention was to confine himself to Beckett as poet and critic, dealing only with published poems, but then he decided to include lengthy discussion of all the Belacqua prose and early unpublished poems as well as *Watt* and the critical writings. In the end he committed what is still supposed to be a capital sin in some Beckett circles by shamelessly interpreting Beckett's work in terms of his inner psychology and deducing aspects of his sexuality from it. Unfortunately, Harvey clouds and occludes almost everything by blanketing it in a fog of academic language, but despite this, and in spite of his willing co-operation in the enterprise, when finally sent the completed manuscript Beckett reacted with extreme dismay, removing some things from the typescript himself before asking Con Leventhal to read it and to excise further biographical references. Either Leventhal found it unreadable (which it very largely is) or he did a careless job, for a lot of biographical material seeped through. Some of Harvey's findings have already been quoted in Chapter Eleven.

A Beckett characteristic which contributed its share to the air of mystery which now surrounded him was his tendency to compartmentalize his friends. Many of those who knew him, and thought they knew him fairly well, were unaware of the existence of others who thought they knew him and knew him very well. Some of them were seen periodically at well-spaced-out intervals on a one-to-one basis and at a specific place and time; and they included several young writers or would-be writers or students of his work who had succeeded in getting an introduction which proved acceptable or perhaps even met him accidentally or on their own initiative.

Francis Evers was a young Dubliner who was living in England and had been hitch-hiking around France and Spain in 1957 when he called on the Irish journalist, Desmond Ryan, who lived in the rue Molière off the avenue de l'Opéra. Ryan told him he had just missed meeting Samuel Beckett, and though Evers knew little about Beckett except that he was the author of *Waiting for Godot*, he expressed interest and disappointment. He was given a telephone number but, having got no answer, decided to call at the address in the rue des Favorites. Beckett opened the door with a look of slight terror on his face, but visibly relaxed when Evers told him who he was and whom he knew. Asked what he would like to have in the way of refreshment, Evers replied that he would like a cup of tea. Shortly afterwards a woman entered the room whom he took to be a maid or home help, nor was this impression altered when she was introduced, not as Madame Beckett but as Suzanne, or when she silently brought the tea, put it down and left the room. As was often the case, in the conversation which followed Beckett asked more questions than his visitor and Evers recounted some of his hitch-hiking adventures, including an incident with a homosexual motorist when he had to fend off a pass. As Beckett was seeing him to the bus stop he asked how he intended to get back to England. When Evers told him that he intended to hitch-hike, Beckett seemed worried that there might be a repetition of the incident and insisted on giving him in old francs the fare to England, as well as a *carnet* of bus tickets.

Evers discovered in that first conversation, as most people did, that Beckett froze slightly if one touched on his work. Yet he wanted to be read. Evers had brought with him the Grove Press edition of *Proust* which he had obtained in Paris. Asked to inscribe it, Beckett complied happily enough, and then asked if Evers possessed any other works of his. When Evers told him he did not, Beckett gave him a note for Jérôme Lindon at Les Éditions de Minuit asking the publisher to give the young man everything that was in print.

When Evers returned to Paris after a sojourn in Australia he became one of those people who were seen periodically on a one-to-one basis. By then Beckett had moved to the boulevard Saint-Jacques and had come to an arrangement with the telephone company whereby he had a phone that did not ring but flashed a light when people rang him. Friends were told to ring him at a certain time, usually between eleven and twelve noon, and sometimes they were instructed to allow the phone to ring a certain number of times, after which they should ring off and ring again. To suggest a meeting Evers would usually send him a note, to which he would reply either with another note or a telephone call. They would meet in a café, often the Closerie des Lilas. The first time they met there Beckett pointed out to him, on the wall behind the banquettes, the plaques to the famous writers who had frequented it in the thirties, mentioning – Evers thought rather proudly – that he had met some of them.

Understandably, Evers spaced out his suggestions for meetings, but even so, when he dropped Beckett a note, the reply would sometimes be in the negative – that he hoped to spend some time in Ussy, or he was going abroad, or he was not well. However, he would promise to be in contact and would faithfully, in his own time, get in touch and suggest a meeting. He was extremely punctual and would arrive always exactly on time at the appointed place. Frequently he would say he had to leave by a certain time and, if it was an hour, he would leave exactly on the hour. Sometimes for these meetings Beckett would nominate his own apartment on the boulevard Saint-Jacques, and on these occasions he would be standing on the landing waiting when his visitor arrived. They would go into his study so there was rarely any contact with Suzanne. Like many of those whom Beckett seemed to like, Evers is a talkative person and this made conversation easier. Beckett, he found, could be a rather intimidating listener, making it clear by small signs and signals that he was now preoccupied and might even be bored, and that the time had come to bring this particular meeting to an end. On these occasions Evers took care to say his farewells quickly. But in general he found his companion attentive to what he was told and thoughtful about it, though Evers did notice that you could secure an even fuller degree of attention and sympathy by discussing a difficulty with him or telling him a tale of woe. Apart from one or two brief encounters with Suzanne at the apartment, Evers never met Beckett with any other person over a period of thirty years. He came to know Barbara Bray separately, through other mutual acquaintances and because they both lived in the same quarter. During the sixties and seventies he saw a good deal of her. She made no secret of her close friendship with Beckett but never discussed the more intimate aspects of their relationship. A few times he met or saw Beckett in the narrow streets around the rue Séguier, evidently on his way to or from Barbara's apartment, but of course he never discussed Barbara with him either.

Jean Demélier was a student at the École Normale in the 1960s when he wrote an essay on Beckett and sent it to him. 'You are not a critic,' Beckett told him, 'And neither am I.' But perhaps because of this, Beckett befriended Demélier, who was also an aspiring writer. For years, living in cheap hotels as he struggled to write his novels, Demélier was often destitute. 'Beckett', he said, 'understood and felt I had nothing. He was so kind, like a second father.' When they met, Beckett would press money on him, a 500f. or 1000f. note and, in an attempt at recompense, Demélier would send him a drawing. Their meetings – lasting an hour, a time-limit rigorously prescribed by Beckett – took place in a café, where they would drink beer. When they did not sit in mutually comprehending silence, they talked of ordinary things, such as Beckett's troubles with his teeth or more personal matters of Demélier's, such as his sexual affairs. Sometimes though, their talk turned to more profound subjects: 'How extraordinary it is to be alive – he would talk of this, full of fire, very precisely.' Demélier was a Huguenot from Poitiers and they shared a reluctant interest in Protestantism. Unlike most people, Demélier thought Beckett was interested in success and fame and that this came from his Protestantism. '*Il est parti du jeu* – it is part of the game,' he said. And he said it with a movement like a sportsman.

Beckett gave Demélier money to travel to Ireland and, when he was planning a trip to Poland, offered him the use of the zlotys he had earned there but could not take out. Once, in the middle of winter, Demélier was so poor that he had holes in his shoes and was driven to appeal to Beckett on the telephone. 'Come immediately,' Sam told him. When Demélier arrived at the flat an entire wardrobe was laid out in readiness for him – two pairs of shoes, socks, shirts, a suit, an overcoat. He went away with a suitcase of 'fine warm things, like a prince'. Later, in the Dôme, Demélier presented him with some drawings and paintings in return and Beckett pleaded, 'Please, don't give me any more drawings.' He found Demélier a publisher for his first novel, *Le Chemin*, sending him to Georges Lambrichs, his own former advocate at Minuit, who was now at Gallimard and went on to publish three more books.

Their intimacy ended badly. As a kind of jape, Demélier had a friend of his secretly take some photographs of himself and Beckett as they sat at a café table on the boulevard Saint-Jacques, and then sent one to Beckett in what he regarded as a spirit of homage. Beckett was shocked, seemed to suspect that Demélier was going to sell the photographs and ended their meetings. For Demélier, the photos episode was 'a terrible event'. He thought he had been suspected of having a commercial motive and the rupture grieved him deeply because of this misinterpret-ation of his character. For several years there was no communication between them. Then one day he got a letter. It expressed forgiveness but Demélier found it cold and hurtful – 'It was like a letter from the Administration.' But when they met by chance on the boulevard at Saint-Germain-des-Prés, he thought Beckett 'full of tenderness' and was very moved by the meeting.

Charles Juliet was another young man whom Beckett saw on a one-to-one basis. Juliet was a friend of Bram van Velde's and when he wrote to Beckett in the late sixties to ask for a meeting he was invited to come to the boulevard Saint-Jacques. After ringing on the inter-phone and going up in the lift, Juliet also found Beckett waiting on the landing and was invited into the study. He sat nervously on the small sofa facing Beckett's work table and a long silence ensued. Since Maurice Nadeau had told him that Beckett was capable of meeting people and parting from them two hours later without having said a word, his nervousness increased; and when their gaze met he was the first to turn away his eyes. The silence was broken eventually by Juliet rather desperately telling Beckett about an obsession with his work which had begun when he had read a few lines of the *Textes pour rien* three years previously when he was twenty-two. This was heard with some impassivity, but the conversation proper only began when Juliet turned to the topic of Beckett's health. He had been ill earlier that year with an inflammation of the lung around the old knife-wound, and he now began to discuss the topic of ageing with his young visitor. He had always hoped, he said, to have a long, active ageing, during which the flame would continue to burn even though the body might become infirm. He often thought of Yeats, whose best poems were written after the age of sixty, he said.

He spoke of his feeling of being 'assassinated' before his birth and went on to tell him of the Jung lecture and the revelation it had been to him. In this version of one of Beckett's many retellings of the story he described Jung saying, as if to himself, 'basically she had never been born'; and then he added that he too felt this. Later in this conversation he told Juliet that he felt like 'a mole in a molehill'. Of his writing he said that he faced a cliff, 'but I must advance'. It was impossible but one gained a few miserable millimetres. Questioned about his reading, he said that while he wrote he read very little and that the two activities were incompatible. After speaking of Bram van Velde, who lived in Geneva and did not write letters, and of whom therefore Beckett had no news, they parted.

Subsequent meetings, at fairly long intervals, took place in the

Closerie des Lilas. Like Jean Demélier, Juliet noted Beckett's punctuality. He arrived always on the stroke of the arranged time and would usually announce how long he was going to stay. At one of these meetings he told Juliet about Bram van Velde's sister, who lived in difficult circumstances in Amsterdam with a younger companion who was paralysed. Juliet asked him if van Velde was aware of his sister's circumstances and Beckett replied 'no, he is outside all that'. Although Beckett himself was keenly aware of family responsibilities he evidently did not feel that his old friend was delinquent in not having the same attitude. On another occasion he said he had been re-reading *Molloy*, and when Juliet asked him what effect this had had on him a long and agonizing silence followed while Beckett stared abstractedly into space and Juliet noted, amongst other things, how badly shaved he was. At length Beckett said, 'It did not seem to be from me.' When Juliet hazarded the opinion that the progress of the artist could only be understood in terms of a rigorous ethical necessity there was a further lengthy silence before Beckett spoke:

> What you say is just but moral values are not accessible. One can not define them. To define them it would be necessary to produce a judgement of value and that can not be done. It is why I have never been in agreement with this notion of theatre of the absurd, because there, that is a value judgement. One can not even speak of the truth that is part of the distress. Paradoxically it is in form that the artist can find a sort of solution. In giving form to the unformed. It is only at this level that there could be a kind of underlying affirmation.

This conversation seems to have concluded abruptly because Juliet was moved to an impassioned statement of his admiration for Beckett's work. In any case, at this point Beckett suddenly excused himself and the meeting came to an end.

512

Chapter Thirty-Two

Much of Beckett's social life in the 1960s was conducted in the Falstaff bar in the rue du Montparnasse, a small place which had dim lighting and less than a dozen tables grouped closely together around a central pillar. To one side was a short counter at which you could sit on a high stool. The whole effect was rather more old-fashioned American than it was Parisian, but the Falstaff was very French in its discreet and carefully off-hand attitude to celebrities, many of whom used it as a retreat from the larger cafés, somewhere they could be fairly sure of remaining undisturbed. Few people, celebrities or otherwise, went there before nine or ten in the evening, and others who went there included Jean-Paul Sartre, often to be seen in the Falstaff in company with the young Algerian girl who became his adopted daughter.

Beckett had been in the habit of using the place for some years, but in the early 1960s began to do so habitually. This was partly because his friend Con Leventhal had now moved to Paris, lived around the corner and used the Falstaff as his local. After Ethna's death in 1959 Con had begun to take stock. He was now of an age at which he could retire from Trinity. Many of his old friends, including Seumas O'Sullivan, the poet who had edited the *Dublin Magazine*, were now also dead; and there was no other journal to which he could contribute the occasional literary and dramatic criticism which had given him a certain minor status in Dublin. The attractions of Paris to a *bon viveur* like himself were considerable; and there was also the presence there of Samuel Beckett, one of his oldest, and certainly now his most famous living friend. When he first mooted the subject of a possible move to Paris to Beckett it received a modified welcome, tinged with caution, but to the coming itself there was a more enthusiastic response and Beckett exerted himself, with the help of Josette Hayden, to find an apartment on the boulevard du Montparnasse. This was an old-fashioned bourgeois Paris apartment of the sort which Beckett himself never occupied, preferring the sort of modern accommodation that he had rented in the rue des Favorites and now had in the boulevard Saint-Jacques.

When Con began to go to the Falstaff they often met there by arrangement; and since his friend's presence there on almost any evening was a virtual certainty, Beckett now began to drop in there more casually too.

The Beckett to be encountered at the Falstaff was a rather different person from the one who gave audience to selected individuals in the Closerie des Lilas or at the apartment. Someone who knew him under both guises, however, was Peter Lennon, a young Dublin journalist who in the late fifties had established a foothold in Paris as a stringer for the *Guardian*. After arriving in Paris, Lennon thought it would be nice to meet Beckett, so he wrote asking if he would like to meet a young Dubliner who had recently come to live in Paris as a freelance journalist. He also thought that an interview with the famous recluse would be something of a coup and would serve to strengthen his newly formed connection with his paper, so he also asked if Beckett would like to give him an interview, adding diffidently that it would be all right with him if this request was refused and they just had a drink together.

Almost immediately he received one of Beckett's cards in an envelope. These cards with the name Samuel Beckett printed across the top were a new departure, allowing the sender to be briefer and more to the point than when writing on notepaper. Lennon was glad to get the card and glad to note that Beckett had put his telephone number on it. His difficulty was that he could not read Beckett's writing, which 'flowed across the white rectangle as if it had been dipped in a stream'. He therefore had to ring Beckett up from a noisy café and enquire nervously whether the response to his request for a meeting was yes or no. Beckett, who seemed to understand the difficulty, said it was yes and suggested he should come to the Closerie des Lilas at 6 p.m. on the following Friday. He would not give an interview but was prepared to meet him.

Lennon's expectations were hazy. He had seen *Godot* and 'sniggered over Girodias's Olympia Press edition of *Watt*'; but that was all. When Beckett arrived one of the first surprises was his Dublin accent; but Lennon was also somewhat taken aback by the idiomatic Dublinese of his discourse. He thought at first that in using this Beckett was seeking to put him at his ease, but concluded later that he was only partly right and that the use of such phrases about mutual acquaintances, as 'a terrible chancer' or 'a decent man' with whom it would be good to 'organise a jar', was a way of displaying credentials and 'dealing at a tangent with relationships'. Lennon saw it as a specifically Dublin way of suggesting that you are one of the lads, while yet making a point of your literacy, which had been given status by Joyce and Flann O'Brien; and as he says, 'there are plenty of examples of it in Beckett's early writing'. But whereas many literate Dubliners were inclined to keep up this kind of thing all night, Beckett's use of it, he thought, was casual, 'like a light salute to the homeland'.

That first meeting was brief. They had 'no more than three glasses' of Irish whiskey but the consumption of it must have been pretty rapid, for shortly after seven Beckett went off. Other meetings followed a ritual pattern. At fairly protracted intervals Lennon would drop him a line so that he would not feel cornered by a telephone call and Beckett would respond by a telephone call, or, if this had not happened, they would arrange a meeting when they came across one another on the boulevard du Montparnasse. The meetings were made always for the same place and usually for the same time. Sometimes, however, they might reach the fourth whiskey. They might well have continued on this basis but for the fact that Lennon too lived in Montparnasse, in fact next door to the Falstaff, which he began to use as his local.

One fairly regular habitué of the Falstaff was the Irish poet John Montague, who lived close by the boulevard Saint-Jacques in the rue Daguerre and had been introduced both to the bar and to Beckett at the same time by their mutual Dublin acquaintance Con Leventhal. Montague thought that the post 10 o'clock Beckett of the Falstaff was the 'Irish' Beckett, whereas the Beckett of other times and occasions was the formal Frenchman or the scrupulously methodical and exacting professional writer. He was also struck by the contrast between the repeated condemnations of Irish obscurantism, bigotry and prejudice and the sometimes sentimental, indeed almost maudlin nostalgia which certain reminders of Ireland and Irish occasions seemed to evoke. This Irish Beckett of the Falstaff took a great interest in Irish sporting activities, particularly the progress of the rugby team. He was also the one who, in Peter Lennon's fine phrase, 'had a weakness for those with a particular Irish desperation and vulnerability', who would participate in the anecdotal maunderings of Jack MacGowran or the somewhat more pungent story-tellings of Pat Magee, and who would exchange endless reminiscences of Dublin and Trinity acquaintances with Con Leventhal, taking such a great interest in Trinity matters that Montague thought of their conversation together as 'Trinners talk'. Montague also noted that the conversations between Beckett and Leventhal often suggested a fairly comprehensive knowledge of such Paris *bordels* as had survived into an age when they were technically illegal. Con certainly had a wide acquaintance among the *demi-mondaines* of the quarter and as he strolled along the boulevard at certain hours they would return his cordial salutes with evident esteem and affection.

The late-night drinking in the Falstaff could be pretty heavy. Leventhal was a toper of an old-fashioned, quiet, methodical Dublin kind, putting away glasses of whiskey with what to less habitual drinkers might seem amazing sang-froid; and so in his own way was Beckett. He usually turned up at the Falstaff by arrangement with Leventhal but sometimes he would arrive with Arikha or with Henri and Josette Hayden. On a few occasions he came with Barbara Bray. And as Sam's nephew Edward settled into Paris and his music studies he too would come to the Falstaff. Beckett's associates in the bar formed a preponderantly, but by no means exclusively, Irish group. He tended to meet his French friends or his Swiss friend, Giacometti, elsewhere – in the American bar of the Dôme, in the Coupole or in the Select. Frequently in the evenings, after dinner and before the Falstaff – or after the Falstaff, if they had been there earlier – he would play billiards with Leventhal or Edward or both together, going to one of the old-fashioned billiard saloons which still survived in the quarter such as the Cluny on the boulevard Saint-Germain, or, more frequently, to the Trois Mousquetaires on the avenue du Maine.

Beckett played with fierce zest and great skill. The game undoubtedly had the added attraction for him that it disposed of the problem of conversation and sociability on those occasions when he was not inclined for either. Usually though, in the sixties, the evening would end in the Falstaff, Beckett showing no great desire to leave before closing time. To some extent he was like Hemingway's old man in the story 'A Clean Well Lighted Place', for he still suffered from the insomnia which had always been an affliction. Once he discussed this problem with Montague, who for a time was similarly afflicted, due, he thought, to a situation he felt he had created which caused him some guilt. Asked what he did to try to put himself to sleep, Beckett replied rather touchingly, 'I play a round on the old course at Carrickmines Golf Club.' Then he added, 'If that doesn't work I play the new course.' The bar closed later than the cafés and the last customers to emerge would find the great boulevard virtually deserted. Beckett would then set off, sometimes a little unsteadily, towards the boulevard Saint-Jacques, or perhaps, if feeling athletic enough, for a nocturnal perambulation. Sometimes, late though it might be, he would decide to call in on Barbara Bray. Very occasionally he might emerge too drunk to be able to walk unaided, in which case someone might see him home and even take him up in the elevator.

These occasions were not common, however. Beckett had a good head for drink and most Falstaff habitués looking back on those nights remember a man whom drink never made boisterous or over-talkative but, if anything, 'more remote'. Reproving Montague for a story which has Beckett escaping a bore by crawling under a table and coming out on his hands and knees to plead for a kiss from a pretty young woman, Lennon emphasizes that, though not morose, he always gave the impression of being 'mentally under some constraint or control' and attributes such stories to 'an Irish need to recruit Beckett to the back bar of Irish literature'.

The Falstaff was in fact a place where formalities and courtesies were quite strictly observed. There was no table hopping and no casual joining of company other than that which you were accustomed to join. In the sixties it was a resort for others besides Beckett who were too well known to be comfortable on a café terrace, particularly in the tourist season. When one night Lennon brought Ionesco there, Beckett and Con Leventhal were sitting just inside the door, Sartre was at another table with his attractive young Algerian admirer, and Jean-Luc Godard happened to be there with a couple of friends. But no salutations or greetings were exchanged. Beckett and Sartre did sometimes exchange nods, what Montague characterizes as 'the farmer's nod', but that was all.

The sixties were Beckett's most sociable and, in many ways, most relaxed decade; and it was also the decade in which he settled into a routine which still, in the eyes of many who knew him, defines his existence. This was partly because of the Falstaff bar, on which a disproportionate amount of reminiscence centres. Yet parts of his customary routine were unknown, or only dimly known, to some who thought they knew him fairly well. Such people knew enough about his life outside the bar to know that he would take telephone calls or make them between eleven and twelve; that he went for long walks in the afternoon; that he might be seen around the Dôme or the Select at the hour of the apéritif; and that there was a fair chance that he would be in the Falstaff later. What was not known was how much of his life revolved round his relationship with Barbara. Oddly enough, this side of his life was far less known about in Paris – where it would not have been considered anything out of the ordinary – than it was in certain BBC circles in London; and in Paris many who were frequently in his company in the Falstaff, and had even perhaps been there when Barbara was also present, knew nothing about it. His natural reserve and well-developed sense of decorum were allied to his fear of giving offence to Suzanne.

Those who did know something about his relationship with Barbara, however, and who met them together, agreed that with her he seemed almost a different person, brighter, more cheerful and altogether more relaxed. An independent woman, who had a *métier* and a life of her own, and was content with things as they now were, Barbara seems to have asked nothing more from her relationship with Beckett than he was willing or able to give. She admired him very deeply, not only as a writer but as a man, believing him to be, whatever his complications, virtually a saint. Refusing to discuss the more physical side of their relationship, she has permitted herself to say that he was saintly in these matters as well, a remark which, whatever construction one puts upon it, at least implies a high degree of selflessness. 'Men do not understand women anyway, but he understood me better as time went by and I understood him better,' she has also said.

Although aware of his bouts of depression, she thought he had a great capacity for enjoyment – a capacity inseparable from his fineness and keenness of perception. This very fineness of perception, amounting, in her view, to hyperaesthesia, or specially heightened consciousness, made him suffer more than most people did in company or circumstances which were antipathetic to him. He was, she thought, like the man Rousseau speaks of, who finds himself out in the tempest without his skin. At the same time his ultra-sensitivity was balanced by humour and insight, so that unlike other hypersensitive people he would not allow himself to go to pieces or to be blown off course because of it. He was immensely strong as well.

Her overall view of his work was that it should be judged as a discourse which does not reveal all of the writer's attitude to life. Her analogy is with a classical architect – that all of his buildings are so formal does not mean that he doesn't like trees – and the fact that Beckett's explorations are of the other side of life does not mean that he had no knowledge of or capacity for happiness. The work is no more than his voice, what came out when he wrote, in a sense no more than the tip of the iceberg. And it can in any case be seen as containing its own contradictions, so that when anything is said, nearly always the opposite is said as well. And the opposite is in any case contained in the humour. 'I regret everything', besides being tragic, is also very funny.

Yet Beckett did regret much, particularly the situation in which he now found himself with Suzanne. When his English publisher John

Calder arrived in Paris in July 1961, they spent the entire evening, which included a very good meal and some choice bottles of wine, talking about suicide. It was shortly after Ernest Hemingway had blown his brains out and Beckett was moved by the news to discuss the many and varied ways in which one could take one's own life. There was, Calder insists, nothing facetious or fantastical about this discussion. Beckett was genuinely concerned about how it might be done and which of the many ways might be the least painful or unpleasant. On another occasion in the Falstaff, however, when the topic of this often discussed solution to the problem of existence was raised, he brusquely replied that such a solution was 'out of the question'; though on a lighter note he went on to say that a staged disappearance, using a boat that could be sunk, would solve many problems – adding that such a solution would have the drawback though of creating testamentary problems for the widow.

The impression Calder formed of Beckett's attitudes on this occasion was borne out by many subsequent conversations. Existence, to which we are condemned without our permission, was something to be endured. Hope was an illusion. Human beings were rapacious and cruel to each other, seemingly by instinct. Calder had no doubt that Beckett believed that life was a pretty awful business. Nevertheless it might be thought odd that he should have been talking so fervently and seriously about suicide in the summer of 1961, just after Barbara had come to Paris. Even in the case of people less complex than Beckett, however, the guilt and suffering caused by an affair, particularly in its early stages, can outweigh whatever happiness it brings, and his guilt about Suzanne was now very great, though he was not prepared to do much about it.

This was partly a matter of his lifestyle. Even when he had not come from Barbara's, he arrived home very late virtually every night, at a time when his wife, who might well have spent a solitary evening, would have gone to bed; and in the mornings there was often very little communication. Through much of the 1960s Suzanne was not well, suffering from persistent colds and flu, headaches and even temperatures, ailments so persistent and unexplained that they were almost undoubtedly psychosomatic, as many of Sam's ailments were at an earlier period of his life. And there were now disagreements and quarrels of a more serious order than before, when the habit of reasonable amity was still strong. In the early days of Barbara's residence in Paris he was not as sedulously discreet as he was to become. Barbara was occasionally to be seen with him in the Falstaff or elsewhere; and though few of the habitués there knew anything about their relationship and any who did were unlikely to be in communication with Suzanne, she heard about Barbara through one party or another. The result, inevitably, was scenes. Beckett was not a maker of scenes, nor was Suzanne – who was almost as reserved and self-contained as he was; but it was equally inevitable that during these outbreaks hurtful things were said and she gave expression to her unhappiness. It was in an attempt to resolve this situation and the misery it caused that he began in 1962 to work on the strange playlet that is called simply *Play*.

Play is about a man, a wife and a mistress, stuck eternally in the same unresolvable situation. As James Knowlson and John Pilling have put it, 'Three stock characters, a man, his wife and his mistress are involved in an only too familiar triangular relationship. Clandestine meetings, domestic squabbles, partings and recriminations have long formed the staple diet of much domestic drama or light comedy. Once again here they are only too clearly the stuff of a 'play'.' Beckett's treatment of the theme is unusual, however, in that there is no movement towards resolution, happy or tragic, farcical or emotional; indeed the last stage direction requires that the play should be repeated in its entirety. '*Repeat* Play *exactly*', it says. The situation is apparently unending.

There is another peculiarity. The three characters in *Play* are encased up to their necks in urns. A light plays upon each of them in turn. When it does, they speak, telling the story of their relationship in fragments, as if the light were an interrogator. So far, so strange, but the real strangeness in this being a Beckett play is otherwise. It is the cliché-ridden nature of the story, of the characters' reactions, of the situation itself. The man, the wife and the mistress are given an English upper-middle-class milieu, English upper-middle-class speech patterns and, seemingly, values and outlooks to go with them. Or rather, they are given these characteristics in terms of what generations of West End theatre-goers had been led to expect. We hear of little jaunts to the Riviera or 'our darling Grand Canary'. There is a butler – called Erskine, like the butler in Mr Knott's house in *Watt*. There is talk of morning rooms, vanity bags, lawn-mowers. The mistress has 'means'. The man prefers Lipton's. One rings for the servants.

If this could be Noël Coward, the way in which the characters speak of crises in the affair could almost be Pinero. The wife says: 'I said to him, "Give her up. I swore by all I held most sacred."' The mistress asserts: 'Fearing that she was about to offer me violence, I rang for Erskine and had her shown out. Her parting words were . . . to the effect that she would settle my hash. I confess this did alarm me a little, at the time.' The man tells us: 'We were not long together when she smelled the rat. Give up that whore, she said, or I'll cut my throat.'

The wife has the husband 'dogged for months by a first-rate man'. He swears to the mistress that they will go away together 'as soon as he had put his affairs in order'.

And there is, furthermore, an almost comic over-emphasis on the man's physicality. Although the wife has her suspicions, he continues to be 'as assiduous as ever' in his love-making. This 'and his horror of the merely Platonic thing' makes her wonder if she 'is not accusing him unjustly'. When the mistress accuses him of still having relations with the wife, he asks her what she takes him for, 'a something machine?' She knows that with him 'there is no danger of . . . the spiritual thing'. When the wife knows all, her first feeling is one of wonderment, 'What a male!'

All this is a very far cry from the Left Bank of Paris, Beckett's way of life and that of the two women with whom he was emotionally involved. It is almost as if the whole conception of the situation in the play is an attempt to point this out. He is not he. They are not they. And yet the pain is real. And the punishment, the circular repetition without any outlet, is real too. The characters cry out together for release from the situation they are in. When they first experienced 'this change' – apparently to some sort of after-life – they thought it would bring peace and rest. It hasn't. When the cruel spotlights play on them they must re-hash the story over and over again. 'Is it that I do not tell the truth,' the wife asks, 'is that it, that some day somehow I may tell the truth at last and then no more light at last, for the truth?' The husband has a rather foolish wish that they might all three 'meet, and sit, now in the one dear place, now in the other, and sorrow together, and compare'. The mistress wants to be listened to, but feels that no one has or will. The wife seems to suffer most – and to have suffered most, even if her suffering has been most comically described; and it is she who articulates the commonest intimation and the commonest wish among Beckett characters – that there should be some statement of a truth which would give release and which they yearn to find. She pities the mistress, and then all three of them. 'Poor creature,' she says. Then adds: 'Poor creatures.'

But the circular, unresolved, repetitious nature of the situation afflicts them all. They are all subject to the same interrogation and they all interrogate themselves over and over again. There is no doubt that for all the theatrical trappings, the parodic and comic descriptions of the characters' class and circumstance – and in a way they merely point it up – the suffering is real; and perhaps stranger still, though the husband talks of 'happy memories', it is the most real thing there is.

But even though there were quarrels, Sam and Suzanne still went together to concerts and opening nights in the theatre, or to dinner with mutual friends. He was a regular and patriotic attender of Irish productions – whether by the Abbey Theatre or other companies at the Théâtre des Nations – though he refused to go to Behan's plays. Throughout the sixties too there were productions of his own plays; and, though he stuck to his rule of not attending on opening nights, Suzanne usually went and her status was, more or less, that of his representative.

One such event at the beginning of the decade that he did attend, however, was a long-awaited and -discussed production of *Waiting for Godot* at the Odéon. He had always wanted to have the play done in French on a bigger stage and now he had his wish. He was acutely aware of the symbolic importance of a production at the Odéon and, for all his disdain of other forms of success, it gave him considerable pleasure to have it performed there, particularly with two of the original cast, Jean Martin as Lucky and Lucien Raimbourg as Vladimir. When it was certain that it would go on he wrote to the old friend and companion of his post-midnight perambulations, Alberto Giacometti, asking if he would make the tree which is the focal point of the décor. Giacometti too was now world-famous, but he accepted the celebrity in much the same spirit as Beckett did. When he was awarded the major prize for sculpture at the Venice Biennale at about this time and his fame and something of his legend began to penetrate the popular consciousness, he declared that he thought it was all based on a 'misunderstanding' – which was what Beckett had said to Alec Reid about the success of *Godot*.

Though Giacometti had never worked for the theatre he accepted Beckett's request eagerly and, with the help of his ever-faithful studio assistant, his brother Diego, set to work at once, making a tall white slim tree which grew straight up until its four branches curved out from it, almost in the manner of a palm tree. Beckett came round to the studio while this was being made and they considered its creation together. 'All one night', Giacometti said, 'we tried to make that plaster tree larger or smaller, its branches more slender. It never seemed right, and each said to the other: 'maybe'.'

From the beginning Beckett stayed close to this production of *Godot*, keeping, as he said, his nose in the theatre day after day and finding rehearsals both absorbing and exhausting. When it was first mooted there had been serious disagreements about responsibility for direction between Blin and Jean-Louis Barrault, who was artistic director of the Odéon and also a friend of Beckett's. Barrault felt to some extent responsible for everything that was done in the Odéon and for the way in which it was done. Since he had the power, a production depended on his agreement and these differences had already led to postponement and what had seemed at one stage to be a cancellation. Beckett was in an awkward position. He wanted the production to go ahead but was somewhat more in favour of the tried and trusted Blin's ideas of who should be responsible than he was of Barrault's. Blin and Barrault were old friends and associates, Barrault's admiration for the slightly older Blin going back to the time when he used to go to productions of the avant-garde Groupe Octobre in the early 1930s; and Barrault also admired Beckett deeply. But he felt that both of them had a profoundly ambiguous attitude to success. Both wanted, he thought, the luxuries and freedoms of marginalization while at the same time being by no means unwilling to have people come and see their productions. When the difficulty was finally resolved, Blin was confirmed as director but unfortunately the postponement now meant a conflict of commitments, so finally Beckett became director in his place.

This was the first time he had actually directed a production, though he had often sat in and made suggestions. As might have been expected, he was meticulous, exact and demanding, but confined himself to movement and delivery of the text without discussing its meaning. The production opened in early May and was such a great success that there was talk of extending the run beyond the scheduled run into July, but some of those who attended felt that Beckett had taken too many pains to nullify the jokes and dampen down the possible laughs. (Yet when he came to London in 1964 for the English Stage Company's production of *Waiting for Godot* directed by Anthony Paine at the Royal Court, he told the actors – Jack MacGowran, Nicol Williamson, Paul Curran and Alfred Lynch – 'All that matters is the laugh and the tear.' He also told them that the play was 'full of implications and every important statement may be taken at three or four levels. But the actor has only to find the dominant one; because he does so does not mean the other levels will be lost.')

An incident which may have taken something of the shine off the opening at the Odéon – a rather official French occasion – was when he was introduced to the novelist Olivia Manning, author of *The Balkan Trilogy*. She immediately began to attack him on the grounds of his vision of things, his pessimism, aridity and lack of hope. Beckett bore this diatribe in silence and made no attempt to reply, but he grew increasingly nervous and finally turned away, asking, 'Where is Con? Has anybody seen Con?' in a tone which clearly conveyed an appeal for help.

One of those who met Beckett for the first time in the early sixties was the playwright Harold Pinter. Pinter was in Paris for rehearsals of the French version of *The Caretaker*, and when he told the director Roger Blin that he would like to meet Samuel Beckett, the word was passed on and it was arranged that Beckett should call and pick him up at his hotel. Beckett strolled briskly into the lobby and in the first exchanges proved so affable and talkative that Pinter was taken by surprise. There were enquiries about mutual acquaintances such as Pat Magee, and the seemingly easy flow of civilities and small talk made Pinter begin to wonder if the long silences and broodings about which he had been told by more than one person were a myth. As he often did with visitors to Paris, Beckett said that he could only stay until a specific time – eight o'clock – and suggested that they should go for a quick one together; but as also often happened with visitors whom he found amenable, this specific time was forgotten and the quick drink lasted until the small hours. The conversation traversed a great deal of ground, encompassing literature, philosophy, even politics and, of course, their great mutual interest, cricket.

Beckett talked at length about Racine, the master in whom all his life he had been able to find both inspiration and renewal. They discussed form and Pinter expressed his admiration for Beckett's, as he saw it, unending struggle to impose form on the mess of experience; but Beckett responded by telling him that he had been in hospital once when a man in another ward had been dying of throat cancer. 'I could hear his screams continually in the night,' he said. 'That's the only kind of form my work has.'

Even when the last bar had closed Beckett wanted to go on drinking and suggested that they should go to the market area of Les Halles, where you could eat and drink in the early hours. There they had more drinks, accompanied by the famous onion soup which in those days was the recourse of so many drunkards; but this dish gave Pinter a piercing attack of indigestion, to which was shortly added a sense of abandonment, for Beckett seemed to have disappeared, leaving him alone and feeling very wretched among the early morning bustle and activity of the market café. Suddenly his companion reappeared. He had gone off to find a pharmacy where he could get some bicarbonate of soda and had traversed a large part of Paris in his quest. After the bicarbonate of soda had been successfully administered, Beckett drove Pinter in the Deux Chevaux back to his hotel and a welcome bed, returning in the afternoon to drive him to the airport. This was the beginning of a friendship with the younger English writer which was to be important to both of them.

Though the compliment was not returned in the same form, Pinter began to send Beckett copies of his plays and take note of any comment he might make. He found that Beckett's perceptions of dramaturgical matters were usually very acute. When Pinter's play, *Silence*, was being produced by Peter Hall Beckett read it and in his reply asked Pinter if he would reconsider a particular speech. Pinter had a look at it and decided there was nothing wrong but a few days later, in a conversation with Peter Hall, the director assured him things were going very well – but would Pinter have another think about such and such a speech, which he thought was a little over-poetical? It was the same speech about which Beckett had reservations, though he had not specified what he thought was wrong. Looking at it again, Pinter saw they were right.

When Peggy Ashcroft was appearing in *Landscape*, Beckett suggested that Pinter should explain carefully to her exactly how he wanted the last lines of the play to be delivered. Pinter replied that he was uncertain himself and thought that Peggy Ashcroft should be trusted to get it right. This attitude was anathema to Beckett, who simply did not trust actors to get it right. For all his gratitude to and occasional enthusiasm for actors, Beckett, like many another playwright, felt that both directors and actors were an unfortunate necessity and that the ideal dramatic executant would be one totally subservient to the author's ideas about production and prepared to abandon his or her personality and character in favour of the text. The ideal actor would be, as a person, invisible.

Beckett's attitude was that actors (or, for that matter, anybody else) should not seek for deep 'meanings' or feel that they should understand the philosophical implications of a play. He was happiest with actors who could *be* the part, not understand it. 'I never knew what was going on, really,' said Jack MacGowran after the first production of *Endgame*. Which is not to suggest that Beckett did not respect or like actors. He was particularly drawn to the combination of egotism and vulnerability which many actors possess and was very evident in the case of Mac-Gowran. In the early sixties MacGowran was doing a one-man anthology of Beckett's work called *Beginning to End*. Beckett was strangely indulgent to this show which, after all, was an amalgam of several disparate works, creating the impression that there were no boundaries between them; that they had a simple autobiographical thread which ran throughout; and that all the various projections of the 'I' figure could be treated as one. He was also strangely indulgent to the liberties which he undoubtedly knew Jackie was inclined to take with the approved text when out of the author's ken. Of course, there were royalties and the show proved to have a wide popular appeal in Ireland,

Britain and the United States, thus making Beckett's prose more accessible to many people, even if it did create misconceptions about the novels.

As time went by, however, and Jackie's weaknesses and excesses became dominant factors in his life, a good part of whatever royalties Beckett had received were generously disbursed again to help the actor out of difficulties. Barney Rosset was instructed to turn royalties from other works over to Jackie when he was in the United States in the early seventies and was in financial trouble, and at an earlier stage Beckett had made his compatriot an extraordinary gift: the Yeats which he had struggled to buy in 1937 was impulsively handed over when Jackie remarked that if there was one thing he wanted it was a work by Jack Yeats. Probably all that Jackie wanted to convey was that he was aware of Yeats's cultural importance, but the repetition of the remark on other occasions resulted in the sudden gift of the painting, to the astonishment of their mutual friends. When Jackie got into difficulties, Beckett was anxious that he should not sell the work and hinted broadly that he would be willing to buy it back.

There is no doubt that he was entranced by both Jackie and Pat Magee and loved drinking with them. A friend who met Suzanne in the street in 1964 and asked after Sam was told, 'I will not see him for weeks now. The Irish are in Paris.' Perhaps it ought to be stressed that they were almost the first not specifically literary and not Protestant Irishmen he had known really well – one is tempted to say 'ordinary Irishmen', but that would be to beg too many questions. Apart from their sensitivity both had deep inner conflicts, but whereas Jackie's conflicts were all too apparent even to the casual observer, the more caustic Magee gave the impression of tensions and conflicts beneath the surface.

Both acted in a production of *Endgame* by Michael Blake in 1964 which was rehearsed in London but took place in Paris, Jackie playing Clov as he had done in the first Royal Court production, and Magee playing Hamm. Beckett came to London for the rehearsals, virtually taking over the direction from the young producer, who was anxious about deeper meanings. 'I only know what's on the page,' Beckett would say when questioned. Used to Beckett though they both were by then, both MacGowran and Magee seemed to him to be worrying unnecessarily about the depths rather than concentrating on what was going on on the surface, but of course they could not, any more than the critics, be entirely oblivious of other things. When, at a certain point, Beckett cautioned the producer against looking for symbols in his plays, Magee grinned and said audibly, 'He means don't play it like symbols.'

Beckett was in London three times that year. He went for these rehearsals of *Endgame* in various locations, including the gaming room of the Establishment Nightclub, then a room over a pub in the Euston Road which seemed normally to be used as a store for the pub's unwanted tables and a battered piano, and for the last few days a rather luridly decorated nightclub over the Royal Court in Sloane Square. Earlier he had been in the English capital to supervise a shortened forty-five-minute version of Jackie's *Beginning to End* for the BBC television programme, *Monitor*, the rehearsals for which took place in the North Kensington Community Centre in the desolate landscape of the Peabody Estate. Towards the close of the year he was back to rehearse *Godot* for the Royal Court production, which of course again included Jackie. Beckett formally took over the direction of the Magee-MacGowran *Endgame* himself before it opened in Paris, where it ran for nine weeks, followed by two shorter runs at the Aldwych in London. Discouraged though the actors were from reflecting on the deeper implications of the play, this experience left Jackie MacGowran with a feeling that the point of the play was inter-dependency and need. Clov and Hamm need each other. Clov had to be needed, which was why he never left.

On these visits to London he usually stayed with John and Bettina Calder. In spite of his desire for periods of solitude, even when he became affluent enough to have the privacy of a hotel bedroom, Beckett preferred to stay with people he knew and liked. He got on very well with Calder's wife, the singer Bettina Jonic, with whom he had a common interest in music. She was one of a number of attractive women he became friendly with, usually married to men with whom he had a professional relationship. Another was Jocelyn Herbert, the stage-designer wife of George Devine, artistic director of the Royal Court, with whom he sometimes stayed in the country in England. Both Bettina and Jocelyn Herbert were artists with careers of their own. They admired Beckett very deeply, both as man and artist, but naturally they tended to notice different aspects of him. Jocelyn Herbert paid particular attention to his perfectionism in all matters of staging, and particularly the visual aspects of it, how his plays were composed visually as well as aurally; while Bettina was interested in his understanding of music. The first time he came to stay he was busy with the London production of *Happy Days* and very fretful about the performance of Brenda Bruce in the leading role. There had been a long search for an actress to play Winnie, now he was unhappy with George Devine's ultimate choice of Bruce; but it is impossible not to feel that some of her shortcomings were the result of Beckett's own interference in the production. He wanted her to speak the lines to a very definite rhythm and, to her consternation, he produced a metronome and plonked it down on the floor in front of her. Each day he gave her extraordinarily detailed notes, which he had composed that morning before coming to the theatre.

Happy Days, with its seemingly optimistic central character, was a surprise to many Beckett critics. It has only two characters, a middle-aged couple called Winnie and Willie; and since not much is seen of Willie, Winnie is almost the sole focus of the audience's attention. She is buried up to her waist in a mound of earth under a burning sun; then, in the second act, up to her neck. But, perhaps because she is a well-brought-up Protestant lady, she is irrepressibly optimistic, not to say complacent; and grateful for everything, apparently to a Supreme Being, believing that there is 'so much to be thankful for', and often repeating the phrase, 'great mercies'. No explanation is given of how she got into her present plight, or what her end is likely to be, if she is to have an end, for time seems to have been abolished, either because we are in the after-life or because her creator is still toying with Bergsonian ideas of time.

Winnie's complacency is boundless. She is delighted to have an occasional word from Willie, who crouches for most of the play behind the mound with only his bald head visible, and to have him as a listener. True, she has realistic ideas about courtship and marriage and about Willie as a life-partner. 'Oh I know you were never one to talk, I worship you Winnie be mine and then nothing more from that day forth but titbits from Reynolds news' is her comment on the level of communication between them. But she is deeply afraid of being left without anyone to reflect her identity or confirm her existence by listening to her. 'That is what enables me to go on, go on talking that is. Whereas if you were to die . . . to speak in the old style . . . or go away and leave me, then what would I do, what could I do, all day long . . . Simply gaze before me with compressed lips.'

But she knows that to ask another human being to listen is to ask a great deal. She even acknowledges that when one human being does anything at all for another, it is cause for gratitude. When Willie, to her delight and astonishment, sings one verse of 'I love you so' from *The Merry Widow*, but refuses to go on, she reflects, 'One does not appear to be asking a great deal, indeed at times it would seem hardly possible . . . to ask less – of a fellow creature – to put it mildly – whereas actually – when you think about it – look into your heart – see the other – what he needs – peace – to be left in peace – then perhaps the moon – all this time – asking for the moon.'

Happy Days is another of Beckett's studies of marriage; and in spite of its chilling realism, it is often almost unbearably moving. It is also another triumphant example of his ability to deal with the entire human predicament with the utmost economy of means. Winnie spills out the contents of her handbag. Except for a revolver it could be the contents of any woman's handbag, but the desperation of our relationship with objects is exposed. She remembers decorous courtships from her middle-class past and the poverty of all emotion becomes clear.

At the end of 1964 Beckett discovered that he had been deeply involved in or directly responsible for no less than six productions of his work in the course of the year. Besides the London-Paris *Endgame*, the *Monitor* programme's *Beginning to End* and the London *Godot*, there was a production of *Fin de partie* which Roger Blin took on tour. Then there was Jean-Louis Barrault's revival of *Fin de partie* in Paris and there were French and English productions of the play called *Comédie* in French and *Play* in English. In all of these he was fully involved and this became a pattern of his life to an increasing extent from now on. Given the fierce concentration that he brought to each production, it was no wonder that he constantly complained of extreme exhaustion and spoke of his unavailing longing for periods of rest and recuperation at Ussy. Little wonder also, perhaps, that he failed to begin an extended prose work and that he was dissatisfied by his inability to do so.

Beckett was a perfectionist where productions of his work were concerned, a brooding presence in the stalls who distrusted actors and insisted on a meticulous obedience to his written stage directions. What makes this devotion to perfection all the more extraordinary is his disclaimer of any interest in the effect of a play on an audience. 'I'm not interested in the effect my plays have on an audience. I simply produce an object. What people think of it is not my concern.' If this is to be believed, the actual production of a play was, in the purest sense, Art for Art's sake as far as Beckett was concerned. But he was interested of course. He read the reviews and he frequently expressed what seemed a genuine pleasure when things had gone right. Still, when combined with the fact that he sometimes never saw these productions and certainly never attended the first night, often actually leaving London on the very day the curtain went up, it does mean that for him they took place in what was a rather extraordinary void.

One of the very small number of occasions on which he is known to have relaxed his rule about first nights was the London New Arts Theatre opening of Jack MacGowran's one-man show. His presence at this was attributed by some, including MacGowran, to the actor's deep insecurity and need for reassurance. This was probably part of the reason, but one cannot entirely rule out Beckett's anxieties about departures from the script and the need to know exactly what Jackie was up to.

This was also the year of the production of the film, *Film*, in which Beckett was also deeply involved. The original suggestion for *Film* had come from Barney Rosset, who had thought it would be a nice idea to get certain eminent authors, also published by Grove Press, to write original film scripts, and was prepared to put some money into the project himself. Some of the other authors, who included Alain Robbe-Grillet and Marguerite Duras, agreed to write a script, but Beckett was the only one who actually produced the goods, setting to work almost immediately. The idea appealed to him partly because a filmed script would be a final, frozen production from which there could be no departures; partly also because he was always willing to experiment with new media and had already done so successfully with radio, as well as making use of the tape-recorder in *Krapp's Last Tape*. The script he produced quite early in 1963 was about perception – the original outline actually had a Latin epigraph (from the Irish philosopher George Berkeley), '*Esse est percipi*' – 'To be is be perceived'. It is possible that the idea of a script about a man who tries to achieve a form of non-being by avoiding perception by others had some part of its origins in the fact that he himself was now experiencing trouble and discomfort from being perceived in public places – particularly in the summer months, when American academics came to Paris in large numbers and made his café life miserable by staring at him and even approaching him. Ironically, their willingness to cast aside natural good manners and approach him was partly caused by his reputation as a recluse. The sight of a famed recluse calmly drinking brandy in the Dôme apparently went to the heads of these normally well-mannered members of various French and English departments. Dazzled by the idea of being able to say when they got home that they had actually exchanged words with the recluse in question, they could not resist breaking in on his reclusion.

It is now firmly enshrined in legend that Beckett chose Buster Keaton to play the principal role in *Film*. He had a profound admiration for Keaton and saw *The General* upwards of a dozen times. Since there is an obvious affinity between the expressionless, enigmatic actor and the expressionless, enigmatic writer, the idea has its attractions. However, the truth is that Keaton was not the first choice. Beckett's first suggestion was Charlie Chaplin, and Barney Rosset wrote to the most famous comedian of all enclosing a copy of the script. There was at first no reply, but when Barney wrote again the reply was a letter stating, 'Mr Chaplin does not read scripts,' to which Barney responded in rather juvenile fashion with another letter saying he was sorry that Mr Chaplin couldn't read. Other suggestions were Zero Mostel, who proved unavailable, and Jackie MacGowran, who had had something of a success as the highwayman in *Tom Jones*. Jackie was anxious to do it and would have, had the production dates not kept changing because of the usual production difficulties, including shortage of money. A point came when he was no longer available because of another film commitment and it was only at this stage that Beckett suggested Buster Keaton, the star of such silent masterpieces as *The Navigator* and *The General*.

In retrospect Keaton, with his expressively inexpressive mask, seems an obvious choice. But in 1964, except for a brief appearance in Charlie Chaplin's *Limelight*, he had not made a film for well over thirty years and, though interest in his work was beginning to burgeon, nobody seemed to know whether he was alive or dead. And there were other doubts. He had been offered the part of Lucky in the first American *Godot* some years before and had turned it down. He was known to have had alcohol troubles. His state of health was unknown.

Some of these doubts having been put to rest, the director of the film, Alan Schneider, flew out to Hollywood to talk to Buster and, if necessary, to woo him. The address he had been given turned out to be a house in an obscure suburb of Los Angeles, furnished much as it must have been in 1927 or even earlier. A sober, inexpressive Buster Keaton, who was in the middle of a card game in the parlour when Schneider arrived, admitted that, yes, he had read the script. He did not think it was very funny but maybe things could be done to 'fix it up'. Maybe he could introduce some special business with his famous walk or a few old gags such as the sharpening-a-pencil-which-would-get-smaller-and-smaller-until-it-eventually-disappeared gag, 'things which might be new to people now'. Schneider told him that one did not normally pad Beckett's material. Then Keaton insisted, as somebody with a lot of experience of the business, that the script he had been given couldn't possibly run for more than about four minutes. He had even timed it. The cat and dog business seemed promising, he said, but even if it was stretched to its limits, four or five minutes was all that could be got out of the whole script. However, as he said, he was willing to supply ideas which he knew would work. They had after all worked before. Finally he agreed to do it, though he made no attempt to disguise his personal bewilderment about the whole project. The money was not bad for three weeks' work and he needed it.

If Keaton was not too enthusiastic about the film, neither was Beckett about New York. When Francis Evers asked him in the 1950s if he had ever been to America his reaction was one of some surprise. Why should he go there? He imagined it, rightly, as being hot and noisy in the summer. He thought he would be subjected to an endless round of cocktail parties. When he arrived late at night on Friday 10 July 1964 he was relieved to find that the first production conference was to be a poolside one at Barney's country residence in East Hampton, Long Island, to which he was to be transferred immediately by a privately hired plane. This turned out to be barely large enough to accommodate his long legs. Over the next few days those involved in the production sat around, talked and played tennis. Barney noted that Sam was somewhat dismayed by his performance on the tennis-court. He had not played for some time and had not realized how much his eyes were deteriorating.

He had worked out the necessary camera positions and angles and he now explained to all concerned that there could be no departure from these preconceived ideas, even when it was explained to him that with hand-held cameras everything could be a good deal more flexible. The script's theme was that even if all 'extraneous perception', animal, human or divine, could be suppressed, self-perception would remain in being and that the search for non-being by flight from extraneous perception would break down. The general notes for the script added that 'no truth value attaches to above, regarded as a merely structural and dramatic convenience'. The script envisages two characters, O and E. For most of the film E is the camera, perceiving O from various angles, but never, so to speak, getting a good look at him, while pursuing O through his efforts to avoid other forms of perception; wearing a long coat with the collar turned up and a hat pulled down, shielding his face with his hand as he hurries along the sidewalk of a busy street; putting the cat and the dog out of his room when he gets there; hanging a rug over the mirror, etc. At the end of the film we realize that O is in fact E, the self perceiving the self, so the camera angles were important.

The weekend conference over, everybody went back to New York to scout locations and await Keaton's arrival. Beckett, who was staying in Barney Rosset's house on Houston Street, decided that he liked New York a little better than he had thought he would, especially the Village. The first meeting between him and Keaton had been anticipated with mixed feelings by the various participants, but it turned out to be worse than anyone had thought. When Schneider and Beckett arrived at his hotel Keaton was drinking a can of beer and watching a baseball game on television. The first few exchanges were awkward, Beckett finding himself in the unusual position of having to make the running and venturing a few general words about his admiration for Keaton's work.

When these petered out everybody sat in silence while Keaton, seemingly unperturbed – though because of the famous mask nobody could be sure – continued to drink his beer and to watch the game. Every now and then, Beckett, almost like somebody meeting Samuel Beckett for the first time, would venture a few words, to which Keaton would respond at best in monosyllables. Did Buster have any questions about anything to do with the script? No, he hadn't. What did he think about the film when he first read it? No reply followed by a long pause. The minutes passed, the situation seemed to get more harrowing and hopeless and the silences to become more interminable. It was, to use Schneider's word, a disaster.

As they were leaving, however, Keaton volunteered a suggestion. He had brought along some of the flattened-down stetson hats which had been his trademark in the early days. Would it not be a good idea if he wore one in the film? Somewhat to Schneider's surprise, a probably demoralized Beckett said yes, then went on to demonstrate how the handkerchief which E would wear inside his hat to hide his face should be worn. On this note the first meeting between two great masters of comedy came to an end.

Shooting began next day in the dreadful heat and humidity of New York in July. There were the usual problems, the only bright spot being Keaton's magnificent professionalism. He was patient, imperturbable and relaxed, doing everything that was asked of him and doing it over and over again in the broiling heat while wearing the long overcoat that the script required. The first day's rushes turned out to be a fiasco. Everything now looked strange to Schneider, different from the way he thought it had looked while he was shooting it; and many of the shots were unusable because of what is called a strobe effect, an undulation of the background during a panning shot. Eventually, amid much recrimination, Beckett quietly proposed that the entire opening sequence with cyclists and cabs passing by be eliminated and that the film should start with Buster Keaton running along by the wall. Everybody agreed that this would in any case be an artistic improvement.

For the next three weeks the heroes of the shoot were Beckett and Keaton. When things went wrong Beckett would blame his script, in other words himself, rather than the director or anyone else. And though Keaton did remark a couple of times that his face had been his livelihood for years and that now these idiots were knocking themselves out to prevent the camera catching a glimpse of it, his behaviour on the set was always co-operative and amenable. Since there is only one piece of dialogue – a whispered 'Sssh!' – *Film* is to all intents and purposes a silent film, and so he was in his element, encouraging Schneider to give him vocal directions during a shot and sometimes coming through with suggestions for solving difficulties which, he would intimate, had cropped up in much the same form back in 1927.

As the three weeks of shooting progressed, Keaton grew a little more talkative, though he still never smiled. Eating lunch with the others, he explained the principles of movie-making as they had been understood in 1927. There had been no script then. All stories began with an idea about a character in trouble. You had a series of gags and improvisations to get him out of trouble, and that was that. End.

In the evenings principals and crew went to a small Italian restaurant in Greenwich Village which had air conditioning, a welcome relief after the torrid heat of the day's shoot. Keaton would sit at a separate table from the others, silently drinking his beer and staring down at the checked cloth. Beckett, who seemed fascinated by him, would steal surreptitious glances at his face, but there were few words exchanged. Beckett had, however, found a friend in the cameraman, Joe Coffey, an old hand, fond of his beer, and with a long list of credits behind him, who had an old Morgan three-wheeler in which Sam liked to be driven through the Manhattan traffic. Like most writers he enjoyed being made privy to the technical processes of another craft and was always there on the set, ready to answer a question or look through the camera.

Most of the filming was now taking place indoors, at a studio location. The principle of the film is that O successfully avoids perception until the very end, when he falls asleep in the rocking chair and wakes up to find himself observed by himself – that is to say, the camera. This is the only bit of real acting the script calls for. The rest could almost have been done by an understudy. It is also the first and only time that the audience sees the famous face. As the observer, he stares at himself with what the script calls a 'very different expression, impossible to describe, neither severity nor benignity, but rather acute intentness'. As the observed, waking up to find himself so, he was required to have a look of horror. These are difficult directions to follow but Keaton was magnificent. As Schneider put it, 'Buster, finally given his chance not only to let us see his face but to see him act, let loose from deep inside somewhere. When we finally saw it, that face paid off – even if we hadn't known it was Keaton's.' When shooting was finished Buster collected his money and went back by train to the West Coast, as puzzled as when he had arrived. In interviews subsequently he said he did not understand what he was doing or what the film was about, but he had, without knowing it, obeyed the Beckett principle – which Schneider paraphrased as 'an actor must not mean but do' –

and Schneider treasured the memory of the half-smile he got when he asked him after the last shoot if he thought the film was worth doing.

The editor, Sydney Meyers, now did a quick rough cut for Beckett to look at before going back to Paris. The editing was contentious, with Meyers attempting to depart from the narrow provisions of the script and Beckett and Schneider trying to keep him inside them. During the few days that remained of his stay Beckett saw some old acquaintances, such as the Reaveys, Horace Gregory and his wife, and Kay Boyle. Boyle had not been above suggesting to various people that she had had an affair in Paris in the 1920s with the now famous Irish author, but the other guests at the dinner party noticed that they had very little to say to each other. In fact when they sat together beforehand, far from giving the appearance of a couple of old flames seizing an opportunity to joke about old times, it seemed rather as if Boyle was desperately trying to keep the conversation going. It was almost as if they had never met before. However, they did correspond subsequently.

Barney Rosset had made a reservation for Beckett on an early-morning flight, so they went to bed early in the little brownstone on Houston Street, where Barney and his then wife, Christine, lived. When Barney and his wife awoke and saw the time, they found to their horror that they had probably missed the plane. When they opened the door of their bedroom they found Beckett sitting outside on the landing in his overcoat and with his suitcase. He had got up in time, but had been too polite to knock on their door and then had fallen asleep himself. Another reservation was made on a flight for that evening and Barney suggested that they might pay a visit to the World's Fair, which was in Flushing Meadows on the way to the airport. It was another very hot and humid New York day but Beckett was still wearing his overcoat as they walked through the exhibits at the fair, all three rapidly getting into a state of torpor and exhaustion. When Barney and Christine were sampling some food at an exhibit they suddenly discovered that Sam was nowhere to be seen. In panic they went back the way they had come, looking to left and right. It took them about half an hour to find him, sleeping again, this time on a bench, where he had sat down to rest while they investigated another exhibit. After he had been woken and somewhat revived, he was driven to the airport and caught his plane. This ended Beckett's only visit to the United States.

In the months that followed the makers had great difficulty in marketing the film. As Schneider put it: 'No one wants shorts anyhow and this one they didn't want (or understand) with a vengeance. Nor did showing it around help us. We stopped showing it. It became a lone, very lone, piece indeed. Which no one ever saw and seemingly very few wanted to see.' Then, almost a year after the making, they were suddenly invited to show it at the New York Film Festival as part of a Keaton revival. The film was becoming Buster's, not Beckett's. In spite of Schneider's reservations it was sandwiched in between two Keaton shorts, an old one and a commercial he had just completed for the railroads. The result was predictable. The audience, which had been delighting in Keaton's artistry, presumed this was going to be more of the same. When the credits started to roll and the close-up of Buster's eyelid appeared on the screen they roared with laughter. As he walked along by the wall with his face carefully hidden from the camera the laugh petered out into a few puzzled titters. Finally they stopped laughing altogether and sat in resentful silence. When the film ended there was loud booing. The reaction of the critics was as bad as that of the audience. No one could understand why Keaton had his back to the camera until the very end.

Gradually, however, *Film* began to be shown at various European film festivals. Though it was never released, it did have sporadic, one-off public showings, usually for university audiences; and a little cult began to develop around it. When, four years after it was made, it was shown in a programme of shorts at the Evergreen Theatre in New York it received favourable reviews. Schneider's comment shows how much people wanted the reception of Beckett's work to fit the by then well-established modernist pattern, the saga of rejection followed by late acclaim which was the history of most of the great modernist works. 'Hard as it is for those involved to appreciate, each time that's par for the Beckettian course. All of his stage plays, radio and TV pieces, first get slammed, derided, ignored. Then, five years later, they are hailed as classics.' In fact, ever since *Godot*, none of Beckett's works had been ignored and mostly they had been hailed with acclaim, even if it was sometimes puzzled or uncomprehending. There were pockets of resistance, particularly in London theatre criticism, and 'Beckettian' had now become a by-word for an obscure and enigmatic work about debility and despair, but each new work was greeted with a great deal of critical respect. A partial exception was *How It Is*, the work he had struggled with so wearily at Ussy for so long. This English version of *Comment c'est* was published in 1964 and met with a very mixed reception. John Updike in the *New Yorker* wrote a review which parodied the style of the book and ended '. . . good the end of review the END of meditating upon this mud and sub-primate sadism NO MORE no more thinking upon it few books have I read I will not re-read sooner SORRY but that is how it is.'

The narrator of *How It Is* is crawling across an endless plain of mud, telling his story in a series of verbal gasps without punctuation or conventional syntax. He promises a system which will enable him to tell it all: he will do it in the 'natural order' and in three parts – 'before Pim with Pim after Pim'. In the first phase the speaker crawls intermittently with a sack of tins tied round his neck, across the mud; in the second, he comes upon the prone and motionless Pim and teaches him to answer certain questions in response to stimuli inflicted by fingernails, fist or tin-opener; in the third the speaker, abandoned by Pim, awaits the coming of Bom, who will torment him as he has tormented Pim, while Pim crawls away to where he will find, in his turn, his victim.

The three-part structure begins to seem insufficient almost as soon as it is established, however. There are certain anomalies and inconsistencies which have to be accounted for. The speaker's narrative clearly demands four parts, not the promised three – journeying to Pim, tormenting Pim, abandoned by Pim and being tormented by Bom. And although we are told that this last phase need not be recorded because it would be merely a repetition of the second, there now seems to be one other figure involved, Pim's victim, Pem; and in fact, after examining various hypotheses to account for the inconsistencies of his story and his inability to tell it as promised, the speaker finally begins to believe there may be an infinite number of individuals involved in this procession across the mud, since justice demands that no man shall be a victim without having a victim of his own, nor a tormentor without having a tormentor of his own.

And besides these figures in the mud and darkness, tormenting and being tormented, there is also a voice. Perhaps the speaker's existence is being observed from another plane and by another person, someone who may be called Kram and who employs the services of a recorder called Krim, though how he knows these names we do not know. And as to the numbers of people involved, a solution 'more simple by far and by far more radical' may be that there is no one at all except himself, all the tormentors and all the victims rolled into one. There may be no one else at all involved, 'only me yes alone yes with my voice yes my murmur yes when the panting stops yes all that holds yes panting yes worse and worse no answer WORSE AND WORSE yes flat on my belly yes in the mud yes the dark yes nothing to emend there no the arms spread yes like a cross no answer LIKE A CROSS no answer YES OR NO yes'.

There had come a time in the careers of some of the other great modernists when they finally parted company from their readers, a point after which comprehension and true enthusiasm were largely supplanted by a sort of puzzled respect. Such to some extent had been the response to *Finnegans Wake* and to the *Cantos* of Ezra Pound.

Re-reading my own review of *How It Is* in the *TLS* at this distance of time I find myself uneasy at my enthusiasm for this final 'novel' or prose work of novel length; and feel I should have suggested that Beckett – or *Mr* Beckett, as I insisted on calling him – was, as far as his extended prose was concerned anyway, perhaps reaching a point when the aesthetic satisfactions, the incidental beauties of sound and sense, even the illuminations of human existence, seemed insufficient reward for the pains and difficulties the reader was asked to undergo.

Chapter Thirty-Three

After the travels, excitements and productions of 1964, it is not surprising that Beckett's New Year resolution for 1965 was to make the year a quieter one, with as much time spent at Ussy as possible, and to try and find out 'if there is any more work in me'. He was finding Paris more and more difficult because of the interruptions and demands which were the accompaniments of fame and were, as he put it, 'difficult to evade and impossible to satisfy'. Suzanne no longer came to Ussy, so he was there – and she in Paris – for much of the time alone; but he was pleased that she had had a trip to Venice with her friend Marthe Gautier to see their mutual friend Madeleine Renaud, starring – with her husband Jean-Louis Barrault – in *Oh les beaux Jours* (*Happy Days*) and that she seemed in better form when she returned. They were moving towards an accommodation. He would make no public display of his relationship with Barbara; but she would accept its existence and its importance to him. In fact she was enjoying his fame rather more than he was, and where he did not attend first nights she was pleased to do so and to take the place of honour, as she did when *Oh les beaux Jours* opened at the Odéon in October 1963. Nevertheless he continued to be afflicted by guilt about her and the quality of her life.

Renaud had been eventually cast in this production by Blin, although she did not correspond to the physical character of Winnie as described by Beckett. Blin was nervous when he first made the suggestion to the author and he was delighted to find him receptive. 'When you have the greatest voice in the French theatre, you do not insist on other, less important qualities,' Beckett said. In fact it was Beckett, not Blin, who really directed this production, for which the designer Mathias had created a bright orange sky rather than a blue one, saying that he wanted to stress the unbearable heat rather than the beauty of it.

François Mauriac, the senior Catholic novelist of France and an immensely influential polemicist through his column in *Le Figaro*, chose this production for an attack on Beckett's work as a 'calumniation of life'. It turned out that Mauriac had not seen the play himself; he had dispatched what his opponents called his minions to see it; but his attack marked out the battle-lines between what his followers called 'despair' on the one hand, and faith, hope and charity – or at least Christian fortitude – on the other. One of the most interesting pro-Beckett contributions to the controversy was that by Alfred Simon in the journal, *L'Esprit*. 'The pessimistic vision of Beckett has nothing arbitrary about it,' he said.

It is merely untenable. And here is the tragic paradox of Beckett: he affirms the untenable. And he forces us to affirm it. Sartre said that life begins on the other side of despair. For the believer in our time, faith begins after the death of God ... We must return to the very foundation of tragedy as it was perceived by the Greeks: the confrontation between an evil God and human Freedom in revolt: the wickedness of God and the revolt of man each being implied by the other. What is the meaning of Winnie on her bare mound? What is the meaning of Job on his heap of ashes, and of Prometheus on his rock? The tragedy of the modern period is that man knows more about God than Job did, and more than Prometheus did about man ... The evil God weighs heavy upon Beckett's universe. It is the dead God. God has killed himself out of wickedness, out of hatred for man. And if he made man in his own image, then God has killed himself out of self-hatred. The sado-masochism of God constitutes the very bedrock of modern tragedy.

This production of *Happy Days* went on tour to many French cities, as well as to Geneva, Vienna and Prague and afterwards to New York. For much of this itinerary Suzanne was present. Though no doubt she was happy to receive the considerable *réclame* which came her way, her admiration for Beckett's work was sincere and her joy in his fame considerable. And besides, her friendship with Renaud was a deep one, and without these involvements her own life would have been even emptier than it was. In spite of these travels and distractions, however, the ailments from which she suffered persisted, seemingly not alleviated by the homeopathic remedies which were the only ones in which she had any faith.

The psychosomatic symptoms from which Beckett had suffered as a younger man seemed to have disappeared, but one ailment from which he suffered in the sixties was real enough. This was an abscess or growth in his mouth, which was at first thought to be possibly cancerous. Typically, he did nothing about it for a while and, when it was found to be benign, postponed any operation even further. When at length he yielded, the operation did not prove very successful. His palate, which had had to be opened, would not heal and in April 1965 he again had to have surgery. This time a small plate was inserted to close the opening in the palette but neither did this work, and at the end of May he had to have further surgery, this time apparently with better results. He now had only his eyes to worry about, for his vision, as he had realized while playing tennis at Barney Rosset's, was certainly darkening. He wondered if he was going blind but again for the time being did nothing about it. Towards the end of the decade his sight had become so bad that he had to squarely face the fact that it did look as if he was losing his sight and at last consulted an oculist, to be told that he was suffering from glaucoma.

Then in the winter of 1967-8 he suffered from a series of illnesses, which he identified as a cold, bursitis and a mysterious fever of some kind. The main link between these complaints was a cough, which caused him pain, and a pain in his chest irrespective of the cough. Eventually these were diagnosed as consequent on an inflammation of the pleura in the area where the knife had penetrated more than thirty years before. The mortal state and the ills that flesh is heir to were much in his mind at this time. His oldest male friend, Tom MacGreevy died in a Dublin hospital in March 1967 of a coronary attack following an operation for a hernia and he felt the loss keenly. MacGreevy had been one of his few real confidants and one of the few who really knew him.

The state of his own health generally and of his lungs in particular became one of the main reasons why he now began to seek out sunny, warm places where he might spend periods of rest and recuperation during the severest part of the French winter. There were lengthy visits to Cascais in 1966-7 and 1968-9. In 1966-7 he went to Porto Santo, Madeira, with Suzanne. Later they would go to North Africa and Malta. Whether these winter visits to sunnier climes or his half-hearted attempts to cut down on his smoking were responsible or not, the discomfort did subside; though, as the doctor had told him after the stabbing, his lung was still a barometer and the pain recurred in certain kinds of weather. As he was predisposed to do, Beckett often entertained fears of the worst kind by imagining that he might have cancer of the lung, the cruel affliction from which his brother Frank had died under his very eyes. Early in 1969 he told the Polish theatre director and critic

Adam Tarn of these fears, and Tarn reassured him that he himself had successfully undergone surgery for lung cancer. When the Pole raised the question of smoking, however, Beckett spoke fatalistically, as people who have failed to give them up often do, about the effect of the cigarettes, claiming that it was now too late to give up, that whatever damage they might do had already been done. There had already been a cancer scare in May 1968 just as the streets were filling with the student barricades. But tests revealed that the suspected growth was only an abscess. The scare had caused Beckett to give up cigarettes and even drink for a while but his renunciation of these comforts was short-lived.

Tarn had been largely responsible for Beckett's reputation in Poland, a country where avant-garde theatre was, oddly enough, flourishing. He had often invited Beckett to visit Poland, and when *Godot* was revived there in 1962 he repeated the invitation. Unfortunately for him, it was just at the time when Beckett was beginning to feel the miseries of being an object of attention in public places and he demanded to know whether he could walk down the street and sit in a café without being recognized or bothered. When Tarn honestly answered that he couldn't guarantee this, Beckett said that in that case he could not come.

When Tarn decided to leave Poland in 1964 and was on his way to Canada, having been refused admission to the United States, they met in Paris. Beckett offered him money or manuscripts to sell, a gift which he often gave or offered instead of money. He also wrote letters of introduction for Tarn which made his life easier in Canada. Giving manuscripts was a favourite Beckett mode of charity and there were other recipients besides Tarn. He had begun to sell manuscripts and other material himself in 1956, when he was approached by a dealer, Jake Schwartz, who contacted many other writers during these years. After meeting Schwartz in Paris, Beckett sold him manuscripts for £200 which were, of course, worth a great deal more. He liked Schwartz, as he did other, equally dubious characters, describing him as 'an entertaining ruffian', and he continued to send him manuscripts almost as soon as the works in question were completed, evidently believing in keeping his study uncluttered.

Much of what is now in the possession of various universities in America was sold on by Schwartz; but although Beckett knew he was a sharp operator, he was still shocked when, some years later, he met the American bookseller and manuscript dealer Henry Wenning and discovered that there was an enormous and ever-widening gap between the cheques which Schwartz sent him and those the dealer was receiving from the United States. He began then to sell to Wenning, forwarding not only manuscripts but, as he had done with Schwartz, asking his publishers to return his corrected galley proofs so that he could sell them too. In 1963, when he began to sell to Wenning, he no longer needed the money, but he continued to sell every scrap which was saleable just the same.

In October 1969 Beckett again had trouble with his lung. The pain had returned, especially when he coughed, which he did frequently. Told that it was the same old complaint and advised to go away if possible, he and Suzanne decided to seek the sun early that year. They also decided to try a new location for the winter break and chose Nabeul, a resort in Tunisia, not far from Tunis itself. They were advised that it was usually sunny, dry and warm there in October and November, but it turned out to be nothing of the kind. It was cold and the rain poured down in torrents every day. On the 23rd October the telephone rang in their hotel room and Suzanne answered it. After listening for a few moments she turned to Beckett with a grave face and exclaimed, 'Quelle catastrophe!' She had just been told that Beckett had been awarded the Nobel Prize. The call confirmed what Jérôme Lindon had already said in a telegram despatched that day from Paris: 'In spite of everything they have given you the Nobel Prize – I advise you to go into hiding.' Suzanne's remark on hearing the news admirably reveals her qualities and her understanding of Beckett.

But the news was not entirely a surprise. Beckett had half expected and equally half dreaded the Nobel Prize for some time. The nominating procedure allows any professor of literature anywhere in the world to put forward a name to be considered by the Swedish Academy, and Beckett knew that for at least five years now his name had been put forward on several occasions by weighty academics, among them the Joycean William York Tyndall, to whom he had written in response in 1966, 'if there is any justice I shall be spurned again and for good', adding however, 'What counts is the thought of the few such as you.' The ambiguity this reveals was typical of his attitude. He had frequently discussed the prize with his friends, saying that he would be inclined to refuse the supposed honour as Jean-Paul Sartre had done but that Sartre had to some extent queered the pitch: to do so now would seem to be imitating him. Besides, he told Lindon, Sartre's refusal had been 'inelegant' and at one point he instructed Lindon that if enquiries came from the Swedish Academy he was to intimate that Beckett would prefer not to be offered the prize. He had known that moves were afoot again in 1969 to have him awarded the prize, but he told people in the Falstaff that he was comforted by the fact that Lindon was going to Sweden and would pass on his feelings on the matter. Montague, who was there, was struck by Beckett's trustfulness in expecting a publisher to actively lobby against one of his writers getting the Nobel Prize. In fact Lindon did not go to Sweden prior to the award. Of course, Beckett was fully aware that the prize was a questionable honour, that it was largely given out on a rotatory basis and often for political reasons. Discussing the money aspect of the prize, he often adverted to the story of Yeats who, on being rung up in the middle of the night by the editor of the *Irish Times* to be given the news of his Nobel Prize, immediately asked, 'How much?'

On 23 October the Swedish Academy had actually sent a telegram to Beckett, care of Les Éditions de Minuit, saying that the decision had been made to award him the prize and asking if it would be possible for him to be present in Stockholm on 10 December to receive it 'from the hands of the king'. When they received no reply the Swedish Ambassador in Paris was asked to make enquiries as to why. Lindon explained that Beckett was in Tunisia, so the academy went ahead and announced the recipient of the prize regardless. The citation said that it was for 'a body of work, that in new forms of fiction and the theatre, has transmuted the destitution of modern man into his exaltation', which is elegant and not inaccurate. Reuters Report said that 'Beckett had been a leading candidate for the prize for many years'.

In the immediate aftermath of the announcement the telephone in the Hotel Riadh never stopped ringing. It seemed as if the whole world wanted to interview him or offer its congratulations, and also as if the whole world knew his whereabouts and telephone number, though he had no notion of how it might have got into circulation. Callers were told by the hotel manager that Mrs and Mrs Beckett were staying at the hotel, but that they 'had left on an excursion'. The incessant rain had flooded the road out of Tunis, almost cutting Nabeul off from the outside world, but nevertheless a determined group of journalists and photographers managed to get to the hotel, only to find that the famous recluse was refusing to leave his bedroom.

But Jérôme Lindon had by now also flown in and a pact was arranged: Beckett would appear and photographs would be taken but there were to be no questions. When eventually he stood, head bowed and looking out from under his brows as he had done so long ago for a school photograph at Portora, his appearance confirmed many of the journalists' preconceived notions about him. He was tall, thin, gaunt. He wore steel-rimmed spectacles and his hair was cropped to his bony skull in a crew-cut. He could have been a convict, a tramp or even a concentration camp victim.

The legend of the eccentric, silent recluse seemed to be fully justified by his appearance, and since journalists are never happier than when the story they have brought turns out to be the one they can take away, the assembled group was, in spite of his silence, hugely gratified. The spectre that stood before them was the one they had imagined. They were also quite intimidated by what they understood to be the silence of greatness. One of them even apologized for the intrusion. 'That's all right. I understand,' Beckett replied. It was something of a nuisance, however, that stragglers were still arriving, so that there were photographers hanging around the hotel for days, hoping for shots or interviews. To leave the hotel for a swim Beckett had to creep down a back stairs and go hurriedly off to a deserted beach. All this of course confirmed again everything that had already been decided about him. One day he met a photographer on the beach who had the temerity to congratulate him. Beckett received the congratulations courteously, saying, 'I regret that my mother is not alive to hear about it.'

Cut off from newspapers and magazines though he largely was, Beckett was now becoming dimly aware of the deluge of publicity that was descending on his head, and he did not like it. He was apparently, he said, 'lacking in Nobel fibre', not the sort of person who regards the award as the summit of all literary aspiration and is delighted that others think so too.

But largely because of the money and the element of competition involved, the Nobel Prize is regarded as important by large numbers of people who know or care nothing about literature. They would soon begin to make Beckett's life impossible, forcing changes of habit on one who was very much a creature of habit, and of lifestyle on one who was very much a *flâneur* in the old sense of the word, who liked to stroll and stare and immerse himself anonymously in the life of the city where he lived. Meanwhile letters and telegrams were pouring in from well-wishers and old friends, which he felt he must answer. Work was out of the question, 'With all this I don't even know when I can *try* to write again,' he said. Paris would be worse than Tunisia, but Tunisia was becoming impossible, so he and Suzanne decided to seek refuge elsewhere. After some deliberation they decided on Cascais, the old-fashioned Portuguese resort where they had been a couple of years before, flying there from Rabat on 13 December.

In Cascais they stayed at the solid and rather resplendent Hotel Cidedela, which Suzanne liked. Though Beckett thought he might have more privacy in a smaller place, the Cidedela was better equipped to deal with intruders and fend off telephone queries, as large establishments usually are. Since very few people now knew where they were, their new abode was reasonably peaceful and the weather was mild. Though very near Lisbon, Cascais, like neighbouring Estoril, was an exclusive place. There were walks that Beckett could take without fear of molestation and, though there was still correspondence to be attended to, they enjoyed a fair degree of privacy, if not the rest that had been hoped for when they set out for Tunisia. Though there might have been times in the past when the receipt of £30,000 would have enabled even Beckett to sleep sounder at nights, those days had long passed. He already had enough money for his needs. For the moment anyway, the prize presented itself to him mostly in the guise of a nuisance. To escape or postpone the consequences of this great accretion of his already considerable fame, he and Suzanne stayed in Cascais for six weeks, adding to the invisibility and mystery which the world's press had now made his story.

Jérôme Lindon went to Stockholm to receive the Nobel Prize for Literature on Beckett's behalf while he and Suzanne were still in Cascais. When an author is unable to be present, the usual practice is that his country's ambassador should receive the prize from the Swedish monarch on his behalf. The Irish Ambassador had been instructed by his government to express his willingness to do this. But Beckett insisted that it should be Lindon and it was to him that King Gustav presented the Gold Medal and the Diploma in the hall of the Swedish Academy, to the applause of an audience which had so impressed Yeats with its status that he had listed all the members of the royal family who sat in the front row; though he did point out ironically that, judging from their uniforms, the Swedish functionaries might have been 'chosen by a London manager staging, let us say, some dramatised version of *The Prisoner of Zenda*'.

The cheque for $73,000 – or roughly speaking £30,000 – was not handed over but followed later in the diplomatic bag, and was delivered to the boulevard Saint-Jacques by special arrangement. Beckett reacted coolly not only to the honour of the Nobel Prize but also to the money it brought him. James Joyce, he remarked, should have got it. 'He would have known how to spend it.' He gave a lot of it away. When he spoke to John Calder about writers who might deserve and need some money, Calder rang back within twenty-four hours with a list. One of the recipients of his generosity was the experimental novelist B. S. Johnson, whose sports-car was said to have been bought with Beckett's Nobel money. Another was Djuna Barnes, now old, lonely and ailing in New York, who got a cheque for $3000. Many celebrations were planned and a few were even held after the new Nobel Laureate's return to Paris, but he did not attend any of them, except those in the

Falstaff, where of course people insisted on congratulating him.

The Becketts returned to Paris from Lisbon on 25 January. Among the more urgent matters to attend to were the repeated and now strident requests from his publishers to be allowed to re-publish *More Pricks Than Kicks*, and to publish for the first time 'Premier Amour', *Mercier et Camier* and *Eleutheria*. It was an irony that publishers should now be pleading to be allowed to publish works that had so often been rejected and that he, who, when they were written, would only have been too delighted to see them in print, should now be resisting the idea. Both Barney Rosset and John Calder had been agitating to be allowed to re-publish *More Pricks than Kicks* for some time but Beckett had been resistant. Then he had weakened, to the extent of giving permission for limited editions 'for scholars' and the Calder and Boyars photo-lithographic edition had appeared in 1966. Richard Seaver, the former editor of *Merlin*, was now working for Barney in New York and he kept up the pressure, which became especially intense when the Nobel Prize was announced. Now Beckett suddenly weakened and gave permission all round, writing to Calder a few days after his return to Paris.

Some days ago I had a letter from Seaver saying 'they could wait no longer to publish M.P.T.K.' I answered that this was against my wish. But in the last few days pressure on all sides has grown so strong – and I was so tired, that I capitulate. You may therefore proceed with trade edition of this juvenilium. I also capitulate for 'Premier Amour' & *Mercier et Camier* – but Not for *Eleutheria*.

He added a characteristically wry little joke: 'I hope you don't realise what this will involve for me.'

More Pricks Than Kicks appeared as part of the collected works of Samuel Beckett in a Grove Press edition in 1970, and that year Les Éditions de Minuit published *Mercier et Camier* and 'Premier Amour'. Beckett's resistance had never been as absolute as it should have been if he was utterly determined not to have these works appear. His usual ambiguities had been on display throughout; and of course when the publishers saw any sign of weakness they redoubled the pressure. From a certain point of view, though, he was right to resist and wrong to give in. The disguises he had adopted in all his later works had been, like the impersonality of Eliot's poems, part of their power. Behind these masks he could be revelatory without being self-revealing. For all that they were more revealing about humanity's weaknesses and failures than anything in previous literature, the author himself was not on display, indeed could scarcely be discerned at all. The mystery of the man, the legend of the recluse had combined with the personal impenetrability of the work to create a literary presence of an extraordinary power. He had been invisible as another exile, James Joyce, had appeared to be in the vast works of his maturity, and had insisted that the author should be. It was a specifically modern stance, a reaction against the chatty discursiveness, the hand-on-heart sincerities and the buttonholing charm of the Victorians and Edwardians. Now, to some extent, he had abandoned it, for Belacqua, as he well knew, was a poor disguise and for all its daringness, *Mercier et Camier* was as chatty, as intentionally charming and embarrassingly personal as the works of the English Edwardians. *More Pricks Than Kicks* and, to a lesser degree, *Mercier et Camier* are explicit and self-revealing books. The voice in them is the author's voice, sometimes embarrassingly so. It took him three years to manage a translation of *Mercier et Camier* and when he received the proofs of *More Pricks Than Kicks* from Grove he succeeded in reading only about thirty or forty pages before giving up in disgust.

His eyes were now seriously troubling him and the Paris traffic was an even more terrifying hazard than usual because of the almost complete loss of peripheral vision which accompanied the blurring effect of the glaucoma. But when he mentioned this to Mary Manning and she wrote back in sympathy, saying that he needed his eyes more than most, he replied, 'Why do I need my eyes more than another? It seems to me they never focused anything. I console myself purely and simply with the thought.' On the way back from Sardinia in 1968 he had seen a specialist in Berne but had been told that the time was not yet right for an operation. He was well aware that Joyce too had suffered from glaucoma and he had daily experience of the melancholy state of the master's eyes in his last years. His own eyes were in better state than Joyce's, however, and the basic problem was less complicated. Though he was seriously incommoded and even reading and writing were difficult, he was assured that an operation would be successful.

The first operation to relieve the pressure of the fluid within the eyeball of his right eye was performed at the Clinique in the rue du Faubourg, off the rue Saint-Jacques, on 15 October 1970; and the second, a similar operation on the left eye, in February 1971. Like many people whose eyes have gradually deteriorated, he had lost the memory of what good eyesight could be and when the results were apparent he was amazed at the light which flooded in from all sides, forcing him to wear dark glasses for a while. Focusing his vision was still a problem, but he noted that he could read the time on the clock of the Palais du Luxembourg from a considerable distance, which he had not been able to do before.

The infirmities of the flesh were further borne in upon him in the month after the second operation on his eyes by the death of his uncle, Dr James Beckett, in Dublin. James and his brother Howard had both suffered terribly from the circulatory problems which afflicted the Becketts. Both had had to have legs amputated and by the time he died James Beckett had lost both legs as well as virtually all his sight. This was a sad end for one who had been, like most of the Becketts, a noted athlete and a swimming champion, a hockey and tennis player. In comparison with these things Sam should perhaps have borne more lightly the 'periarthritis', a discomfort affecting both the shoulders and the arms, of which he was now complaining. A more serious matter though was a good deal of pain in the lung that had given trouble before; and a very uncomfortable ailment, an outbreak of the cysts that had so often plagued him before on various parts of his body. About this time he also began to suffer from Dupuytren's contracture in his right hand. This is a gradual closing inwards of the fingers of the hand, usually beginning, as it did in Beckett's case, with the outer or little finger. It is called after a French surgeon, Baron Dupuytren, who gave it a name in the early nineteenth century, and its effects can be clearly seen in some of the later photographs of Beckett in which his right hand is visible. Until he was told what it was it puzzled him and made shaking hands awkward, but it did not affect the holding of a pen and the fingers never clenched fully in a fist as they might have done. Hoping for warm weather that might alleviate his lung problem at least, he went to Rapallo with Suzanne in May 1971, but a cold wind blew off the Ligurian Sea at Santa Margherita Ligure, where they stayed, and rain fell intermittently for a couple of weeks. When the weather improved in early June they decided to stay on until July, when they returned to Paris. Visits to warmer and sunnier climes where fresher air might be expected had now become a regular part of Beckett's life and twice a year he and Suzanne would go off together for a month or more to such places. Later in 1971 they went to Malta, 'a bare old battered rock', which they liked enough to return to the following year. In January 1973 they visited Morocco, not returning until 10 April; and in the closing months of that year it became again their place of escape from the autumnal mists of Paris. In the spring of 1974 it was Tangier which attracted them.

These places had seemingly little to offer except the walks, bathing, the evening meal in a hotel dining room and much reading of the thrillers – French (the Gallimard series), English, and American (Ellery Queen) – to which he was mildly addicted. Beckett would bring a backlog of letters with him to be answered but he did little serious work.

Often the expectation of fine weather was not realized. Nevertheless, it was Tangier again 'in spite of all' in the autumn of 1974 and, in the winter of 1975-6, Morocco. There were few excursions, little sightseeing, not many buildings or pictures for the connoisseur to gaze at; and there were very few encounters with other people of any note. Going away with Suzanne like this, frequently to places where they knew nobody and were shut off from the outside world, was, in part, a way of restoring and re-establishing their relationship. It was also a way of making amends.

Of course, these sojourns with Suzanne were sometimes hard for Barbara to bear. Before one of the trips to Morocco in the early seventies Beckett had had a bad cough and his lung had been very painful. Harold Pinter had taken a cottage in the English countryside near Stratford to get away and do some writing. Very few people had the telephone number, but one night Barbara rang. She explained that Sam was in Morocco but that she was very worried about him. She had a telephone number. Would Pinter like to ring and enquire how he was? Somewhat surprised but understanding her dilemma, Pinter rang the number she gave him and got Sam. He said he believed he hadn't been too well before he left and asked how he was, saying he hoped he was better. In spite of the unlikelihood of Pinter ringing him up in these circumstances, or even having the telephone number, Beckett did not seem at all surprised; and Pinter thought he understood that the enquiry was being made on behalf of someone else. Beckett assured him that he was much better and that the sun had done him good. Pinter relayed the result to a thankful Barbara.

These sojourns had a beneficial effect on the condition of his lung but even after the operations on his eyes his physical troubles were by no means eliminated. Apart from the 'periarthritis' which still bothered him, his remaining teeth were giving trouble and only the difficulties posed by the cyst in his mouth had prevented him from having them extracted *en masse*. Suddenly, before going to London to be present at rehearsals of a double bill – Albert Finney in a revival of *Krapp's Last Tape* and Billie Whitelaw in a new play, *Not I* – in December 1972, he decided to have the job done. The extractions went well but he had been staying in the Hyde Park Hotel for almost a week before he could begin to think of eating food that needed chewing.

Billie Whitelaw was for Beckett one of the discoveries of the sixties, just as Magee had been of the fifties. Her voice was sweet and soft, intensely feminine without loss of timbre, and it haunted him as no other voice except Magee's had ever done. He had first heard this voice when she appeared in *Play* in 1963. On the occasion when they first met, Billie had walked into the rehearsal room and found a man in a raincoat sitting there. His grey hair 'looked as if it had been crew-cut by some backstreet barber' and he wore steel-rimmed spectacles on the end of his nose, behind which she noticed his very pale blue eyes. She received, as many other people did, an impression of gentleness and quietness. 'I soon realised Beckett was a most gentle person, a gentle *man* – kind, quiet and private.' After rehearsal Beckett said he would come to each of the four actors' dressing rooms in turn. When he came to hers he sat in complete silence for the first ten minutes before asking her to make a series of minute changes in her script, including the substitution of two dots for three between certain words. They had no further contact for nine years, but Beckett continued to be fascinated by her voice. When he came to write *Not I* for her he could hear it in his head as he had heard Magee's when writing *Krapp*. But it must be said that he gave its more obviously musical qualities less chance than he might have in this play of a disembodied mouth, suspended eight feet above the stage and lit in the surrounding darkness.

Both Finney's and Whitelaw's were solo performances and both were to be directed by Anthony Page. Having watched Finney in *Not I* and discovered that, when he made a suggestion, the actor, rather rudely but in accordance with protocol, would turn to Page for confirmation, Beckett concentrated instead on Whitelaw's performance, virtually taking it out of Page's hands. In Whitelaw he had malleable material. His demands on her were tyrannical but she responded willingly and all the more so because they were gently and courteously couched. She had a strong sense of safety under Beckett's direction, almost as if he had taken her by the hand and was leading her towards what he wanted. She had felt this sense of safety emanating from his mere presence during the production of *Play*, when he had had a less directorial role. Now she felt it intensely, sensing that he knew exactly what he wanted and was confident that she could give it to him. Whitelaw had already decided that the words had to be taken fast and had already written 'speed and not gabble' on her script. She found that Beckett also wanted it to be said at speed, even if this meant a loss of intelligibility. During rehearsals he would say, 'Too much colour, too much colour,' which she correctly interpreted as 'For God's sake, don't act.'

Not I is one of the stream of Beckett plays, beginning with *Happy Days*, in which women are the central characters. It is a monologue, the voice coming from an illuminated mouth which is the only thing the audience sees. In this production it was vitally necessary that Whitelaw did not move her head even a fraction of an inch so that the illuminated mouth did not move at all, up and down or from side to side. The first solution for this was to stand her on a platform holding a bar, with only her mouth lit. During rehearsal she began to be afraid that she was about to tumble off this raised platform into the dark body of the theatre. She felt like an astronaut in space, somehow out of the safety of the capsule. This fear became so strong that eventually she broke down. 'I'm sorry, Sam,' she kept saying. 'I'm so sorry.' Beckett was deeply concerned. 'Oh Billie,' he said. 'What have I done to you.' The difficulty was resolved by strapping her into a sort of armchair affair with her head firmly wedged between two blocks.

Like many who came to know Beckett she found him much more human than she had initially expected, very gentle, courteous and solici- tous for others, enquiring always after her small son Matthew, who had been ill. He came to her home in Camden Town several times during rehearsals and a very warm and close relationship developed between them. Billie Whitelaw was a wonderful discovery for Beckett, inspiring him creatively, as well as bringing him a new and undemanding friend- ship. With her light but sensuous voice, and her fulsome figure, she touched some chord in his feeling for womanhood. For her he would write two other plays, *That Time* and *Footfalls*, whose central character is given his mother's name, May. The impact of her performance in *Not I*, a torrent of words continuing even after curtain fall, was huge and most of the praise and discussion which the double bill received was accorded to *Not I*.

It is evidence of the way in which Beckett was now regarded that this eighteen-minute drama received a 1,750-word review in the *New Statesman* – in other words, almost 100 words for every minute of the play. Beckett claimed to be somewhat surprised by this reception, which he gratefully acknowledged was due to 'the incomparable Billie'; but strangely enough, in all their relationship she only once heard him speak a word of praise about her work and that was by accident. It was after the television production of *Not I*, directed by Tristram Powell, in which the mouth filled the entire screen, in Whitelaw's words, 'strangely sexual and glutinous, slimy and weird, like a crazed, oversexed jellyfish'. She watched this in a dark viewing room, crowded in with Beckett, Powell and the cameraman, wanting to get out, trying not to look at the object on the screen, closing her eyes or looking at her knees. At the end she heard Beckett say just one word in his soft Irish accent: 'Miraculous.' It was enough for her.

The illuminated mouth which is the centre of the audience's attention belongs to a foundling, a woman whose father had vanished as soon as he had buttoned up his breeches and whose mother similarly disappeared into thin air immediately after giving birth. Thus she was 'spared' the 'love such as [is] normally vented on the ... speechless infant ... in the home', or 'indeed for that matter' spared 'love of any kind ... at any subsequent stage'. The idea that to be loved is to have a burden imposed on one and is a calamity to be endured is frequent in Beckett. In spite of this absence of love the woman in question claims that her life has been a 'typical affair' with 'nothing of any note till coming up to sixty' – or it may have been seventy – when one April morning, wandering in Beckett's old childhood haunt, Croker's Acres, she finds herself 'in the dark', able to see only 'a ray of light', coming and going like the moon drifting 'in and out of cloud', and unable to hear anything except a buzzing sound. Since like the other waifs in her orphanage she had been brought up to believe in a merciful God, she imagines she is being punished for her sins. Then it occurs to her that she is not suffering; in fact she cannot remember offhand a time when she suffered less. Perhaps, she thinks, she is meant to be suffering or thought to be suffering, just as on certain occasions in her life, 'when clearly intended to be having pleasure ... she was in fact ... having none ... not the slightest'.

The oddest thing, though, is that she who has been almost totally silent all her life – except, it would seem, about twice a year, usually in winter, when she was so assailed by the need to give utterance that she would even rush out and talk to complete strangers, babbling nonsense, 'mad stuff', and afterwards would crawl back into her home and almost die of shame – now cannot refrain from utterance, feeling a compulsion to go on talking because of 'something she had to tell' or 'something that would tell ... how it was'. Though she finds this need for utterance unbearable, she knows she must go on, keep on trying until she has said something or answered something, she does not know what. Thus she joins the long line of Beckett characters who are compelled or commanded to give utterance though they wish only for silence. She is also one of a number of Beckett characters whose sense of identity is not strong enough for them to say 'I' and who therefore refer to themselves in the third person. Among the memories that come to her is one of an evening when she was on her way home (home!) across Croker's Acres and found herself sitting on a little mound in the dusk, staring at her hand which is in her lap and appears to be wet with tears.

This reference to Croker's Acres is the only circumstantial geographical detail present in Not I, but the genesis of the play had been much more explicit in various ways. Some authors love detail: Joyce, for example, could not have enough of it and it all had to be accurate, part of the real world. Beckett's method was entirely the opposite. In what appears to be a version of *Not I*, written as far back as 1963, a female voice describes a move to Kilcool, which is a real place, albeit with a final 'e' – Kilcoole. There, like other Dubliners, Willie Beckett had once rented a house and installed his family for the summer. The girl in the unfinished fragment had lost both parents, her father first, then her mother 'perhaps of grief or partly grief'. The memory of her mother's death, 'the darkened room, the sleepless days', still brings tears and she even offers prayers to some sort of supreme being, 'Time the father', who gives light and takes it away. The woman in *Not I* is, incidentally, apparently also a deist, for in her desolation she too offers gratitude to such a being, echoing Winnie's phrase in *Happy Days*, 'great mercies', but calling them 'tender mercies'.

Left bereaved, the girl in 'Kilcool' goes to stay with an old aunt who lives in that village and her monologue is enlivened by accurate, circumstantial details of the journey by the old Dublin and South Eastern Railway, the Damn Slow and Easy, which runs by the sea and therefore has a 'beautiful view all bare and glitter' and 'tunnels through the Head', which is Bray Head. But in successive drafts of this important fragment Beckett eliminated almost all the naturalistic detail, while of course retaining some substantive matter, as in the woman's memory of lying in the field with her face in the grass on an April morning when the larks are singing. Such eliminations undoubtedly gave his work a power and a mystery which it would not otherwise have. In his Trinity lectures decades before he had even defined modernism in terms of such eliminations with their resulting obscurities, knowing very well apparently, even at that stage, what a source of power they were.

Writing of *Not I* at this time – when of course he knew nothing of the unfinished fragment, 'Kilcool' – the Irish critic, Alec Reid, who was also a friend of Beckett's, speculated brilliantly about the composition of the play. Probably, he thought, Beckett had, like many other dramatists, started by defining for himself the basic dramatic situation and deciding what the audience would have to be told to establish the central character in that context.

Next, perhaps, he set down these essential facts in a coherent sequence which might, however, have included such flashbacks and digressions as would be found in any spontaneous narrative. So far the sentences would have been simple, the syntax conventional. Then, deliberately, he would have proceeded to smash these sequences. The earlier, complete sentences are slashed to telegraphese and the narrative progression interrupted by exclamations and repetitions, phrases are bent back upon themselves echoing what has gone before even as they in their turn form the material for future echoes. Thus, when C———ing broken or out of order, he was now made to say that it is 'no

Reid suggested that the first sentences of such a draft might have read: 'This baby, a tiny girl, had come into the world before her time, at a godforsaken hole the name of which doesn't matter. Her parents were unknown, her father having vanished into thin air immediately after the one casual encounter with her mother, who likewise had disappeared on the birth of the child precisely eight months later.' This had become in due course: '. . . out . . . into this world . . . this world . . . tiny little thing . . . before its time . . . in a godfor- . . . what? . . . girl? . . . yes . . . tiny little girl . . . into this . . . out into this . . . before her time . . . godforsaken hole called . . . called . . . no matter . . .' And so on and so forth. It was not quite true of Beckett's method of composition, for in his case even the first drafts are elliptical enough in themselves to puzzle the reader, but it was a brilliant guess at the basis of Beckett's method and its progression towards a potent and fascinating obscurity, a process which had of course, after Ezra Pound's intervention, been paralleled almost exactly half a century before the composition of *Not I* in that of the most seminal of all modern works, T. S. Eliot's *The Waste Land*.

The month Beckett spent in London rehearsing Whitelaw was only a small part of the time he gave to theatrical productions in the early seventies. In August 1971 he went to Berlin to direct Eva-Katarina Schultz and Rudy Schmidt in a German version of *Happy Days*, *Glückliche Tage*, at the Schiller Theatre, where he had already directed *Endgame* in 1967 and *Krapp's Last Tape* in 1969. This production of *Endgame* was the first time that Beckett's name had appeared in a programme as a director of his own work. He approached the directing very seriously and methodically, preparing a director's notebook containing detailed directions for the action at every line of text – in fact a set of expanded and minutely detailed stage directions. Perhaps the most illuminating thing about it is its division of the play into sixteen sections, suggesting a very tightly organized, premeditated dramatic structure indeed, from establishment to exposition to action to resolution. He also annotated a copy of the 1956 tri-lingual edition of the play, which includes Elmar Tophoven's German translation. In this he made certain changes in the text and stage directions, deleting, for example, the stage direction indicating that both hands and Hamm's and Clov's faces should be notably red and changing the Spitz dog to a poodle in honour of Schopenhauer, who liked poodles.

The changes to the German text suggest both a more than adequate knowledge of that language and long thought about the play he had written. Thus, when Clov says that the world outside is '*kaputt*', meaning broken or out of order, he was now made to say that it is '*aus*', which has the meaning of expended or extinguished. The line, '*Der Spass ist zu Ende*', meaning 'the fun is over', was changed to '*Das Fest ist jetzt zu Ende*', which is a deliberate echo of Friedrich Schlegel's well-known translation of the line from *The Tempest*, 'Our revels now are ended.' He also removed the rather stagy references to the audience which the characters make during the play, such as Clov's sarcastic 'a multitude in transport of joy', thus making the play more claustrophobic and enclosed.

During this 1967 visit he stayed in what was to be his abode in Berlin over the years, the Akademie der Kunst, which was a spartan place, not much above the hostel level. He arrived at the theatre the first morning determined that there should be no nonsense about meaning and, pulling the text out of the thin black portfolio, he announced, 'I don't want to talk about my play, it has to be taken purely dramatically to take shape on the stage. There's nothing in it about philosophy, . . . the only interest of the play is as dramatic material.' But in spite of his evident determination to be in command, he appeared to the actors rather shy and ill-at-ease, though the shyness vanished when they went to work and he took up his stance near the footlights, with a cigarette butt glowing in his left hand, observing closely through his glasses and making his comments softly but with emphasis. From this point on he kept a courteous distance, but his reserve and the sureness of his directions lent him a natural authority. His production assistant, Michael Haerdter, thought that he deliberately created a certain tension and that he knew how to deflect questions of a kind he did not like. Everyone respected his reserve. Asked if Nell really died, he disclaimed any special knowledge: 'So it seems, but no one knows.' But at the point where Nell came to the line, 'nothing is funnier than unhappiness', Beckett suddenly broke in to say, 'That for me is the most important sentence in the play.'

A few mornings later he told the actors, 'I would like as much laughter as possible in this play. It is a playful piece.' He meant internal laughter; that it ought to sound plausible when Hamm said to Clov, 'Ah, great fun we had, the two of us, great fun.' He also insisted a few days later that he would like to get as much comedy as possible into the play, though only within the limits of what he thought was proper and allowable. He would not allow the actors to 'pause for effect' and try to get laughs by this sort of timing; but when Horst Böllmann, who was playing Clov, delivered sadly the line, 'Sometimes I wonder if I am in my right mind', and then more cheerfully the following one, 'then it passes and I am as lucid as before', he told him to reverse the procedure, ending the first line cheerfully and the second sadly, assuring him rightly that it was funnier that way.

Of course, the actors did ask questions. Ernst Schroeder, who was playing Hamm, had some difficulty imagining himself into the role until he asked Beckett whether the mighty Hamm did not, after all, have a bad conscience. In response Beckett looked at him with a crafty expression, seeming surprised and a little unhappy; and then asked softly, 'Do you think so?' Schroeder did not know any author or director who would have answered that way, and thought that at that moment Hamm became a real person for him. He was no longer a role; he was 'an acquaintance of Beckett's and mine'. Böllmann wanted to know why Clov must learn to suffer better, to be there better? Beckett answered, 'the notion is that when one has given the tyrant his full account of suffering, he lets his victim go. Only when one has given life its full share can one leave it.' This reply is, in its way, a response to those who ask why an artist with Beckett's vision would not decide to commit suicide.

He was still playing around a little with his own text, inserting one 'No!' and one 'Clov!' into Hamm's last monologue, but when Schroeder asked who this 'No!' was addressed to, he replied only, 'Hamm says the No! against the nothingness.' As they were leaving the rehearsals one day and he seemed relaxed, Haerdter asked him, 'What does theatre mean to you?' and Beckett replied, 'Theatre for me is mainly a recreation from working on the novel. One has a given space to deal with and people in that space. That is relaxing.' To the question whether the directing was also relaxing, he replied with a laugh saying, 'No, not so much, that's hard work.' He wanted, he said, to get back to writing novels but he was writing absolutely nothing. Then he grew expansive. One could scarcely call them novels either. It was something different; 'We don't write novels any longer.' When Haerdter enquired whether all writing, including novels, did not grow out of a preoccupation with one's self, with one's own past, he replied, 'No, hardly that, pure force of imagination . . . naturally memory plays a part but it's a matter of imagination; of the attempt to escape from the tangle.'

Over lunch he expatiated on his own theme of the impossibility of knowing. 'The crisis started with the end of the eighteenth century. The encyclopedists were all mad, *tous fous . . . ils déraisonnent!* They give reason a responsibility which it simply can't bear, it's too weak. The encyclopedists wanted to know everything . . . but that direct relation between the self and – as the Italians say – *lo scibile*, the knowable, was already broken.' Staring down at his plate he continued: 'Leonardo da Vinci still had everything in his head, still knew everything, but now!' Then, looking up with a smile that was between bitterness and resignation, he continued, 'Now it's no longer possible to know everything. The tie between the Self and Things no longer exists. One must make a world of one's own in order to satisfy one's need to know, to understand one's need for order.' Almost on a more cheerful note, he concluded, 'There for me lies the value of the Theatre. One turns out a small world with its own laws, conducts the action as if upon a chess board ... yes, even a game of chess is too complex.' There is a sort of implication that he found the ordering of the confusion he was talking about easier in the theatre than in non-theatrical, imaginative prose composition, which of course it is.

Another resident in the Akademie residential building was Mira Avrech, a dark-haired, attractive Israeli journalist. After formal greetings had been exchanged a few times in the corridors, the acquaintance developed. She sent him a note inviting him to her room for a drink and that evening he arrived at her door with a bottle of Jameson and two glasses. What followed has been described as an affair; but in her account Ms Avrech makes no such claim, saying only that they spent evenings together talking of Kafka, the Holocaust and the state of the world. In any case Suzanne was arriving, and perhaps awkwardly for him, Barbara Bray was also coming to Berlin to see the production. Suzanne arrived after four weeks of rehearsal. The actors now had an audience for the first time, and under what Haerdter called her 'critical glance and amused participation', the pace got quicker. It seemed all a little more colourful than before. When Hamm demanded his painkiller and Clov replied with sadistic enjoyment, 'There is no more pain-killer,' Suzanne called out, '*C'est formidable!*' with straightforward enthusiasm; but Beckett, a trifle surly, interjected, '*Il y a encore du travail*' – 'there is still work'. Some days before, he had sought to explain the relationship of Hamm and Clov by saying it was the love-hate relationship, the typical relationship of marriage partners; and instanced the Latin proverb '*Nec tecum possum vivere nec sine te*' – 'I cannot live with you or without you.' Almost immediately he removed himself so obviously from further discussion by taking a step backwards that Schroeder, amused, promised not to ask any more questions.

Chapter Thirty-Four

Happy Days was produced in Berlin in 1971 as part of a festival at which MacGowran also performed his one-man show, *Beginning to End*. On the day of his expected arrival in Berlin, Beckett went to the airport to meet him but, after two annoying and frustrating days spent trying to locate him, the actor failed to arrive. Beckett discovered at last that he was marooned in New York with an out-of-date passport and in financial difficulties. Through the intervention of the Schiller Theatre management money was dispatched to New York; the Irish Consulate renewed his passport; and MacGowran arrived very much the worse for wear. Fearing that he would be unable to perform, Beckett took him to his hotel. But Jackie was one of those actors who always go on and who always seem to be capable of giving their best; indeed the often-repeated damage done to the nervous system by stretching it tight enough to perform with a hangover and withdrawal symptoms may have been one of the things which hastened his early death. Later that year he was in serious difficulties again in the United States, and Beckett asked Barney Rosset to advance him $2,000. Meanwhile, in response to further pleas, Beckett himself had sent money in Sterling to the account of Jackie's wife, Gloria. Honourably, Jackie returned Rosset's cheque to Beckett.

After *Happy Days* in 1971 Beckett returned four more times to Berlin to direct: *Godot* in 1975, *Footfalls* in 1976, *Krapp's Last Tape* in English with the San Quentin Drama Workshop in 1977, and *Play* in 1978. The Krapp of the 1977 production was an American, Rick Cluchey, who as a convict in San Quentin organized and directed a production by the San Quentin Drama Workshop of *Waiting for Godot*. He corresponded with Beckett, who helped him as best he could, and when he was released the two met in Paris. Cluchy was a short, burly, masculine fellow with tattooed arms, curly hair, sideburns and a moustache. Beckett found him engaging, as he did other people who had something of the rogue or the criminal in their make-up. He thought he was a good actor, an opinion not always shared by others, and he encouraged him also as a director, smoothing his path for a touring production of *Endgame*. Described as 'short on humour and long on existentialism', Cluchy's was a slow *Endgame*, lasting for three hours instead of the hour and a quarter which the author had specified to Alan Schneider. But Beckett did not seem to disapprove, even though he had told Böllmann in Berlin that to take it slowly was 'very dangerous for the play'. So engaging did Beckett find Cluchey that when he went to Berlin again in 1977 to do *Krapp* he took Cluchey with him as his assistant director and he also instructed Barney Rosset to pay over sums to him from royalties.

The career as the director of his own work on which Beckett had embarked was both wearing and time-consuming. In 1975-6, for example, he directed *Godot* at the Schiller; directed Madeleine Renaud in *Pas Moi (Not I)* on the same bill with *Krapp's Last Tape* at the Théâtre d'Orsay; was involved in the first broadcast of *Rough* for Radio Two, which was directed by Martin Esslin, with Harold Pinter, Billie Whitelaw and Pat Magee; directed Magee in *That Time* at the Royal Court and Whitelaw in *Footfalls*; directed *Krapp* at the Akademie der Kunst in Berlin; and was closely involved in the BBC broadcast of the *Ghost Trio* and . . . *but the clouds* . . . , though this was directed by Donald McWhinnie.

All this would have been taxing and exhausting enough for a professional director, but for Beckett, with his perfectionist concern for his own work, his production notebooks, his considerations of the texts and his exposure to probings, questions and interviews, it was much more so. Meanwhile his protestations that he was done with directing became ever more frequent and vehement, like an alcoholic's forswearings of alcohol. And, besides the demands of directing, there were also rights, permissions and translations to be attended to. There were his three primary publishers, eager for every scrap, including hitherto unpublished or abandoned work, plus a multitude of other people eager to produce plays or reprint other work. Although he had an agent, Curtis Brown, he still did much of the work himself, taking an intense interest in at least the major productions and publications. And of course there were inevitable differences and frictions.

When *Not I* was premiered at Lincoln Center in New York on a double bill with *Krapp's Last Tape*, Alan Schneider flew to Paris along with Jessica Tandy and Hume Cronyn, who were to play the mouth and Krapp. None of them had seen the script of *Not I*; it was only known that Beckett had written a new play and they were very excited about it. When they were all assembled for dinner, Beckett produced typewritten copies of this new work, five pages long, in single spacing – a monologue for a woman's voice, or rather a woman's mouth. These were passed around at the dinner table and, to their dismay, when Jessica Tandy asked what might be thought a pertinent question, 'What happened to the woman in the field?' there was a distinct awkwardness. Beckett never liked being questioned about circumstance and life histories not given in the text. 'Was she raped?' she persisted. 'No, no, not at all – it wasn't that at all,' he answered with some vigour. 'How could you think of such a thing!'

Before the first night in New York there was a row with Cronyn about *Krapp's Last Tape*. There is a point where the author indicates that Krapp 'curses', without specifying the actual words used, rather as a novelist might. Cronyn had to elaborate his own expletives, and naturally they were American. Harold Pinter went to the preview and, rather disturbed by the swear-words Cronyn thought might fit the case, communicated his unease to Beckett, who promptly cabled Schneider and asked him to make a change. With the opening imminent, this naturally upset Cronyn, who cabled Beckett rather tartly, 'We know what you don't want us to say. What *do* you want us to say?' Quoting one of his own expletives, he added, 'this minuscule point is balls and rubbish, especially at this time'. Beckett stuffily replied that when he had anything to communicate to the actors, he would do it through his director, Mr Alan Schneider.

Among the permissions sought in 1973 was one by Estelle Parsons and Shelley Winters for an all-woman *Godot*, starring themselves. This was refused immediately and, in a letter to Barney Rosset, Beckett elaborated on his refusal, saying, 'I AM AGAINST WOMEN PLAYING GODOT, and wrote to Miss Parsons to tell her so. Theatre sex is not interchangeable and *Godot* by women would sound as spurious as *Happy Days* or *Not I* played all by men. It was once performed in Israel without our authorisation by an all-female cast with disastrous effect.' Zero Mostel's request for permission to make a film of *Waiting for Godot* was likewise refused at this time. It was not, Beckett thought, suitable for filming, and it had not been written for that medium. Sometimes people secured permission from agents for productions of which the author would have disapproved, had he known their nature. If money had been invested and actors and theatre hired, however, he always felt himself in a quandary, mainly because of his general sympathy with actors who might be in need of a job. When he heard about an

André Gregory production of *Endgame* with a set made of chicken wire he was put in one such dilemma. The production was quite far advanced when he learned of its nature. After a hectic interchange of letters and cables he told Barney Rosset, 'Frankly, I don't know what the answer should be. I only know I'll regret it in either case.' In the end he decided for the actors' sake to let it run. When Nicol Williamson used material without permission in a one-man show, he too was treated indulgently, Beckett remarking to Rosset that he thought of Williamson as a friend whose speaking of the closing section of *How It Is*, which he had heard, had been very remarkable.

On top of these worries about theatrical productions there was the chore of translations – in the seventies of shorter pieces such as *Sans*, translated as *Lessness*; *Le Dépeupleur*, translated as *The Lost Ones*; 'Premier Amour', translated as 'First Love'; of *Pour Finir encore* as *For to End Yet Again*. There were also the shorter pieces which were included in *Pour Finir encore et autres foirades*, published by Les Éditions de Minuit in Paris in 1976, which became *For to End Yet Again and other fizzles*, published in London in the same year. He explained to Magee, who wanted recently published work for a one-man show he was devising – to avoid covering altogether the same ground as MacGowran had – that these fizzles would be appearing towards the end of 1976 in an art edition with etchings by Jasper Johns. They were all, with two exceptions, translated by him from 'aborted' French pieces dating from the early sixties. There were two more recent fizzles, 'farted in the last few years': *Stirrings Still*, which had been written in English, and *Pour Finir encore*, his most recent work, in French. He told Magee that *The Oxford English Dictionary* defined fizzle as '1. – the action of breaking wind quietly; the action of hissing or sputtering. 2. – a failure or fiasco.' This, he said, was exactly the sense of *foirade* in French. *Fizzles* had a mixed reception, American critics being on the whole less enthusiastic than their English counterparts. 'Since they really are what they claim to be, the publication of the *Fizzles* is puzzling,' said J. D. O'Hara in the *Nation*. But English critics were now prepared to see profundities even in the least of Beckett's works. For some time an impression had been growing that he was steadily advancing towards silence. Although Valentine Cunningham was prepared to see in these later works a greater expansiveness than he had expected, and signs even of a resurrection, he spoke of 'that remorseless tearing away, steely accumulating of lessnesses, that pincer movement on the void, that must logically end in silence'. And others had spoken of an epic search for release from utterance, which would end finally in a work only one sentence long. But – in spite of the repeated protestations of his various mouthpieces that they speak only in the hope of being released from speech by their mysterious masters – none of the evidence suggests that Beckett was in fact consciously striving towards silence in the way that these critics assumed.

Nearly everything he said on the subject suggests that it remained a matter of regret to him that he was unable to find the time or the concentration or creative inspiration for longer works. Of course, there was a difficulty. By stripping away the circumstantial matter that keeps most fictions going and collapsing the pretences that sustain the bulk of imaginative works, he had made extended creative prose composition almost impossible for himself. But, just the same, that he was unable to engage in it was partly his own fault. Whatever inspiration he might find in Morocco or elsewhere, Ussy was his place of composition, where such works might mature undisturbed in a silence broken only by birdsong or the distant noise of a passing train, and where they might finally be with much difficulty accomplished. But for long periods during the seventies Ussy lay unvisited; and when he got there his time was as often taken up with translation as with original creation. When in December 1971 he returned for the first time for over a year, it was to find the house reeking of damp which a blazing stove would not dispel. He also found it dark and gloomy and, though he declared that the peace and quiet made it worthwhile to be there, this did not seem to outweigh the disadvantages of a winter stay for very long and he retreated to Paris almost immediately. Other sojourns there during the seventies were by comparison with the previous decade short and far between. Of course, he continued to work, as the published results show, but the distractions that beset him were endless and largely of his own creation. When he did succeed in working, his joy was evident.

The aphorist and essayist E. M. Cioran met him by chance on the rue Guynemer in the early seventies. Beckett enquired if he were working and Cioran told him that he had lost his taste for work, that writing was too much of an ordeal for him and that he did not see the point or the necessity of going through it any longer. Beckett seemed astonished by this but Cioran was even more surprised when, 'precisely in reference to writing', Beckett 'spoke of joy'. Afterwards Cioran asked himself if he had really used that word, but yes, he was sure he had. At the same time he recalled that when they had met for the first time at the Closerie des Lilas in 1961, Beckett had confessed to him 'his great weariness, the feeling he had that nothing could be squeezed out of words anymore'.

Cioran, who has been described as the last of the great French moralists, was the son of a Greek Orthodox priest who was born in Romania and, after studying philosophy in Bucharest, settled in Paris in 1937, a time when there was much interchange between French and Romanian cultures. He had many things in common with Beckett. One was an interest in the sentence as what Hugh Kenner has called 'the unit of effect'. As Kenner says, the unit of effect for most writers – especially English writers – 'when it has been something briefer than the episode, for nearly two centuries has been the phrase'. The sentence has been used almost to legitimize the phrase. Beckett's works, more than most, are composed in sentences which have their own interior order and dependencies. Cioran too wrote in carefully formed, often beautiful and haunting sentences.

He shared with Beckett also an interest in philosophy as a sort of underpinning of the writer's work, and his general outlook, with its advocacy of a 'will to powerlessness', was not unlike Beckett's. When the two first met Cioran was fascinated by Beckett's reticence. He himself came, as he puts it, from a corner of Europe where 'effusiveness, lack of inhibition, immediate unsolicited shameless avowals are the rule'. He felt at once Beckett's power to be himself, to stand apart, but with no apparent pride, 'none of the stigmata inherent in the awareness of being unique'. He was also struck by the fact that Beckett never disparaged anyone, that he was apparently 'ignorant of the hygienic function of spite – its salutary virtues, its usefulness as an outlet'; and he even thought – probably with truth – that this must make him suffer unconsciously, that to deny oneself the outlet of maligning other people could cause agitation and unease. If he had known the pre-war Beckett, he would have known a different man, not spiteful perhaps as spitefulness goes, but not averse to the occasional spiteful remark either. When the present writer asked another friend, who had known him well in earlier times, about the post-war, or at least post-success Beckett's absence of spite or malignity, his utter refusal to say anything adverse about anybody, the reply was, 'Why should he bother?' To Barbara at least he often spoke in terms which, if more amused than malign, showed quite clearly that he knew the score about certain people, particularly certain habitués of the Falstaff.

Cioran and he almost never spoke of literature together, Beckett seeming to find more conversational sustenance in everyday incidents, difficulties and annoyances. The usual writer's talk of comparative reputations did not interest him and he would make his displeasure felt when others insisted on engaging in it. Of his own work he would speak only in terms of its realization and how he visualized it on stage. In a prolonged discussion of the requirements to be satisfied by the actress who played Not I, his eyes seemed to gleam as he saw that mouth, 'insignificant and yet invading, omnipresent'. Attempts to probe behind the work to its meaning were of course rebuffed in one way or another. Cioran remembered a dinner party at which people insisted on asking such questions until finally Beckett took refuge in complete silence and at length almost turned his back on the company, leaving before dinner was over. Yet Cioran himself speculated about the relationship between Beckett and his characters, asking himself whether Beckett's own existence was bathed in that 'leaden light' which is described in *Malone Dies*. He thought that in another time Beckett would have been drawn like an anchorite to the desert and that one could very easily imagine him in a bare cell whose walls were stripped of all decorations, even a crucifix. Once when they talked of Swift, Beckett told him that he had been re-reading *Gulliver's Travels* and that he had particularly admired the country of the Houyhnhnms, especially the scene in which the approach of a female Yahoo drives Gulliver mad with terror and disgust. Discourage such intrusions though he might, however, and with all the means available to him, Beckett could not always avoid being asked about his view of the human condition and the ultimate implications of his work. His comments on these occasions are sometimes revealing and they should be distinguished from the kind of guyings of his own pessimism in which he frequently indulged in letters and elsewhere. The impromptu responses when cornered deserve perhaps to be taken more seriously than the humorous asides, with their deliberately self-mocking tone. There are examples of both in his acquaintance with the Irish writer, Aidan Higgins. Higgins was acquainted with Beckett's cousin, John Beckett, the musician, and after reading *Murphy* decided, in spite of John's warning that he was most unlikely to reply, to write the author a fan letter. He did get a reply, but he found the handwriting unreadable and was forced to take it to John Beckett's mother for deciphering. The final phrase of the letter was a piece of advice. 'Despair young and never look back,' it said.

Later, when John Beckett was living in London and Sam came over on a visit, a meeting was arranged. Higgins wondered if the man he was to meet would turn out to be Democritus or Heraclitus, the man who says nay or the man who says yea. In fact he turned out to be neither. Higgins was very nervous, but Beckett was polite, courteous, considerate, 'devastatingly so'. So great was Higgins's nervousness, however, that at one point he retched into the wash-basin in the bathroom. He felt that this was the only appropriate response to meeting Beckett, who in his courtesy and consideration was almost too great to be endured.

After dinner Sam moved chairs so that he could sit beside his youthful admirer and he gave Higgins a signed copy of *Textes pour rien*, even though he knew the recipient of the gift did not read French. It was agreed that there should be a further meeting while he was still in London, perhaps an outing of some sort, and after the dinner there was much discussion between John and Aidan about where the great man might be taken as a diversion. His fondness for the music-hall was known about, so in the end they decided on the last of the London music-halls – Collins, in Islington. The party included Higgins's wife, Jill, as well as John's companion, Vera. They went by trolley-bus and Higgins noticed as further evidence of Beckett's scrupulous politeness that he insisted on dividing his attentions between the two women during the journey, resisting Vera's attempts to monopolize him. The show they saw included a comedian who delivered a long and hilarious monologue about cod liver oil as a cure for all ills. It reminded Higgins of Lucky's speech in *Godot* and at a certain point he looked along the row at the faces of his companions. The only one not registering great amusement was the bony visage of the author of *Godot*, who at one point turned to Higgins with a sort of blankly questioning look, as if to say, 'What's so amusing about all this?' But in the bar afterwards, when the comedian came through and the others stopped him to offer congratulations, Beckett, on being introduced, joined in the general assent, saying politely, 'I loved your stuff.'

During subsequent meetings in Paris Higgins found that there were frequent silences. He knew that one did not ask direct questions, and he felt he could equally not offer opinions about Beckett's work, which was of course what he had most in mind. 'It was like being with the Queen of England, you should not presume too much or ask questions. It is for the Queen to ask questions, but if she is not prepared to ask them, then you are stuck with silence.' The resulting nervousness often caused Higgins to drink too much and on one occasion he found himself showing Beckett photographs of his children, which he had an uneasy feeling he had shown him before, but which were still received and gazed at with scrupulously polite interest. In the years that followed Higgins corresponded with Beckett at intervals and sent him a typescript of the stories in his first book *Felo da Se*. Beckett was sufficiently impressed by this to recommend it to his publisher, John Calder, and to Barney Rosset for publication by Grove, in both cases the beginning for Higgins of a long connection. Their last meeting took place in 1969 in Berlin, where Higgins was then living and where Beckett was rehearsing Martin Held in *Krapp's Last Tape* for six weeks. It was arranged that Higgins should call at the Akademie der Kunst where

Beckett was staying, after which they would go to a restaurant he had picked out in Klopstockstrasse, where Jill was to join them. Going to the Akademie by U-Bahn, Higgins got on the wrong train. It was the sort of thing that happened whenever he went to meet Beckett, due, he thought, to a kind of terror. 'One admired him so much that one felt a fragility in the relationship, as if it might go away.'

In his little room in the Akademie building the heating was turned on full blast and Beckett was in his shirt sleeves. With its reading lamp and little desk, the room struck Higgins as like 'an operating theatre of the mind'. He also had the feeling that any room that Beckett was in would be Beckett's room, but this one certainly was. They drank whiskey, copiously poured from a bottle. Higgins felt the need to water it, but also felt that in Beckett's company one did not take water in one's whiskey, one took the pure spirit. So after several stern measures in the extremely hot room, he began to be a little drunk. He found Beckett's politeness almost intolerable and the inhibition about saying anything about his work devastating, so that one wound up feeling that it would be better not to say anything at all; but with the resulting silence came a sense of painful inadequacy. His sense of constraint in the great man's presence was rendered the keener by consciousness of the fact that he had contributed a piece to a recent *Festschrift, Beckett at Sixty,* compiled by John Calder. In this he had ventured some criticism of Beckett's latest work, *How It Is.*

> The wind has died down in his work; no faces, no time of year, no time of day – only friction, for the time being at least. No rivers, no animals, no tides, no days, no faces; only movements, painful movements, sad accomplishments, common nouns manipulated by overbearing intelligence . . . Into this plain mist of words the most accomplished Master of English Prose since Joyce is disappearing . . .

Now he felt the distance between them was even greater than it had been on other occasions. Criticism was not welcomed. He sensed a definite coldness in Beckett's manner, though he later thought that this might have been subjective, the result of guilt. The restaurant in Klopstockstrasse was called the Giraffe, and was Beckett's regular haunt, but in spite of this cheerful name things did not improve when they got there, even after Jill arrived. When suddenly, having had a whiskey or two, Jill asked the questions which were on many people's minds when they met Beckett, Higgins was appalled and feared their host might immediately leave the table. The questions were, 'Why do you despair so much? Why is there no expression of hope in your work?'

In the silence that followed Beckett picked up a crumb of bread from the tablecloth and looked at it for a long moment. Then he replied with what Higgins later identified as a quotation from Sébastien Chamfort, eight of whose maxims Beckett translated into doggerel English verse in 1975-6, incorporating what he of course knew to be a quotation from Dante. 'I would have written over the gates of Heaven what is said to be written over Hell – abandon all hope ye who enter here.' Then he dropped the crumb and said: 'That's what I think of hope.'

From the late 1970s on Beckett adopted the Hôtel Saint-Jacques, afterwards the PLM, more or less opposite where he lived on the boulevard Saint-Jacques, as a rendezvous. Founded and financed by Rothschild money and frequented by an aristocratic clientele when he started going there, the entrance foyer and the lounge were *belle époque* in style with deep red banquettes and discreet widely spaced tables. It was a four-star *hôtel de luxe*. Then in 1981, when the election of François Mitterrand and a socialist government caused the French monied classes to behave as if revolution had broken out in the streets and the guillotine had been re-erected in the place du Carrousel, the hotel was sold and down-graded, becoming a three-star catering for Japanese and Americans in Paris on package tours. The change of management and re-orientation towards a new type of customer brought sweeping changes in its interior decoration. *Belle époque* was replaced by the usual nondescript modern fittings. This did not seem to worry Beckett and he continued to use the PLM in spite of its duty-free shops and signs in Japanese and English; though a further change to the décor in the late eighties involving fountains and neon lights did perturb him a little, and caused him to retreat to the café, which was separate from the lounge. But he still continued to use the PLM; and, though he usually took one past the fountains to this rather featureless little glass-fronted café with marble-topped tables and straight-backed chairs, the PLM was a fairly intimidating introduction for those who had never met Beckett before and might have achieved a rendezvous with him with difficulty and apprehension. Upstairs on the first floor was a cafeteria with numbered dishes which were illustrated on the large menu. Here he would very occasionally take visitors, and those who accompanied him to lunch there saw that he was familiar with it. In fact, he and Suzanne often crossed the crowded street to eat one of the prepared plates at lunchtime.

A new young friend of the 1980s whom he frequently met in the PLM was André Bernold. Bernold, like Jean Demélier, had written to Beckett as a schoolboy from Alsace. His letter had posed a query which could be replied to with a yes or a no, and he had asked only for this.

Beckett took him at his word and replied, 'Yes', but Bernold wrote again and more often than not Beckett would answer his letters, however briefly. When Bernold came to Paris to study at the École Normale, he was too shy to make direct contact, but began to haunt the 14th arrondissement hoping to see Beckett in the street. One day he did see him but not where he had expected. It was in the rue d'Ulm and Beckett was standing outside the École gazing up at a second-floor window. When he moved on Bernold followed him for a while and then, at a pedestrian crossing, decided to introduce himself and declare his identity. They fell into step together, Beckett on the outside and walking slowly. He told Bernold that it was ten years since he had been in the rue d'Ulm. The student was struck by his Irish accent before Beckett's embarrassment and what Bernold felt to be a certain comedy in their ensuing silent stroll together impelled him to leave him. Then, a smile brightening his face, Beckett took his hand. 'If you need anything signal to me.'

Three months later Bernold asked for a meeting by letter and received one of the now usual cards by return indicating a day, time and place. An hour was allotted to this first interview, but it passed almost entirely in silence. Nevertheless a further meeting was arranged and soon Beckett was meeting his young friend every few days. Silences were frequent during these meetings too, during which Beckett 'gazed with abandon at the table'. Bernold describes him as leaning on his elbows smoking in silence and raising his eyes from distance to distance to see if the solution was to be seen on the horizon, careless of scattering ash on the marble table or in his coffee cup, the sleeves of his jacket littered with burn marks, the subtle movement of the head contrasting with the immobility of the body. Sometimes, seemingly for conversation's sake, Beckett would encourage his young friend to retail small circumstances, asking if he had come on foot or by some form of transport, passing on to other trivia of the day; but in fact he seemed to take a 'tireless interest' in the elementary things of existence, in sleep, comings and goings, what form of heating Bernold had in his room, and what winter clothes would keep out the cold. When a coat Bernold had bought in the flea market was discussed, Beckett felt the material and repeated its name in incredulous tones, '*Kamelhaar*'. When Bernold got another coat which Beckett approved of, he tried out the vocable 'alpaca'. He himself always wore a blue jacket with a tartan lining or a lumber-jacket with a large fur collar, sometimes a voluminous blue scarf and in winter a little brown woollen hat that Suzanne had found at the flea market twenty or thirty years before and which he wore pulled down over his ears. Because of his lungs he had yielded to the advice of the doctors in the seventies and given up the cheap cigarettes he had always smoked in favour of small cigars or, as they are sometimes called, cigarillos. Of these he would smoke at least one, leaving the packet on the table. Nearly always during their meetings he would have a coffee, sometimes a double or sometimes a small black; but Bernold drank whiskey 'to drown the tension always possible', allowing him to feign surprise – 'you are taking to the drink in the morning?' Sometimes, absentmindedly, Beckett would drop the wrapping-paper from the sugar-cube into the coffee instead of the sugar itself or light his cigar at the wrong end. Bernold admired his beauty and, during the silences or when Beckett was talking, would study his face, changeable and sometimes, in his way of turning his head, eagle-like. During the frequent silences he felt that Beckett was studying his face too – 'I don't know what he thought of my youth – but he examined the climate of my face precisely enough to sometimes divine my thoughts,' he said. Asked, with reference to appointments, why he saw so many people, Beckett replied that it was not to talk but to know still what it was to 'meet others . . . to verify still if this might not be in vain'.

They talked of birds – of an owl Beckett had heard, saying he used to hear them often at Ussy as well as in Paris near the Luxembourg and that he found the cry of the owl very moving; of dreams – Bernold telling him of a dream in which he had to get across a bridge, though if he succeeded the world would end. 'God expects you,' said Beckett. He said that in his own dreams there were no words, only images and forests. They spoke of hands, to which Beckett seemed to devote great attention. He would like to have given an entire play to them, he said, and expressed compassion for people's hands 'at rest after all they have done', remarking on the hands of the old woman on the stairs in *Film* and the lovely hands of Buster Keaton, which he said were his only good memory of the shoot – he and Keaton had not had much in common. Once they spoke of Wittgenstein and of his sojourn in Dublin, where nobody knew him. Beckett seemed interested in the philosophy lectures which Bernold was attending, asking about Derrida and Deleuze, the style of their lectures, their voices and what kind of men they were. When Bernold replied that they were nobles, *edle Menschen*, Beckett seemed happy with the response.

Chapter Thirty-Five

Throughout the seventies Beckett's creative output was small, consisting of minimalist works for theatre or television, frequently written for actors such as Whitelaw, Magee or, a new enthusiasm, David Warilow, and often for specific occasions. He continued to be involved in directing his own work or supervising other people's productions, an always time-consuming and frequently exhausting activity. After several weeks spent directing *Play* in Berlin in 1978 he told Barney Rosset that he 'seldom if ever had felt so tired in my tired old life'. Besides his London and Paris commitments, the Akademie der Kunst and the Schiller Theatre, from the early eighties on there was a new connection with the German television service Süddeutscher Rundfunk, which involved him in trips to Stuttgart for productions of the television pieces *Quadrat 1 & 2*, *Nacht und Träume*, and an adaptation of his last piece for theatre, *What Where*. Seeing *Quadrat* on to the screen took seven hours a day for ten days, working from nine to five with one pause, for a twelve-minute production. He made no attempt to analyse the philosophy of this or why he thought such effort was worth it. *Quadrat 1 & 2* (now called *Quad* in English) is an original television work which utilizes the technical possibilities of the medium to the full. This production still exists on tape and is still sometimes transmitted by various networks and channels. One could argue that he was creating an original work of art. But the effort still seems disproportionate. Sometimes two or even three years would go by without any new written creative work at all, and throughout the seventies there were scarcely any prose works, only three or four of the *Fizzles* which were published in 1976.

Then, in 1980, besides writing *Rockaby*, one of the most moving of his monologues for women, he wrote a prose work, *Company*, in

English. This was followed the next year by another, slightly shorter prose work, *Mal vu mal dit*, shortly to be translated into English as *Ill Seen Ill Said*. Then, after a gap of a year, much of it taken up with Süddeutscher Rundfunk productions, there came *Worstward Ho*, the ironic title of which is derived from Charles Kingsley's famous tale of adventure on the Spanish Main with which Beckett had been familiar as a boy. As the title implies, it was written in English and he found it very difficult to translate into French. He hoped, or said he hoped, that *Worstward Ho* had 'achieved him', telling his young friend Bernold that 'It is the final point'; and a year after its composition remarking with some wonder, 'I see myself in it still.' His other works, he said, had become strange to him and he remarked of *The Unnamable*, 'I do not know this author.'

Worstward Ho begins with the familiar Beckett imperative to continue to give utterance: 'On. Say on', followed by the familiar sense of being both spoken through and even created by the imperative, 'Be said on. Somehow on' until silence is achieved, 'Till no how on. Said no how on.' Of course, in this respect it is a failure and though the speaker finds himself at the end 'at bounds of boundless void. Whence no farther', he has to admit that he has not arrived at silence, only at a further impasse: 'Nohow nought. Nohow on.' The very last sentence repeats the end of the opening paragraph, but ambiguously, 'Said nohow on.'

Its beauty is late Joycean, the auditory beauty of *Finnegans Wake*.

Longing the so-said mind long lost to longing. The so mis-said. So far so mis-said. Dint of long longing lost to longing. Long vain longing. And longing still. Faintly longing still. Faintly vainly longing still. For fainter still. For faintest. Faintly vainly longing for the least of longing. Unlessenable least of longing. Unstillable vain least of longing still.

These three works, separately published in the 1980s, were collected together in the year of Beckett's death in a single volume entitled *Nohow On*. *Nohow On* is one of Beckett's most moving expressions of the belief that failure is the inevitable outcome of all attempts at artistic creation. Over thirty years before he had said in the Duthuit dialogues that 'to be an artist is to fail as no other dare fail'; that failure was the world of the artist and the retreat from it, 'desertion, art and craft, good housekeeping, living'. His conclusion there had been that the artist should make of 'this admission, this fidelity to failure, a new occasion, a new term of relation'; that it could itself be 'an expressive act, even if only of itself, of its impossibility, of its obligation'. To this harsh and stoic artistic creed he had been faithful ever since. The trilogy, *How It Is* and these late works had all been expressions of this attitude.

The three pieces finally composing *Nohow On* were written, like much else of great beauty, at Ussy, his place of creativity, all too seldom visited during recent years, but where he was now again finding refuge from distraction. Now that his sight had been restored, he could also enjoy watching the birds through his binoculars again. One day he found a tit nesting in his letter-box, the nest made of twigs and lined with moss to make it soft. Inside there were little balls of bronze with gaping mouths. He was much moved and told the postman not to leave letters in the box, but to bring the letters up to the door or leave them with the neighbours.

In 1981, the year of publication of *Mal vu mal dit – Ill Seen Ill Said* – the Paris Festival d'Automne was largely given over to his works in celebration of his seventy-fifth birthday. There were performances, presentations, round-table discussions and video displays at the Centre Pompidou, at the Théâtre du Rond-point, the Théâtre Gérard Philipe de Saint-Denis, and at the American Center.

Perhaps because of all this official recognition and the many complimentary sentiments expressed he was now beginning to speak rather more of the past than had been his wont, except perhaps with Leventhal. To Bernold he spoke of the days before all this recognition had descended on him, of Joyce with undiminished reverence, of Marcel Duchamp, of his early days in Paris. *Company* is strangely nostalgic, composed mainly of early memories – the hedgehog he had foolishly shut up in a box, falling from the tree in the garden, diving at the Forty Foot, his father in the summer-house reading *Punch*, the beggar woman fumbling at the gate of Cooldrinagh – complicated by a device, a voice heard in the darkness, in which someone lying on his back is telling the someone these stories while another voice seems to be commanding it to do so. If you abstract these voices, what is left is a straightforward, almost sentimental evocation of childhood such as many authors could have written, though perhaps few so well. It is the voices, easily separable from the rest of the material, which give the composition its modernist, Beckettian character.

But he was frequently in nostalgic vein. Discovering that the Irish poet Derek Mahon had been to Belfast Academical Institute and Trinity College, he spoke of going up to Belfast on the train from Portora to play the 'Inst' first fifteen 'and how the Portora chaps were all 'shit scared'.' When Mahon, who had sought an interview apropos of a

573

piece he was writing about the Irish poets of the thirties – Devlin, Coffey et al – but who, having got little mileage out of this subject, still found he had to make nearly all the conversational running, asked in desperation, with 'cold sweat forming' behind his ears, about playing the piano, Beckett said 'not anymore'. He held up his 'arthritic hands, which were purple and gnarled like trees in the wind' and then, having 'contemplated the paws a minute', relieved Mahon of further conversational effort by launching into a remarkable paean to old age, 'how your memory goes, your vocabulary goes. He had been looking forward to it all his life, he said; and the interesting thing was that he did not seem to be conscious of self-parody, of doing a Malone on me. He was perfectly sincere, a bit abstracted, chuckling faintly.'

Yet he was showing few signs of ageing. Many observers, including Mahon, noted his general athleticism, his brisk arrivals and departures. He still went for his long contemplative walks around the Left Bank; and about this time or even a little later, he began the physical labour of taking down a palisade which stretched along the edge of the wood at Ussy, consisting of more than 100 stakes sunk in concrete and more than a kilometre of wire. However, he had given up driving in Paris and now kept the car in a garage near the train station at La Ferté-sous-Jouarre, the country town near Ussy. In spite of what he told Mahon he did still play the piano, around this time especially the sonatas of Haydn. Playing the piano passed the time, he told Bernold, but then added more enthusiastically, 'It's marvellous and it's beautiful.' Insomnia, a lifelong affliction, continued to make his nights long but he continued to deal with it as he always had, shortening the hours during which other people slept by staying up late, listening to the BBC World Service and trusting to the paradox of insomnia, that sufferers are inclined to doze off in the morning when the light is returning and others are stirring.

During the night too he sometimes switched on the light and read the French or American detective stories that he kept by his bed. How out of touch he was with contemporary names in literature was demonstrated when Mahon asked him what he thought of Golding getting the Nobel Prize for Literature, just announced when they met. '"Did Louis Golding get the Nobel?" he asked incredulously. No, no, sez I, William Golding. Oh yes, sez Beckett – "the man who wrote The King of the Rings".' He still went to the bars and ate out in the evenings, usually in his favourite restaurant, the Iles Marquises. Though he dreaded ordinary tourist or literary academic intrusion, in the bars he was by no means unapproachable, particularly by people who did not know or care who he was or by those who were not awestruck and retained a casual friendly demeanour – and of course who did not want to talk about literature. The American novelist Randall Greene fell into conversation with him in a bar one evening and they wound up having a game of billiards together, the first of many. An Irishman, Johnny Glanville, was asked by a stranger in a bar if he could borrow his copy of the *Irish Times*. The tall, gaunt stranger turned out to be Samuel Beckett and they bought each other drinks and had a pleasant conversation before they parted.

His guilt about Suzanne was lessened by the long trips abroad, by the evident pleasure she got from such occasions as the Festival d'Automne and by the occasional official or semi-official function they attended together. His dislike of formal openings, launchings and literary gatherings is well known. These artificial occasions increased his feeling of being 'not there'. The necessity to make smalltalk pained him and the tediums and intrusions one risked at such gatherings caused him apprehension. He had largely given up drinking during the daytime and early evening. Gone were the days when several whiskeys would be consumed during a six o'clock rendezvous in the Closerie des Lilas. That he now commonly drank several large espressos instead when he met someone cannot have helped his insomnia.

Many of the infirmities that had plagued him in the past had gone. His teeth and his eyes were no longer giving him trouble and even the cysts on various parts of his body which had caused him so much discomfort from his early twenties on seemed to be dormant. But from 1978 on he had an intermittent cough and attacks of dyspnoea or difficulty in breathing. This was partly the effect of the stabbing, partly emphysema, very probably caused by a lifelong addiction to the cheap cigarettes which had only recently given way to cigarillos. He continued to direct, however, to sit in attentively at other people's productions and even to travel to England and Germany for productions of his works, particularly if they involved Billie Whitelaw or David Warilow. In 1982 he 'hobbled in on' rehearsals for Alan Schneider's production of Billie Whitelaw in *Rockaby* in London, remarking with his usual enthusiasm for her that Billie was 'great as always'.

In the following year he went to Stuttgart to oversee the first television broadcast of *Nacht und Träume* and he returned to the television station on the hill in the park there in 1985 to supervise a television adaptation of *What Where*. This was to be his last trip abroad. He liked Stuttgart: part of its attraction was that his favourite composer, Schubert, had spent a happy interval there during his short and generally unhappy life. Beckett's interest in other artists, musical included, always had a biographical dimension. He was interested in how they had fared in life and drawn to those like Schubert whose lives had been bleak and unfulfilled.

He went to London in 1984 to supervise productions of *Waiting for Godot*, *Endgame* and *Krapp's Last Tape* by Rick Cluchey's San Quentin Drama Company. These productions went on an extended tour of Australia and elsewhere, and such was his devotion to Cluchey that he allowed them to be billed collectively as *Beckett Directs Beckett*, a truly extraordinary indulgence. While it was true that he had participated and made some suggestions during these London rehearsals, the general style of the productions had been arrived at before that, and of course he had no further control over them once they went on the road. *Beckett Directs Beckett* was, to make no bones about it, a somewhat fraudulent billing and would certainly never have been allowed to any other company or company manager, but Cluchey needed the label and Beckett acquiesced. More than that, though Barney Rosset had raised serious objections to the inclusion of the Cluchey version of *Krapp's Last Tape* in a projected documentary film also to be called *Beckett Directs Beckett*, Sam overruled him, writing more than once to insist that the permission should be given without delay. 'I beg of you now again', he wrote, 'to stand no longer in the way of this production, so important for Rick Cluchey and fully approved by me.'

What made this all the more strange was that he was still trying to prevent productions that he did not approve of, or were by people of whose credentials he thought he did not know enough. Though he often repeated that he was getting tired of the necessity to do this, it did not seem to occur to him that some opening of the flood-gates was inevitable. Most of his views on such matters were expressed on cards, in the minimalist form to which his epistles had shrunk. His handwriting had improved. It was now more spidery and far more legible. It was on one of these cards that I received in 1982 his assent to membership of Aosdana, an affiliation of artists in all disciplines set up by the Irish Arts Council, with government backing and approval. I had urged him to become a member because it seemed more than likely that the scheme, which provided stipends for up to 150 artists, might incur political and public disapproval. Beckett was not a clubman and, apart from sporting clubs and the French Society at Trinity during his student days, Aosdana was the only association or society of any kind – certainly the only official state body – of which he became a member. Even more forthcomingly, a couple of years later he agreed to become a Saoi, one of a very small number of artists in all disciplines who have been singled out by their fellow members for special honour. 'The news of my election to the dignity of Saoi has moved me deeply,' he wrote to Adrian Munnelly, the registrar. 'Please convey to the members of Aosdana how sensible I am of this great honour.' It was almost as if through Aosdana he was making his peace with Ireland. Certainly he seemed to recognize that its mere institution had altered the relationship of the country to its artists for ever and that the bad old days of censorship and semi-starvation were now over.

When I invited him to become a member I also asked him to come to Ireland for the James Joyce centenary celebrations in 1982. This he declined with a 'Sorry can't be with you on Bloomsday'; but two controversies which stirred him in these years involved James Joyce, to whose memory he had a fierce loyalty. A central figure in Irish cultural circles, the architect Michael Scott, put a Joyce death-mask up for sale at Sotheby's. Beckett was outraged, more particularly since it seemed that the mask did not belong to Scott but had been given to him by Frau Giedion-Welcher in trust for the Joyce Museum at the Martello Tower in Sandycove where *Ulysses* begins. The other, a little earlier, had been the publication in 1976 of some letters that Joyce had written to Nora from Dublin in December 1909. The publication of these letters, with their intimate and loving obscenities, horrified Beckett; and he held all the parties involved – Richard Ellmann as editor, the Columbia and Oxford University presses as publishers and the Society of Authors as Joyce's Trustees – to be equally guilty. He was also inclined to blame the greed and insensitivity of literary widows for what he considered to be an unwarrantable intrusion into the privacy of the dead, for the letters had been left behind in Trieste and had been sold to Columbia by Stanislaus Joyce's widow. In a rare outburst he declared, 'Literary widows should be burned alive like Hindus.'

After the publication of *Worstward Ho* he thought he had finished with writing and said so to many people. 'I have written nothing for a long time and there is nothing on its way,' he wrote to Antoni Libera, the Polish dissident and translator of *Godot*, but in the previous year, 1984, he had begun to be haunted by the sentence: 'One night as he sat by his table head on hands he saw himself rise and go.' When Louis le Brocquy and Anne Madden talked to him in the PLM at the end of May 1984 he told them about this sentence, that it kept running through his head but that he could not progress any further. He told them that he knew that his character watched himself from behind, but he did not know why and said that 'there must be some very good reason or some very bad one'. Talking to Louis and Anne he went through the sentence several times, saying, 'If you take away 'his head on hands' it does not work.' The sentence eventually became the opening of his penultimate piece of writing, the tripartite prose piece, *Stirrings*

Still, eventually published in a de-luxe edition, limited to 200 copies, illustrated and signed by Louis le Brocquy, which finally appeared on the last day of 1988.

Le Brocquy was one of three distinguished illustrators of Beckett in his own lifetime, the others being Jasper Johns, who illustrated *Textes pour rien*, and Avigdor Arikha. A near-miss was Sydney Nolan, who met Beckett in Paris to talk about a project. Unfortunately Sydney had been reading Deirdre Bair's biography and, with a strange lack of tact, mentioned Beckett's difficulties with his mother. Beckett froze, the conversation tailed away into an embarrassed silence and the projected illustrations were never asked about again. 'I simply wanted to tell him about my own difficulties with my mother,' Sydney said.

Louis le Brocquy and Anne Madden were both Irish painters who divided their year between the South of France and Ireland and were frequent visitors to Paris, where they had an apartment. They had been friends with Jean-Paul Riopelle and others whom Beckett knew, and frequented the Falstaff. But they did not meet Beckett himself until 1978 when, after Louis had done a series of Beckett heads, Con Leventhal expressed surprise that they had not met and suggested that they 'contact him without delay'. From now on they saw Beckett whenever they were in Paris. On this occasion in May 1984 he seemed downcast and, after telling them of his insomnia and the night fears which still assailed him, he spoke of being 'tormented by what he called his "cruelty to people"'. As Anne Madden puts it, 'An unsparing judge of himself, Sam obviously suffered intricate moral torment.' Who were these people to whom he thought he had been so cruel? His mother? Suzanne? Barbara? Perhaps all of them and perhaps also he was thinking of the portrait of Peggy he had drawn in *More Pricks Than Kicks*, published in 1934 while Boss and of course Cissie were still alive.

A more minor but more immediate reason for guilt in 1986 was his reneging on the permission he had given to Barney Rosset to publish his still unpublished and unproduced play *Eleutheria*, a permission that seemed to be accompanied by a promise to translate. On 25 June 1986 he wrote to Barney that he had now read it again 'with loathing. I cannot translate it let alone have it published. Another rash promise.' He felt particularly bad about this because Barney was in difficulties and the permission had been given, as he said, with intent to lighten the publisher's burden. Now he had added to it. 'It goes to my heart to break this bad news but I must,' he continued; and he went on to promise to write something that would be worth having for his American publisher, even if only a few pages. 'I feel unforgivable. So please forgive me,' he wrote, ending 'much love from guilt ridden Sam'. It does not demean it to say that *Stirrings Still* was in part at least the result of this guilt and the fulfilment of this promise.

In 1986 the elaborate celebrations which had been planned by the French theatrical and cultural establishment for Beckett's eightieth birthday caused him very mixed feelings. They were to be on a very large scale, beginning in April with lectures and symposia in the Centre Pompidou and there were also to be private celebrations, organized by Stephen Joyce. He wanted nothing to do with these events, either public or private, at least in so far as those in which he was expected to be present in person were concerned; but at the same time he did not want to convey the appearance of being rude or ungrateful. He thought the best thing to do would be to leave Paris, but Suzanne was not well and he felt he could not do this. He had contracted a viral infection in his lungs during his visit to London in 1984 for the rehearsals of the Rick Cluchey shows which considerably aggravated his cough and his difficulty in breathing, besides weakening him generally. Now these breathing difficulties had increased and he was depressed.

In the aftermath of the celebrations, however, he cheered up. Able to get away to Ussy, he began to make a little progress with *Stirrings Still*, the piece of writing beginning, 'One night as he sat at his table head on hands he saw himself rise and go.' At last he knew something about the character and his situation and soon two of the fragments which finally composed the triad called *Stirrings Still* were brought into existence. When they were dispatched to Barney, he was told that if he liked them he could rely on a third and that the suggested title would be *Stirrings Still*.

'Seen always from behind whithersoever he went', the figure in the story wore the 'Same hat and coat as of old when he walked the roads. The back roads', which are still quite evidently the back roads of Foxrock and the foothills of the Dublin mountains. Walking thus he hears a clock strike and a series of cries in the distance, coming and going on the breeze, although he has no idea where these sounds are coming from. He ventures forth, but continues to hear the cries and bells as before until he begins to wonder if he is in his right mind. 'Then he sought help in the thought of one hastening westward at sundown to obtain a better view of Venus and found it of none.' He finds himself in a field of grass, which though 'not the green grass he seemed to remember eaten down by flocks and herds but long and light grey in colour verging here and there on white' is still somehow evocative of Croker's Acres. Moving on through this long hoar grass he becomes resigned, not knowing where he is or how he got there. Strangely there seems no end to this field or pasturage, but on he goes 'unknowing and what is more no wish to know nor indeed any wish of any kind nor therefore any sorrow save that he would have wished the strokes to cease and the cries for good and was sorry that they did not'.

In spite of his earlier optimism about a third section, the story hung fire at this point. He knew there would be more, but did not know what it was. Through the early part of 1987 he was nevertheless in good spirits and when he met the Le Brocquys in the PLM in February he seemed cheerful, urging Anne, whose brother had recently died, not to be depressed and to keep on trying to work. When he wrote to her afterwards, he spoke as usual of the work having an urgency of its own, of its speaking through the sayer. 'Don't be depressed, it is trying to be said,' he wrote to her a few days after they had parted. When he saw Richard Seaver in Paris in March he was still confident that there would be a third piece to *Stirrings Still* and he had begun to think that it would end with a rhyme incorporating a word of lamentation of which Irish dramatists had been so fond in the early years of the Abbey Theatre that it had become something of a joke: '*Ochone, ochone*, Dead and not gone.'

He was still taking an active interest in productions of his work and was pleased at the reception of Madeleine Renaud's performance in *Happy Days*, directed by Jean-Louis Barrault at the Rimini Festival, and curious about productions of *Godot* at the National in London and Lincoln Center in New York. He had also authorized yet another one-man show drawing on various texts, arranged by Gerry Dukes and interpreted by Barry McGovern. Those who met him saw little evidence of infirmity and in the early autumn he and Suzanne returned from a few weeks in Morocco tanned and reasonably fit. But before going abroad he wrote to Walter Asmus that physically he was 'more or less all right – but the mind in a bad mess'; and at Ussy in July he had begun to be troubled by a new ailment – dizzy spells, accompanied naturally enough by a fear of falling. The likely cause was a circulatory problem. As the attacks became more frequent and the dizzy spells more acute, he began to be more apprehensive about the consequences of having such an attack in the street or even in the middle of Paris traffic as the long walks he took required many crossings at busy intersections. For the time being, however, he continued to walk and to cross the road to the PLM through the busy traffic of the boulevard Saint-Jacques without benefit of any intersections or traffic lights at all, as Barney Rosset saw him do while waiting for him there. But this began to seem a more and more hazardous undertaking.

When Barney arrived in Paris towards the end of 1987 he was accompanied by the American film-maker John Reilly, who wanted to make a film of Beckett's life. They had brought with them several videos of American productions of the plays, some of which, such as *Ohio Impromptu*, Beckett had never seen produced in any form. They took a room in the PLM to show him the videos and on the way up in the lift Barney mentioned casually that Reilly would like to make a biographical film, to which Beckett assented without demur. Whenever he was supposed to join them in the hotel they would watch for him from the window and, to Reilly's amazement, he would walk slowly across the street through the racing traffic, apparently oblivious of danger. Reilly observed that he had difficulty in walking and this was beginning to be the case, for his legs were now affected by his circulatory problems.

Beckett took an intense interest in the productions he saw on video, some of which had been directed and the videos made by S. E. Gontarski, who was a professor at the Georgia Institute of Technology and editor of the *Journal of Beckett Studies*, a publication in which academic views of aspects of Beckett's work were now regularly aired. In the film Riley made there is a shot of Beckett leaning forward to watch Gontarski's production of *What Where* on the monitor in the hotel bedroom. He is watching it so intensely that he is reading some of the words with his lips and seems to be conducting the tempo of the speeches with his hands.

He was still struggling intermittently to provide the third section of *Stirrings Still*. In the early part of 1988 the third part came to him. Still apparently walking across the endless pasture on the hoar grass, the figure who had seen himself rise and go, he knew not where, so troubled by the bells and the cries, now begins to hear 'from deep within oh how and here a word he could not catch it were to end where never till then'. He does not know if it is telling him 'to press on regardless, or to stir no more', whether it is a warning or a direction. 'Such and much more such the hubbub in his mind so-called till nothing left from deep within but only ever fainter oh to end. No matter how no matter where. Time and grief and self so-called. Oh all to end.'

Chapter Thirty-Six

In Paris in February 1988 the Le Brocquys found him thin and much reduced in his little woollen cap, but lively and animated when he began to talk. *Godot* was to be produced as part of a Beckett Festival at the Gate Theatre in Dublin and Louis was designing the set. For his benefit Sam drew a sketch of the tree on a sheet of paper, a cross with slightly raised arms on a gently undulating ground. When Anne raised the question of the cross and its deep Christian symbolism, Beckett shrugged and chuckled, implying that since the cross had so much symbolism, why not use it? He spoke of Didi and Gogo waiting for 'salvation or whatever'. But a short while before, when Louis had fallen and broken some ribs, Sam had written, 'Distressed to hear of your fall. Thank God no worse.' As had been common in Ireland among Catholics and Protestants when he was growing up, he had always sprinkled his letters with 'Thank Gods' and 'Please Gods', even the occasional DV, short for 'Deo Volente', or 'God willing', which is more of a Catholic practice. He had ended this one with 'Old mind in a bad way so no more for now', repeating what he had written to Walter Asmus.

But in spite of his momentary animation, he would suddenly, according to Anne Madden's account, 'put his head in his hands and look a hundred'. Speaking of the many insensitive or unauthorized productions of his works which were taking place around the world, he said he could not keep track of them. 'I am too tired. I give up.'

But it was not quite true. He was still keeping an eye on things. Sent the script of an operatic version of *Watt* which was about to go into production, he immediately wrote back to the author to know who, if anybody, had given permission. 'Did you get permission for this adaptation from me or from Barney Rosset who holds US performance rights? Not to my knowledge. Correct me if I am wrong.'

He had in fact written one original opera in response to a request from the American composer, Morton Feldman. Feldman, seeking him out in Paris in 1962, was told he would be at rehearsals of the Barrault production of *Happy Days*. He found Beckett sitting alone in the darkened stalls and during a break he spoke to him, introducing himself and asking whether there was any chance that Beckett would write a libretto for a short opera. At this point Beckett interrupted him, saying, 'I hate opera,' to which Feldman replied, 'So do I.' The result was *neither*, a libretto of less than twenty lines of text, which, as an opera with music by Feldman, received its world premiere in Amsterdam in 1990.

But as usual he found himself in a quandary where actors had been engaged and a production embarked on without permission, what he called 'the now alas familiar *fait accompli*', concluding 'in consideration of the pains you and your collaborators have taken, I cannot refuse permission for this project which normally would not have been given'.

For much of the early months of the year he was unwell with a cough, chest pains and difficulty in breathing as well as the problem of his legs, and he left the house very little. To Walter Asmus, who was in Paris in September 1988 to make videos of three short Beckett plays, he said that his worst affliction was his difficulty in walking. Without the solace of his long swings through the streets of Paris he was disconsolate. And of course he had used the excuse of these long walks for many afternoons spent in the rue Séguier. Suzanne too was unwell and friends suspected that they were eating very little at home.

In June when the Le Brocquys came to meet him in the PLM he seemed to Anne to be 'transparent looking'. Before leaving them he held both their arms and, quoting his own work, said, 'In *Endgame* Clov says 'When I fall I'll weep for happiness.' And as I go out I say, If I fall do not pick me up.' He took pleasure in this rueful joke and said it to others as well, including Gerry Dukes, the deviser of the actor Barry McGovern's one-man show. In August he did fall, collapsing in the kitchen of the apartment in boulevard Saint-Jacques. It was the culmination of a series of falls in the street, or, once in the park, and on this occasion Suzanne telephoned for an ambulance as well as their doctor, who lived near by, and he was taken to hospital. Nothing was found except what was already known – that he had circulatory and breathing difficulties. When he was released Suzanne declared that she could no longer look after him. His doctor found him a place in a *maison de retraite* in the rue Rémy Dumoncel, in the 14th arrondissement, off the lower regions of the boulevard Raspail. Some people were surprised, even upset, to find him in Tiers Temps, as the *retraite* was called. Others found it not unfitting.

Tiers Temps was a municipal old people's home. Beyond the small lobby one passed through a room which served as both dining area and television lounge into a small grassed courtyard with a single small tree. Beckett's room opened directly off this courtyard, which might have been designed by the fates for the author of *Godot*, in which the single tree means so much. His room measured sixteen feet by twelve. It was painted dark blue and furnished with an iron bedstead, a wardrobe, and two tables – one a small writing table under the window. In the mornings he would feed the pigeons which strutted round the tree outside, enjoying the ceremony of crumbling bread and watching them peck at it.

When he arrived, staff at Tiers Temps were told that he had not been eating well and was suffering from anaemia. They found him courteous and co-operative and he made special friends with Madame Isabelle Jernand, the *directrice*, a young woman of some culture and understanding. She thought it would be difficult to tell his age from his appearance and general bearing and noticed the vitality and youthfulness of his blue gaze. She also found he could be very amusing, though sometimes he would tell her that he wanted to be undisturbed and would remain in his room. He had no telephone in his room and would come to the desk in the lobby to answer the numerous calls that came for him; but there were days when he said that he did not want to answer any and the staff would take messages.

The other residents also found him courteous and, impressed by his presence in their midst, were always glad to go to his room with a message or receive a few words in passing. He still went out often, either to the post office in the avenue du Général Leclerc or in the other direction to the Parc Montsouris. Since it was only a few minutes walk away along the rue de la Tombe Issoire, he also sometimes went to the apartment in the boulevard Saint-Jacques, which was, after all, still his home. But as time went by he seemed to become increasingly reluctant to do this, as if it were part of another life that he did not wish to resume. To begin with Suzanne came almost every day to Tiers Temps or he went to her. Madame Jernand found them a very bizarre couple and had a distinct impression that though they were in a sense inseparable they could not live together. She thought Madame Beckett very authoritarian and found that her visits disturbed Sam – though he was also unhappy if she did not come. There were frequent arguments when Suzanne would shout at him, but he likewise would be annoyed and speak angrily to her. Suzanne's memory for ordinary day-to-day things was becoming uncertain and there were often misunderstandings and confusions on this account which caused him great annoyance.

Madame Jernand noticed a resemblance between them, particularly in profile, and thought Mme Beckett, who usually wore trousers, neither very elegant nor very feminine, though there was undoubtedly about her a sense of being somebody, a person of consideration. Numerous other visitors included Edward Beckett, Jérôme Lindon, and of course Barbara Bray, whose visits were arranged by telephone and timed so as not to coincide with Suzanne's. Edith Fournier, a devoted Beckett scholar who had become a good friend of both Sam and Suzanne was a frequent visitor, bringing him such things as he wanted from the apartment and elsewhere; and she also looked after Suzanne's needs.

Unlike the other inmates, Sam took his meals in his room. He had a television set which was pushed in under the bed when not in use. On this he watched sports programmes and, in the evening, the news and public affairs programmes which were all he had ever watched on television. Most of his mail, which was considerable in volume, eventually began to come to the rue Rémy Dumoncel, and he was kept busy answering it, though later in 1988 there was a French postal strike which impeded delivery. On 8 November he wrote to Barney Rosset that he was still in this 'old crock's *retraite*' and having physiotherapy on his 'cautious legs'. He was improving slowly and said that he hoped to resume at Number 38 before very long. He ended with a little joke about the title of his most famous play: 'Postal strike still on here so God knows when this will reach you. May He find you *en attendant*, with love from your old Sam.'

When Barney came to Paris to visit him in the home he found him cheerful and philosophical. Among other visitors from abroad were Hugh and Mary-Ann Kenner. Like most people, the Kenners had never heard Beckett quote or recite anything at any length. And like others they knew that, though it had changed, his opinion of Yeats's later work had in the past not been of the highest. Now he suddenly surprised them by reciting in full Yeats's lines about death from 'The Tower'.

> Now shall I make my soul
> Compelling it to study
> In a learned school
> Til the wreck of body,
> Slow decay of blood,
> Testy delirium
> Or dull decrepitude,
> Or what worse evil come –
> The death of friends, or death

Of every brilliant eye
That made a catch in the breath –
Seem but the clouds of the sky
When the horizon fades;
Or a bird's sleepy cry
Among the deepening shades.

There had been much mortality among his friends and acquaintances. Jack Yeats was gone and Tom MacGreevy. So were Henri Hayden and George Reavey. Roger Blin, a key figure in his career, had died in 1984; and so had his eager young American director, Alan Schneider, who had been struck by a motor-bike in the street. Both Ethna and Con Leventhal were dead, Con ten years previously, in 1979.

Sam's young doctor – who was also Suzanne's – came almost every day, frequently bringing cigarillos, the little cigars that he still smoked, and a bottle of Jameson's Irish whiskey, which he consumed at a steady rate and poured for visitors with a fairly lavish, if not always very certain hand, sometimes with the comment, 'It helps.' These comforts aside, his life had now attained an elemental monastic simplicity, that of the single room, table, bed and chair – not quite as elemental as Malone's but getting closer to that state. Of course, he could have afforded a more luxurious place and some of his friends, Barbara included, considered Tiers Temps an unsuitable place and his presence there the result of mismanagement. Jean Demélier, the young novelist whom he had encouraged and supported in the 1960s, was sorry when he came to find Beckett in such a 'stupid, horrible place'. But perhaps he would have been lonelier in another sort of establishment. With one side of his nature, he had always had a surprising need of people; and it is possible that he found the rather intimate arrangements and little rituals of the small and personal Tiers Temps, including the presence of the other inmates when he left his room, soothing and comforting. The simply furnished room and simple regimen, the small courtyard with its single tree and the pigeons he was fond of feeding were somehow in tune with his rejection of the vanities of the world, and with the spirit of his work.

As 1988 gave way to 1989 he began to think of Tiers Temps as less of a temporary expedient, more of a home for the foreseeable future, perhaps more especially after Suzanne was taken ill and had also to be hospitalized. In November 1988 he still liked to tell people that he was there for physiotherapy of the legs 'as tired of carrying me as me of being carried'. He still walked almost every day to the Parc Montsouris, along the avenue René Coty. But he felt that his walking was 'not improving', though he was determined to 'keep on trying gingerly to and fro on the Coty strip'.

But eventually there began to creep into his comments on his presence in Tiers Temps a note of reconciliation to the prospect of being there indefinitely and even a sort of grim satisfaction at the thought. 'Come whenever you like, I won't be leaving here,' he now told friends. In addition to the risk-fraught, somewhat painful journey to the Parc Montsouris he still went every day to the post office. He was reading Dante's *Divina Commedia* yet again; and yet another biography of his lifelong hero, Samuel Johnson. When he read Richard Ellmann's *Oscar Wilde*, he was struck by what he called 'the hypocrisy of the English', but thought the book too long and too detailed, as indeed most modern biographies are.

He was also writing. *Stirrings Still* had been delivered and over Christmas 1988–9 he translated it, giving it the title *Soubresauts*, he did not quite know why. Then he began a piece in French, 'Comment dire', finally translated into English as 'What Is the Word'. This short piece, no more than three pages, went very slowly and he felt most of the time that his heart was not in it. It is a play on certain repeated phrases broken by dashes, and seems to be a natural continuation of the closing sentences of *Stirrings Still*, whose central figure is tortured by the sense of a missing word of great importance.

> folly –
> folly for to –
> for to –
> what is the word –
> folly from this –
> all this –
> folly from all this –

it begins. And it ends:

> glimpse –
> seem to glimpse –
> need to seem to glimpse –
> afaint, afar away over there what –
> folly for to need to seem to glimpse afaint,
>   afar away over there what –
> what is the word –
>
> what is the word.

The last line in the French version, 'Comment dire', is much more ambiguous in English: 'what is the word', what being the possible answer. The uncertainty about everything which had been given expression in *Watt*, the first work of his maturity, is echoed in his last, for this was Samuel Beckett's last piece of writing.

On 17 July 1989 Suzanne died. They had been together for over fifty years in a relationship of great intensity. If there had not always been a full communion of mind or heart, they had none the less, to use a musical term, always vibrated in sympathy with each other, albeit in a way which was sometimes very painful. There was no service, only a burial attended by Sam and a few of her close friends in the Cimitière du Montparnasse just off the boulevard Raspail, after which he returned to Tiers Temps. He was now free to marry Barbara if they both wished but there was very little time left; and the marriage, though discussed, did not take place. But Barbara was close to him during his time in Tiers Temps.

He had been reluctant to visit 38 boulevard Saint-Jacques when Suzanne was alive; now he was doubly so. Even when there were books or papers which it would have been useful to refer to and which could have been fetched from the apartment, he would not go there or encourage anybody else to do so. When the suggestion was made that something which he wanted could be collected from there, he would discourage it, and he certainly would not go into the building himself even when out walking in that direction. Tiers Temps was now the place where he was making his soul. But shortly after Suzanne's death he arranged with his young doctor to make a last visit to Ussy.

It was a beautiful late August summer's day and the Marne valley looked its best from the hill on which the house stood. Ussy had been Beckett's place of retreat from the world of engagements and affairs, his place of work and self-communing for almost thirty years. He had probably lived more intensely there than anywhere else. He knew he would never see it again and when he told the neighbouring woman who looked after it for him that this was his last visit, there were tears in his eyes.

As summer ended refurbishment work was being carried out at Tiers Temps which meant he had to change rooms for a while. There was constant noise and the little courtyard where he fed the birds was cluttered with workmen's planks and ladders. The new room had floral wallpaper which he did not like and the furniture was stained an ugly mahogany brown. His breathing difficulties were increasing and he was now given an oxygen tank which was kept in the corner of his room. Visitors thought him very frail and weak. He had finished the French translation of *Stirrings Still* and the English of 'Comment dire' and he felt that all writing of any sort was at an end. When Louis le Brocquy visited him he spoke of his recent visit to Ussy.

> Visibly moved he recounted how beautiful he had found it on that Summer day up there on the hill alone, the countryside all around and no one to be seen. 'I never realised how beautiful it was until I lost it. I will never go back again.' Of the apartment on the boulevard Saint-Jacques he declared that he would never return there either, of his writing – and we felt it was of his life too he murmured again – 'Nothing. Nothing left.'

One small task remained. He had agreed to copy out a poem on parchment for *The Great Book of Ireland*, a bound volume initiated by Poetry Ireland in which many Irish poets were represented by a manuscript page decorated by a painter. The poets and the painters were supplied with a loose vellum page which would eventually be stitched into the book. Sam's page was to be decorated by Louis le Brocquy and he discussed the choice of poem with Louis, reciting both 'The Vulture':

> dragging his hunger through the sky
> of my skull shell of sky and earth . . .

and 'Da Tagte Es':

> redeem the surrogate goodbyes
> the sheet astream in your hand
> who have no more for the land
> and the glass unmisted above your eyes

before deciding on the latter, which is a poem of both love and death written after the death of his father.

The Irish poet John Montague, once a habitué of the Falstaff, now one of the directors of Poetry Ireland, who was in Paris in November, brought the parchment sheet to Beckett at Tiers Temps. When he asked after Beckett's state of health the reply was, 'I'm done.' A moment later this was vehemently repeated with the addendum, 'But it takes such a long time.' He adverted to his father's death: 'I sat beside my father when he was dying. Fight, fight, fight, he kept saying. But I have no fight left.' Montague's account describes 'a gesture of resignation and disappointment' and continues:

Most of our conversation had been circumspect, cautious and courteous, but now, sensing an opening, I feel brave enough for a direct personal question, even without the ritual glass before me.

'And now it's nearly over, Sam, can I ask you, was there much of the journey you found worthwhile?'

The blue eyes briefly ignite.

'Precious little.' And in case I did not hear or comprehend, he repeats it again with redoubled force, 'Precious little,' and adds, 'For bad measure, I watched both my parents die.'

After a drink or two they turned their attentions to the transcribing of the poem on parchment. It proved difficult and Montague noticed that Beckett's hands were shaking as he struggled with curling parchment and inked pen. He made some mistakes, but fortunately spare sheets had been provided and at length the task was accomplished; whereupon, 'with a gesture of finality', he swept the lot, including the ink-bottle and the pen, into the wastepaper basket.

After this they settled to the whiskey, finishing what remained in the bottle. Then Beckett shuffled to a cupboard inside which several more were stacked and came back clutching a bottle of Bushmills. The conversation after this point was desultory, prodded on mostly by Montague's mentions of people such as Leventhal, but it included a reference to Suzanne – 'We were friends for fifty years' – and to a recent reading of Keats's 'Ode to a Nightingale' – 'It's very beautiful,' he declared. Montague was surprised, but he did not know that Beckett had learned it by heart almost seventy years before, under a tree in a field outside Enniskillen with his friend Geoffrey Thompson. A certain mellowness was induced by the Bushmills, so much so that, to his guest's surprise, Beckett suddenly broke quaveringly into a Church of Ireland hymn, the one that he had made Krapp's favourite:

> Now the day is over,
> Night is drawing nigh,
> Shadows of the evening
> Steal across the sky.

It was not the first time he had been known to sing a verse or two of this hymn. When he did, he sang it according to the air used in the Church of Ireland, not that in the English Hymnal.

In November 1989 the Berlin wall came down. It was the end of an era. Beckett had been watching a news programme on television in his room as he did every night. He emerged very agitated and exclaimed to Madame Jernand. '*Ça va trop vite*' – 'It's happening too quickly' – he kept repeating. It was almost as if he foresaw the problems that would follow the break-up of the Soviet Empire.

Another Irish poet who came to see him that November was Derek Mahon. Like Montague, he was in Paris with a group of Irish writers for a series of readings; and he too had been charged with an errand: to deliver a letter to Beckett from Professor Terence Brown of Trinity.

> There was no correspondence this time and no phone call; I simply sought out the nursing-home, rang the bell, and asked a high-spirited nurse to deliver the letter for me. To my surprise she told me to go through to the back and knock on his door; so I traversed a room full of startlingly convivial old folks watching television and found myself in a courtyard enclosed by a square of low, chalet-like quarters with names on the doors.

When Beckett opened his own door, he seemed to Mahon 'no longer the rather soigné figure of ten years before, but a sort of scarecrow with protruding ears and patches of white stubble where, old-man fashion, he hadn't shaved very attentively'. They sat together for about half an hour over a glass of Jameson. In spite of Beckett's frail appearance, Mahon remembers being impressed by his self-sufficiency.

> He seemed buoyant, if unsteady on his feet, and tremendously relaxed, as if to demonstrate the logic of his previous *paean à l'outrance*: the older you get the better it felt. It seemed to me he was unquestionably having *fun* as the corporeal envelope disintegrated and the end drew nigh. He positively twinkled. He read the letter I'd brought, told me to tell T. Brown he'd be happy to see him, and again, off I went, through the pleasant courtyard and the cackling TV room.

But however cheerful he may have seemed, he was undoubtedly failing and his breathing difficulties were increasing. One morning in December he got up and dressed himself as usual. At noon, when the other residents were lunching, he came into the dining area and walked up and down by the tables, gesticulating and talking rapidly. He was making no sense and when Madame Jernand was called it was clear to her that he was raving. She contacted his doctor and then the hospital. As he was carried out on the stretcher he called out that he was going away but that he would be back.

Chapter Thirty-Seven

Samuel Beckett was taken to hospital from the Tiers Temps nursing home early in December 1989. He recovered rational consciousness only fitfully and shortly after his arrival sank into a coma. On 19 December rumours of his illness began to circulate throughout the world; but Jérôme Lindon had now assumed control of the situation. Enquirers were informed that 'M. Beckett does not like publicity' and that no information would be made available. Those of his associates who, hearing that he was mortally ill, rushed to Paris were not allowed to see him. Only his doctors, Lindon and his closest relatives were by his bedside. Barbara Bray did not go to the hospital. To have done so, she has said, would in any case have been pointless. He died on Friday 22 December. No announcement was made and the silence which had descended around him continued until the day after Christmas.

On that day, known as St Stephen's Day in Ireland, his mortal remains were buried in Montparnasse cemetery in the same grave as Suzanne's. Lindon's office rang a small number of people, some of them that morning, and only they were present. There was no priest and no speeches were made. Afterwards a brief announcement said that Samuel Beckett had died of 'respiratory failure'.

'Whatever opinion we may be pleased to hold on the subject of death, we may be sure that it is meaningless and valueless. Death has not required us to keep a day free,' he had said in the little book on Proust, published fifty-five years before, at the start of his writing life.

It is doubtful if he believed in any sort of survival of consciousness, or disbelieved in it either, since belief – or disbelief – was not something he permitted himself. He thought that all the guides were poor ones and that it was better to live, and to admit to living, in complete uncertainty: better because more honest; better in his own case because it was only after he had decided to do this that he had written *Waiting for Godot* and his trilogy of great novels. Better perhaps also because convictions and the desire for them had led to a great deal of trouble and unhappiness in the world.

His characters live in confusion, and the possibility of an after-life is part of it. 'Yes, the confusion of my ideas on the subject of death was such that I sometimes wondered, believe me or not, if it wasn't a state of being even worse than life,' says Molloy. The characters in *How It Is*, in *Happy Days* and in *Play* all appear to exist in some sort of after-life. But we cannot deduce anything about his own beliefs from this. It may be only a 'structural device', a use of Bergsonian ideas about time as neither divisible nor measurable, a scaffolding for works of art. The last words of his last story, 'Oh all to end', express, as Christopher Ricks has pointed out, 'both a hope and a regret: they could mean 'if only the whole thing would be over', or 'how sad that it should stop'.'

How sad indeed. 'Yes, there is no good pretending. It is hard to leave everything,' Malone, dying, had said. Many are not fully at home in the world. Samuel Beckett was less at home than most. Yet he too had found some final partings difficult; and his work, like all great art, is in some sense a celebration of existence.

BIBLIOGRAPHY

Fortunately for the general reader for whom this book is intended, most of Beckett's work in English is currently in print and easily obtainable. The dramatic works, with the exception of *Eleutheria*, are contained in the *Complete Dramatic Works* (Faber, 1986). The majority of Beckett's shorter prose pieces are contained in *Collected Shorter Prose, 1945–1980* (John Calder, 1984). Those pieces not included in the *Collected Shorter Prose* can be found in *Nohow On* (John Calder, 1989) and *As the Story Was Told* (John Calder, 1990). The following are the principal details of first publication. Except where otherwise stated English translations are by Samuel Beckett.

Samuel Beckett's Works in English:

NARRATIVE PROSE

More Pricks Than Kicks. London: Chatto and Windus, 1934.

Murphy. London: G. Routledge, 1938.

Molloy. Paris: Les Editions de Minuit, 1951. English translation as *Molloy* by Samuel Beckett and Patrick Bowles. Paris: Olympia Press, 1955.

Malone Meurt. Paris: Les Editions de Minuit, 1951. English translation as *Malone Dies*. New York: Grove Press, 1956. London: John Calder, 1958.

L'Innommable. Paris: Les Editions de Minuit, 1953. English translation as *The Unnamable*. New York: Grove Press, 1958.

Watt. Paris: Olympia Press, 1953.

Nouvelles et textes pour rien. Paris: Les Editions de Minuit, 1955. Translated into English by Richard Seaver, Anthony Bonner and Samuel Beckett as *Stories and Texts For Nothing*. New York: Grove Press, 1967.

Comment C'est. Paris: Les Editions de Minuit, 1961. English translation as *How It Is*. New York: Grove Press, 1964.

Imagination Morte Imaginez. Paris: Les Editions de Minuit, 1965. English translation as *Imagination Dead Imagine*. London: Calder and Boyars, 1965.

From an Abandoned Work. Reprinted in *No's Knife. Collected Shorter Prose 1945–1966*. London: John Calder, 1967.

Sans. Paris: Les Editions de Minuit, 1969. English translation as *Lessness*. London: Calder and Boyars, 1970.

Mercier et Camier. Paris: Les Editions de Minuit, 1970. English translation as *Mercier and Camier*. New York: Grove Press, 1974.

Premier Amour. Paris: Les Editions de Minuit, 1970. English translation as *First Love*, included in *First Love and Other Stories*. New York: Grove Press, 1974.

Le Depeupleur. Paris: Les Editions de Minuit, 1970. English translation as *The Lost Ones*. New York: Grove Press, 1972. London: Calder and Boyars, 1972.

All Strange Away. New York: Gotham Book Mart, 1976.

Pour Finir Encore et Autres Foirades. Paris: Les Editions de Minuit, 1976. English translation as *Fizzles*. New York: Grove Press, 1976.

Company. New York: Grove Press, 1980. London: John Calder, 1980.

Mal Vu Mal Dit. Paris: Les Editions de Minuit, 1981. English translation as *Ill Seen Ill Said*. New York: Grove Press, 1981.

Worstward Ho. London: John Calder, 1983.

L'Image. Paris: Les Editions de Minuit, 1988. English translation as *The Image*, included in *As the Story Was Told*. London: John Calder, 1990. New York: Riverrun Press, 1990.

Stirrings Still. New York: Grove Press, 1988. London: John Calder, 1988.

Dream of Fair to Middling Women, edited by Eoin O'Brien and Edith Fournier. Dublin: Black Cat Press, 1992.

PRINCIPAL DRAMATIC WORKS

En Attendant Godot. Paris: Les Editions de Minuit, 1952. New York: Grove Press, 1954. First production, Theatre De Babylone, 5th January 1953. English translation as *Waiting For Godot*

All That Fall. New York: Grove Press, 1957. Produced BBC, 1957.

Fin De Partie. Paris: Les Editions de Minuit, 1957. Translated into English as *Endgame*. New York: Grove Press, 1958. London: Faber and Faber, 1958. First production of English translation, London, 1958.

Krapp's Last Tape. London: Faber and Faber, 1959. New York: Evergreen, 1960. First Production, London, 1958.

Embers. London: Faber and Faber, 1960. New York: Grove Press, 1960. First production, BBC, 1959.

Happy Days. New York: Grove Press, 1961. First production, New York, 1961.

Play. London: Faber and Faber, 1964. First production, New York, 1964; London, 1964.

Film. New York: Grove Press, 1969. First production, 1969.

Not I. London: Faber and Faber, 1973. First production, New York, 1972.

Footfalls. New York: Grove Press, 1976. First production, London, 1976.

Rockaby. London: Faber and Faber, 1982. First production, Buffalo, New York, 1981.

What Where. London: Faber and Faber, 1984. First production, New York, 1983.

Eleutheria. Paris: Les Editions de Minuit, 1996. English translation as *Eleutheria* by Joseph Brodsky. New York: Foxrock, 1995.

POETRY

Whoroscope. Paris: The Hours Press, 1930.
Echo's Bones and Other Precipitates. Paris: Europa Press, 1935.
Poems in English. London: John Calder, 1961.
Collected Poems 1930–1978 (including poems in French). London: John Calder, 1984.

CRITICISM

Proust. London: Chatto and Windus, 1931.
Disjecta. Miscellaneous Writings and a Dramatic Fragment, edited by Ruby Cohn. London: John Calder, 1983.

Abbreviations used:

S.B.	Samuel Beckett
T.MacG.	Thomas MacGreevy
M.M.	Mary Manning
G.R.	George Reavey
B.R.	Barney Rosset
A.C.	Anthony Cronin

CHAPTER ONE

p2 'Even before the foetus . . .': S.B. in interview with John Gruen, *Vogue*, February 1970.

p2 'retained a terrible memory . . .': Peggy Guggenheim, *Out of This Century: Confessions of an Art Addict* (Deutsch, London, 1980).

p3 'and then, in a bold flashback . . .': S.B., *All That Fall* (Faber & Faber, London, 1986).

p9 'The males and some of the females . . .': Vivian Mercier, *Beckett/Beckett* (Souvenir Press, London, 1977).

p11 'This is a charming Tudor style . . .': quoted by Vivian Mercier, op. cit.

p12 'a fragrance in which . . .': S.B., *Dream of Fair to Middling Women* (Black Cat Press, Dublin; Calder, London, 1992).

CHAPTER TWO

p14 'motionless . . . under the singing lamp . . .': S.B., *Dream*.

p14 'his big belly . . .': S.B., 'The Calmative'.

p14 'the adventures of one . . .': S.B., ibid.

p15 'for Jesus Christ's sake . . .': S.B., *Dream*.

p15 'The waist of his trousers . . .': S.B., *Company*.

p15 'You are alone in the garden . . .': ibid.

p16 'You stand at the tip . . .': ibid.

p17 'lacerated by . . .': S.B., *Watt*.

p17 'No – no – too high . . .': S.B., *Eleutheria*.

p17 'The fear of dreaming . . .': Lawrence Harvey, *Samuel Beckett, Poet and Critic* (Princeton University Press, 1970).

p18 'It is precisely . . .': S.B., *Malone Dies*.

p18 'For some reason you could never . . .': S.B., *Company*.

p18 'A small boy, stretching . . .': S.B., 'The End'.

p18 'It's a miracle . . .': S.B., *Malone Dies*.

p18 'This was largely due . . .': Beatrice Lady Glenavy, *Today We Will Only Gossip* (Constable, London, 1964).

p19 'a very moody person': Mary Manning, interview with A.C.

p19 'as long as he could kneel': S.B. to Tom Driver, 'Beckett By the Madeleine', Columbia University Forum, Summer 1961.

p19 'no religious feeling . . .': ibid.

p19 'her huge head hatted . . .': S.B., *How It Is*.

p20 'alone with her maker . . .': S.B., *All That Fall*.

p21 'My brother and mother . . .': S.B. to Driver, op. cit.

p21 'absurd prohibitions . . .': S.B. to Charles Juliet, quoted in Juliet, *Rencontres avec Samuel Beckett* (Éditions Fata Morgana, Montpelier, 1978).

p22 'You glowed easily . . .': S.B., *Company*.

p23 'I'll be her sweet pet . . .': S.B., *Texts for Nothing*.

p23 'she must have a strawberry nose . . .': S.B., letter to Thomas MacGreevy, 4.8.1932.

p23 'mother of thousands': S.B., *Dream*.

p23 'There wasn't much love . . .': Mary Manning, interview with A.C.

p24 'by one of these mind doctors . . .': S.B., *All That Fall*.

p24 'Behold Belacqua . . .': S.B., *Dream*.

p25 'and went singing away . . .': S.B., *Watt*.

p25 'horrible slum . . .' and following: S.B., *Dream*.

p25 'the clicking of mallet . . .': S.B., *Molloy*.

p26 'there were so many of those old crones . . .': S.B. to Alan Schneider, quoted in S. Gontarski, *The Intent of Undoing in Samuel Beckett's Dramatic Texts* (Indiana University Press, 1985).

p26 'On the way home . . .': S.B., *Company*.

p28 'These latter were springing . . .': S.B., *More Pricks Than Kicks*.

p28 'now one voice . . .': S.B., 'That Time'.

p28 'It is here that one . . .': S.B., *Mercier and Camier*.

p29 'walked himself . . .': S.B., quoted in Eoin O'Brien, *The Beckett Country* (Black Cat Press, Dublin, 1986).

p29 'It cuts across vast . . .': S.B., *Mercier and Camier*.

p30 'It was evening and . . .': S.B., 'The End'.

p30 'One day I told him . . .': S.B., 'From an Abandoned Work'.

p30 'That was always the way . . .': S.B., *Embers*.

p31 'reduced almost to incontinence . . .': S.B., letter to T.MacG., 18.10.1935.

p32 'That's where that man . . .' and following: M.M., interview with John Reilly for *Waiting for Beckett* video, Global Village, New York.

p35 'outrageous riot': *Irish Times Handbook of the Rebellion* (Dublin, 1916).

p38 'The crocuses and the larch . . .': S.B., *Watt*.

CHAPTER THREE

p40 'natural, unconventional': Douglas Seal, tape in RTE archive.

p42 'some really stylish strokes': *Portora*, vol. XV, no. 3, 1921.

p42 'All this is involved...': Brian Coffey, tape in RTE archive.

p43 'The one tradition...': V. Mercier, 'Portora', *The Bell*, October 1949.

p45 'the affection...': *Portora*, vol. XV, no. 3, 1925.

p45 'Tetley has gone...': anon., quoted in Eoin O'Brien, *The Beckett Country* (Black Cat Press, Dublin, 1986).

p46 'But I stilled my cringing...': S.B., 'For Future Reference', *transition*, June 1930.

p47 'violent and eloquent speeches': *Portora*, Michaelmas 1921, quoted in O'Brien, op. cit.

p48 'When a bit of sunshine...': Tom Cox, tape in RTE archive.

p48 'friending': A. E. Housman, *A Shropshire Lad*, LXII.

p49 'trained like a racehorse': Vivian Mercier, *Beckett/Beckett*.

p49 'a fine cheery duet': ibid.

p50 'no invitation...': Mary Rogers, the Beckett file, Portora School, 1969.

CHAPTER FOUR

p54 'regarded as one of the...': Ulick O'Connor, *Sport is My Lifeline* (Pelham Books, London, 1984).

p58 'is a master of knockabout...' and following: S.B. reviewing Sean O'Casey, *Windfalls* in *Disjecta*, ed. Ruby Cohn (Calder, London, 1983).

p58 'I wouldn't suggest that...': S.B. in letter to Cyril Cusack, quoted by A. J. Leventhal in *Beckett, An Exhibition* (Turret Books, Reading, 1971).

p60 'approach to language learning': Roger Little, 'Beckett's Mentor, Rudmose-Brown, Sketch for a Portrait', *Irish University Review*, Spring 1984.

p63 'neither Fascist nor...' and following: extracts from an unpublished memoir by T. B. Rudmose Brown, *Dublin Magazine*, January–March 1956.

p64 'I have, I am fully...': Rudmose-Brown, op. cit.

p65 'But I angered...': James Joyce, included in *Poems and Shorter Writings*, ed. Richard Ellmann, A. Walton Litz and John Whittier Ferguson (Faber & Faber, London, 1991).

p65 'There subsisted as much...' and following: S.B., 'Dante and the Lobster', *More Pricks Than Kicks*.

CHAPTER FIVE

p67 'that rare, almost unique...': V. Mercier, op. cit.

p69 'crescendo of...': S.B., *Dream*.

p69 'a wake of objurgation...': S.B., 'Love and Lethe', *More Pricks Than Kicks*.

p70 'She had only to unleash...': S.B., 'A Wet Night'.

p70 'Seeing a girl home...': Denis Johnston, *Orders and Decorations* (Lilliput, Dublin, 1992).

p75 'for all the poets...': Beatrice Lady Glenavy, op. cit.

p75 'the loveliest little...': S.B., *Dream*.

p76 'very vanguardful...': ibid.

CHAPTER SIX

p78 'One by one, I have beheld . . .': Henry de Montherlant, *Nouvelle Revue française*, June 1927.

p79 'But then at once comes . . .': Marcel Arland, *Nouvelle Revue française*, June 1924.

p80 'There are no reasons . . .': Jacques Rigaut, quoted by Jacques Émile Blanche in *Les Nouvelles littéraires*, December 1929.

p80 'To live, that is to say . . .': Jacques Rigaut, *La Révolution surréaliste*, December 1929.

p82 'they aimed their . . .': Germaine Bree, *Twentieth Century French Literature* (University of Chicago Press, 1983).

p86 'When, for a time . . .': T.MacG., unpublished memoir, quoted in Susan Schreibman, Introduction to *Collected Poems of Thomas MacGreevy* (Anna Livia, Dublin, 1991).

p87 'thousands of celibates . . .': Anthony Cronin, *No Laughing Matter* (Grafton, London, 1989).

p90 'was not a very good way . . .' and following: Dougald McMillan, *transition* 1927–38; *The History of a Literary Era* (Calder, London, 1975).

p92 'had a great deal of esteem . . .' and following: Adrienne Monnier, 'Gazettes et Écrits Divers', *Les Lettres nouvelles*, March 1953.

p93 'Your belle amie . . .': Arthur Power, quoted in Brenda Maddox, *Nora* (Houghton Mifflin, Boston, 1988).

CHAPTER SEVEN

p95 'clearly adapted by . . .' and following: S.B., 'Dante . . . Bruno. Vico . . . Joyce'; *Our Exagmination Round His Factification for Incamination of Work in Progress* (Paris, 1929).

p97 'In the silence of his room . . .' and following: S.B., 'Assumption', *transition* 16/17, June 1929.

p100 'directed towards . . .': Richard Ellmann, *James Joyce* (Faber & Faber, London, 1959).

p101 'Day after day . . .': S.B., *Ohio Impromptu*.

p101 'I can justify . . .': James Joyce, quoted by S.B. in conversation with A.C.

p101 'Why does nobody . . .': ibid.

p101 'certain aspects of . . .': Bree, op. cit.

p101 'Looks like rain . . .': Ellmann, op. cit.

p102 'deeply under the influence . . .' and following: James Joyce, letter to Valery Larbaud, 30.7.1929.

p102 'had a weakness for . . .': T. S. Eliot in conversation with A.C.

CHAPTER EIGHT

p103 'At last I find . . .': S.B., *Dream*.

p105 'Scalded the eyes . . .': S.B., *Krapp's Last Tape*.

p105 'Because her body . . .': S.B., *Dream*.

p106 '. . . it was not when he . . .': ibid.

p107 'So she had been . . .': ibid.

p107 'I said again . . .': S.B., *Krapp's Last Tape*.

p109 'I don't love anyone . . .': Ellmann, op. cit.

p111 'offensively fastidious...' and following: S.B., letter to T.MacG., undated (summer 1929).

p112 'jewel thief': Brian Coffey in conversation with A.C.

p112 'submitted themselves...': S.B., *Bookman*, August 1934.

p112 'great trouble with...': S.B., letter to Patricia Hutchins, 18.12.1953.

p112 'a sorrow much deeper...': T. S. Eliot to Geoffrey Faber, quoted in Philip Herring, *Djuna: The Life and Work of Djuna Barnes* (Viking, New York, 1995).

p112 'appeal mainly to readers...': T. S. Eliot, Preface to Djuna Barnes, *Nightwood* (Faber & Faber, London, 1936).

p114 'the work of a very young...': Harvey, op. cit.

p114 'The conflict in the...': ibid.

p114 'the most interesting...' and following: *European Caravan*, ed. S. Putnam, M. C. Darnton, G. Reavey, J. Bronowski (Brewer Warren & Putnam, New York, 1931).

p116 'What remarkable lines...': Nancy Cunard, *Those Were the Hours* (Southern Illinois Press, Carbondale, Illinois, 1969).

p117 'caught up with the international...': V. Mercier, op. cit.

p118 'a lovely young Irishman': Nancy Cunard, quoted in Anne Chisholm Davie, *Nancy Cunard* (Sidgwick and Jackson, London, 1979).

p118 'Christ in Heaven...': Sir Harold Acton to Anne Chisholm Davie, quoted in Deirdre Bair, *Beckett* (Jonathan Cape, London, 1978).

p118 'What terrible instinct...': S.B., letter to T.MacG., undated (probably May 1930).

p119 'The acceptance of this thing...': ibid.

p119 'disgusted by the chasm...': S.B., letter to T.MacG., 7.7.1930.

p120 'looking forward to...': ditto, undated (1930).

p119 'I can't do the fucking thing': ditto, 25.8.1930.

p120 'as little variety...': ibid.

p120 'I have a copy box...': George William Russell, quoted by S.B., letter to T.MacG., undated (1930).

p121 'Heidegger's most important...': Geert Lernout, *The French Joyce* (University of Michigan Press, Ann Arbor, 1992).

p121 'intellectual justification...' and following: S.B., letter to T.MacG., undated (1930).

p121 '... has always been less to...': Bertrand Russell, *A History of Western Philosophy* (Unwin, London, 1988).

p121 'that of explaining evil...': except where indicated, the quotations from Schopenhauer are from *Essays and Aphorisms*, trans. R. J. Hollingdale (Penguin, London, 1970), and *The World As Will and Idea*, ed. David Berman (Everyman, London, 1993).

p121 'The major sin is...': S.B., quoted in Driver, op. cit.

p122 'There is no salvation...': Iris Murdoch, *Metaphysics As a Guide to Morals* (Chatto and Windus, London, 1992).

p122 'Van Gogh's shoes...' and following: ibid.

CHAPTER NINE

p125 'silly little wife...': S.B. to T.MacG., undated (1930).

p125 'this life is terrible...': S.B., letter to T.MacG., 5.10.1930.

p125 'an approximation of . . .': ibid.

p126 'with an excess of Picabia . . .': A. J. Leventhal, editorial in the *Klaxon*, Winter 1923–4.

p129 'be co–ordinated with . . .': S.B., letter to T.MacG., 5.10.1930.

p129 'Bel Bel . . .' and following: S.B., *More Pricks* (there is a slightly different version in *Dream*).

p133 'I admit . . . Beatrice . . .': S.B., *Dream*.

p135 'a marvel surpassing . . .': James Lord, *Giacometti* (Faber & Faber, London, 1986).

p136 'want to be pawed . . .': S.B., *Dream*.

p137 'It's all over.': Georges Belmont (Georges Pelorson), in interview with A.C.

CHAPTER TEN

p139 'One left filled . . .': Mervyn Wall, interviewed by Michael Smith, *Lace Curtain*, No. 4., 1971.

p140 'No one creates . . .': Jack B. Yeats to Joseph Hone, quoted in Hilary Pyle, Catalogue notes to 'Images of Yeats' exhibition, Dublin and Monte Carlo, 1990.

p141 'in the sense that . . .': T.MacG., 'Jack B. Yeats: An Appreciation and an Interpretation', Victor Waddington Gallery, Dublin, 1945.

p141 'art criticism of . . .': S.B., 'MacGreevy on Yeats', *Irish Times*, 4.8.1945.

p141 'very beautiful and . . .': S.B., letter to T.MacG., 25.1.1931.

p141 'two entirely delightful . . .': ditto, 3.2.1931.

p142 'The Island . . .': S.B., review of Jack B. Yeats's *The Amaranthers*, *Dublin Magazine*, July–September, 1936.

p142 'So, unlike Miranda . . .' and following: S.B., *Proust* (Calder, London, 1965).

p142 'at its best . . .': S.B., letter to T.MacG., 11.3.1931.

p143 'Although the essay . . .': A. Alvarez, *Beckett* (Fontana, London, 1973).

p143 'There is no escape . . .' and following: S.B., *Proust*.

p143 'Of all human plants . . .' and following: ibid.

p145 'everything is lost . . .' and following: Marcel Proust, *The Captive*, trans. C. K. Scott Moncrieff (Chatto, London, 1929).

p147 'All poetry . . .': S.B., review of T.MacG., *Poems*, *Dublin Magazine*, July–September 1934.

p147 'pale grey sandpaper . . .' and following: S.B., letter to T.MacG., 11.3.1931.

p148 'very pessimistic . . .': Georges Belmont, symposium in *TCD*, October 1991.

p152 'I never think . . .': S.B., letter to T.MacG., undated (probably late June 1930).

p153 'That girl will have . . .': ditto, 5.10.1930.

p153 'poor Lucia . . .': ditto, 5.1.1931.

p153 'another very calm . . .' and following: ibid.

p154 'Went out to Camden . . .': S.B., letter to T.MacG., 14.11.1931.

p155 'comedy . . .': ibid.

p156 'cadging drinks . . .': S.B., letter to T.MacG., 9.10.1931.

p156 'the best of company . . .': Arland Ussher in conversation with A.C.

p156 'the life went . . .': Sean O'Sullivan in conversation with A.C.

p157 'poets and peasants . . .': S.B., *More Pricks*.

p157 'A lowly house . . .': ibid.

p159 'before morning...' and following: S.B., 'Alba', *Collected Poems* (Calder, London, 1984).

p159 'Smoke less...' and following: S.B., *Dream*.

p160 'can settle down...': Lawrence Harvey, op. cit.

p161 'But it would not do...': ibid.

p162 'mooching about...': S.B., letter to T.MacG., 9.10.1931.

p162 'appear au courant...': ibid.

p162 'this terrible Dublin...' and following: S.B., letter to T.MacG., 11.4.1931.

p163 'somewhere into Germany...': ibid.

CHAPTER ELEVEN

p167 'a classic, pious...' – Maddox, op. cit.

p170 'immature and unworthy': Harvey, op. cit.

p170 'a slob of...' and following: S.B., *Dream*.

p170 'valuable for...': Harvey, op. cit.

p171 'assume that he...': Raymond Federman and John Fletcher, *Samuel Beckett. His Works and His Critics* (University of California Press, Berkeley, 1970).

p172 'new expressionist...': McMillan, op. cit.

p172 'the hegemony of...' and following: 'Poetry Is Vertical', *transition* 21, 1932.

p176 'that foul fucker...': S.B., letter to T.MacG., 4.8.1932.

p176 'Or perhaps...': ditto, 13.8.1932.

p176 'a proper pudding': S.B., letter to T.MacG., 18.8.1932.

p177 'thought he might possibly...' and following: ibid.

p177 'perfecting my...': S.B., letter to T.MacG., 4.8.1932.

p177 'hideous': ibid.

p177 'très émouvant': ibid.

p178 'mined-granite' and following: ibid.

p178 'somehow ludicrous': S.B., letter to T.MacG., 4.8.1932.

p178 'too tired and...' and following: ditto, 18.8.1932.

p179 'and now look...' and following: ditto, 4.8.1932.

p179 'private school or...': ditto, 18.8.1932.

p179 'no use insisting...': ibid.

p179 'I am too unbelievably...': ibid.

p179 'concerts, cinemas...': S.B., *Dream*.

CHAPTER TWELVE

p180 'anaesthetic of...' and following: S.B., letter to T.MacG., 30.6.1932.

p181 'in nervous comfort...': ibid.

p182 'the guts to be...': ibid.

p182 'I never see nor...': S.B., letter to T.MacG., 8.10.1932.

p183 'a grand little magic...' and following: ibid.

p184 'immeasurably and...': S.B., letter to T.MacG., 18.10.1932.

p184 'something unapproachably...': ditto, undated.

p184 'Pretending to like...' and following: ditto, undated (1932).

p184 'The fact of myself...': ditto, 21.11.1932.

p184 'My parents do...': ibid.

p185 'they had clean forgotten...': S.B., 'Yellow', *More Pricks*.

p186 'hardly a coat': S.B., letter to T.MacG, 5.1.33.

p186 'subtly unbearable' and following: S.B., letter to T.MacG., 27.1.1933.

p187 'Yes, that's what I like . . .': S.B., *Malone Dies.*

p188 'The thought of taking root . . .': S.B., letter to T.MacG., 23 (probably March) 1933.

p189 'sans conviction' and following: ditto, 13.5.1933.

p189 'all jigsaw': ditto, 22.6.1933.

p189 'quite pleasant . . .': ditto, 13.5.1933.

p190 'it was an ill cyst . . .': ibid.

p191 'I can't write . . .': S.B., letter to T.MacG., 12.7.1933.

CHAPTER THIRTEEN

p193 'antiquarians' and following: S.B. (under the pseudonym Andrew Belis), *Bookman*, August 1934.

p195 'I'm afraid I didn't . . .' and following: S.B., letter to T.MacG., 9.10.1933.

p195 'rather detached . . .': Francis Stuart in conversation with A.C.

p195 'on the principle of . . .': S.B., letter to T.MacG., 25.7.1933.

p197 'It's a frail life . . .': ditto, 3.11.1933.

p198 'I don't know a damn . . .': ditto, 6.12.1933.

p200 'the enormity of . . .': James S. Grotstein, *Do I Dare Disturb the Universe? A Memorial to W. R. Bion* (Caesura Press, Beverley Hills, 1981).

p200 'He has lost . . .': Wilfred R. Bion, *The Long Weekend* (Abingdon, 1982), quoted in Bennett Simon, 'The Imaginary Twins: The Case of Beckett and Bion', *International Review of Psychoanalysis*, 1988.

p201 'invested in . . .': Bennett Simon, article cited.

p201 'screaming' and following: Didier Anzieu, 'Beckett and Bion', *International Review of Psychoanalysis*, 1989.

p201 'shocking family . . .': Wilfred R. Bion, 'The Case of the Imaginary Twins', *Second Thoughts* (New York, 1974).

p202 'narcissistic . . .': Anzieu. op. cit.

p202 'intervention and . . .': S.B., 'Proustian Pieces', *Spectator*, 23.6.1934.

p203 'He has the fidgets . . .': S.B., review of J. B. Leishman's translation of Rilke's poems, *Criterion*, July 1934.

p203 'I might be made . . .': S.B., quoted in Harvey, op. cit.

p203 'all right' and following: James Joyce, letter to Stuart Gilbert, 22.7.1932.

p204 'dry, harsh': Gerald Gould, *Observer*, 10.6.1934.

p204 'capable of . . .': A. Calder-Marshall, *Spectator*, 1.6.1934.

p204 'a book very . . .': Edwin Muir, *The Listener*, 4.7.1934.

p204 'a dangerous model . . .': *The Times Literary Supplement*, 25.7.1934.

p206 'Dublin costumes . . .': S.B., letter to T.MacG., 8.8.1934.

p206 'essence of . . .' and following: S.B., 'Censorship in the Saorstat', published in *Disjecta*.

p208 'every day I hear . . .': S.B., letter to T.MacG., 8.1.1935.

p208 'one moment in . . .': S.B., *Murphy.*

p208 'It was so still . . .': S.B., letter to T.MacG., 1.1.1935.

p209 'I wish . . .': ibid.

p209 'Dublin was . . .': ibid.

p210 'My daughter is not...': James Joyce, quoted in Ellmann, op. cit.

p210 'his own anima...' : C. G. Jung, letter to Patricia Hutchins, quoted in P. Hutchins, *James Joyce's World* (Thames and Hudson, London, 1957).

p210 'Father, if ever...' and following: Lucia Joyce, letter to James Joyce, quoted in Ellmann, op. cit.

p211 'she was a very...': C. G. Jung, letter to Giorgio and Helen Joyce, 5.2.1935, quoted in Maddox, op. cit.

p211 'to think that...': Ellmann, op. cit.

p212 'consciously and...' and following: S.B., letter to T.MacG., 10.2.1935.

CHAPTER FOURTEEN

p215 'Please God...': S.B., letter to T.MacG., 26.4.1935.

p217 'shabby depravity': ditto, 15.5.1935.

p217 'a riot of priests...': ibid.

p217 'Up to the...' and following: S.B., letter to T.Mac.G., 25.7.1935.

p217 'as little to write...': ibid.

p218 'frightfully deeply...': Geoffrey Thompson, tape in RTE archive.

p219 'craved': S.B., *Murphy*.

p220 'stifled giggles...': S.B., 'First Love'.

p220 'cuttle fish's...': S.B., letter to T.MacG., 8.9.1935.

p221 'Jung had rather...': Geoffrey Thompson, RTE tape.

p221 'one of these new...' and following: S.B., *All That Fall*.

p223 'with pleasure and...': W. R. Bion, letter to S.B., quoted in S.B., letter to T.MacG., 10.1.1936.

p223 'keep the death...': S.B., letter to T.MacG., undated (early 1936).

p224 'there was a happy land...': S.B., *Whoroscope*.

p225 'are apt to leave a reader...': Hugh Kenner, *A Reader's Guide to Samuel Beckett* (Thames and Hudson, London, 1988).

p225 'The reader sees...': ibid.

p225 'revolted silence': S.B., letter to T.MacG., 21.12.1935.

p226 'an unconscious tribute...': T. S. Eliot, lecture in University College Dublin, January 1936, quoted by S.B. in letter to T.MacG., 23.1.1936 (incorrectly dated 1935 by S.B.).

p227 'always morning...': S.B., letter to T.MacG., 7.5.1936.

p229 'abhorred gates': S.B., letter to T.MacG., 23.1.1936 (incorrectly dated 1935 by S.B.).

p229 'in the water closet': S.B., 'The End'.

p229 'Nothing is to be done...' and following: Arnold Geulincx, 'Ethica', *Sämtliche Schriftlichen*, ed. H. J. Vleeschaueur (Stuttgart, 1968), trans. here by Ronald Begley, quoted in Dougald McMillan and Martha Fehsenfeld, *Beckett in the Theatre* (Calder, London, 1988).

p231 'I am not a philosopher...': S.B., quoted in Driver, op. cit.

p232 'I take no sides...': S.B. to Alan Schneider, quoted in Alan Schneider, 'Waiting for Beckett', *Beckett at Sixty* (Calder, London, 1967).

p232 'The stoics aspired...': S.B., *Whoroscope Notebook*.

CHAPTER FIFTEEN

p233 'from bad to . . .': S.B., letter to T.MacG., 25.3.1936.

p233 'held up over . . .': ditto, 9.4.1936.

p233 'again on Croghan . . .': S.B., *Krapp's Last Tape*.

p234 'Anything would be better . . .': James Joyce, quoted by Arthur Power in conversation with A.C.

p235 'Why were you not . . .': version copied out by S.B. in MacGreevy Papers TCD. When the poem was published in the *Dublin Magazine* (October–December 1934) he made certain changes, numbering the sections and adding the line 'Unless they love you' as a third section and last line. He changed the poem slightly again for the *Collected Poems*.

p237 'That is worse . . .': S.B., letter to M.M., 14.11.1936.

p237 'I do not feel . . .': S.B., letter to T.MacG., 26.7.1936.

p238 'drunken vainglorious': W. B. Yeats, '1916', *Collected Poems* (Macmillan, London, 1954).

p239 'you needed a degree . . .': J. C. Mays, Preface to Brian Coffey, *Poems and Versions* (Dublin, 1991).

p239 'its saturation in the conviction . . .': S.B., letter to T.MacG., 9.9.1936.

p240 'I think we all feel . . .': ibid.

p241 'strangling his . . .' and following: S.B., letter to T.MacG., 19.9.1936.

p241 'She sschwarms . . .': ditto, 18.1.1937.

p243 'Well, the Lüneburg . . .': S.B., 'The Expelled'.

p243 'remoted, unfriended . . .': Oliver Goldsmith, *The Traveller*.

p243 'voice crackling . . .': S.B., letter to T.MacG., 18.1.1937.

p244 'indescribably lovely . . .' and following: ibid.

p244 'with a more definite . . .': S.B letter to T.MacG., 16.2.1937.

p245 'journey *from* . . .': S.B., letter to M.M., 13.12.1936.

p245 'both fully . . .': S.B., letter to T.MacG., 16.1.1937.

p245 'I shall never . . .': ibid.

p245 'Truly an architecture . . .': ibid.

p245 'like all the others . . .': S.B., letter to T.MacG., 16.2.1937.

p246 'knows them all . . .': ibid.

p246 'blue with blood': S.B., letter to T.MacG., 16.2.1937.

p246 'all the usual . . .': S.B. letter to T.MacG., 14.5.1937.

p246 'the usual bilge . . .': S.B., letter to M.M., 22.5.1937.

p248 'extreme aversion . . .' and following: S.B., letter to G.R., 13.11.1936.

p248 'the comic exaggeration . . .': ibid.

p248 'go further . . .': S.B., letter to G.R., 20.12.1936.

p249 'mentally as usual . . .': S.B., letter to T.MacG., 25.3.1937.

CHAPTER SIXTEEN

p251 'I feel now that . . .': S.B., letter to T.MacG., 20.4.1937.

p252 'the brute . . .': ditto, 4.8.1937.

p254 'art has nothing to do . . .': S.B., review of Denis Devlin's *Intercessions*, *transition*, April–May 1938.

p255 'need to suffer' and following: S.B., letter to T.MacG., 11.8.1937.

p255 'absurdly in love': ditto 20.4.1937.

p255 'the fake rage...': ibid.

p256 'being so old...': S.B., letter to G.R., 27.7.1937.

p256 'weeping on...': M.M., in an interview with A.C.

p256 'or any nonsense...' and following: S.B., letter to T.MacG., 4.8.1937.

p257 'Mrs D. He is late...' and following: S.B., *Human Wishes*, published in *Disjecta*.

p257 'And from the poop...': S.B., *Molloy*.

p257 'the impotent...': S.B., letter to M.M., 11.7.1937.

p257 'the time I spent...': S.B., letter to M.M., 2.1.1959.

p259 'An old usurer...' and following: Oliver St John Gogarty, *As I Was Going Down Sackville Street* (Hutchinson, London, 1962).

p260 'I am quite convinced...': S.B., letter to T.MacG., 7.7.1937.

p260 'I feel riddled...': ditto, 23.8.1937.

p261 'cordial fucking...': ditto, 23.5.1937.

p262 'deaf and dumb...': ditto, 25.8.1937.

p262 'a useless kip...': ditto, 28.9.1937.

p263 'had enough of mothers': ibid.

p263 'peace in the heart...': ditto, 6.10.1937.

p265 'unsuccessful abortion...': S.B., quoted in Enoch Brater, *Why Beckett* (Thames and Hudson, London, 1989).

p265 'Oh no...' and following: S.B. to Martin Esslin, quoted in report of Beckett Symposium, University of Reading, May 1971; *New Theatre Magazine*, Vol. 11, No. 3.

p266 'You know the kind...': S.B. to Israel Shenker, quoted in Shenker, 'Moody Man of letters', *New York Times*, 5.5.1956.

p266 'I just slipped away...': S.B., quoted in Brater, op. cit.

p268 'I purchased a copy...': affidavit sworn out by S.B. 13.9.1937, quoted in Bair, op. cit.

p269 'only the *Pickwick Papers*...': Charles Graves, quoted in O'Connor, op. cit.

p270 'vilified the...': excerpts from trial transcripts from Ulick O'Connor, *Oliver St John Gogarty* (Mandarin, London, 1990); and the *Irish Times*, 23–27.11.1937.

p275 'had been a thorough...': S.B., letter to T.MacG., May 1938.

CHAPTER SEVENTEEN

p277 'tied to the cart-tail...': S.B., review of Denis Devlin's *Poems*, *transition* 27.

p277 'James Joyce is completing...': Eugene Jolas, editorial, *transition* 27.

p278 'I am so hard up...': S.B., letter to M.M., December 1937.

p278 'no jubilation...': S.B., letter to T.MacG., 10.12.1937.

p279 'No more slop-emptying...': ibid.

p279 'it was a friendship...' and following: Lord, op. cit.

p281 'tall lanky Irishman...' and following: Peggy Guggenheim, op. cit.

p284 'like the stock exchange...': James Joyce, letter to Helen and Giorgio Joyce, 12.1.1938.

p285 'his glasses sparkling...' and following: Nino Frank, 'L'Ombre qui avait perdu son Homme', *Mémoires brisées* (Calman–Lévy, Paris, 1967), trans. quoted in Willard Potts, *Portraits of the Artist in Exile* (Wolfhound, Dublin, 1979).

p287 'gusts of affection': S.B., letter to T.MacG., 12.1.1938.

p288 'they come . . .' and following: Guggenheim, op. cit.

p290 'exchanged amiabilites': S.B., letter to T.MacG., 31.2.1938.

p291 'His mail . . .': S.B., letter to Arland Ussher, 5.4.1938.

p291 'the obscenity . . .' and following: S.B., letter to T.MacG., 21.2.1938.

p292 'go through . . .' and following: Guggenheim, op. cit.

p293 'a sort of . . .': ibid.

p295 'Murphy swept . . .' and following: Kate O'Brien, review of *Murphy*, *Spectator*, 25.3.1938.

p296 'energy, hilarity . . .': Dylan Thomas, *New English Weekly*, 17.3.1938.

p296 'One might explain . . .': *The Times Literary Supplement*, 12.3.1938.

p297 'to be yearning . . .': Guggenheim, op. cit.

p297 'Every step . . .': ibid.

p298 'This was all . . .': ibid.

p299 'diabolical . . .': ibid.

p299 'waited for me . . .': ibid.

p299 'he managed to get off . . .': ibid.

p300 'unaccountably avaricious': S.B., letter to T.MacG., 26.5.1938.

p301 'all of the old songs . . .': S.B., letter to T.MacG., 26.5.1938.

p302 'apparently inspired . . .': Harvey, op. cit.

p303 'grown tedious . . .': S.B., letter to T.MacG., 5.8.1938.

p303 'terribly in love . . .': Guggenheim, op. cit.

p304 'with my own handful . . .': S.B., letter to G.R. 27.9.1938.

p304 'getting a whole people . . .': James Joyce, quoted by S.B. in Ellmann, op. cit.

p304 'I done me best . . .': James Joyce, *Finnegans Wake* (Faber & Faber, London, 1938).

p306 'from the window . . .' and following: S.B., letter to A. Ussher, 20.12.1938.

CHAPTER NINETEEN

p306 'if the second half . . .': S.B., letter to T.MacG., 18.4.1939.

p306 'If there is a war . . .': ibid.

p308 'extraordinarily good': S.B., letter to T.MacG., 26.5.1938.

p308 'ignominious': S.B. to T.MacG., 26.5.1938

p308 'I know on my side . . .': ditto, April 1939.

p308 'I did not feel . . .': ditto, 18.4.1939.

p310 'I preferred France . . .': S.B., quoted in Shenker, op. cit.

p312 'My daughter-in-law . . .': James Joyce, letter to Jacques Mercanton, 9.1.1940.

p312 'What is the use . . .': James Joyce, Ellmann, op. cit.

p313 'We're going downhill . . .': James Joyce, quoted by S.B. in conversation with A.C.

p320 'insofar as reconciliation . . .': Ellmann, op. cit.

p321 '*C'est le bouquet* . . .': James Joyce, quoted in Ellmann, op. cit.

p322 'It seemed to me . . .': Lucie Léon, *James Joyce and Paul Léon, The Story of a Friendship* (New York, 1950).

p325 'The choice between . . .': Bree, op. cit.

p325 'I was so outraged . . .': S.B. to John Kobler, in unpublished article written for the *Saturday Evening Post*. Quoted in Bair op.cit.

p325 'I was fighting against . . .': S.B. to Alan Simpson, quoted in Simpson, *Beckett and Behan and a Theatre in Dublin* (Routledge and Kegan Paul, London, 1962).

p327 'The campaign of . . .': Alfred Cobban, *A History of Modern France*, Vol. 3 (Penguin, London, 1970).

CHAPTER TWENTY-ONE

p334 'not the most courageous' and following: Aimé Bonhomme, interview with John Reilly for *Waiting for Beckett* video, Global Village, New York.

p334 'anyone could get it' and following: ibid.

p336 'fragility of . . .' and following: S.B., *Watt*.

p338 'ALL WELL . . .': S.B., telegram, April 1943, Department of External Affairs archive, National Archives, Dublin.

p340 'BROTHER ANXIOUS . . .': telegram from Frank Beckett, loc. cit.

p341 'a gratuitous allowance . . .': form E1A, completed by Frank Beckett, loc.cit., 25.11.1942.

p341 'It was boy scout stuff': S.B., quoted by Sean O'Sullivan and Niall Montgomery in conversation with A.C.

p343 'My friends eat sawdust . . .': S.B. to Freda Young, quoted in Bair, op. cit.

CHAPTER TWENTY-TWO

p348 'the city was like . . .': Raymond Lelièvre, *Terre normande*, quoted by Eoin O'Brien, op. cit.

p349 'a frightful job': S.B., quoted in Harvey, op. cit.

p350 'Some packets weighed . . .': Alan Thompson, report to Irish Red Cross Society, quoted in Eoin O'Brien, op. cit.

p350 'met by Colonel . . .': letter from James Gaffney to his sister Maureen, 14.9.1945, quoted in Eoin O'Brien, op. cit.

p350 'It was novel . . .': ibid.

p350 'in lovely weather . . .': letter from James Gaffney to his sister Nora, quoted in Eoin O'Brien, op. cit.

p351 'who would prefer . . .': S.B., 'The Capital of the Ruins', talk on Radio Éireann, 10.6.1946; published in *As None Other Dare Fail*, ed. John Calder (Calder, London, 1986).

p353 'la vie des saints . . .': S.B., 'Mort de A. D.', *Collected Poems* (Calder, London, 1984).

p355 '"Provisional is not . . ."': S.B., 'The Capital of the Ruins', loc. cit.

p355 'Vire will wind . . .': S.B., 'Saint-Lô', *Irish Times*, 24.6.1945. A slightly revised version is in *Collected Poems*.

CHAPTER TWENTY-THREE

p357 'it seems absurd . . .': S.B., 'Peintres de l'empêchement', *Derrière le Miroir*, Galérie Maeght, Paris, June 1948, trans. in *Disjecta*.

p357 'are at bottom . . .': S.B., 'La Peinture des van Veldes ou le Monde et le pantalon', commissioned by *Cahiers d'art*, published in *Disjecta*.

p358 'immense mental vitality...': fragment of letter to A. P. Watt from unknown publisher, MacGreevy Papers TCD.

p360 'Perhaps only the French...': S.B., *Dream*.

p360 'it was easier...': S.B. to Nicholas Gessner, quoted in Enoch Brater, op. cit.

p360 'the right weakening...': S.B. to Herbert Brow, quoted in Brater, op. cit.

p360 'you couldn't help...': S.B. to Richard Coe, quoted in Brater, op. cit.

p360 'for him, an Irishman...': Harvey, op. cit.

p361 'And more and more...': S.B., letter to Axel Kaun, 9.7.1937, published in *Disjecta*.

p361 'The journey of...' and following: S.B., *Mercier and Camier*.

p364 'siege in the room': S.B. to various people at various times.

p365 'a literature of...': Jean-Paul Sartre, editorial in *Les Temps modernes*, Vol. 1, No. 1., 1945.

p367 'At that point...': S.B., *Eleutheria*, trans. Michael Brodsky (Foxrock, N.Y., 1995).

p367 'my own translation...': S.B., letter to G.R., 14.5.1947.

p370 'an unsatisfactory book...': ibid.

p370 'unless you feel...': S.B., letter to G.R., 26.6.1947.

p371 'perhaps, to encourage him...': ditto. 16.8.1947.

CHAPTER TWENTY-FOUR

p373 'the day I became aware...': S.B., quoted in Ludovic Janvier, *Beckett par lui-même* (Seuil, Paris, 1969).

p373 'the still, rich...': Herman Melville, quoted in Charles Olson, *Herman Melville* (New York, 1955).

p374 'What tedium...': S.B., *Malone Dies*.

p375 'a labyrinthine torment...': S.B., *The Unnamable*.

p375 'It is difficult...': ibid.

p377 'I have spoken...': S.B., *Molloy*.

p377 'We are getting on...': S.B., *Malone Dies*.

p377 'If this continues...': ibid.

p378 'a wealth of filthy...': S.B., *Molloy*.

p380 'Morning is the time...': ibid.

p380 'ram this code': ibid.

p381 'Yes, I once took...': ibid.

p381 'some insight into...': S.B., *Malone Dies*.

p381 'summoning to their aid...': ibid.

p382 'Such are the advantages...': S.B., *Molloy*.

p382 'goes back to Killarney': S.B., *The Unnamable*.

p382 'How do you know they hadn't?': S.B., quoted by V. Mercier, op. cit.

p382 'more shabby...': V. Mercier, op. cit.

p383 'and if I had met...': S.B., *Molloy*.

p383 'it was then I saw...': S.B., *Malone Dies*.

p384 'a little clearly at last': S.B., letter to T.MacG., 8.3.1948.

p384 'The feeling of...': ibid.

p385 'a long book, the second last...': S.B., letter to T.MacG., 4.1.1948.

p385 'This time, then once more...': S.B., *Molloy*, trans. from French, 1951.

p385 'This time, then once more . . .': S.B., *Molloy* (English edition, 1955).

p388 'It is hard sometimes . . .': S.B., letter to T.MacG., 4.1.1948.

p388 'a presence, embryonic . . .': S.B., quoted in Harvey, op. cit.

p388 'images of getting down . . .': ibid.

p389 'They turned towards the sea . . .': S.B., *Molloy*.

p389 'the authentic weakness ..': S.B. to Lawrence Harvey, quoted in Harvey, op. cit.

p389 'What complicates it all . . .': S.B., quoted in Harvey, op. cit.

p390 'Yes all my life, I think . . .': S.B., *Molloy*.

p390 'the last I hope . . .': S.B., letter to G.R., 8.2.1948.

p390 'I am preparing him for burial': ibid.

p390 'back into the light': S.B., quoted in Colin Duckworth, 'The Making of *Godot*', *Casebook of Waiting for Godot*, ed. Ruby Cohn (New York, 1967).

p390 'the wildness and rulelessness . . .': S.B., quoted in John Fletcher, Afterword to *Waiting for Godot* (London, 1971).

CHAPTER TWENTY-FIVE

p391 'a great music hall sketch . . .': Jean Anouilh, *Arts spectacles*, 5.3.1953.

p392 'Estragon: That wasn't such a bad . . .': these quotations from *Waiting for Godot* are from the first Grove edition, 1954. There were so many textual changes by the author over the years that it would be difficult to choose between later editions.

p393 'Godeau . . .': S.B., letter to Hugh Kenner, 27.6.1959.

p394 'dragging around somewhere': S.B., letter to T.MacG., 27.3.1949.

p396 'a lost and surprised child' and following: Patrick Waldberg, quoted in Bram van Velde, catalogue to the Exhibition, Centre Georges Pompidou, October 1989.

p396 'Bram is my great familiar . . .': S.B., letter to Marthe Arnaud, 25.3.1952, quoted in ibid.

p397 'Masson bears the scars . . .': all the following quotations are from S.B., 'Three Dialogues with Georges Duthuit', *Transition*, December 1949, published in *Disjecta*.

p400 'All my life . . .': S.B., *The Unnamable*.

CHAPTER TWENTY-SIX

p402 'nothing like breathing your last . . .' and following: S.B., *Texts for Nothing*.

p406 'a solidarity': Roger Blin, 'Conversation avec Linda Peskine', *Revue d'esthetique*, 1990.

p407 'wishing she were gone': S.B., *Krapp's Last Tape*.

p409 'I read *Molloy* . . .': Jérôme Lindon, 'First Meeting with Samuel Beckett', *Beckett at Sixty*.

p411 'Your letter of yesterday . . .': S.B., letter to Jérôme Lindon, 18.4.1951, quoted in ibid.

p412 'Though not a rich man . . .': Jérôme Lindon, interview with Anne Haverty and A.C.

p413 'an astonishing play': Georges Neveaux, quoted by Blin, op. cit.

p413 'Now that we are embarked . . .': S.B., letter to R. Blin, 15.1.1951.

p413 'swim, talk, void . . .': ditto, 26.8.1952.

p415 'I never want to be cold again': S.B., quoted by Francis Evers in interview with A.C.

p416 'really good-for-nothing...': S.B., letter to Bram van Velde, 18.1.1952.

p416 'I have never understood...': S.B., letter to Marthe Arnaud and Bram van Velde, 25.3.1952.

p419 'with a kind of infinite distress': Roger Blin, quoted in 'Conversation avec Linda Peskine', loc. cit.

p421 'I couldn't take any more': S.B., letter to R. Blin, 9.1.1953.

p421 'one of today's...': Sylvain Zegel, *Libération*, 7.1.1953.

p421 '*Godot* is a masterpiece...': Jean Anouilh, loc. cit.

p422 'One thing troubles...': S.B., letter to R. Blin, 9.1.1953.

p422 'the snobs': Jean Martin, interview with A.H. and A.C.

p423 'made out of nothingness': Alain Robbe-Grillet, *Critique*, February 1953.

CHAPTER TWENTY-SEVEN

p427 'stunning': Richard Seaver, op. cit.

p427 'Not *Molloy* or...': Seaver, op. cit.

p428 'a broad bad-boy grin': ibid.

p429 'elegant, suave': ibid.

p430 'With regard to...': S.B., letter to Barney Rosset, 25.6.1953.

p430 'Sometimes things...': B.R., letter to S.B., 30.7.1953.

p430 'a clear understanding...': S.B., letter to B.R., 8.7.1953.

p431 'had no particular program...': S. E. Gontarski, 'Dionysus on Publishing', *Review of Contemporary Fiction*, New York, Fall 1990.

p432 'a tough lady...': John Ashbery, catalogue note to Joan Mitchell exhibition, Robert Miller Gallery, New York 1989.

p432 'there was a deep interest...': B.R., interview with A.C.

p433 'Stick to Riopelle...': Anne Madden, interview with A.C.

p434 'a new book...': Richard Seaver, *I can't Go On, I'll Go On: A Samuel Beckett Reader* (Grove, New York, 1976).

p435 'bound to be...': S.B., letter to B.R., 1.9.1953.

p435 'with all the adventure gone': S.B., letter to T.MacG., 27.9.1953.

p435 'the old team...': ibid.

p436 'an experiment...': Maurice Blanchot, *Nouvelle Revue française*, October 1953, trans. Richard Howard in *Samuel Beckett, The Critical Heritage* (Routledge & Kegan Paul, London, 1979).

p436 'gave the impression...': Maurice Nadeau, *Mercure de France*, March 1952.

p436 'I feel more and more...' and following: S.B., letter to T.MacG., 12.12.1953.

p437 'I don't think...': ibid.

p437 'Beckett came into...': Ralph Richardson, *New Yorker*, 21.2.1977.

p438 'He wanted the low-down...': S.B., letter to M.M., 14.10.1954.

p439 'Scraps come through...': S.B., letter to T.MacG., 10.11.1951.

p439 'impossible to have health...': ditto, 27.3.1949.

p440 'Mr Beckett is...': Thomas Hickey, letter to T.MacG., 7.10.1952, quoted in Bair, op. cit.

p441 'I'm horribly tired...': S.B., letter to B.R., 11.2.1954.

p442 'Bram van Velde...': S.B., letter to T.MacG., 1.3.1954.

p442 'I have dreadful difficulty...' and following: S.B., letter to T.MacG., 1.3.1954.

p443 'almost smiled' and following: Desmond O'Grady, 'On Beckett in Paris', *Poetry Ireland*, Winter 1992–3.

p444 'It is sometimes difficult...': S.B., letter to M.M., 26.8.1954.

CHAPTER TWENTY-EIGHT

p447 'most of them were...': Peter Bull, *I Know the Face But ...* (Peter Davies, London, 1959).

p448 'no more profound...': Milton Shulman, *Evening Standard*, 4.8.1955.

p448 'shopkeepers' and following: S.B., letter to T.MacG., 7.8.1955.

p448 'Mr Beckett...': Harold Hobson, *The Sunday Times*, 7.8.1955.

p448 'a play ... is basically...': Kenneth Tynan, *Observer*, 7.8.1955.

p449 'in the attentive...': anon., *The Times*, quoted in Lawrence Graver, *Landmarks of World Literature: Beckett: Waiting for Godot* (Cambridge University Press, 1989).

p449 'Every night also ... we would carefully...': Alan Schneider, 'Waiting for Beckett', Beckett at Sixty (Chelsea Review, London, 1967) *Chelsea Review*, Autumn 1958.

p449 'expected it to go so far': S.B., quoted by Arland Ussher in conversation with A.C.

p450 'the dwindling phalanx...': Philip Toynbee, *Observer*, 18.12.1955.

p451 'conquered the audience...': René Lalou, *Nouvelles littéraires*, 22.12.1955.

p452 'like a Swiss guard': Avigdor Arikha, interview with E. Jane Dickson, *Independent on Sunday*, 7.5.1994.

p452 'siege laid again...': S.B., catalogue note quoted by Avigdor Arikha, ibid.

p453 'It is time now...': S.B., letter to T.MacG., 8.11.1955.

p453 'Perhaps some day...': S.B., letter to Patricia Ryan, 12.12.1961.

p454 'very plush' and following: Alan Schneider, 'Waiting for Beckett', loc. cit.

p455 'had billed the production...': John Lahr, *Notes on a Cowardly Lion* (Ballantine Books, New York, 1969).

p455 'for the moment...': S.B., letter to Alan Schneider, quoted in 'Beckett's Letters on Endgame', *Village Voice Reader*, ed. Daniel Wolf and Edwin Fancher (Doubleday, New York, 1962).

p456 'in a few weeks...' and following: Lawrence Grover, op. cit.

p456 'in what seems to be...': S.B., letter to T.MacG., 27.4.1956.

p457 'The end ... is to give...': S.B., quoted by Alec Reid, *All I Can Manage, More Than I Could* (Dolmen, Dublin, 1964).

CHAPTER TWENTY-NINE

p458 'no, *mon pigeon*...': S.B., draft quoted in Gontarski, *The Intent of Undoing*.

p459 'Struggling...' and following: S.B., letter to Alan Schneider, *Village Voice Reader*.

p459 'straighten up the statue': Glenavy, op. cit.

p460 'Hamm is a king...': S.B., quoted in Cohn, *Back to Beckett* (Princeton University Press, 1973).

p461 'never thought about...': S.B., letter to Nancy Cunard, 4.7.1956.

p461 'extremely keen...': John Morris, letter to Val Gielgud, quoted in Brater, op. cit.

p461 'with some scepticism': Barbara Bray in interview with A.C.

p462 'queer to be struggling...': S.B., letter to Nancy Cunard, undated (early autumn 1956).

p462 'Rather difficult and...': S.B., letter to Alan Schneider, *Village Voice Reader*.

p462 'With *Godot* after all...': S.B., letter to T.MacG., 18.10.1956.

p463 'on some kind of...': S.B., letter to Alan Schneider, *Village Voice Reader*.

p463 'It is not easy...': S.B., letter to T.MacG., undated (autumn 1956).

p463 'The trouble with palliatives...': S.B., letter to B.R., undated (autumn 1956).

p463 'with the help of dope...': S.B., ditto.

p463 'Everything is drenched...': S.B., letter to T.MacG., 18.10.1956.

p464 'It is a text written...': S.B., quoted by Alec Reid, *All I Can Manage*.

p464 'standard realism': Donald McWhinnie, letter to S.B., quoted in Federman and Fletcher, op. cit.

p464 'radio classic': Roy Walker, *Listener*, 24.1.1957.

p464 'Radio triumphant': Christopher Logue, *New Statesman*, 14.9.1957.

p465 'derivative slapstick': Donald Davie, *Spectrum*, Winter 1958.

p465 'rather grim...': S.B., letter to Alan Schneider, *Village Voice Reader*.

p466 'Here is life...': Kenneth Tynan, *Observer*, 7.4.1957.

p466 'There is no drama...': S.B., quoted by Roger Blin, Peskine, loc. cit.

p467 'mostly depending on...': S.B., quoted by Schneider, loc. cit.

p467 'more like what...': S.B., letter to Nancy Cunard, undated.

p467 'only one of a spate...': V. Mercier, op. cit.

p468 'Well, they are not...': Harold Hobson, *The Sunday Times*, 15.7.1973.

p468 'the intellectual...': Marc Bernard, *Nouvelles littéraires*, 9.5.1957.

p469 'a fabulist of...': Robert Welch, *The Kilcolman Notebook* (Brandon, Dingle, 1994).

p469 'At a party...': S.B., quoted in Driver, op. cit.

p471 'a serious attempt...': S.B., letter to T.MacG., 31.3.1958.

p471 'come back to me...': ibid.

p472 'a text which...': S.B., letter to Alan Schneider, *Village Voice Reader*.

p472 'undecantable...': S.B., letter to B.R., 6.4.1957.

p472 'all dried up...': S.B., quoted by V. Mercier, op. cit.

p472 'I just don't happen...': S.B., letter to T.MacG., undated.

p473 'to recuperate...': S.B., letter to B.R., 11.8.1955.

p474 'It is lonely...': S.B., letter to T.MacG., 27.11.1957.

p474 'for the few years...': ditto, 9.4.1956.

p474 'last years...': S.B., letter to T.MacG., 6.4.1957.

CHAPTER THIRTY

p475 'a blind old broken man': S.B., letter to T.MacG., 2.5.1938.

p475 'a brief stage...': S.B., letter to B.R., 5.3.1958.

p475/6 'getting more and more...': ditto, 20.11.1958.

p476 'The only chance...': ditto, 23.11.1958.

p476 'they are a sorry...': S.B., letter to T.MacG., 2.5.1958.

p476 'very sober...' and following: S.B., letter to B.R., 29.7.1958.

p477 'astonishment and delight': John Calder, interview with A.C.

p478 'As you know, I am...': S.B., letter to B.R., 16.4.1957.

p478 'Sorry again, Dear Barney...': ditto, 2.2.1958.

p480 'seems a little in doubt...': anon., 'Samuel Beckett the Messenger of Gloom', *Observer*, 1958.

p481 'exercises in peevish despair' and following: Alan Brien, *Spectator*, 4.11.1955.

p482 'It was about this time...': S.B., *Watt*.

p482 'Not even the Westminster...': Harold Pinter, 'Pinter on Beckett', *New Theatre Review*, Vol. XI, No.3.

p483 'In *Waiting for Godot*...': T. C. Worsley, *Listener*, 4.11.1958.

p483 '*Godot* is rollicking...': S.B., letter to Nancy Cunard, undated (1955).

p484 'Beckett's gloomy...': V. Mercier, op. cit.

p484 'Is that all the review...': Kenneth Tynan, *Observer*, 2.1.1958.

p484 'in no way senile' and following: Michael Haerdter, 'Production Account', McMillan and Fehsenfeld, op. cit.

p486 'Krapp seems to think...': S.B. in conversation with A.C.

p486 'I thought of writing...': S.B. to Pierre Chabert, quoted in Gontarski, *The Intent of Undoing*.

p488 'perished with cold...': S.B., letter to T.MacG., 9.12.1958.

p488 'I won't try...': ditto, 29.3.1958.

p488 '*besoin d'un frère*' and following: ditto, 21.4.1958.

p488 'Ah well!': ditto, 19.10.1958.

p489 'one of those queer compulsions': S.B., letter to B.R., 20.3.1959.

p489 'This is just...': S.B., letter to Ethna McCarthy, 25.5.59.

p489 'I rely a lot on...': ditto, 5.5.1959.

p489 'I hear their siren...': ibid.

p489 'and the sea out as far...' and following: S.B., *Embers*.

p490 'technical weaknesses...' and following: Karl Miller, *Encounter*, September 1959.

p490 'the dramatic which...': John Whiting, *London Magazine*, May 1960, quoted in Federman and Fletcher, op. cit.

p491 'Cronin delivered his...': S.B., letter to B.R., 20.3.1959.

p492 'unworthy' and following: S.B., letter to H. O. White, 3.2.1959.

p493 'talking him round': S.B., letter to B.R., 8.4.1959.

p493 'morally bound to...': S.B., letter to T.MacG., 10.5.1959.

p494 'for a while...' and following: Public Orator's Address, trans. A. J. Leventhal, quoted in Bair, op. cit.

p494/5 'to see to some things...': S.B., letter to T.MacG., 17.5.1959.

p495 'great absorption' and following: ibid.

p495 'quite horrifying' and following: S.B., letter to T.MacG., 19.9.1959.

p496 'I see nothing else for it': S.B., letter to B.R., 20.3.1959.

p497 'Beckett went to Brendan's...': Ulick O'Connor, *Brendan Behan* (Hamish Hamilton, London, 1970).

p497 'he paid them...': Alec Reid, 'The Reluctant Prizeman', loc. cit.

CHAPTER THIRTY-ONE

p501 'We simply must...': S.B to M.M, 22.12.59.

p501 'pretty flattened' and following: S.B., letter to T.MacG., 15.8.1960.

p501 'troubled by the...' and following: ditto, 9.1.1961.

p502 'tea à toothpaste': Edward Beckett, interview with Peter Lennon, *Guardian*, 30.3.1996.

p502 'In any case . . .': Barbara Bray, interview with A.C.

p502 'to deal with . . .': S.B., letter to T.MacG., 10.3.1961.

p503 'whatever his movements . . .': ibid.

p503 'Thank God it's . . .' : S.B., letter to T.MacG., 27.3.1961.

p505 'One of Sam's . . .': Barbara Bray, interview with A.C.

p507 'for talk I hope . . .': S.B., letter to Lawrence Harvey, 28.10.1963.

p507 'a score of . . .': Harvey, op. cit.

p507 'My feeling about . . .': S.B., letter to T.MacG., 7.4.1962.

p507 'greatly touched by . . .': S.B., letter to Lawrence Harvey, 30.6.1965.

p510 'You are not a critic . . .' and following: S.B., quoted by Jean Demélier, interview with A.C.

p510 'Beckett . . . understood . . .' and following: Jean Demélier, loc. cit.

p511 'assassinated' and following: Charles Juliet, op. cit.

CHAPTER THIRTY-TWO

p514 'flowed across the white . . .' and following: Peter Lennon, *Foreign Correspondent* (Picador, London, 1994).

p515 'like a light salute . . .' and following: ibid.

p517 'more remote': ibid.

p518 'Men do not . . .': Barbara Bray, interview with A.C.

p520 'Three stock characters . . .': James Knowlson and John Pilling, *Frescoes of the Skull, The Later Prose and Drama of Samuel Beckett* (Grove, New York, 1981).

p520 'our darling Grand Canary' and following: S.B., *Play*.

p522 'All one night . . .': Alberto Giacometti, quoted in Lord, op. cit.

p523 'All that matters . . .' and following: S.B., quoted in Michael Haerdter, Rehearsal Diary, in Dougald McMillan and Martha Fehsenfeld (eds.), *Beckett and the Theatre* (Calder, London, 1988).

p524 'I could hear . . .': S.B. to Harold Pinter, quoted by Pinter in interview with A.C.

p525 'I never knew . . .': Jack MacGowran in conversation with A.C.

p526 'I only know . . .': S.B., quoted by Pat Magee in conversation with A.C.

p528 'so much to be . . .' and following: S.B., *Happy Days*.

p531 'fix it up' and following: Buster Keaton, quoted by Alan Schneider in 'On Directing *Film* by Samuel Beckett', Samuel Beckett, *Film* (Grove, New York, 1969).

p534 'Buster, finally . . .': ibid.

p535 'No one wants . . .': ibid.

p536 '. . . good the end . . .': John Updike, *New Yorker*, December 1964.

CHAPTER THIRTY-THREE

p539 'if there is any more . . .': S.B., letter to T.MacG., 30.1.1965.

p539 'When you have . . .': S.B., quoted by Roger Blin, Peskine, loc. cit.

p540 'The pessimistic vision . . .': Alfred Simon, *Esprit*, December 1963.

p542 'an entertaining ruffian': S.B., letter to Nancy Cunard, 11.5.1956.

p543 'in spite of everything': Jérôme Lindon, telegram to S.B., 23.10.1969.

p543 'if there is any justice . . .': S.B., letter to William York Tindall, 20.1.1963.

p545 'lacking in Nobel fibre': S.B., letter to Dr Müller-Freienfels, 12.11.1969.

p545 'With all this I . . .': ibid.

p546 'chosen by a London . . .': W. B. Yeats, 'The Bounty of Sweden', *Autobiographies* (Macmillan, London, 1955).

p546 'He would have known . . .': Jean Demélier, interview with A.H. and A.C.

p547 'Some days ago . . .': S.B., letter to John Calder, 31.1.1970.

p548 'Why do I need . . .': S.B., letter to M.M., 7.1.1969.

p549 'a bare old battered rock': S.B., letter to Charles Monteith, 14.9.1972.

p550 'in spite of all': S.B letter to B.R., undated (autumn 1974).

p551 'looked as if it had been . . .' and following: Billie Whitelaw, *Who He?* (Hodder & Stoughton, London 1995).

p552 'I'm sorry, Sam . . .': ibid.

p552 'the incomparable Billie': S.B., letter to Peter du Sautoy, 8.2.1973.

p553 'spared . . . love . . .': S.B., *Not I*.

p554 'Next, perhaps . . .' and following: Alec Reid, 'Impact and Parable in Beckett', *Hermathena*, Winter 1986.

p556 'I don't want to talk about . . .' and following S.B., quoted in Haerdter, loc. cit.

CHAPTER THIRTY-FOUR

p560 'very dangerous for . . .': ibid.

p561 'We know what . . .': Hume Cronyn in cable to S.B., quoted in Bair, op. cit.

p561 'I AM AGAINST . . .': S.B., letter to B.R., 11.7.1973.

p562 'Frankly, I don't know . . .': ibid.

p562 'farted in the last few years' and following: S.B. to Pat Magee, 15.12.1975.

p562 'Since they really . . .': J. D. O'Hara, *Nation*, 19.2.1977.

p562 'that remorseless . . .': Valentine Cunningham, *New Statesman*, 29.10.1976.

p563 'precisely in reference to writing' and following: E. M. Cioran, *Partisan Review*, XLIII, February 1976.

p564 'the unit of effect' and following: Hugh Kenner, *A Reader's Guide to Samuel Beckett* (Thames and Hudson, London, 1973).

p564 'effusiveness . . .' and following: Cioran, loc. cit.

p565 'Despair young . . .': S.B., letter to Aidan Higgins, quoted by Higgins in interview with A.C.

p565 'devastatingly so' and following: Aidan Higgins, interview with A.C.

p567 'The wind has . . .': A. Higgins, tribute in *Beckett at Sixty*.

p568 'I would have written . . .': S.B. to A. Higgins and Jill Higgins, quoted by A. Higgins in interview with A.C.

p569 'If you need anything . . .': S.B. to André Bernold, quoted in A. Bernold, *Amitiés de Beckett* (Hermann, Paris, 1992).

p569 'gazed with abandon . . .' and following: Bernold, op. cit.

CHAPTER THIRTY-FIVE

p571 'seldom if ever . . .': S.B., letter to B.R., 5.10.1978.

p572 'achieved him': Bernold, op. cit.

p572 'It is the final point' and following: S.B. to André Bernold, quoted in Bernold, op. cit.

p572 'On. Say on' and following: S.B., *Worstward Ho*.

p572 'to be an artist . . .' and following: S.B., 'Three Dialogues with Georges Duthuit'.

p573 'and how the Portora . . .': Derek Mahon, letter to A.C., 10.12.1994.

p574 'It's marvellous . . .': S.B. to André Bernold, Bernold, op. cit.

p574 'Did Louis Golding . . .': Derek Mahon, letter to A.C., 10.12.1994.

p576 'I beg of you . . .': S.B., letter to B.R., 30.1.1986.

p576 'The news of my . . .' and following: S.B., letter to Adrian Munnelly, Registrar of Aosdana, 5.12.1984.

p577 'Sorry can't be . . .': S.B., letter to A.C.

p577 'Literary widows . . .': S.B., letter to Adam Tarn, quoted in Bair, op. cit.

p577 'I have written nothing . . .': S.B., letter to Antoni Libera, 21.7.1985.

p577 'One night as he . . .' and following: S.B. to Anne Madden and Louis le Brocquy, quoted in Anne Madden le Brocquy and Louis le Brocquy, *A Painter Seeing His Way* (Gill & Macmillan, Dublin, 1994).

p578 'I simply wanted . . .': Sydney Nolan in conversation with A.C.

p578 'tormented by what . . .': Madden, le Brocquy, op. cit.

p578 'with loathing . . .': S.B., letter to B.R., 25.6.1986.

p579 'Seen always from behind . . .' and following: S.B., *Stirrings Still*.

p580 'Don't be depressed . . .': S.B., letter to Anne Madden le Brocquy, quoted in Madden, le Brocquy, op. cit.

p580 'more or less all right . . .': S.B., letter to Walter Asmus, 8.9.1987.

p581 'from deep within . . .' and following: S.B., 'As the Story Was Told'.

CHAPTER THIRTY-SIX

p582 'salvation or whatever' and following.: S.B. to Anne Madden le Brocquy, quoted in Madden, le Brocquy, op. cit.

p582 'Distressed to hear . . .' and following: S.B., letter to Louis le Brocquy, quoted in ibid.

p582 'Did you get . . .' and following: S.B., letter to Grose, 26.11.1988.

p583 'transparent . . .': Anne Madden le Brocquy, op. cit.

p585 'Now shall I . . .': W. B. Yeats, 'The Tower'.

p586 'stupid, horrible . . .': Jean Demélier, interview with A.H. and A.C.

p586 'as tired . . .': S.B., letter to B.R., 19.11.1988.

p587 'keep on trying . . .': ditto, 17.12.1988.

p586 'Come whenever you like . . .': ibid.

p587 'folly: . . .': S.B., 'As the Story Was Told'.

p589 'Visibly moved . . .' and following: Madden, le Brocquy, op. cit.

p589 'I'm done' and following: S.B. to John Montague. The following account is based on Montague's article in the *Irish Times* (22.12.1994) and on an interview with A.C.

p591 'There was no correspondence . . .' and following: Derek Mahon, letter to A.C., 18.12.1994.

CHAPTER THIRTY-SEVEN

p592 'Whatever opinion . . .': S.B., *Proust*.

p593 'Yes, the confusion . . .': S.B., *Molloy*.

p593 'both a hope . . .': Christopher Ricks, *Beckett's Dying Words* (Clarendon Press, Oxford, 1993).

Acknowledgements

The author and publisher are grateful to the Samuel Beckett Estate, The Calder Educational Trust, London, and Grove Press, US, for permission to quote from the prose and poetry of Samuel Beckett; Faber and Faber and Grove Press, US, for permission to quote from the dramatic works; the Samuel Beckett Estate for permission to quote from *Assumption* and *Kilcool*; and the Samuel Beckett Estate and Eoin O'Brien to quote from the Black Cat edition of *Dream of Fair to Middling Women*. All works copyright © Samuel Beckett and the Samuel Beckett Estate.

Illustrations

Illustrations

Index

627

634

635

pornography, 291, 429
Port Chasseur, Théâtre de, 413
Porter, Cole, 82
Portora Royal School, 39-51, 73
Portora (school magazine), 43, 45
*Portrait of the Artist as a Young
    Man, A* (Joyce), 110, 266
Portugal, SB visits, 541, 545-6
Pound, Ezra, 83, 85, 110-12,
    193, 555
Pourrat, Henri, 60
Powell, Tristram, 552
Power, Arthur, 88, 93, 100,
    167, 234
Prague, 306
Prayers and Meditations
    (Johnson), 257
'Premier Amour' *see* 'First Love'
Prentice, Charles, 110-11, 120,
    125, 155, 174-6, 195,
    197-8, 209, 234
primal myths, 391
Prince William's Seat (moun-
    tain), 29, 31
Prix Rivarol, 394
Prosper (resistance group), 328
prostitutes, 106, 134-6, 171,
    219-20, 280, 284
Protestant
    business community, 9
    education, SB's, 34-5, 39-51
    middle class outlook, 10-11
Protestantism, 19, 21, 376, 510
Proust, Marcel, 111, 115, 119-
    20, 127, 143-7, 182, 202,
    271, 381, 424
Proust (Beckett), 111, 118-20,
    125, 127, 139, 142-3,
    147, 155, 378, 508, 592
Prudent, Robert Jules, 284,
    286-7, 289-2
psychoanalysis, 199-202, 204-5,
    208-9, 212-14, 217, 220-2
Pudovkin (film director), 228
Purgatorio (Dante), 169
Putnam, Jack, 416
Putnam, Samuel, 114, 155

Quadrat (Beckett), 571
Quatre Nouvelles (Beckett), 359,
    361-2, 365, 368, 370,
    378, 385, 390, 441
    publication, 451, 453
Queen, Ellery, 549
Queen's Theatre, 57
Quennell, Peter, 204
quietism, 229-31

Racine, Jean, 59-60, 149, 463-4, 524
Radio Éireann, 351, 355
Ragg, Tim, 278
Raggle Taggle (Starkie), 62
Raimbourg, Lucien, 418-19,
    435, 522
Rapallo, SB visits, 549
Raphael, 244
Rathdowne, Lord, 252
Ray, Man, 116-17
Réaction (review), 277
Read
    Herbert, 278, 281, 342, 357
    and *Watt*, 358
Reader's Digest, 363
realism, 379
Reavey
    George, 112-14, 151, 207,
        215, 224, 238-9, 256,
        268, 298, 304, 307, 535,
        586
    and *Murphy*, 234, 239, 246-8,
        254, 278, 287, 367, 390
    SB lack of optimism, 417
    war years, 314, 317, 358
    and *Watt*, 358, 370-1
Red Cross, 348-55
Reddin, Kenneth, 267
Reid, Alec, 457, 464, 522, 554-
    5
Reilly, John, 580-1
Rembrandt, 141, 244
Renard, Jules, 148
Renaud, Marthe, 416
Renaud, Madeleine, 504, 539-
    40, 560, 580
repatriation, 340-1
Rescke, Jean de, 65
Résistance, 322
resistance groups, 319, 322-7,
    328-9, 333-5
    Gloria, 323-4, 328, 341
    Prosper, 328
Restaurant Drouot, 168
Resurrection (Yeats), 206
Révolution surréaliste, La, 79-80,
    83, 302
Revolution of the Word mani-
    festo, 171-2
Reynolds, Mary, 283, 311, 324
*Richard Aldington: An
    Englishman*, 111
Richardson, Ralph, 437-8
Rickman, John, 200
Ricks, Christopher, 593
Rickword, Edgell, 181
Riding, Laura, 117
Riff war, Morocco, 82, 276

Rigaut, Jacques, 80
Rilke, Rainer Maria, 202-3
Rimbaud, Arthur, 71, 112, 127,
    147-8, 173, 181, 193
Riopelle, Jean, 395, 432-3, 451,
    578
Robbe-Grillet, Alain, 423-4,
    530
Robert Hale, 254
Robert Miller Gallery, New
    York, 432
Roberts, Ellis, 177
Robinson, Lennox, 253
la Rochelle, Drieu, 404
Rockaby (Beckett), 571, 575
Rodgers, W.D., 460
Rodker, John, 95
Roe, Edward Price, 32
Roe, Jack, 32-3
Roe, Maria *see* Beckett, May
Roe, Mollie, 32-3, 195, 205,
    309-11
Roe, Samuel Robinson, 6-7, 10
Roe, Sheila *see* Page, Sheila
Rogers, Mrs Mary, 50
Rolland, Romain, 78
Romains, Jules, 102, 72
Romeo and Juliet
    (MacLiammóir), 183
Ronsard, Pierre de, 68
Rosebud bar, 433, 451-2
Rosset, Barney, 430-2, 441,
    454, 463, 472, 474-8,
    481, 489, 491, 496, 530-
    5, 547, 559-60, 562, 566,
    571, 576, 578, 580, 582,
    585
Rouault, Georges, 141, 345,
    356
Rough, 560
Roussillon, SB in, 329-39
Route des Flandres, La, 491
Routledge and Kegan Paul,
    278, 281, 287, 341, 357
Royal Court theatre, 465, 468-
    72, 478, 480-1, 523, 527,
    560
Royal Dublin, Portmarnock, 53
Royal Dublin Society, 3
Royal Hibernian Academy, 74
Royal (music-hall), 57
Rubens, 244
Rubinstein, Artur, 183
Rubinstein, Helena, 114
Rudmose-Brown
    R.B. 'Ruddy', 239
        alcoholism, 228-9
        French lectureship, 72-3, 84

645